Electroporation and Electrofusion in Cell Biology

Electroporation and Electrofusion in Cell Biology

Edited by
Eberhard Neumann
University of Bielefeld
Bielefeld, Federal Republic of Germany

Arthur E. Sowers
American Red Cross
Rockville, Maryland

and

Carol A. Jordan
Science Applications International Corporation
McLean, Virginia

Plenum Press • New York and London

Library of Congress Cataloging in Publication Data

Electroporation and electrofusion in cell biology / edited by Eberhard Neumann, Arthur E. Sowers, and Carol A. Jordan.

p. cm.

Dedicated to the memory of Herbert Acland Pohl.

Includes bibliographies and index.

ISBN 0-306-43043-6

1. Electroporation. 2. Electrofusion. 3. Cytology—Technique. I. Neumann, Eberhard, 1939- . II. Sowers, Arthur E. III. Jordan, Carol A.

QH585.5.E48E44 1989 89-30872

574.87—dc20 CIP

The preparation of this book was partially supported by
Office of Naval Research, Contract # N00014-83-C-0008 (CJ)

Printed in the United States of America

Dedication

To Herbert Ackland Pohl, Ph.D.

Born of American parents in Lisbon, Portugal, in 1916, Herb Pohl was one of the true geniuses of this century. He was a man of tremendous foresight, unbounded creativity, and a most unusual sense of humor. A true scientist, he was dedicated to the pursuit of truth and to the alleviation of human suffering. As a good actor on the stage of life, he was always "on," for science was his life and he was always prepared to enter into a scientific debate or discuss a philosophical issue. It was his untiring pursuit of knowledge and dedication to science that led to the findings that have made electrofusion and electroporation the scientific tools that they are and this book possible.

Pohl did his undergraduate and graduate work at Duke University and received his Ph.D. in physical chemistry in 1939. During World War II, he worked as a senior chemist at the prestigious Naval Research Laboratory in Washington, D.C. He spent the remainder of his career in academia, including Princeton University, the Brooklyn Polytechnic Institute, and Oklahoma State University, where he was professor of physics from 1964 until his retirement in 1981. He also held visiting professorships at the University of Uppsala (Sweden), Cambridge University (England), the University of California at Riverside, and the Massachusetts Institute of Technology. It was at the latter institution, to which he had gone as a visiting professor after his formal retirement, that he died in his laboratory on June 21, 1986.

Although trained as a physical chemist, Pohl had a tremendous breadth of knowledge that spanned biology, medicine, and biophysics. He was editor of the *Journal of Biological Physics* and co-editor of *Digest on Dielectrics* and was on the editorial board of the *Journal of Electrostatics*. A fellow of the American Institute of Chemists, Pohl was a consultant to the World Health Organization and to a number of industries. An active researcher all his life, he published over 200 research papers, as well as a textbook on quantum mechanics that was translated into French, Italian, and Japanese. He also authored a monograph on dielectrophoresis (DEP) and edited a number of books and conference proceedings.

The vision and scientific creativity of Herb Pohl are best exemplified in his discovery of DEP, the motion caused by a nonuniform electric field on neutral matter. DEP is the basis for xerography, electrofusion, and electroporation. As this book amply demonstrates, the implications of electrofusion and electroporation are very far-reaching. Pohl's interest in cellular phenomena drove him to study the biophysical basis of cellular processes and function in the hope of someday being able to better control cancer. He worked tirelessly for the

Oklahoma Laboratory of the National Foundation for Cancer Research and established the Pohl Cancer Research Laboratory in Stillwater, Oklahoma.

We could easily devote an entire chapter of this book to Herb Pohl's scientific accomplishments. Suffice it to say that his impact on cellular biophysics, polymer chemistry, and a host of other areas of science and technology will be felt for many years to come. Pohl was indeed a man of bold vision whose ideas and imagination, charm and wit, and friendly and constructive criticism are sorely missed by his many friends throughout the world. We have all suffered a great loss.

T. C. Rozzell
National Academy of Sciences

Contributors

W. Asche, Krüss GmbH, 2000 Hamburg 61, Federal Republic of Germany

C. Blangero, Center for Biochemistry and Cellular Genetics, CNRS, 31062 Toulouse, France

Donald C. Chang, Department of Physiology and Molecular Biophysics, Baylor College of Medicine, Houston, Texas 77030, and Marine Biological Laboratory, Woods Hole, Massachusetts 02543

Leonid V. Chernomordik, The A.N. Frumkin Institute of Electrochemistry, Academy of Sciences of the USSR, 117071 Moscow, USSR

Yuri A. Chizmadzhev, The A.N. Frumkin Institute of Electrochemistry, Academy of Sciences of the USSR, 117071 Moscow, USSR

Bernhard Deuticke, Department of Physiology, Medical Faculty, RWTH Aachen, D-5100 Aachen, Federal Republic of Germany

Angel Domínguez, Department of Microbiology, Genetics, and Preventive Medicine, University of Salamanca, Salamanca, Spain

Daniel L. Farkas, Department of Physiology, University of Connecticut Health Center, Farmington, Connecticut 06032

Walter Förster, Faculty of Chemistry, Department of Physical and Biophysical Chemistry, University of Bielefeld, D-4800 Bielefeld 1, Federal Republic of Germany

Mark C Glassy, Brunswick Biotechnetics, 4116 Sorrento Valley Boulevard, San Diego, California 92121

Gunter A. Hofmann, BTX-Biotechnologies and Experimental Research Inc., San Diego, California 92109

Sek Wen Hui, Electron Optics Laboratory, Biophysics Department, Roswell Park Memorial Institute, Buffalo, New York 14263

Francisco J. Iglesias, Department of Applied Physics, University of Salamanca, Salamanca, Spain

P. R. Imrie, Institute of Cancer Research, Royal Cancer Hospital, Sutton 2M2 5PX, United Kingdom

Barbara Junker, Max Planck Institut für Züchtungsforschung, D-5000 Köln 30, Federal Republic of Germany

D. E. Knight, M.R.C. Secretory Mechanisms Group, Department of Physiology, Kings College London, London WC2R2LS, United Kingdom

Hans-Ulrich Koop, Botanical Institute, University of Munich, D-8000 Munich 19, Federal Republic of Germany

Ingolf Lamprecht, Institute for Biophysics, Free University of Berlin, D-1000 Berlin 33, Federal Republic of Germany

Mathew M. S. Lo, Molecular Biology and Genetics Unit, Neuroscience Branch, Addiction Research Center, National Institute on Drug Abuse, Baltimore, Maryland 21224

M. Carmen López, Department of Microbiology, Genetics, and Preventive Medicine, University of Salamanca, Salamanca, Spain

Horst Lörz, Max Planck Institut für Züchtungsforschung, D-5000 Köln 30, Federal Republic of Germany

Benjamin F. Matthews, Plant Molecular Genetics Laboratory, USDA, ARS, Beltsville, Maryland 20705

Grigory B. Melikyan, The A.N. Frumkin Institute of Electrochemistry, Academy of Sciences of the USSR, 117071 Moscow, USSR

Paul D. Miller, Vegetable Laboratory, USDA, ARS, Beltsville, Maryland 20705. *Present address:* DNA Plant Technology Corporation, Cinnaminson, New Jersey 08077

Maja Mischel, Institute for Biophysics, Free University of Berlin, D-1000 Berlin 33, Federal Republic of Germany

Eberhard Neumann, Faculty of Chemistry, Department of Physical and Biophysical Chemistry, University of Bielefeld, D-4800 Bielefeld 1, Federal Republic of Germany

M. J. O'Hare, Institute of Cancer Research, Royal Cancer Hospital, Sutton 2M2 5PX, United Kingdom

Takako Ohno-Shosaku, Department of Physiology, Kyoto University Faculty of Medicine, Kyoto 606, Japan. *Present address:* Department of Physiology, Osaka Medical College, Osaka 569, Japan

Yasunobu Okada, Department of Physiology, Kyoto University Faculty of Medicine, Kyoto 606, Japan

M. G. Ormerod, Institute of Cancer Research, Royal Cancer Hospital, Sutton 2M2 5PX, United Kingdom

J. H. Peacock, Institute of Cancer Research, Royal Cancer Hospital, Sutton 2M2 5PX, United Kingdom

Huntington Potter, Department of Neurobiology, Harvard Medical School, Boston, Massachusetts 02115

Kevin T. Powell, Harvard–MIT Division of Health Sciences and Technology, Massachusetts Institute of Technology, Cambridge, Massachusetts 02139

Michael Pratt, UCSD Cancer Center, University of California at San Diego, La Jolla, California 92903

M. Pröls, Max Planck Institute for Plant Breeding, D-5000 Köln 30, Federal Republic of Germany

M. P. Rols, Center for Biochemistry and Cellular Genetics, CNRS, 31062 Toulouse, France

Carlos Santamaría, Department of Applied Physics, University of Pais Vasco, Bilbao, Spain

James A. Saunders, Germplasm Quality and Enhancement Laboratory, USDA, ARS, Beltsville, Maryland 20705

J. Schell, Max Planck Institute for Plant Breeding, D-5000 Köln 30, Federal Republic of Germany

H. P. Schwan, Institute for Biotechnology, University of Würzburg, Würzburg, Federal Republic of Germany, and Department of Bioengineering, University of Pennsylvania, Philadelphia, Pennsylvania 19104

Karl Schwister, Department of Physiology, Medical Faculty, RWTH Aachen, D-5100 Aachen, Federal Republic of Germany

Arthur E. Sowers, Jerome H. Holland Laboratory for the Biomedical Sciences, Rockville, Maryland 20855

Germán Spangenberg, Max Planck Institute for Cell Biology, D-6802 Ladenburg, Federal Republic of Germany

H.-H. Steinbiß, Max Planck Institute for Plant Breeding, D-5000 Köln 30, Federal Republic of Germany

David A. Stenger, Electron Optics Laboratory, Biophysics Department, Roswell Park Memorial Institute, Buffalo, New York 14263

Istvan P. Sugar, Departments of Biomathematical Sciences and Physiology and Biophysics, Mt. Sinai Medical Center, New York, New York 10029, and Institute of Biophysics, Semmelweis Medical University, 1444 Budapest, Hungary

J. Teissié, Center for Biochemistry and Cellular Genetics, CNRS, 31062 Toulouse, France

Tian Yow Tsong, Department of Biochemistry, University of Minnesota, St. Paul, Minnesota 55108

James C. Weaver, Harvard–MIT Division of Health Sciences and Technology, Massachusetts Institute of Technology, Cambridge, Massachusetts 02139

Foreword

Cells can be funny. Try to grow them with a slightly wrong recipe, and they turn over and die. But hit them with an electric field strong enough to knock over a horse, and they do enough things to justify international meetings, to fill a sizable book, and to lead one to speak of an entirely new technology for cell manipulation.

The very improbability of these events not only raises questions about why things happen but also leads to a long list of practical systems in which the application of strong electric fields might enable the merger of cell contents or the introduction of alien but vital material. Inevitably, the basic questions and the practical applications will not keep in step. The questions are intrinsically tough. It is hard enough to analyze the action of the relatively weak fields that rotate or align cells, but it is nearly impossible to predict responses to the cell-shredding bursts of electricity that cause them to fuse or to open up to very large molecular assemblies. Even so, theoretical studies and systematic examination of model systems have produced some creditable results, ideas which should ultimately provide hints of what to try next.

But I expect it will be the try-it-and-see-what-happens experiments that will grab headlines. It's hard to compete for attention with big bucks and new forms of life. Already electroporation is a major element in any attempt at gene transfer. Already one expects in principle to be able to fuse any two or three kinds of cells. Already one hears of the hope of stimulating fertilization where hope was lost before.

Where will it lead? One thinks of the next decade with optimism. A careful read through this volume should give at least a glimpse of what to expect.

Adrian Parsegian

Physical Sciences Laboratory
Division of Computer Research and Technology
National Institutes of Health
Bethesda, Maryland

Preface

The use of electric pulse techniques in cell biology, biotechnology, and medicine has attracted enormous interest. Pulsed electric fields can cause electroporation, the permeabilization of the membranes of cells and organelles, or electrofusion, the connection of two separate membranes into one. Electric fields can also induce pearl chain formation and the rotation of cells in rotating electric fields.

Membrane field effects are of great potential not only for a variety of applications in biotechnology, plant and animal genetic engineering, gene therapy, or monoclonal antibody production, but also for fundamental biophysical studies of structural rearrangements of membrane components such as membrane fusion or pore induction. Indeed, the electroporation technique of gene transfer is often the method of choice for cell transfection and transformation.

The intent of the editors has been to secure an international collection of contributions on electroporation and electrofusion and associated phenomena such as dielectrophoresis and cell rotation. The volume is an attempt to review, unify, and consolidate the research and developments in these new and challenging areas of study. It covers basic, applied, and instrumentation aspects of electroporation and electrofusion and presents discussions of biological and biophysical model systems. The contributions of the various authors represent individual views and interpretations and provide the reader with a useful, single reference source. As in any new area of study, a standard lexicon of nomenclature, definitions, and interpretations has yet to be agreed upon.

The book is dedicated to the memory of Herbert Pohl (1916–1986), a pioneer in the investigation of electric field effects on cells.

Eberhard Neumann
Bielefeld, Federal Republic of Germany
Arthur E. Sowers
Rockville, Maryland
Carol A. Jordan
McLean, Virginia

Contents

Part I

Cells in Electric Fields

Chapter 1

Dielectrophoresis and Rotation of Cells

H. P. Schwan

1. INTRODUCTION

Electrical fields exert forces on particles and biological cells. These field-induced forces are well known to physicists and many textbooks discuss these so-called ''ponderomotoric'' forces. Usually this is done for the DC case only. The forces can be shown to be proportional to the square E. Since the forces are proportional to E^2, an alternating field will produce forces of the form $\cos^2\omega t = 0.5(1+ \cos 2\omega t)$ (ω = angular frequency, t = time) and therefore must have a time-averaged positive value.

These forces can demonstrate themselves in various ways. Just as two capacitor plates attract each other when an alternating field is applied, so will two particles exposed to a field move toward each other and form the beginning ''pearl chains'' linking in time many suspended particles. The formation of ''pearl chains'' was probably first observed by Kerr nearly 100 years ago as mentioned by O'Konski (1981). Liebesny (1938) was the first to demonstrate the field-induced alignment of erythrocytes. The movement of biological cells toward each other with the aid of alternating fields is, of course, the first step toward their fusion, an important topic of this volume. Pearl chain formation can be considered a special case of the movement of particles in an inhomogeneous field since one particle distorts the field acting on the other and vice versa. Movement in an inhomogeneous field will take place if the gradient of E^2 does not vanish and if the field force is strong enough to overcome random thermal motion. It was termed ''dielectrophoresis'' by Herbert Pohl (1978) who spent a lifetime exploring this effect. Other force effects include particle shape changes, orientation of nonspherical particles, and rotation. Interest in these effects acting on biological cells has rapidly grown during the past decade since it was realized that it is possible to manipulate cells with the aid of electrical fields for all sorts of purposes. As mentioned above, cell fusion can be achieved by a clever combination of alternating and pulsed fields. Rotation of cells is caused by rotating electrical fields and can be used to extract cellular properties such as membrane capacitance and cytoplasmic conductivity. Shape changes are caused by fields and may be used to determine mechanical properties of cell membranes. Shape changes may also cause cellular volume changes if

H. P. SCHWAN • Institute for Biotechnology, University of Würzburg, Würzburg, Federal Republic of Germany, and Department of Bioengineering, University of Pennsylvania, Philadelphia, Pennsylvania 19104.

the membrane surface cannot change. This can yield water exchange and rupture of the membrane. These effects, as well as electrical "breakdown," may participate in the complex cellular fusion process.

More detailed summaries of all these effects have been published (Schwan and Sher, 1969; Schwan, 1985a,b) as well as extended reviews on dielectrophoresis (Pohl, 1978) and cell fusion and electroporation (Zimmermann, 1982; Arnold and Zimmermann, 1984; Neumann and Rosenheck, 1972; Neumann *et al.*, 1982; Neumann, 1986). No attempt will be made to survey the large literature. Instead, we concentrate on principles and relevant physics for pearl chain formation, cell rotation, and certain field "coupling" considerations which are relevant.

2. PEARL CHAIN FORMATION OF CELLS

2.1. Earlier Theories

To our knowledge, Krasny-Ergen (1936, 1937) was the first to present a theory of pearl chain formation. He applied the principle that a state of minimal potential energy is approached by any physical system provided that the system is permitted to seek it. He furthermore assumed that the particles have metallic properties, in order to simplify the complexity of the problem. No comparison with experimental data was made. Schwan and Sher (1969), Saito and Schwan (1961), and Takashima and Schwan (1985) were probably the first to study, theoretically and experimentally, this effect. As mentioned, a minimal field strength is needed to cause whatever field effect. We term this minimal field value, the threshold field strength E_{th}. Saito and Schwan (1961) presented an equation for E_{th} for pearl chain formation which was based on an expression derived for the potential electrical energy of a particle suspended in a medium of different dielectric properties by Schwarz (1963). This equation was also used by Sher (1968) to calculate the dielectrophoretic force acting on a particle exposed to an inhomogenous field

$$F = 2\pi R^3 \operatorname{Re}\left(\epsilon_2{}^* \frac{\epsilon_1 - \epsilon_2}{\epsilon_1 + 2\epsilon_2} \right) \nabla (E^2) \rightarrow -4.6 R^3 \epsilon_1 \nabla (E^2) \tag{1}$$

where the expression following the arrow was shown to result for biological cells. The subscripts 1 and 2 indicate particle and medium, * denotes complex conjugate, ϵ's are complex dielectric constants $\epsilon = \epsilon' - j\epsilon''$, and R is the particle radius.

This equation states that the force acting on cells in proportional to the dielectric constant of the particle instead of that of the medium, a result which was entirely unexpected. It could not be reconciled with experimental data casting doubt on the validity of the Schwarz equation and the Saito–Schwan expression for E_{th}.

Another simple equation was proposed by Schwan for pearl chain formation and based on the Langevin criteria for significant orientation of polar dipoles (Schwan, 1985a,b). This principle states that preferential orientation of dipoles emerges above the random orientation caused by thermal motion if the field strength is greater than the threshold E_{th} given by

$$\mu E_{th} = KT \tag{2}$$

where μ is the induced dipole moment, K is the Boltzmann constant, and T is the absolute temperature. Introducing the value for μ yields

$$E_{th}^2 \, R^3 \epsilon_2{}' \, \frac{\epsilon_1 - \epsilon_2}{\epsilon_1 + 2\epsilon_2} = KT \tag{3}$$

However, the applicability of the Langevin criterion to induced dipoles may be questioned. It is also not clear to what phenomena E_{th} in Eqs. (2) and (3) refer, pearl chain formation or orientation of nonspherical particles, etc. Another disadvantage of Eqs. (2) and (3) is that the distance between two particles is not included. Furthermore, it is not clear from Eqs. (2) and (3) what part of the ratio $(\epsilon_1 - \epsilon_2)/(\epsilon_1 + 2\epsilon_2)$ needs to be considered. Is it its magnitude, its real or imaginary part? In practice, the threshold values for pearl chain formation and orientation appear not to differ too much and Eq. (3) appears to describe the experimental data as long as it is used to estimate approximate values (Figs. 1 and 2).

2.2. Recent Theoretical Advances

The potential energy equation given by Schwarz (1963) has been criticized since the meaning of potential energy in the presence of dissipative and dispersive properties for particles and surrounding medium is questionable (Sauer, 1983). Sauer instead used the Maxwell stress tensor to calculate the forces acting on the particle. His rigorous theory correctly predicts trajectories passed by particles if their original separation is not in line with the field. Sauer and Schloegl (1985) also gave an equation for the threshold field strength for pearl chain formation as follows:

$$E_{\text{th}}^2 R^3 \epsilon_2' \left[\text{Re} \left(\frac{\epsilon_1 - \epsilon_2}{\epsilon_1 + 2\epsilon_2} \right) \right] = \frac{KT}{6p} \quad (4)$$

It considers the original separation of the two particles by introduction of the particle volume concentration. Sauer's result was derived from a series development of the potential. However, convergence of the series may be poor at small particle distances. We also note that Eqs. (3) and (4) are nearly identical, at least for p values near 16% and $(\epsilon_1 - \epsilon_2)/(\epsilon_1 + 2\epsilon_2)$ ratios near 1.

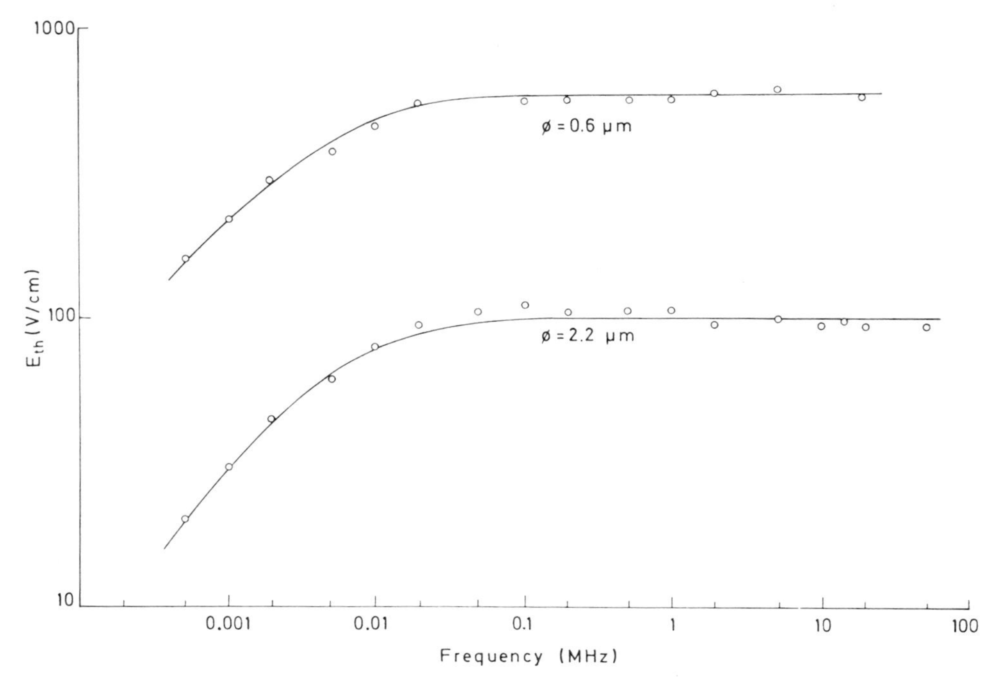

Figure 1. Threshold field strength for pearl chain formation as a function of frequency. Silicon particles with two different diameters. From Sher (1961) and Takashima and Schwan (1985). The high-frequency data are well explained by theory. The low-frequency decrease may be caused by a change in the medium's dielectric constant (see text).

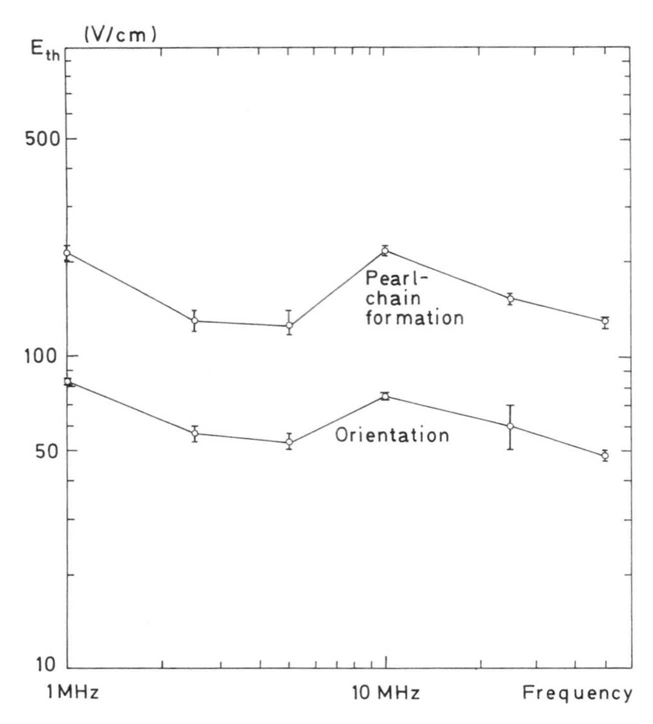

Figure 2. Threshold field strength values E_{th} for orientation and pearl chain formation for *E. coli*. A similar frequency dependence is indicated as anticipated from theory (Sher, 1961). The variations with frequency probably reflect corresponding variations of the *u*-factor.

2.3. Experimental Data

Experimental data are largely confined to those published by Schwan and Sher (1969) and Schwan (1985a). These data fit theoretical predictions based on the Sauer equation (4) as well as Eq. (3) fairly well, at least at high frequencies. Schwan (1985b) gives a summary of such data and Figs. 1 and 2 some frequency dependences. At low frequencies, E_{th} decreases about tenfold, which may be due to the medium's dielectric constant. This value increases substantially if the particles observed are surrounded by other particles and consequently is characterized by the bulk dielectric constant of the medium. Its large value at low frequencies is caused by polarization of the ionic atmosphere surrounding the particles (Schwan *et al.*, 1962; Schwarz, 1962). There is no quantitative confirmation of this effect since precise volume fractions used in the experiments are not available.

2.4. Conclusions

The basic physics of the pearl chain formation effect has been well established for quite some time. Recently, significant advances have been made, yielding precise expressions for the force between two cells and their response. Earlier experimental threshold data fit fairly well. More data are needed to explain the low-frequency behavior shown in Fig. 1. Threshold values for various particle concentrations should be established, while at the same time measuring the dielectric properties of the bulk suspension.

3. CELLULAR COUPLING CONSIDERATIONS

In order to understand how electrical fields interact with cells, we shall briefly discuss how electrical fields "couple" into cells. The analytical tools are provided by the solution of the Laplace equation for the potential caused by an external alternating field. The general solution of this problem is given in the Appendix of this chapter.

3.1. The Membrane Potential

We consider a spherical cell with a homogeneous isotropic cytoplasm surrounded by a thin membrane as indicated in Fig. 10. The general solution for the membrane potential is given in the Appendix. It simplifies for small thickness values of the membrane to Eq. (A15). If cytoplasm and medium are purely resistive so that the influence of their dielectric constants can be neglected, this equation reduces to

$$V_m = \frac{1.5\,ER}{1 + RG_m(\rho_i + 0.5\,\rho_a)}\;\frac{1}{1 + j\omega T}\cos\delta \tag{5}$$

Here ρ_i and ρ_a are resistivities in and outside the cell, R the cell radius, and δ the angle between E and radius vector. T is a time constant given in Eq. (8) and G_m the membrane conductance. If the membrane conductance can be neglected, the following equation results:

$$|V_m| = \frac{1.5ER}{\sqrt{1 + (\omega T)^2}}\cos\delta \tag{6}$$

If the cytoplasmic and extracellular dielectric constants cannot be neglected, more complicated equations result. Comparison of Eqs. (5) and (6) readily permits estimates under what circumstances the membrane conductance may be neglected. Figures 3 and 4 show typical frequency dependences of the membrane potential. The frequency-independent low-frequency limit value is given by

$$V_0 = \frac{1.5\,ER}{1 + RG_m(\rho_i + 0.5\,\rho_a)}\cos\delta \tag{7}$$

Above the "characteristic" frequency $f_0 = 1/2\pi T$ with T given by

$$T = \frac{RC_m(\rho_i + 0.5\,\rho_a)}{1 + RG_m(\rho_i + 0.5\,\rho_a)}\,* \tag{8}$$

the membrane potential is declining and for neglibible dielectric constants ϵ_1 and ϵ_3 and for $\omega T >> 1$

$$V_\infty = \frac{1.5E}{\omega C_m(\rho_i + 0.5\,\rho_a)}\cos\delta \tag{9}$$

It is inversely related to frequency until a high-frequency limit can be derived from Eqs. (A2) and (A9) which is for $d << R$ and $\epsilon_1 = \epsilon_3$

$$V_\infty = \frac{\epsilon_1}{\epsilon_2}\,dE\,\cos\delta \tag{10}$$

*This time constant is identical to the time constant which defines the center of the dielectric dispersion of a suspension of cells modeled as stated above (Schwan, 1957; Pauly and Schwan, 1959).

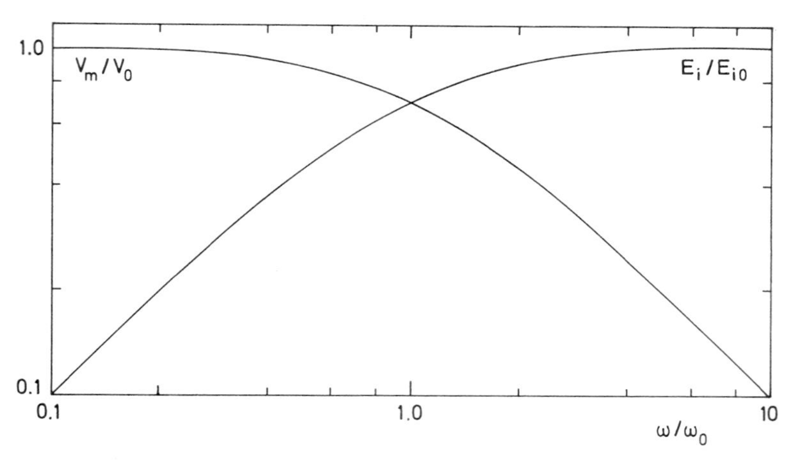

Figure 3. Frequency dependence of the membrane potential V and cytoplasmic field strength E_i induced by a frequency-independent extracellular field. V and E_i are normalized with regard to their limit values V_0 and E_{i0}, respectively. Membrane potential and internal potential drop E_iR add vectorially to the total potential applied to the cell.

It is reached when the potential is only related to the dielectric constants of the internal and external media and that of the membrane and orders of magnitude smaller than V_0. For cells in a natural environment, internal and external fluids have a conductivity of about 10 mS/cm. Hence, membrane potentials in the microwave frequency range are about 1000 times lower than at low frequencies. Therefore, membrane interactions are less likely to occur at such high frequencies.

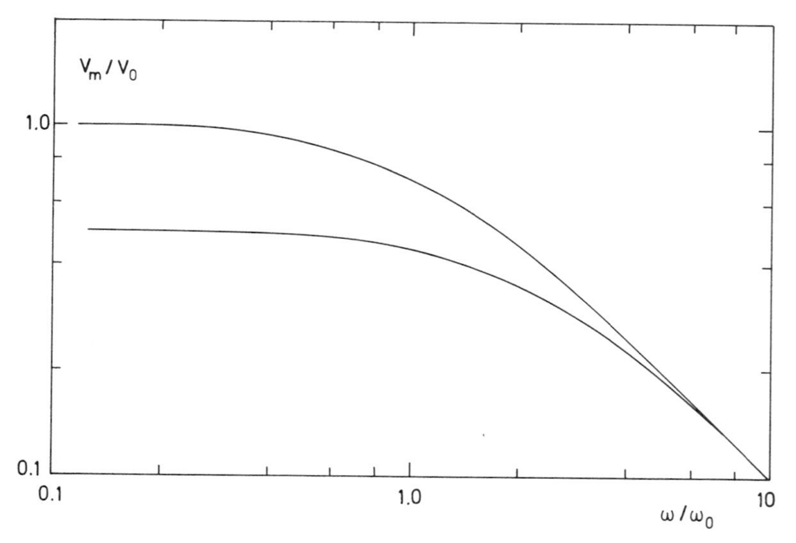

Figure 4. Frequency dependence of the membrane potential V for two different membrane conductances G_m. Upper curve, $G_m = 0$. Lower curve, $G_m = 1/R(\rho_i + 0.5\,\rho_m)$. The curves are normalized with regard to the low-frequency-limit potential $V_0 = 1.5ER$ for the case $G_m = 0$.

3.2. Internal Field Strength

The internal field strength is derived in the Appendix. For a small membrane thickness, it is given by:

$$E_i = \frac{3K_1K_2\,E}{K_2(K_3+2K_1)-2\,\frac{d}{R}\,(K_2-K_1)(K_3-K_2)} \tag{11}$$

This differs from V_m by the ration K_2/K_3d as seen from Eqs. (A15) and (A16). Equation (11) reduces for negligible influence of the cytoplasmic and medium dielectric constant's ϵ_1 and ϵ_3 to

$$E_i = \frac{1.5ER\,\rho_i}{1+RG_m(\rho_i+0.5\rho_a)}\;\frac{G_m+j\omega C_m}{1+j\omega T} \tag{12}$$

and for negligible membrane conductance G_m to

$$|E_i| = \frac{1.5\;E}{1+0.5\;\rho_a/\rho_i}\;\frac{\omega T}{\sqrt{1+\;(\omega T)^2}} \tag{13}$$

Typical frequency dependences are presented in Fig. 3. At very low frequencies, the field strength is zero, at least if the membrane conductance is neglected. Then the cell interior is shielded by the capacitive membrane. Thus, the membrane receives the total potential applied to the cell, thereby multiplying the external field strength E to the membrane field strength level E_m by the factor R/d. On the other hand, at very high frequencies the value of E_i becomes constant. The membrane capacitance is short-circuited and the total potential applied to the cell by the external field is available to the cytoplasm.

As stated before, external and internal fluid compartments may be considered as purely resistive under physiological circumstances. However, the external resistivity is often substantially increased in order to reduce heat development detrimental to the manipulation of cells for various purposes. In such cases, the dielectric constant of the medium must be taken into account and more complex equations are appropriate.

3.3. Nuclear Membrane Potential

The field strength of the nuclear membrane and that of any other organelle inside the cell can be obtained readily from the above equations. We assume that the nucleus or organelle is small compared to the cell itself. Then the field acting on the nucleus is given by E_i, i.e., Eqs. (12) and (13), instead of E. We thus obtain the nuclear membrane potential by replacing E in Eq. (5) by E_i and obtain for the case that the fluid dielectric constants can be neglected

$$V_n = \frac{2.25R_nRE\;\rho_i\cos\delta}{[1+RG_n(\rho_n+\frac{1}{2}\rho_i)][1+RG_m(\rho_i+\frac{1}{2}\rho_a)]}\qquad\frac{G_m+j\omega C_m}{(1+j\omega T)(1+j\omega T_n)} \tag{14}$$

with

$$T_n = R_nC_n(\rho_n+\tfrac{1}{2}\rho_i) \tag{15}$$

If $G_n = G_m = 0$

$$V_n = \frac{2.25\;R_nRE\;\rho_i}{(1+j\omega T)(1+j\omega T_n)}\;j\omega C_m\;\cos\delta \tag{16}$$

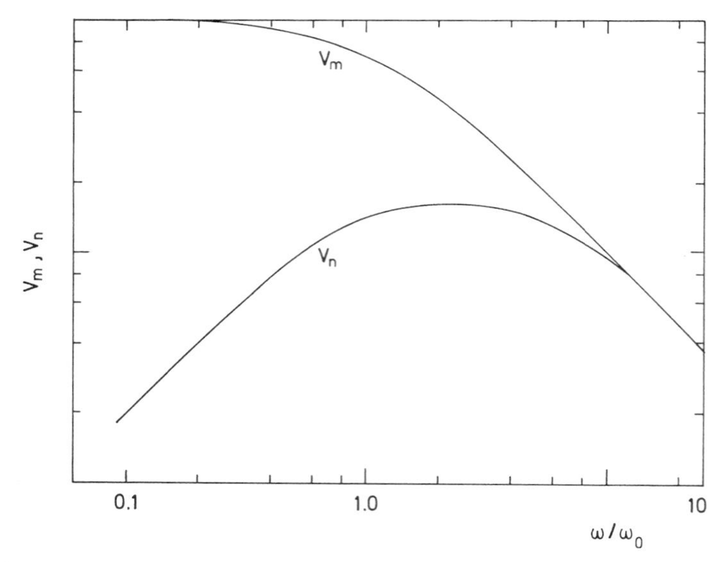

Figure 5. Frequency dependence of cellular and nuclear membrane potential and V_n, respectively, for an extracellular field equal to $1/1.5ER$. The nuclear membrane potential displays two cutoff frequencies which are identical to that of the cell and that of the nucleus. Identical resistivities of all fluid media and the two membrane capacitances C_m and C_n are assumed. The ratio of cellular to nuclear radius is assumed to be 5. In the chosen case, the potentials become identical at high frequencies. In most cases this is not the case and the nuclear potential may be larger or smaller than the potential across the cell.

where subscript n indicates the nucleus or organelle and ρ_n, ρ_i, ρ_a are resistivities in the nucleus, cytoplasm, and extracellular space. Figure 5 shows typical frequency dependences for V_m and V_n. In the same fashion we obtain the field strength inside the nucleus or organelle by replacing E in Eqs. (12) and (13) by E_i. The more general solution is

$$E_n = \frac{2.25\, R_n R E\, \rho_i\, \rho_n}{[1+RG_n(\rho_n+\frac{1}{2}\rho_i)][1+RG_m(\rho_i+\frac{1}{2}\rho_a)]} \quad \frac{(G_n+j\omega C_n)(G_m+j\omega C_m)}{(1+j\omega T)(1+j\omega T_n)} \tag{17}$$

The equation

$$E_n = \frac{2.25\, R_n R E\, \rho_i\, \rho_n}{(1+j\omega T)(1+j\omega T_n)} \quad (-\omega^2 C_n C_m) \tag{18}$$

holds if membrane conductances can be neglected. Equations (17) and (18) are valid if dielectric constants of internal and external fluids can be neglected. In all of the above equations, the membrane thickness values are assumed to be small compared to the radius of the cell and organelle.

We next compare the cellular and nuclear membrane potentials at high frequencies. Dividing Eqs. (6) and (16) with ωT, $\omega T_n > 1$ yields

$$\frac{V_n}{V_m}\,(\omega T,\ \omega T_n>1) = 1.5\,\frac{C_m}{C_n} \quad \frac{\rho_i}{\rho_n+0.5\,\rho_i} \tag{19}$$

Thus, the nuclear membrane potential can be larger or smaller than the cellular membrane potential, depending on membrane capacitances and fluid resistivities. If the membrane capacitances C_n and C_m are equal and the resistivities ρ_i and ρ_n are equal, the two potentials become identical.

3.4. Conclusions

Simple relations hold for membrane potential, cytoplasmic field strength, and nuclear membrane potential for many cases of practical interest. The cell membrane potential decreases above the characteristic frequency of the β dispersion which characterizes the RF behavior of cell suspensions (Schwan, 1957). The internal field strength then reaches its optimal value. At lower frequencies the interior of the cell is well shielded. The nuclear membrane potential has two cutoff frequencies. They are given by the time constants for the β dispersions for cell and nucleus. Only in this limited frequency range can a significant nuclear membrane potential be induced by external fields. At high frequencies the nuclear membrane potential may be larger or smaller than the membrane potential itself, depending on membrane capacitances and resistivities of the various fluid compartments. No detailed calculations have been carried out for the case that the volume of the nucleus occupies a noticeable fraction of the cell interior. Such calculations are suggested in order to determine in greater detail if and how the nuclear membrane may be selectively targeted by the external field in order to achieve fusion of several nuclei.

4. ROTATION OF CELLS

4.1. Early and Recent Theories

Particles immersed in a medium of different dielectric properties will experience a torque if exposed to a rotating electrical field. This torque causes them to rotate around an axis perpendicular to the plane defined by the rotating electrical field vector. This was recognized very early and recently summarized by Fuhr (1985) and Glaser and Fuhr (1986). Early theoretical work used simple assumptions such that the properties of particle and medium are purely resistive or dielectric and frequency independent. Particles of spherical shape and isotropic properties were assumed in order to derive simple closed form solutions for the torque.

During the past decade the rotation of biological cells has become of particular interest. Arnold and Zimmermann (1982) observed that the speed of rotation is a function of frequency as indicated in Fig. 6 and they showed this frequency to be a linear function of medium conductivity as in Fig. 7. They proposed that the time constant, T, which determines the charging of cell membranes also characterizes the peak frequency of rotation $f_0 = 1/2\pi T$. This time constant was derived by Schwan (1957) and is given by Eq. (8). It is identical to that of the "β dispersion" which characterizes the frequency dependence of the dielectric properties of cell suspensions in the radio frequency range. Zimmermann and Arnold (1983) discussed how the β-dispersion equations might be applied to cell or vesicle rotation.

Sauer and Schloegl (1985) presented an equation for the torque experienced by a sphere:

$$L = 4\pi\epsilon_m R^3 E^2 I_m \left[\frac{K'_p - K_m}{K_p + 2K_m} \right] = 4\pi\epsilon_m R^3 E^2 u'' \qquad (20)$$

Here, E is the strength of the rotating electrical field, ϵ_m is the real dielectric constant of the medium surrounding the sphere, and K_m and K_p are complex admittivities $K = \sigma + j\omega\epsilon$. The subscripts indicate particle and medium properties, and u is the Clausius–Mosotti ratio of K_p and K_m. The Sauer–Schloegl expression applies for lossy dielectric properties of particle and medium.

Fuhr (1985), Fuhr *et al.* (1986), and Glaser and Fuhr (1986) derived equations for the torque acting on biological cells. Their model assumed a membrane surrounding an isotropic cytoplasm of spherical shape with dielectric constants and conductivities of all phases assumed frequency independent. The theory predicts two peaks of the torque versus frequency plot, usually of opposite sign. The complicated set of equations reflect equally complicated equations developed by Pauly and Schwan (1959) for the dielectric frequency response of the same cellular model suspended in a medium. The

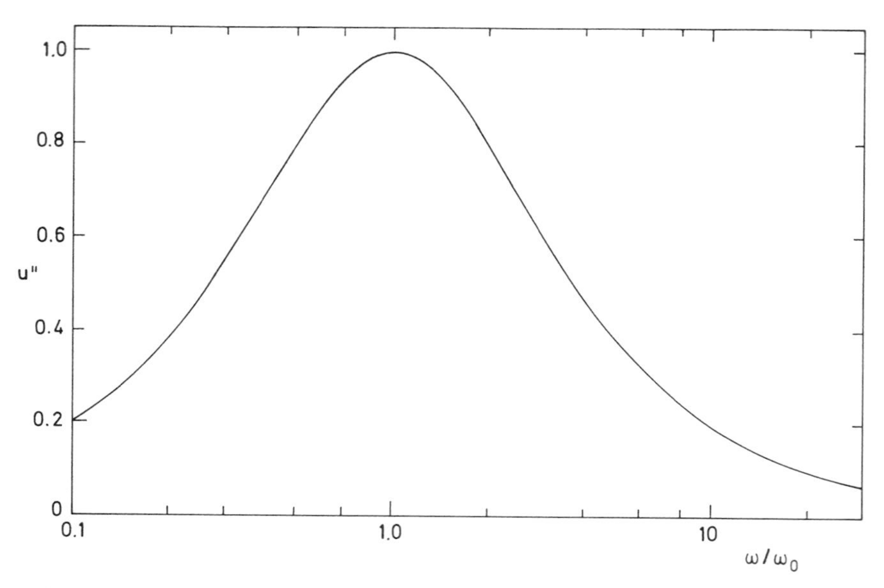

Figure 6. Frequency dependence of the torque experienced by a particle in a rotating field. Biological cells experience at least two such peaks, usually of different sign (corotation and antirotation). Frequency independence of the medium's dielectric constant is assumed, assuring proportionality of torque and u''.

equations do not permit development of closed form expressions for the cellular parameters. Therefore, iterative numerical methods are needed to extract these values. Furthermore, the model does not consider additional relaxation effects caused by protein-bound water and partial molecular rotation in the VHF region, nor counterion relaxation effects responsible for low-frequency dependences (Schwan, 1957; Schwan *et al.*, 1962; Schwarz, 1962). However, with some simple assumptions, the general solution simplifies considerably. Long ago, this was demonstrated for dielectric spectroscopy (Schwan, 1957) and more recently for electrorotation (Fuhr, 1985; Schwan, 1985c).

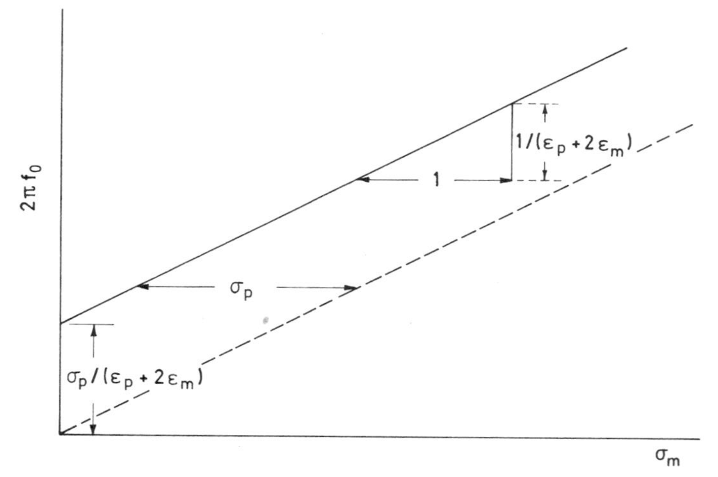

Figure 7. Peak rotation frequency as a function of medium conductivity for particle of frequency-independent properties. Slope and intercept serve to determine these properties (ϵ_p and σ_p).

4.2. Electrorotation of Cells: A Simple Approach

We first consider electrorotation of a spherical particle of frequency-independent dielectric constant ϵ and conductivity σ. Subscript p indicates particle and subscript m the surrounding medium. In this case, the torque is readily calculated from Eq. (20):

$$L = 4\pi\epsilon_m R^3 E^2 \times 3\omega \frac{\epsilon_p\sigma_m - \epsilon_m\sigma_p}{(\sigma_p+2\sigma_m)^2+\omega^2(\epsilon_p+2\epsilon_m)^2} \tag{21}$$

Optimal rotation is achieved when the resonance condition

$$f_0 = 1/2\pi T_0 = \frac{1}{2\pi} \quad \frac{\sigma_p + 2\sigma_m}{\epsilon_p + 2\epsilon_m} \tag{22}$$

is fulfilled and the peak amplitude of u'' is

$$A_0 = 1.5 \frac{\epsilon_p\sigma_m - \epsilon_m\sigma_p}{(\sigma_p+2\sigma_m)(\epsilon_p+2\epsilon_m)} \tag{23}$$

Equation (21) is very similar to that of the imaginary component of the complex dielectric constant of a suspension of particles (Schwan, 1957).

We next consider a particle whose complex conductivity undergoes a dispersion as stated by

$$K_p = \sigma_p + J\omega\epsilon_p = \sigma_\infty \frac{j\omega T}{1+j\omega T} \tag{24}$$

Usually, electrorotation work is carried out at frequencies below some MHz. Then the medium's conductivity is frequency independent. Furthermore, the dielectric constant does not add much to the specific admittance of the electrolyte medium. Therefore, the term K_m in Eq. (20) can be assumed to be frequency independent. Equation (20) then contains only one frequency-dependent property, i.e., the particle's complex dielectric constant or admittivity. Then the torque is given by (Schwan, 1985c)

$$L = 4\pi\epsilon_m R^3 E^2 \times 3\omega T \frac{\sigma_p\sigma_\infty}{4\sigma^2+(\omega T)^2(\sigma_\infty+2\sigma_p)^2} \tag{25}$$

and L peaks when $f_0 = 1/2\pi T_0$ with

$$T_0 = T(1 + \sigma_\infty/2\sigma_p) \tag{26}$$

where the time constants (T) in Eqs. (24) and (26) are identical. We also note that T_o is identical to the time constant of the β dispersion given in Eq. (8) for the case $G_m = 0$.

Next we introduce an additional conductance term, σ_0, in Eq. (24). This conductance term is given by the low-frequency conductivity of the particle or cell and is related to the membrane conductance or surface conductance as stated before (Schwan, 1957; Schwan *et al.*, 1962):

$$K_p = \sigma_0 + (\sigma_\infty - \sigma_0) \frac{j\omega T}{1+j\omega T} \tag{27}$$

The membrane conductance (G_m) contribution to σ_0 is equal to RG_m and the surface conductance (G_s) contribution is $2G_s/R$ with R the radius. Frequently, the membrane conductance contribution and that of surface conductance can be comparable.

The following equations are obtained. The torque is equal to

$$L = 4\pi\epsilon_m R^3 E^2 \frac{3\omega T\sigma_m(\sigma_\infty - \sigma_0)}{(\sigma_0 + 2\sigma_m)^2 + (\omega T)^2(\sigma_\infty + 2\sigma_m)^2} \quad (28)$$

The peak amplitude of u'' is

$$A_0 = \frac{1.5\ \sigma_m\ (\sigma_\infty - \sigma_0)}{(\sigma_0 + 2\sigma_m)(\sigma_\infty + 2\sigma_m)} \quad (29)$$

The peak frequency is $f_0 = 1/2\pi T_0$ with

$$T_0 = T\,\frac{\sigma_\infty + 2\sigma_m}{\sigma_0 + 2\sigma_m} \quad (30)$$

We consider briefly the case that the dispersive particle is a biological cell, assuming for simplicity that membrane and surface conductances are negligible and that the effects of the dielectric constants of the fluid compartments inside and outside the cell may be neglected. Then $\sigma_0 = 0$, σ^∞ is identical to the inverse of the cell's interior resistivity ρ_i, and the cell's time constant is equal to $T = RC_m\rho_i$ as may be seen from Schwan (1957), Eq. (15). Thus, the previously quoted value for the time constant $T_0 = RC_m[\rho_i + 0.5\rho_a]$ is obtained.

Equations (29) and (30) may be used to study other dispersive effects of cells and particles than the β dispersion which characterizes the membrane charging process. They provide additional insight into surface conductance values of charged particles and possibly the mechanism responsible for the large dielectric increment caused by counterion movement.

4.3. Discussion of the Simple Equations

We first discuss to what extent several rotation peaks may affect each other. We consider three cases of particular interest: the case of a particle with surface charges and the case of a cell without and with surface charges.

1. Particles with surface charges exhibit a low-frequency dispersion which may be modeled as stated in Eq. (24) (Schwarz, 1962). The corresponding rotation peak is expected at low frequencies if the particles themselves are poorly conducting. In addition, they display the Maxwell–Wagner rotation peak stated in Eqs. (21) and (22). This peak occurs in the 100 MHz range. Clearly, the two ranges are very far apart and should not affect each other. Experimental evidence supports this conclusion.
2. A biological cell without surface charges will exhibit two RF-rotation peaks. They correspond to the β dispersion and the Maxwell–Wagner peak given by Eqs. (21) and (22). The electrical properties of cell suspensions in the RF range can be expressed as a sum of two relaxation effects (Pauly and Schwan, 1959):

$$\epsilon^* = \epsilon_\infty + \frac{\Delta\epsilon_1}{1 + j\omega T_1} + \frac{\Delta\epsilon_2}{1 + j\omega T_2} \quad (31)$$

Here $\Delta\epsilon_1$, $\Delta\epsilon_2$, T_1, and T_2 are the magnitudes and time constants of the two dispersions. For small membrane thickness values, the mechanism indicated by subscript 1 characterizes the charging of the cell membrane through cytoplasmic conductivities. The mechanism indicated by subscript 2 is identical to the Maxwell–Wagner effect of the cytoplasmic particle surrounded by a short-circuited membrane and suspended in the medium, which must prevail at high frequencies. This effect occurs above 100 MHz. This is orders of magnitude higher than the frequency range for term 1, the β-dispersion range as discussed earlier. Since the torque equation (20) depends on particle dielectric properties, it is expected that the two dispersions in Eq. (31) cause two rotation functions of the type given in Eqs. (21) and (25). This has been shown in detail by Fuhr (1985) and Glaser and Fuhr (1986). The high-frequency peak is about 100 MHz, since the cell conductivity is near 10 μS/cm at these frequencies. On the other hand, the low-frequency peak is well below 1 MHz, since rotation experiments are usually carried out with media of some μS/cm conductivities. Hence, the frequency ratio is large enough to permit the decoupled treatment suggested by the use of the above-quoted equations.

3. A cell with surface charges is anticipated not only to display the two rotation peaks discussed under item 2. In addition, there will be the low-frequency peak discussed in item 1 (α peak). Little experimental data are available to permit judgment if the decoupled treatment, i.e., use of the above simple equations, is permissible for α and β peaks. Some sample calculations by Sauer and Schloegl (1985) indicate that α and β peaks may affect each other. The low-frequency dispersion time constant T_α is (Schwarz, 1962)

$$T_\alpha = \frac{R^2}{2\mu K T_c} \rightarrow f_\alpha = a/R^2 \tag{32}$$

with μ the ionic mobility, K the Boltzmann constant, and T_c the absolute temperature. The constant a has a value of 10^{-6} Hz cm² (Schwan *et al.*, 1962). The β-peak time constant is given in Eq. (8). Therefore, the ratio T_α/T_β is proportional to the radius R. Usually, the medium electrolytes are very diluted to avoid heating. Then the ratio

$$T_\beta/T_\alpha = 10^{-6}\pi C_m \rho_m/R \tag{33}$$

with ρ_m the medium resistivity and C_m the membrane capacitance per area unit. This ratio is near 10^{-3} for resistivities about 10^5 ohm/cm, a membrane capacitance of 1 μF/cm², and a radius of 3 μm. For lower resistivities, it is even smaller. Thus, the two peak frequencies appear to be far apart, suggesting the applicability of the decoupled approach. However, Eq. (33) is based on the Schwarz equation (32). It fits available particle data fairly well (Schwan *et al.*, 1962), but not necessarily biological cells (Schwan, 1957). For biological cells, low-frequency dispersions occur at significantly higher frequencies than suggested by Eq. (32), indicating that the low-frequency and β dispersions are not as widely separated as suggested by Eq. (33).

A final comment concerns dispersion equation (27). It does not include a high-frequency-limit dielectric constant ϵ_∞ and is therefore incomplete. Inclusion of such a high-frequency dielectric constant unfortunately complicates the problem considerably, preventing simple closed form expressions. Use of Eqs. (21) through (30) is only possible if the influence of the high-frequency-limit dielectric constant ϵ_∞ is small or if the dielectric increment $\epsilon_0 - \epsilon_\infty$ is large compared to ϵ_∞. More detailed calculations confirm this statement and justify the use of the dispersive admittivity model.

4.4. Methods of Evaluation

We conclude with a brief summary of techniques used to evaluate particle and cell properties from electrorotation experiments.

First we indicate the technique used to extract particle parameters. In Fig. 7 the peak rotation frequency is plotted against the medium conductivity. From Eq. (22), a straight line is anticipated provided that particle properties do not change with σ_m. The intercept of the line with the ordinate gives $\sigma_p/(\epsilon_p + 2\epsilon_m)$. The slope is equal to $1/(\epsilon_p + 2\epsilon_m)$. Thus, both ϵ_p and σ_p can be extracted.

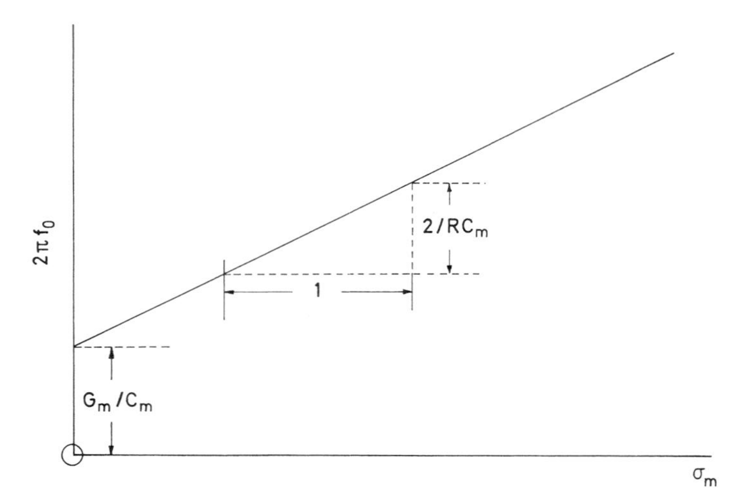

Figure 8. Peak rotation frequency as a function of medium conductivity for a spherical biological cell. Slope and intercept serve to determine membrane capacitance C_m and conductance G_m.

A similar technique was introduced by Arnold and Zimmermann (1982) to extract cell parameters as indicated in Fig. 8. Now the intercept and the slope give the membrane capacitance C_m and conductance G_m.

Finally we plot in Fig. 9 the peak amplitude of rotation against σ_m. A hyperbola is obtained as stated by Eq. (23). Numerical discussions reveal that this hyperbola has limit values which are only

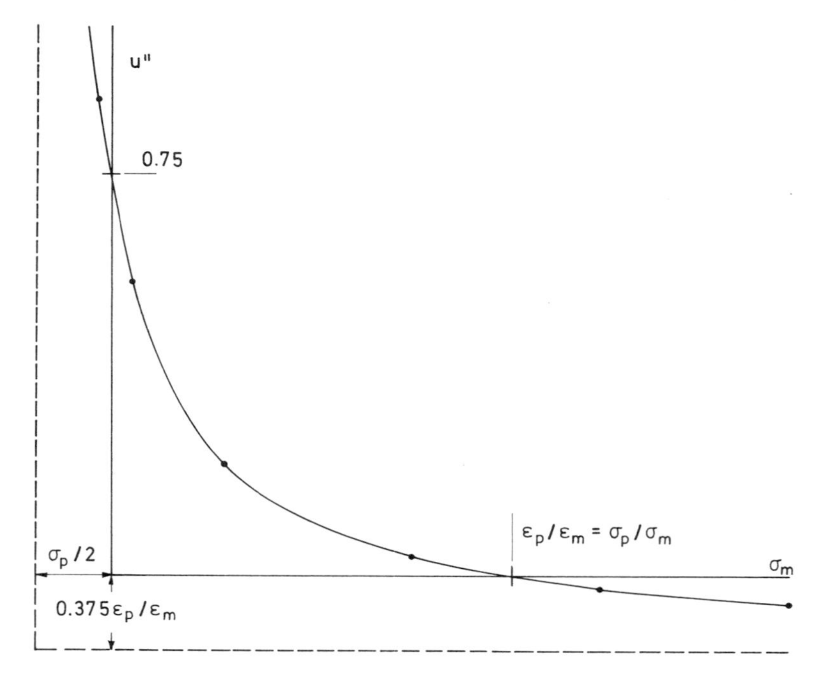

Figure 9. Amplitude of torque and u'' as a function of medium conductivity for a particle of frequency-independent properties. The function is a hyperbola slightly displaced from the σ_m and u'' axes as indicated.

slightly displaced from the axes. Hence, the inverse amplitude is almost a straight line. Its intercept with the ordinate and slope can also be used to extract particle parameters.

Another technique to study cells combines measurement of the β-peak frequency, Eq. (8), with the Maxwell–Wagner high-frequency-peak equation (22). This technique, however, assumes that the high-frequency peak is solely due to the interaction of the cytoplasm with the medium. However, protein-bound water adds another dispersion in this frequency range (Schwan, 1957, 1965; Grant, 1965). This limits the applicability of the approach.

5. CONCLUSIONS

We have discussed pearl chain formation as a prerequisite tool to achieve cell fusion. The cellular coupling considerations, presented here in somewhat greater detail than before, provide cellular potentials and field strength values in the membrane, cytoplasm, and nucleus. They are of use in electroporation applications. They furthermore provide insight into the detailed cellular responses to rotating fields. For example, the rotation of a cellular organelle can be readily predicted by inserting the cellular internal field into the equations for rotation. We finally concentrated on the electrorotation technique which complements dielectric spectroscopy.

We did not discuss other manifestations of field effects on particles and cells. For example, from the equations in the Appendix we derive equations which pertain to membrane compression and cell elongation which can be caused by alternating fields. But relevant experimental work is not discussed. Nor did we survey other ample experimental data, since such data are provided by other contributions to this volume.

6. APPENDIX: POTENTIAL INSIDE AND OUTSIDE A SPHERICAL CELL

We solve the Laplace potential differential equation for a shell-surrounded sphere. A spherical particle of radius R_1 is assumed, surrounded by a membrane of thickness $d = R - R_1$ as indicated in Fig. 10. Specific admittivities $K = \sigma - j\omega\epsilon$ specify the electrical properties of the three phases. The subscripts 1, 2, and 3 indicate external medium, membrane properties, and core, respectively.

We write the potential ϕ in the three regions:

$$\phi_1 = E(r - \alpha/r^2) \cos \delta \tag{A1}$$

$$\phi_2 = (Ar - B/r^2) \cos \delta \tag{A2}$$

$$\phi_3 = Cz = Cr \cos \delta \tag{A3}$$

E is the external unperturbed field, r the radius vector, and δ the angle between E and r. Coefficients A, B, C, and α are to be determined from boundary conditions.

Each equation states a sum of a homogeneous and a dipole field. The dipole term in Eq. (A3) vanishes in order to maintain the field finite at the center. The homogeneous field term in Eq. (A1) is chosen to provide a field strength identical to E for large values of r. The four constants α, A, B, and C are determined from two pairs of boundary conditions, each pair existing at each of the two interfaces 1–2 and 2–3, respectively. The boundary conditions are that the current density component normal to the surfaces is continuous and that the potentials cannot jump. A unique solution is thus obtained, obviating the need to introduce higher-order terms into the series expressions (A1) through (A4).

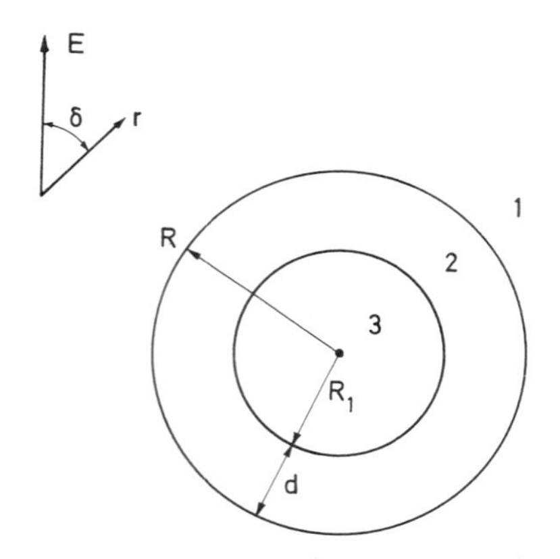

Figure 10. Geometry of the shell-surrounded sphere.

The boundary conditions to be applied to the two surfaces are:

$$K_1(\partial\phi/\partial r)_R = K_2\ (\partial\phi/\partial r)_R \tag{A4}$$

$$K_2(\partial\phi/\partial r)_R = K_3(\partial\phi/\partial r)_R \tag{A5}$$

$$E(R - \alpha/R^2) = AR - B/R^2 \tag{A6}$$

$$AR_1 - B/R_1{}^2 = CR_1 \tag{A7}$$

Solving the resulting set of four linear equations yields the following coefficients:

$$AR = 3K_1E(K_3 + 2K_2)/D \tag{A8}$$

$$B = 3K_1E(K_3 - K_2)/D \tag{A9}$$

$$C = 9K_1K_2E/D \tag{A10}$$

$$\alpha = [R^3(2K_2 + K_3)(K_1 - K_2) + R_1{}^3(2K_2 + K_1)(K_2 - K_3)]/D \tag{A11}$$

with the common denominator D

$$D = (K_2 + 2K_1)(2K_2 + K_3) + 2(R_1/R)^3(K_2 - K_1)(K_3 - K_2) \tag{A12}$$

The more complex Eqs. (A11) and (A12) can be simplified by writing

$$(R_1/R)^3 = (1 - d/R)^3 = 1 - 3d'/R \tag{A13}$$

where $d'/d = 1 - d/R + (d/R)^2/3$ and d' is identical to the membrane thickness for $d << R$. We need only the simplified expression for D:

$$D/3 = K_2(K_3 + 2K_1) - 2(K_2 - K_1)(K_3 - K_2)d'/R = D' \quad \text{(A14)}$$

Now we can write the membrane potential by taking the difference of the two potentials ϕ_2 at the inner and outer membrane surface. For small values of d/R, we obtain for $\cos \delta = 1$

$$V_m = 3K_1K_3dE/D' \quad \text{(A15)}$$

The internal field strength is obtained from Eqs. (A3), (A10) and (A14):

$$E_i = 3K_1K_2E/D' \quad \text{(A16)}$$

with $D' = D/3$ given by Eq. (A14).

We finally discuss the optimal membrane field strength E_m which is obtained if $\cos\delta = 1$. From Eq. (A2)

$$E_m = -\partial\phi/\partial r = A + 2B/r^3 \quad \text{(A17)}$$

Introducing the coefficients A and B [Eqs. (A8) and (A9)] we calculate the difference in field strength values E_m at the inner and outer membrane surface:

$$E_1 - E_2 = \Delta E = 6K_1E(K_3 - K_2)d'/RD' \quad \text{(A18)}$$

This reduces for small K_2 (small membrane conductance and low frequencies) and small membrane thickness values to

$$\Delta E = 3E \quad \text{(A19)}$$

which is $R/2d$ times smaller than the average membrane field strength $E_m = 1.5ER/d$ derived from the low-frequency membrane potential quoted above. The cell membrane therefore experiences several forces. First, there is a compressional force given by the product of a membrane dielectric constant and membrane field strength squared, $\epsilon_m E_m^2/2$. Second, each membrane surface experiences a slightly different force, $\epsilon_m E^2/2$, yielding a net outwards-directed force, $\epsilon_m(E_1^2 - E_2^2)/2 = \epsilon_m 3EE_m$. This force note will tend to elongate the cell in the field direction. It is $R/4d$ times smaller than the compressional force.

Note: The above derivation was based on the Laplace equation. That equation is valid for all three phases provided that the condition of electroneutrality is fulfilled—that charges accumulate only at the phase boundaries, i.e., at the outer and inner membrane surfaces. This is not correct since excess charges are distributed near the phase boundaries over distances comparable to the Debye screening length. Thus, the appropriate treatment of the potential in the neighborhood of the membrane necessitates use of the Poisson instead of the Laplace equation. The corresponding dielectric problem has been solved by Garcia *et al.* (1985). Preliminary estimates indicate that the correction is small under physiological circumstances, but can be substantial if the medium is very diluted.

ACKNOWLEDGMENT. The author appreciates support for this article and related work provided by the University of Wüzburg during his affiliation with the Institute for Biotechnology in 1986–1987.

REFERENCES

Arnold, W. M., and Zimmermann, U., 1982, Rotating-field induced rotation and measurement of the membrane capacitance of single cell *Avena sativa, Z. Naturforsch.* **37C:**908–915.

Arnold, W. M., and Zimmermann, U., 1984, Electric field induced fusion and rotation of cells, in: *Biological Membranes,* Volume 5 (D. Chapman, ed.), Academic Press, London, pp. 381–444.

Fuhr, G. R., 1985, Ueber die Rotation dielektrischer Koerper in rotierenden Feldern, Dissertation, Humboldt-Universitat, Berlin, DDR.

Fuhr, G., Glaser, R., and Hagedorn, R., 1986, Rotation of dielectrics in a rotating high frequency field; model experiments and theoretical explanation of the rotation effect of living cells, *Biophys. J.* **49:**395–402.

Garcia, A., Barchini, R., and Grosse, C., 1985, The influence of diffusion on the permittivity of a suspension of spherical particles with insulating shells in an electrolyte, *J. Phys. D.* **18:**1891–1896.

Glaser, R., and Fuhr, G., 1986, Electrorotation of single cells—A new method for assessment of membrane properties, in: *Electric Double Layers in Biology* (M. Blank, ed.), Plenum Press, New York.

Grant, E. H., 1965, The structure of water neighboring proteins, peptides and amino acids as deduced from dielectric measurements, *Ann. N.Y. Acad. Sci.* **125:**418–427.

Krasny-Ergen, W., 1936, Nicht-thermische Wirkungen elektrischer Schwingungen auf Koloide, *Hochfreguenztech. Elektroakust.* **48:**126–133.

Krasny-Ergen, W., 1937, *Hochfrequenztech. Elektroakust.* **49:**195.

Liebesny, P., 1938, Athermic short-wave therapy, *Arch. Phys. Ther.* **19:**736.

Neumann, E., 1986, Chemical electric field effects in biological macromolecules, *Prog. Biophys. Mol. Biol.* **47:**197–231.

Neumann, E., and Rosenheck, K., 1972, Permeability changes induced by electric impulses in vesicular membranes, *J. Membr. Biol.* **10:**279–290.

Neumann, E., Schaefer-Ridder, M., Wang, Y., and Hofschneider, P. H., 1982, Gene transfer into mouse lyoma cells by electroporation in high electric fields, *EMBO J.* **1:**841–845.

O'Konski, C. T., 1981, A history of electro-optics, in: *Molecular Electro-Optics* (S. Krause, ed.), Plenum Press, New York, pp. 1–25.

Pauly, H., and Schwan, H. P., 1959, The impedance of a suspension of spherical particles surrounded by a shell, *Z. Naturforsch.* **14b:**125–131.

Pohl, H. A., 1978, *Dielectrophoresis,* Cambridge University Press, London.

Saito, M., and Schwan, H. P., 1961, The time constants of pearl-chain formation, in: *Biological Effects of Microwave Radiation,* Volume 1, Plenum Press, New York, pp. 153–176.

Sauer, F. A., 1983, Forces on suspended particles in the electromagnetic field, in: *Coherent Excitations in Biological Systems* (H. Froehlich and F. Kremer, eds.), Springer-Verlag, Berlin, pp. 134–144.

Sauer, F. A., and Schloegl, R. W., 1985, Torques exerted on cylinders and spheres by external electromagnetic fields: A contribution to the theory of field induced cell rotation, in: *Interactions between Electromagnetic Fields and Cells,* Volume 97 (A. Chiabrera, C. Nicolini, and H. P. Schwan, eds.), Plenum Press, New York, pp. 203–251.

Schwan, H. P., 1957, Electrical properties of tissue and cell suspensions, *Adv. Biol. Med. Phys.* **5:**147–209.

Schwan, H. P., 1965, Electrical properties of bound water, *Ann. N.Y. Acad. Sci.* **125:**344–354.

Schwan, H. P., 1985a, Biophysical principles of the interaction of ELF-fields with living matter. II. Coupling considerations and forces, in: *Biological Effects and Dosimetry of Static and Electromagnetic Fields,* Volume 19 (M. Grandolfo, S. M. Michaelson, and A. Rindi, eds.), Plenum Press, New York, pp. 243–271.

Schwan, H. P., 1985b, EM-field induced force effects in: *Interactions between Electromagnetic Fields and Cells,* Volume 97 (A. Chiabrera, C. Nicolini, and H. P. Schwan, eds.), Plenum Press, New York, pp. 371–389.

Schwan, H. P., 1985c, Dielectric properties of the cell surface and biological systems, *Stud. Biophys.* **110:**13–18.

Schwan, H. P., and Sher, L. D., 1969, Alternating-current field-induced forces and their biological implications, *J. Electrochem. Soc.* **116:**170–174.

Schwan, H. P., Schwarz, G., Maczuk, J., and Pauly, H., 1962. On the low-frequency dielectric dispersion of colloidal particles in electrolyte solution, *J. Phys. Chem.* **66:**2626–2635.

Schwarz, G., 1962, A theory of the low-frequency dielectric dispersion of colloidal particles in electrolyte solution, *J. Phys. Chem.* **66:**2636–2642.

Schwarz, G., 1963, General equation for the mean electrical energy of a dielectric body in an alternating electrical field, *J. Phys. Chem.* **39**:2387–2388.

Sher, L. D., 1961, Mechanical effects of AC-fields on particles dispersed in a liquid: Biological implications, Ph.D. thesis, University of Pennsylvania, Philadelphia.

Sher, L. D., 1968, Dielectrophoresis in lossy dielectric media, *Nature* **220**:695–696.

Takashima, S., and Schwan, H. P., 1985, Alignment of microscopic particles in electric fields and its biological implications, *Biophys. J.* **467**:513–518.

Zimmermann, U., 1982, Electric field-mediated fusion and related electrical phenomena, *Biochim. Biophys. Acta* **694**:227–277.

Zimmermann, U., and Arnold, W. M., 1983, The interpretation and use of the rotation of biological cells, in: *Coherent Excitations in Biological Systems* (H. Froehlich and F. Kremer, eds.), Springer-Verlag, Berlin, pp. 211–221.

Chapter 2

Cellular Spin Resonance

Ingolf Lamprecht and Maja Mischel

1. INTRODUCTION

Among the different effects living cells exhibit under the influence of external electric fields, rotation of cells is one of the most predominant ones. Teixeira-Pinto *et al.* (1960) were supposedly the first to report about the spinning of parts of amoebae in a high-frequency field; Füredi and Ohad (1964) observed rotation of erythrocytes and Pohl and Crane (1971) of yeasts. It became clear in the following years that rotation in external fields is a general phenomenon of cells, live as well as dead ones, and that the way of spinning could provide information on the biological and physiological state of the cell under investigation. As spinning often started at a very sharp frequency of the applied field and ceased in the same manner, Pohl and Crane (1971) coined the descriptive name "cellular spin resonance" (CSR) which will be used throughout this chapter.

Rotation of inanimate particles in static, AC, or rotating fields was observed nearly 100 years ago. Theories were developed which remain fundamental for all modern theories concerning CSR. A brief historical survey is presented in this chapter.

1.1. Historical Survey

The first papers on the rotation of bodies like spheres, cylinders, or plates in an electric field were published by Hertz (1880, 1881). As a special case, he described a rotating dielectric hollow sphere with surface charges which rotate with the supporting sphere, so that the external field slows down the spinning rate. By the turn of the century, it was known that a rotating electric field exerts a torque on a resting dielectric sphere. Quincke (1896) observed the rotation of dielectric bodies attached to silk threads hanging between two capacitor plates in an isolating liquid. Von Schweidler (1897) calculated the torque on a conducting isotropic sphere in a homogeneous electrostatic field rotating with a constant angular velocity and compared his theoretical results with the experimental ones of Quincke (1896).

The results of Arno (1892), Quincke (1896, 1897), and von Schweidler (1897) were used by

Ingolf Lamprecht and Maja Mischel • Institute for Biophysics, Free University of Berlin, D-1000 Berlin 33, Federal Republic of Germany.

Lampa (1906) who showed that the torque on dielectric spheres in rotating electrostatic fields linearly increases with the frequency to a maximum and then decreases hyperbolically again, depending on the conductivities and the dielectric constants of the medium and the content of the sphere.

Arno (1892) was the first to apply a four-plate electrode chamber similar to that used by Arnold and Zimmermann (1982) rendering a nonuniform field. With uniform fields, Lertes (1921a,b) proved a theory of Born (1920) for the mobility of ions in an electrolyte. He postulated that a dipole would follow an external rotating field with a lag due to the frictional drag of the surrounding liquid. This torque-producing effect occurs at frequencies in the GHz range and is difficult to measure as it is screened by other influences like polarization and conductivity.

1.2. CSR in AC Fields

In 1960, Teixeira-Pinto *et al.* observed the spinning of paramecium, amoeba, and euglena in an electric field. Füredi and Ohad (1964) reported on a reversible elongation of human erythrocytes in high-frequency fields accompanied by a rotatory motion. Pohl and Crane (1971) observed the phenomenon of cell spinning in a suspension of baker's yeast. Further publications in this field appeared in the early 1980s (Mischel and Lamprecht, 1980, 1983; Zimmermann *et al.*, 1981; Pohl, 1981a,b, 1982; Holzapfel *et al.*, 1982). These authors described the spinning of different cell types (yeast, bacteria, erythrocytes, plant cells, mammalian cells) in AC fields between two electrodes in a frequency range from 100 Hz to 5 MHz.

1.3. CSR in Rotating Fields

The first papers on cell spinning in rotating external fields and chambers with three or four electrodes were published by Arnold and Zimmermann (1982), Mischel *et al.* (1982), Glaser *et al.* (1983), Mischel and Pohl (1983), and Pohl (1983a,b). CSR was established as a simple and sensitive tool to characterize cell parameters. CSR spectra proved to be characteristic for different cell types and for dead or living cells. Mischel *et al.* (1982) were the first to show that the spinning direction of dead and living cells can be opposite in special frequency ranges.

1.4. Cell Screening and Cell Sorting by CSR

CSR is a delicate discriminative method to distinguish between different cells without staining; e.g., treated from untreated, or live from dead cells. It has been found that CSR can differentiate between normal and malignant lines of human lung cells, and that CSR appears to be an age-sensitive method as well (Rivera and Pohl, 1986). This technique may be used to separate cells by their different rotational properties such as characteristic resonance frequency, spinning rate, and spinning sense. Separation of cells was performed under the microscope by means of a micromanipulator—a time-consuming and ineffective procedure.

Fuhr (1985) and Fuhr *et al.* (1985a) developed a vertical four-pole chamber, partly filled with solid agar to render a smooth surface diagonal to the electrodes at the bottom of the chamber. In a rotating field, cells start to roll on the agar surface with different speeds depending on their spinning rate so that a separation is possible for cells differing in their response to the external field. If the cells exhibit different direction of rotation at a chosen frequency, separation is very effective. Another possibility is to apply consecutively two different frequencies at the resonance maxima of the cells and to alter the rotation sense of the field. Although improvements of this technique are necessary, it seems to be a very promising application of CSR in microbiology and biotechnology.

1.5. Electrorotation of Macroscopic Specimens

Fuhr (1985) and Fuhr *et al.* (1985b) reintroduced the old method to investigate macroscopic bodies of some centimeter diameters in rotating fields. They compared the results of model spheres with those of different cells. The specimens hung on thin threads in a four-pole chamber equipped with vertical plate electrodes and filled with either water or air. The model systems consisted of glass spheres. The spheres were used unmodified or alternatively were covered with several layers of material of changing conductivity and permittivity values and a liquid of high conductivity inside the cell to simulate the structures of different cell types. The geometric relations were chosen to meet those of real cells. The unmodified glass spheres resembled cells with a membrane like vacuoles or erythrocytes. The glass spheres having a conductive layer resembled plant cells with membrane and cell wall. Finally, spheres with a glass wall, a conducting agar, and an insulating lacquer layer behaved like protoplasts containing a central vacuole, the cytoplasm, and the two insulating shells of the tonoplast membrane and plasmalemma. Choosing appropriate values of the different conductivities and permittivities, these probes rendered frequency spectra which were in good agreement with those of the model cells. Moreover, Fuhr (1985) showed that the number of frequency extremes is always higher by one than the number of shells of the sphere.

2. THEORIES ON CSR

Besides the above-mentioned theories about spinning of inanimate particles in static, oscillating, or rotating fields, further theoretical development followed the interest in the rotation of biological specimens in electric fields. As the research on CSR started with experiments in a two-pole chamber with AC fields by several investigators (Teixeira-Pinto *et al.*, 1960; Füredi and Ohad, 1964; Pohl and Crane, 1971; Mischel and Lamprecht, 1980; Zimmermann *et al.*, 1981), theories concerning this type of rotation shall be cited first.

2.1. CSR in AC Fields

Different explanations for CSR are presented in the literature: (1) polarization effects at the cell membrane and within the cell due to the external field; (2) the interaction of charges at the cell surface with external electric fields; and (3) naturally occurring oscillating dipoles inside the cell and their interaction with the applied external field. The last one was elaborated by Pohl and Crane (1971) and Pohl and Braden (1982) who observed in a few cases lone cells spinning far away from the electrodes, other cells, or inhomogeneities in the field. As other explanations fail to describe these observations, the theory of spinning of such lone cells assumes (Pohl, 1981b, 1982) that living cells are sources of RF oscillations (Fröhlich, 1968; Phillips *et al.*, 1985). Such rotating cells are not stable in their place but move rather quickly to the electrodes to form pearl chains. Therefore, it is quite tricky to measure their spinning rate.

The second type of CSR in AC fields is described by Holzapfel *et al.* (1982). Cells rotate because of dipole–dipole interactions with other cells or with inhomogeneities of the electrode surface. The rotation occurs in the plane constituted by the connecting line between the two cells and the electric field vector. Speed and direction of the rotation depend on the relative position of the two cells in the applied field. No rotation appears when the connecting line of the cells is parallel or perpendicular to the field lines. The maximum clockwise spinning rate is observed at an angle of 45° between the field and the connection. If this angle amounts to −45°, the spinning sense is counterclockwise. This dipole–dipole interaction leads to a torque with a maximum value at $\omega\tau = 1$ (ω = frequency of the AC field, τ = relaxation time). The torque tends toward zero for lower and higher frequencies so that there exists an optimum frequency range which is determined by the relaxation times of the polarization processes of the cell membranes.

The third type of CSR due to current-carried charge deposition on the particle surface occurs only if the applied field is a DC or a slowly oscillating AC field [$\omega_{par} > \omega_{field}$]. Thus, this type does not fit the experimental conditions presented here where frequencies are usually in the kHz or MHz range and seldom below 1 kHz. The mechanism behind this type of rotation depends on the action of ambipolar currents in producing surface-charge dipoles on the cells having a conductivity and a dielectric constant different from those of the suspending medium (Pohl, 1978). These newly formed surface-deposited dipoles disturb the surrounding electric field. Additionally, Brownian motion eventually introduces a disturbance in the alignment so that rotation starts. Then rotation is maintained by the continuous deposition of ions on the cell surface out of the medium. In this case, the relatively slow spinning rate is influenced by the viscous drag and by the rate of charge deposition on the cell surface. This effect can only be observed in low-frequency fields of a few hertz and can be completely neglected in high-frequency fields where the first and the second type of CSR still exist.

2.2. CSR in Rotating Fields

In the external rotating field, a dipole is induced inside the suspended particle. Since this polarization is time dependent, an angular lag between the dipole and the field appears which leads to a torque on the cell. This torque acts against the friction in the liquid and establishes a constant spinning in the steady state. The spinning rate as a function of the frequency of the external rotating field gives information on a number of biological parameters, such as the age of the cell, the position in the cell cycle, the physiologic state (live or dead, healthy or poisoned by different noxes), and the type of the cell. It is influenced by several physical parameters of the experiment, such as the chosen frequency, the field strength, the viscosity and conductivity of the surrounding liquid, the presence of cations and anions of different valence, the cell density, and the pH value.

Calculations of Fuhr (1985) for the torque on spheres consisting of a different number of shells are based on the theories of Hertz (1881), von Schweidler (1897), and Lampa (1906). He showed that the formula of Arnold and Zimmermann (1982) derived from the time constant of the charging process of the membrane is a special case which generally does not hold. Sauer and Schlögl (1985) pointed out that the early theories of Lampa (1906) and Lertes (1921a,b) are only correct if the surrounding medium with a dielectric constant of unity suffers no losses. Actually, one has to apply the Maxwell stress tensor for a correct description of cell rotation.

3. EXPERIMENTAL APPROACH

Since our first experiments on CSR in 1979, different instrumental setups were tested both with AC and with rotating fields. In all cases, a microscope with an attached TV camera and video recorder was applied. Each run could be repeated several times to determine mean values for the spinning rates of particular cells.

In many cases, it proved to be advantageous not to determine the spinning rate immediately during the experiment. Rather, the changing experimental parameters were recorded verbally on the magnetic tape and analyzed later. With these precautions, the conditions for the cells changed only slightly during the experiment because of the short time they had to remain in the electric field in the chamber. Moreover, one should switch off or reduce the voltage between the individual determinations.

3.1. Chambers for CSR

For AC investigations, chambers similar to the dielectrophoretic ones were used with pin–pin or wire–wire configurations. Special care was taken for a perfect smoothness of the electrode surface to avoid local field inhomogeneities. CSR in rotating fields was observed in four-pole chambers corresponding to those published by Mischel *et al.* (1982) or in the so-called hanging-drop chamber as presented in Fig. 1. These chambers offer a lower friction to the rotating cells and a higher stability in the field as gravitation drives the cell back to the lowest point of the drop even at small field inhomogeneities. In four-pole chambers, only the central part guarantees pure rotational fields so that observations should concentrate on this region. Platinum electrodes are preferred over gold-covered brass plates.

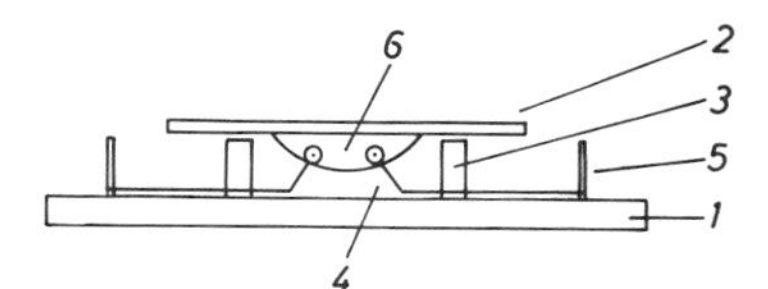

Figure 1. Hanging-drop chamber. 1: microscope slide; 2: coverslip; 3: supporting Perspex ring; 4: electrodes (two shown) with ball-shaped tips; 5: connectors to the generator; 6: hanging drop.

In some experiments, the density of the medium was increased so that the cells were floating between the electrodes. Alternatively, one can fill the chamber up to the electrodes with a denser liquid and then carefully add the medium with the cells. In both cases, friction between the cells and the bottom of the chamber is avoided.

3.2. Frequency Generators for CSR

There are different ways to produce rotating fields. The superposition of two sinusoidally oscillating fields of equal strength separated by 90° leads to a rotating field with the same instantaneous amplitude as the original fields. A combination of phase shifters and phase splitters produces four sinusoidal signals with phase angles of 0, 90, 180, and 270° from the original signal of a frequency generator (Arnold and Zimmermann, 1982; Mischel and Lamprecht, 1983; Fuhr, 1985).

In some experiments we applied trains of square-topped pulses to the usual four-pole chambers. The signal from a voltage-controlled oscillator was transformed in TTL technique to four signals with 90° phase shift each so that a short rectangular pulse of one fourth of a period traveled around the electrodes. In a second mode, the initial frequency was halved before treatment in TTL technique so that an overlap of one fourth of a period existed between two adjacent electrodes. This device could be used up to 4 MHz with a maximum voltage of 15 V. Pilwat and Zimmermann (1983) proposed a combination of two generators for square-topped pulses in connection with a three-pole chamber, while Fuhr *et al.* (1984) applied a four-step ring scaling circuit for the production of the wanted phase-shifted pulses.

Recently, Hölzel and Lamprecht (1987) developed a circuit for frequencies up to 140 MHz with at least 2 V in all ranges. As different electric pathlengths of the four signals result in transient time differences and thus in disturbances of the phase relationships, four identical channels with the same components were established, fed from two slightly inductively coupled LC oscillators (Fig. 2). By this trick, they vibrate with the same frequency and trigger each other. This circuit was

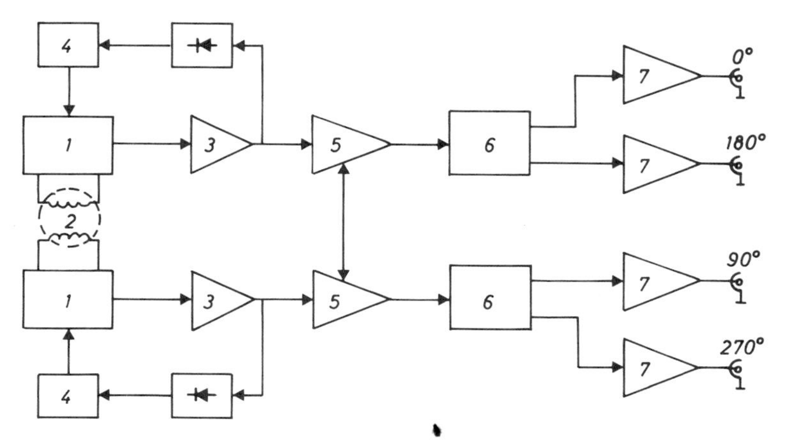

Figure 2. Block diagram for a sinusoidal four-pole generator up to 140 MHz. 1: voltage-controlled oscillator; 2: magnetically coupled coils for a slight inductive coupling of the two oscillators; 3: buffer; 4: PI regulator; 5: voltage-controlled amplifier; 6: phase splitter; 7: amplifier. Note the chamber has four electrodes.

specially used in the frequency range from 10 to 140 MHz to determine the predicted rotation maximum below 100 MHz (Hölzel and Lamprecht, 1987).

3.3. Biological Material

Although other cells like CHO (Chinese hamster ovaries), human melanoma, and algae were also investigated, most experiments were run with the baker's yeast *Saccharomyces cerevisiae*, strain RXII, with a round to elliptic shape, and sometimes with the fission yeast *Schizosaccharomyces pombe* with a very prolonged, nearly rodlike form. Yeasts have some advantages over other cells as they need no osmoticum and can be investigated even in bidistilled water or solutions with low salt concentrations for longer times. Moreover, many tests with other methods were performed on yeasts so that the results can easily be compared. Yeast cells were grown for 1 day in a shaken liquid medium at 30°C, harvested at the end of the logarithmic phase, and fractionally centrifuged to obtain a more homogeneous cell population. The cells were washed several times or just diluted 1 : 1000 with bidistilled water rendering conductivities around 5 μS/cm.

Best results are obtained when the distance between two neighboring cells is larger than five cell diameters so that no interference occurs. This leads to a cell density of less than 5×10^5 cells/ml. In experiments with the hanging drop, single cells were isolated and transferred into the drop by means of micropipettes. In a few experiments, the density was strongly increased to monitor the mutual influence of the cells.

Inactivation of cells was performed by keeping them for 5 min at 70°C. Subsequently, the cell state was tested by vital staining with methylene blue. Because staining changed the medium conductivity, usually unstained cells were further investigated. Mammalian cells were examined at low conductivity but with 0.3 M sorbitol to guarantee a correct osmolarity.

4. RESULTS

4.1. General Observations

During dielectrophoretic experiments dedicated to pearl chain formation, cells are often seen spinning near other cells, near the electrodes, and in very rare cases, free in the medium. It is reasonable that such cells are drawn near others by the effect of "mutual dielectrophoresis" and lose their isolated state. Rotation is easily induced by forming pearl chain bridges between the electrodes at, e.g., 1 MHz and then lowering the collection frequency by a factor of 10. A large percentage of cells spin at high velocities even if they are still lying in pearl chains. When the voltage is decreased, spinning becomes slower and many cells stop spinning at low voltages and only start again at much higher values due to frictional effects which mimic a threshold behavior.

Yeast cells resemble rotational ellipsoids with two different axes. In general, an ellipsoid of revolution can be supposed to be the most common form of a single cell not integrated into a tissue. Completely spherical or cylindric shapes may be deduced from this type. Therefore, discussion about the rotational axis in CSR shall concentrate on such cells. When cells settle on the glass support of the chamber, they orient with the longest axis parallel to the glass surface (Fig. 3). In this case, rotation in four-pole fields (CSR) appears around a short axis perpendicular to the glass (a). In

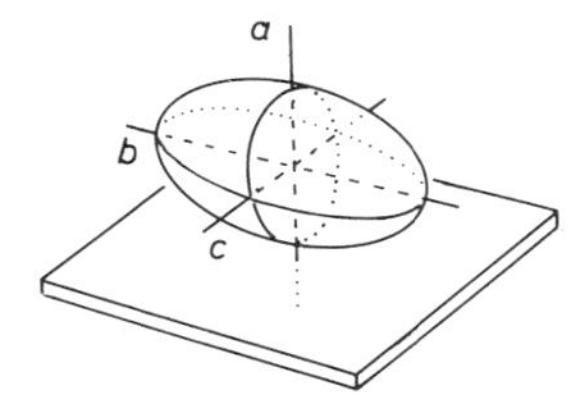

Figure 3. Model cell in the form of an ellipsoid of revolution on the glass support of an experimental chamber. In the case of CSR, rotation occurs only around axis "a," while in DEP experiments, rotation around "b" may be observed also. Near the electrodes or in pearl chains, axes "a," "b," and "c" can be involved.

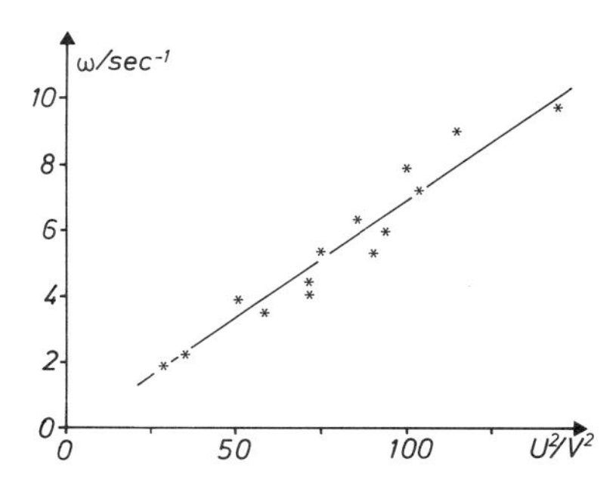

Figure 4. Spinning rate of yeast cells (*Saccharomyces cerevisiae* RXII) as a function of the square of the field strength of an AC field at 1.000 Hz and a conductivity of 4.4 μS/cm.

dielectrophoresis (DEP) experiments, very elongated cells like algae are sometimes seen to rotate at a high rate around the long axis (b), but preferentially also around the shorter ones (c). If the cells are attached to the electrodes or aligned in pearl chains during DEP investigations, all types of orientation and rotation are observed even such that yeast cells with smaller buds perform bumpy spinning over the bud. In conclusion, one can state that rotation may occur around any axis through the cell defined by the geometrical setup or the electrical field. In the medium, the orientation of the larger axis is randomly distributed in space. It turns into the plane of the electrodes when the field is switched on. Rotation then happens around the smaller axis perpendicular to the plane of the field.

When perfectly round cells without any detectable surface structures are investigated, rotation is hard to determine quantitatively. Yeast cells, with small buds as used by Mischel and Lamprecht (1980), or elongated cells facilitate the observation although stronger frictional forces are introduced. To overcome the friction at the bottom of the chamber, the density of the suspension may be increased by substances such as Percoll, Ficoll, or sucrose which do not enter the metabolism of yeast cells. Moreover, cells remain longer in the focus level of the microscope and are more readily investigated.

In all CSR experiments, the spinning frequency of cells is smaller by several orders of magnitude than the frequency of the external AC or rotating field. Typical spinning rates at moderate field strengths are from one cycle per several minutes to several cycles per second. Typical resonance effects are observed at external frequencies between 20 and 100 kHz and around 50 MHz.

4.2. CSR in AC Fields

The first descriptions of spinning cells were made for AC fields with different cell types (e.g., see Pohl, 1978; Mischel and Lamprecht, 1980; Zimmermann *et al.*, 1981; Holzapfel *et al.*, 1982). Optimal frequencies were detected in the range from 20 to 100 kHz for most cell types and from 1 to 5 MHz for malignant human melanocytes (Mischel *et al.*, 1982). Usually, different cells from the same culture spin in a broad frequency range while particular cells just rotate in very small, resonancelike frequency intervals (Pohl and Braden, 1982).

As supposed from elementary power considerations, the rotation rate, ω, is proportional to the square of the field strength, E, as shown in Fig. 4. The first results of Mischel and Lamprecht (1980) and Zimmermann *et al.* (1981) rendering threshold voltages for the onset of rotation were influenced by frictional forces on the bottom of the chamber. Later experiments with floating cells (Mischel and Lamprecht, 1983) were in complete agreement with the E^2 relationship following from the different rotation theories.

4.3. CSR in Rotating Fields

A short time after the revival of rotation experiments in AC fields, rotating fields were introduced by Arnold and Zimmermann (1982) and Mischel *et al.* (1982). As these experiments are

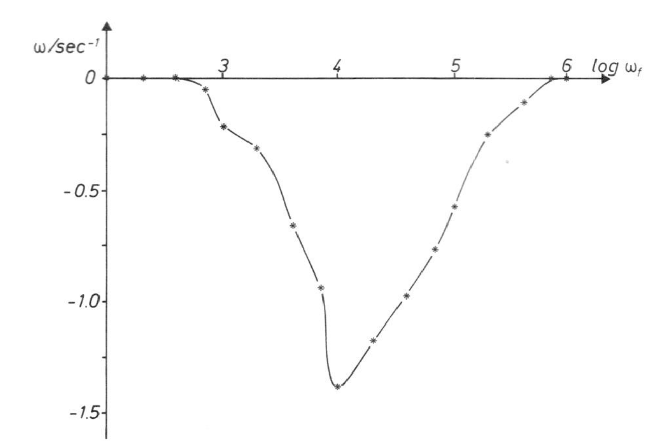

Figure 5. Frequency spectrum of CHO cells suspended in a 0.3 M sorbitol solution at low conductivity. The spinning is antifield.

easier to perform and exhibit better reproducibility, most attention focused on this technique and its application in various scientific areas such as mineralogy, biophysics, and microbiological cell sorting.

The first interesting results are the resonance frequencies at which maximum rotation occurs and the sense of spinning. Such a rotation spectrum of a mammalian cell from a culture of CHO cells in a medium of 0.3 M sorbitol is shown in Fig. 5. The rotation is antifield with a distinct maximum at 10 kHz. In Fig. 6, three yeast cells in different cell states are compared. Again, the spinning direction is opposite to the field rotation except for the single cell at low external frequencies with cofield rotation. Clear differences can be observed in the resonance frequencies and in the spinning rates which are of biological origin, because in steady-state conditions, the cell size should be without influence. With the high-frequency function generator (Hölzel, 1988), the usual frequency spectra could be extended to more than 100 MHz. This frequency range is interesting because it is where living cells should exhibit a further resonance peak (Fuhr, 1985). Figure 7 presents such spectra for live and heat-inactivated yeast cells at cofield spinning at high frequencies. The expected resonance at 30 MHz and an unexpected and yet unexplained second one around 80 MHz were observed.

As for AC fields, theory predicts a linear dependence of the spinning rate on the square of the field strength. This was indeed observed in all frequency ranges, with all cells and under all chosen conditions. In the steady state when the Stokes force of friction equals the torque on the cell, the spinning rate should be inversely proportional to the viscosity of the surrounding liquid. This

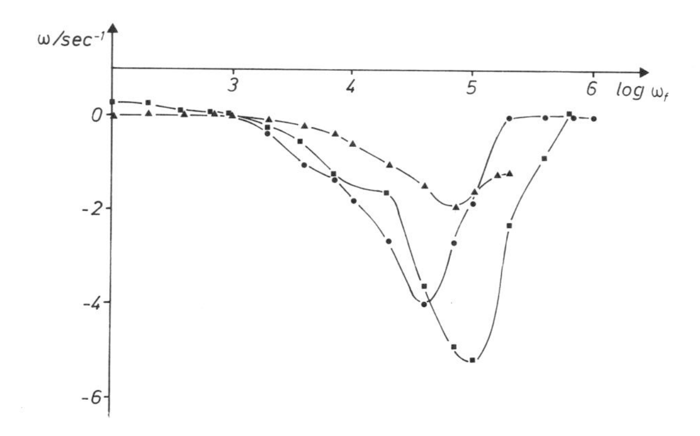

Figure 6. Frequency spectra of three yeast cells (large single cell: ■; cell with a small bud: ●; double cell: ▲) from different stages in the cell cycle. *Saccharomyces cerevisiae* RXII. Conductivity: 8 μmS/cm.

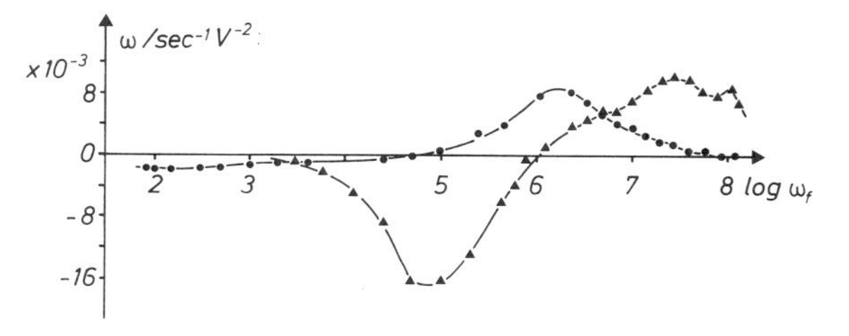

Figure 7. Frequency spectra of a live (▲) and a heat-inactivated (●) yeast cell (*Saccharomyces cerevisiae* strain 211) at a conductivity of 4 μS/cm. Spinning rate is given as reduced to the square of the external field strength.

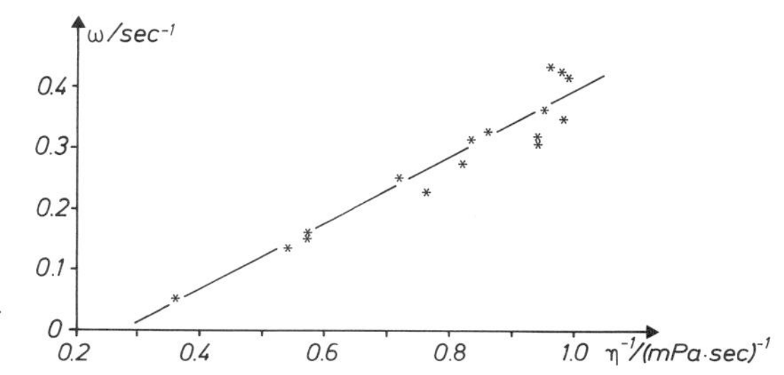

Figure 8. Spinning rate of yeast cells (*Saccharomyces cerevisiae* RXII) as a function of the reciprocal viscosity at 100 kHz.

relationship was established for yeasts in a medium of low conductivity and increasing concentrations of glycerol (Fig. 8) at 100 kHz.

All theories concerning CSR predict a spinning rate strongly dependent on the conductivity of the surrounding liquid. For yeast cells, Fig. 9 presents the increase of ω with increasing conductivity in the low range, a broad maximum between 60 and 80 μS/cm, and a slow decrease to higher values. If possible, experiments were run at low conductivity to avoid heating of the medium due to the electric current. If not, a conductivity in the maximum was chosen for investigations of the pH, the viscosity, and the ion concentrations.

Two further parameters with possible influences on the spinning rate are the pH value and the ion concentration of the surrounding liquid. It is well known from electrophoretic experiments that the pH of the suspending liquid has a strong impact on the surface charges of the cell and thus on its electrophoretic mobility (James, 1979; Beezer and Sharma, 1981). At low pH the surface of yeast cells is positively charged due to the predominance of amino groups while negative charges increase with higher pH (carboxyl groups important). The β dispersion of the frequency spectrum is connected with these surface charges and recharging processes in the membrane. Thus, it was worth-

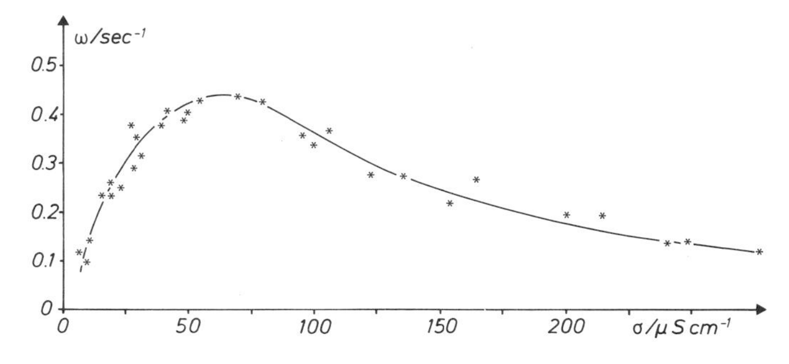

Figure 9. Spinning rate of yeast cells (*Saccharomyces cerevisiae*) as a function of the conductivity of the surrounding medium.

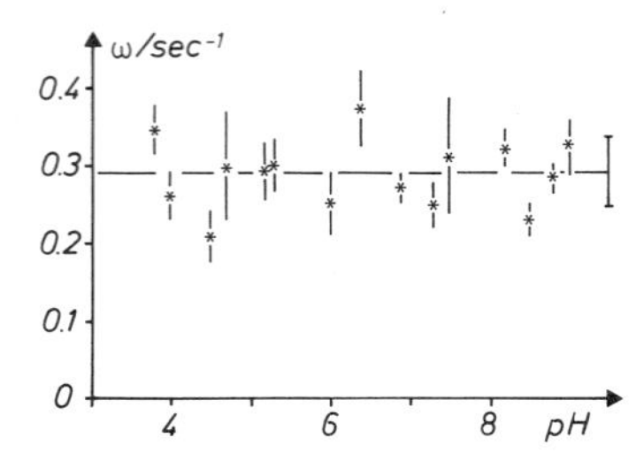

Figure 10. Change of the spinning rate of yeast cells (*Saccharomyces cerevisiae* RXII) with changing pH at 100 kHz and a conductivity of 80 μS/cm.

while to study the dependence of the spinning rate on the pH at a suitable frequency in the β range. A pH interval from 3.8 to 9.0 was obtained with three buffer systems brought to the same conductivity by adding $CaCl_2$. Within this interval, no significant changes in the spinning rate could be detected, just fluctuations around a mean of 0.291 ± 0.046 sec^{-1} (Fig. 10). This indicates that surface charges are of minor influence on the cell rotation in this frequency range.

Similar considerations hold for cations and anions in the liquid around the cell as they alter the dielectrophoretic collection rate in yeasts (Lopez *et al.*, 1984). Their effect can be understood as neutralizing surface charges of the cell membrane. In contrast to the collecting yield, the spinning rate increases with the valence of both cations and anions. A maximum change of 42% is seen for NaH_2PO_4 versus NaCl. In all cases, ion concentrations of 40 μM were adjusted regardless of a slightly changing conductivity (around 12 μS/cm) while Lopez *et al.* (1984) adjusted to a constant conductivity of 20 μS/cm with varying amounts of ions.

Soon after establishing CSR, it was seen to be a very sensitive tool for monitoring environmental poisons such as heavy metals. Figure 11 shows the influence of two concentrations of cadmium chloride on yeast cells. The most striking difference from untreated cells is seen at low frequencies where the cofield rotation of living cells (see also Fig. 6) is significantly changed to antifield spinning with even increased rates. This holds for all Cd concentrations. The changes in the β dispersion are not so clear and should be investigated in more detail.

Figure 12 shows the dependence of the spinning rate on cell density and thus the mutual distance of the cells. At the lowest concentration of 1×10^6 cells/ml, the cell distance varies from four to six cell diameters. At 10×10^6 cells/ml, this distance amounts to 1.5 to 2.5 diameters. The figure clearly exhibits the increase of the spinning rate with higher cell density as expected from mutual influences. The dent around 22×10^6 cells/ml (distance approximately half a diameter) appeared in all experiments but is unexplained as yet. It cannot be due to friction of the cells because at higher densities the effect disappears. To avoid mutual interactions, most experiments were performed with low cell densities of 5×10^5 cells/ml.

The torque as well as the Stokes friction on a rotating cell are proportional to the volume of the cell and thus for a spherical body to the third power of the radius. In the steady state when both

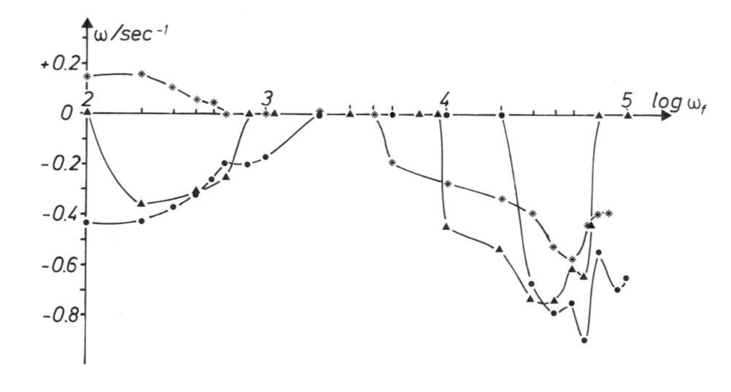

Figure 11. Frequency spectra of yeast cells (*Saccharomyces cerevisiae* RXII) in dependence upon increasing concentrations of cadmium chloride. *: Untreated cells; ●: 1μg/ml; ▲: 100 μg/ml.

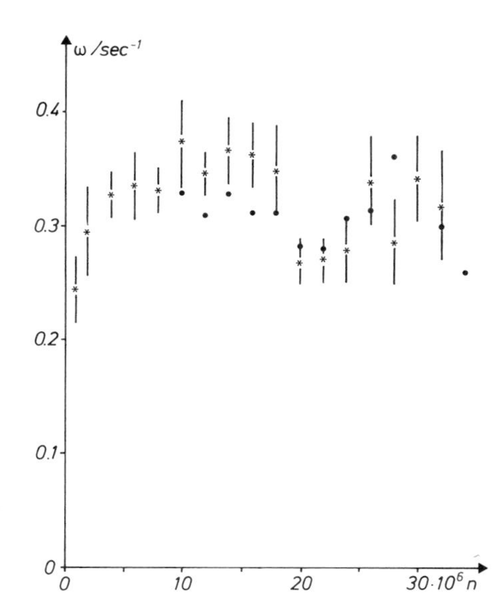

Figure 12. Spinning rate of yeast cells (*Saccharomyces cerevisiae* RXII) as a function of cell density and thus the mutual distance of the cells. * and ● indicate independent experiments.

moments are equal, the volume simplifies out and the spinning rate should be independent of the size of the cell. This prediction was tested with a homozygotic yeast strain system with cells from haploid to hexaploid with volumes varying by a factor of approximately six. Within the range of experimental deviations normally observed with biological material, no dependence on the volume could be detected. This also holds true for specially selected cells of the very elongated yeast *Schizosaccharomyces pombe* with the ratio of axes varying by a factor of four.

5. DISCUSSION

Although CSR is a relatively young field of bioelectricity and bioelectrochemistry, it has already rendered many interesting results. These indicate differences between various types of cells (frequency spectra) as well as physiological states of the same cells (e.g., live or dead, poisoned, cell cycle, benign versus malign). Besides inorganic material, biological particles such as bacteria, yeasts, algae, semen, ova, spores, protoplasts, and even parts of cells were investigated. CSR has proved to be an extremely sensitive method for detecting substances down into the ppm range (Rivera and Pohl, 1986).

CSR is a useful supplement to DEP. The latter is the response to the total volume polarization. CSR is sensitive to the outer periphery of the cell and thus to interfacial polarizations. Therefore, it is specially applicable to investigations of the cell membrane and changes in surface properties after appropriate treatment.

Special advantages may come from CSR chambers which are no longer horizontally, but vertically oriented (Fuhr, 1985; Wicher *et al.*, 1986) and enable long-lasting experiments on particular cells. The device proposed by Wicher *et al.* (1986) could be developed into a flow system for continuously monitoring the action of environmental poisons in increasing concentrations on single cells or to follow a cell through its whole cell cycle.

ACKNOWLEDGMENTS. We express our gratitude to Mrs. R. Ackermann for skilled technical help, Mr. G. Bjeske for constructing the various chambers and instruments, Mrs. C. Umlauf for preparing the figures, and A. Beck, R. Hölzel and G. Schmidt-Hagius for many fruitful discussions.

REFERENCES

Arno, R., 1892, Campo elettrico rotante e rotazioni dovute all'isteresi elettrostatica, *Atti Accad. Lincei Rend. Roma* **I**(5):284–289.

Arnold, W. M., and Zimmermann, U., 1982, Rotating-field-induced rotation and measurement of the membrane capacitance of single mesophyll cells of *Avena sativa, Z. Naturforsch.* **37c**:908–915.

Beezer, A. E., and Sharma, P. B., 1981, On the uptake of nystatin by Saccharomyces cerevisiae. 3. Electrochemistry of the yeast cell surface, *Microbios* **31**:71–82.

Born, M., 1920, Über die Beweglichkeit der elektrolytischen Ionen, *Z. Phys.* **1**:221–249.

Fröhlich, H., 1968, Long-range coherence and energy storage in biological systems, *Int. J. Quant. Chem.* **2**: 641–649.

Fuhr, G., 1985, Über die Rotation dielektrischer Körper in rotierenden Feldern, Dissertation, Humboldt-Universität, Berlin.

Fuhr, G., Hagedorn, R., and Göring, H., 1984, Cell rotation in a discontinuous field of a 4-electrode chamber, *Stud. Biophys.* **102**:221–227.

Fuhr, G., Hagedorn, R., and Göring, H., 1985a, Separation of different cell types by rotating electric fields, *Plant Cell Physiol.* **26**:1527–1531.

Fuhr, G., Hagedorn, R., and Müller, T., 1985b, Simulation of the rotational behaviour of single cells by macroscopic spheres, *Stud. Biophys.* **107**:109–116.

Füredi, A. A., and Ohad, I., 1964, Effects of high-frequency electric fields on the living cell, *Biochim. Biophys. Acta* **79**:1–8.

Glaser, R., Fuhr, G., and Gimsa, J., 1983, Rotation of erythrocytes, plant cells and protoplasts in an outside rotating electric field, *Stud. Biophys.* **96**:11–20.

Hertz, H. R., 1880, Inaugural-Dissertation, in: *Gesammelte Werke von Heinrich Hertz,* Volume 1 (J. A. Barth, ed.), Leipzig, 1895.

Hertz, H. R., 1881, IV. Über die Verteilung der Electrizität auf der Oberfläche bewegter Leiter, *Ann. Phys. Chem.* **13**:266–275.

Hölzel, R., 1988, Sine quadrature oscillator for cellular spin resonance up to 120 MHz, *Med. Biol. Eng. Comput.* **26**:102–105.

Hölzel, R., and Lamprecht, I., 1987, Cellular spin resonance of yeast in a frequency range up to 140 MHz, *Z. Naturforsch.* **42c**:1367–1369.

Holzapfel, C., Vienken, J., and Zimmermann, U., 1982, Rotation of cells in an alternating electric field: Theory and experimental proof, *J. Membr. Biol.* **67**:13–26.

James, A. M., 1979, Electrophoresis of particles in suspension, *Surf. Colloid Sci.* **11**:121–185.

Lampa, A., 1906, Über Rotationen im elektrostatischen Drehfelde, *Sitzungsber. Akad. Wiss. Wien* **115**(IIa): 1659–1690.

Lertes, P., 1921a, Untersuchungen über Rotationen von dielektrischen Flüssigkeiten im elektrostatischen Drehfeld, *Z. Phys.* **4**(3):315–336.

Lertes, P., 1921b, Der Dipolrotationseffekt bei dielektrischen Flüssigkeiten, *Z. Phys.* **6**:56–68.

Lopez, M. C., Iglesias, F. J., Santamaria, C., and Dominguez, A., 1984, Dielectrophoretic behaviour of Saccharomyces cerevisiae: Effect of cations and detergents, *FEMS Microbiol. Lett.* **24**:149–152.

Mischel, M., and Lamprecht, I., 1980, Dielectrophoretic rotation in budding yeast cells, *Z. Naturforsch.* **35c**: 1111–1113.

Mischel, M., and Lamprecht, I., 1983, Rotation of cells in nonuniform fields, *J. Biol. Phys.* **11**:43–44.

Mischel, M., and Pohl, H. A., 1983, Cellular spin resonance: Theory and experiment, *J. Biol. Phys.* **11**:98–102.

Mischel, M., Voss, A., and Pohl, H. A., 1982, Cellular spin resonance in rotating electric fields, *J. Biol. Phys.* **10**:223–226.

Phillips, W., Pohl, H. A., and Pollock, J. K., 1985, Evidence for AC fields produced by mammalian cells, Proc. 1985 Annu. Conf. Institute of Electrostatics, Japan.

Pilwat, G., and Zimmermann, U., 1983, Rotation of a single cell in a discontinuous rotating electric field, *Bioelectrochem. Bioenerg.* **10**:155–162.

Pohl, H. A., 1978, *Dielectrophoresis—The Behavior of Neutral Matter in Nonuniform Electric Fields,* Cambridge University Press, London.

Pohl, H. A., 1981a, Cellular spin resonance, *J. Theor. Biol.* **93**:207–213.

Pohl, H. A., 1981b, Natural electrical rf oscillation from cells, *J. Bioenerg. Biomembr.* **13**:149–169.

Pohl, H. A., 1982, Natural cellular electrical resonances, *Int. J. Quantum Chem.* **9**:399–409.

Pohl, H. A., 1983a, Cellular spinning in pulsed rotating electric fields, *J. Biol. Phys.* **11**:59–62.

Pohl, H. A., 1983b, The spinning of suspended particles in a two-pulsed, three-electrode system, *J. Biol. Phys.* **11**:66–68.

Pohl, H. A., and Braden, T., 1982, Cellular spin resonance of aging yeast and mouse sarcoma cells, *J. Biol. Phys.* **10**:17–30.

Pohl, H. A., and Crane, J. S., 1971, Dielectrophoresis of cells, *Biophys. J.* **11**:711–727.

Quincke, G., 1896, Über Rotationen im constanten electrischen Felde, *Ann. Phys. Chem.* **59**:417–486.

Quincke, G., 1897, Die Klebrigkeit isolierender Flüssigkeiten im constanten electrischen Felde, *Ann. Phys. Chem.* **62**:1–13.

Rivera, H., and Pohl, H. A., 1986, Cellular spin resonance (ĈSR), in: *Modern Bioelectrochemistry* (F. Gutmann and H. Keyzer, eds.), Plenum Press, New York, pp. 431–444.

Sauer, F. A., and Schlögl, R. W., 1985, Torques exerted on cylinders and spheres by external electromagnetic fields: A contribution to the theory of field induced cell rotation, in: *Interactions between Electromagnetic Fields and Cells* (A. Chiabrera, C. Nicolini, and H. P. Schwan, eds.), Plenum Press, New York, pp. 203–251.

Teixeira-Pinto, A. A., Nejelski, L. L., Jr., Cutler, J. L., and Heller, J. H., 1960, The behavior of unicellular organisms in an electromagnetic field, *Exp. Cell Res.* **20**:548–564.

von Schweidler, E. R., 1897, Über Rotationen im homogenen elektrischen Felde, *Sitzungsber. Akad. Wiss. Wien* **106**(IIa):526–532.

Wicher, D., Gündel, J., and Matthies, H., 1986, Measuring chamber with extended applications of the electrorotation. α- and β-dispersion of lipsomes, *Stud. Biophys.* **115**:51–58.

Zimmermann, U., Vienken, J., and Pilwat, G., 1981, Rotation of cells in an alternating electric field: The occurrence of a resonance frequency, *Z. Naturforsch.* **36c**:173–177.

Chapter 3

Dielectrophoresis

Behavior of Microorganisms and Effect of Electric Fields on Orientation Phenomena

Francisco J. Iglesias, Carlos Santamaría, M. Carmen López, and Angel Domínguez

1. INTRODUCTION

In the 1950s, Pohl (1951, 1958) described dielectrophoresis as the movement of neutral particles under the influence of nonuniform electric fields. In recent years, dielectrophoresis has developed rapidly and its present-day field of application is broad and diverse both in inanimate and in biological systems. For example, using dielectrophoresis it may be possible to filter nonconductive liquids (Fielding *et al.*, 1975), to separate mineral powder mixtures into their component parts (Verschure and Ijlst, 1966), and to gain understanding of the behavior of disperse systems or colloids. This could prove to be a powerful aid in the interpretation of the anomalous polarization which occurs in heterogeneous systems (Santamaría *et al.*, 1985). Regarding the properties of biological systems, dielectrophoresis has been used for analyzing phenomena such as: the frequency spectra of bacteria and yeast (Pohl and Crane, 1971); the behavior of living and nonliving cells (Pohl, 1978; Iglesias *et al.*, 1984); and the response of cellular organelles (Ting *et al.*, 1971). The application of dielectrophoresis to biological materials is also useful in analyzing phenomena of natural electric oscillation (Pohl, 1983), cell rotation (Zimmermann and Arnold, 1983), cellular spin resonance (Mischel and Lamprecht, 1980), cellular orientation (Teixeira-Pinto *et al.*, 1960; Iglesias *et al.*, 1985), and cell fusion or electrofusion (Neumann *et al.*, 1980; Bates *et al.*, 1987). However, although in most of the published papers consensus exists concerning the experimental setup and control of physical parameters, we have encountered few papers relating how the biological material was obtained. Accordingly, in the last few years we have attempted to study the behavior of

Francisco J. Iglesias • Department of Applied Physics, University of Salamanca, Salamanca, Spain. Carlos Santamaría • Department of Applied Physics, University of Pais Vasco, Bilbao, Spain. M. Carmen López and Angel Domínguez • Department of Microbiology, Genetics, and Preventive Medicine, University of Salamanca, Salamanca, Spain.

microorganisms, mainly yeast, by dielectrophoresis. We carefully standardized biological parameters such as growth conditions, influence of shape, effect of the cell wall and so forth.

2. EFFECT OF NONUNIFORM ELECTRIC FIELDS ON CELL SUSPENSIONS

A neutral particle, such as a microbial cell, when subjected to a nonuniform electric field will become polarized. Due to the nonuniformity of the field, a net force will act on the particle. This force will produce movement of the suspended cell, a phenomenon known as *dielectrophoresis* (DEP) (Pohl, 1978) or *dipolophoresis* (Petkanchin and Stoilov, 1975). Unlike what happens with charged particles and uniform fields, movement is maintained when the field is AC. The use of AC voltage has the advantage that movement is only caused by DEP and does not interfere with electrophoretic phenomena. In the long run, the characteristics of this movement, i.e., of DEP, are determined by the polarization of the biological material in suspension. Polarization most often occurs due to the formation of dipoles or the orientation of permanent dipoles in the membrane or at the interface. When the applied field is AC, the movement and orientation of the dipoles do not vary in accordance with the changes in the field; electrical dispersion occurs. Associated with this dispersion is an absorption of energy from the AC field by the dielectric body. Living systems exhibit remarkable dispersions known as α, β, and γ dispersions which occur in different intervals in the frequency spectrum. The β dispersion occurs in the 10^4 to 10^7 Hz range and is the so-called Maxwell–Wagner dispersion. Because this coincides with the frequency range studied by us, it is of interest in the present chapter. Dispersion or interface polarization mechanisms is caused by processes of change in the electric properties of the particle and suspension medium.

Besides the translational movement of neutral matter, one must take into account the possible interactions among cells. This phenomenon is known as mutual dielectrophoresis. The presence of a particle in an electric field distorts the lines of the field, causing the particles nearby to be attracted to the original particle and giving rise to the formation of pearl chains. When the field is nonuniform, these pearl chains are attracted by the electrodes to form a deposit (the dielectrophoretic deposit, Fig. 1). With nonspherical particles, phenomena of reorientation with respect to the field lines can occur. Our research has focused mainly on the study of these two phenomena—DEP and

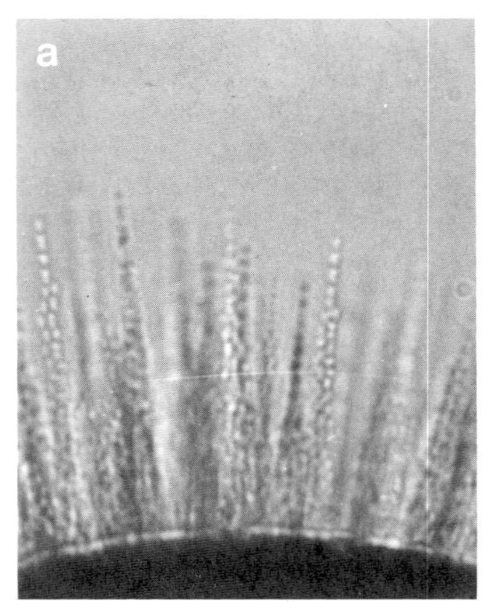

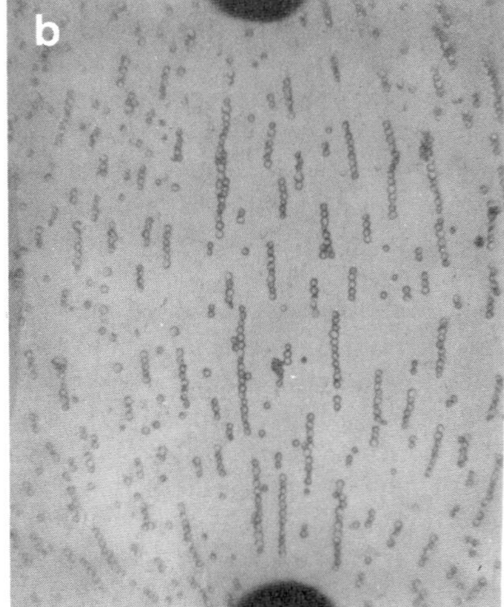

Figure 1. Photomicrographs of the dielectrophoretic behavior of (a) *Escherichia coli* and (b) spores of *Phycomyces blakesleeanus*.

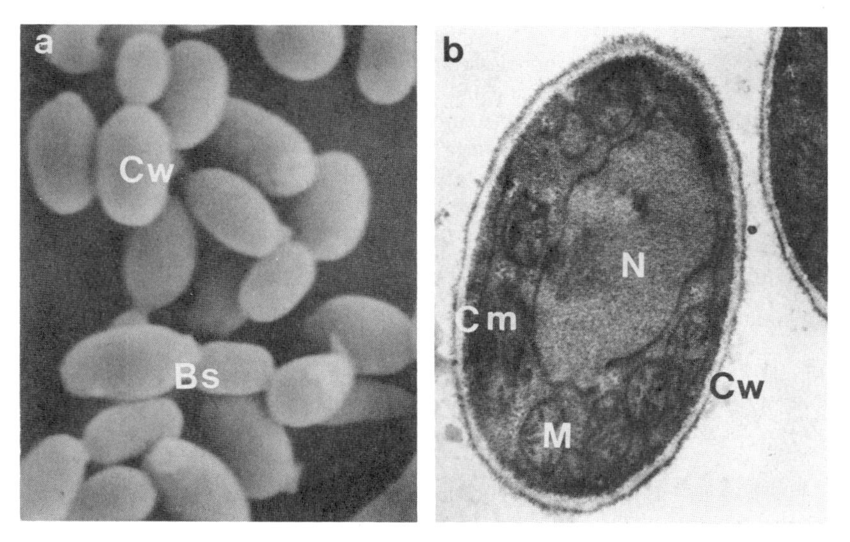

Figure 2. (a) Scanning and (b) transmission electron photomicrograph of *Yarrowia lipolytica*. Cw, Cell wall; Bs, bud scar; M, mitochondria; N, nucleus; Cm, cell membrane.

cell orientation—in yeast and spores. We have also observed the existence of cell rotation or spinning (Holzapfel *et al.*, 1982).

Probably the first successful experiments on the use of DEP to collect living cells were reported by Pohl and Hawk (1966). They showed that yeast cells could be made to collect at a reliable rate in the region of highest field intensity. We therefore decided to work with yeast due to our previous experience and to the fact that yeast seem to offer a good model as an experimental organism.

The external and internal structures of yeast can be examined by electron microscopy. The micrographs (Fig. 2a)of *Yarrowia lipolytica* and of all yeast (Domínguez *et al.*, 1982; Rodríguez and Domínguez, 1984) demonstrate by scanning the uniformity of the culture. It is also seen that yeast have a homogeneous porous surface, the cell wall (Fig. 2a). By transmission electron microscopy, one observes (Fig. 2b) a cell wall surface with an outer, electron-dense layer and an inner electron-transparent layer. Adjacent to the cell wall is the plasma membrane composed of a lipid bilayer and proteins with a width of 7.5 nm. The plasma membrane is responsible for the selective permeability of the cell and for the phenomenon of polarizability of the lipid layers due to the ions bound to its surface.

The intracellular organelles of yeast are typical of eukaryotic cells. Yeast contain ribosomes, mitochondria with double membranes, vacuoles, lipid bodies and polyphosphate granules, and a double membrane nucleus containing chromosomes and mitotic spindle (Rose and Harrison, 1969; Odds, 1979). Although Pohl (1978) suggested that the dielectrophoretic behavior of yeast could be explained by a three-layer model, our results presented later in this chapter disagree to a certain point with such conclusions. However, since yeast are eukaryotic microorganisms, there is the advantage that many of the results obtained can be extrapolated to mammalian cells.

According to Pohl, the value of the dielectric force acting on a small spherical particle of radius a and volume v is:

$$\vec{F} = 2\pi a^3 \, \mathrm{R_e} \left\{ \tilde{\epsilon}_1^* \frac{\tilde{\epsilon}_2 - \tilde{\epsilon}_1}{\tilde{\epsilon}_2 + 2\tilde{\epsilon}_1} \right\} \vec{\nabla}|\vec{E}|^2 = \frac{3}{2} v \, \chi \, \vec{\nabla}|\vec{E}|^2 \tag{1}$$

where $\vec{E}$ is the external electric field applied in the solution at a distance from the particle; $\tilde{\epsilon}_1$ and $\tilde{\epsilon}_2$ are the complex permittivities of the medium and the particle, respectively; $\tilde{\epsilon}_1 = \epsilon_1' - i$

ϵ_1'',$\tilde{\epsilon}_2 = \epsilon_2' - i\epsilon_2''$, $\epsilon_1'' = \sigma_1/\omega$, $\epsilon_2'' = \sigma_2/\omega$. The terms ϵ_1', ϵ_2', ϵ_1'', and ϵ_2'' are the in-phase permittivities and the dielectric losses of the surrounding medium and the particle, respectively. The in-phase conductivity of the medium and particle and the angular frequency are given by σ_1, σ_2, and ω, respectively. R_e is the real part of the complex number, and the asterisk denotes the complex conjugate. The use of complex permittivities is due to the fact that neither the medium nor the particle are perfect dielectrics. χ is known as the effective excess permittivity of cells in the presence of the surrounding medium. Accordingly, it is possible to obtain precise information through the dielectrophoretic study concerning the polarization mechanism present in the frequency range studied and, eventually, information concerning the dielectric properties of the cell.

In Eq. (1), it is seen that the value of the dielectrophoretic force acting on the cell depends on the nonuniformity of the field through the term $\vec{\nabla}|\vec{E}|^2$. The value of this force is independent of the sign of the field, which justifies the use of AC. Taking this into account, the way of achieving nonuniformity in the field, i.e., the geometry and configuration of the electrodes, is decisive in the dielectrophoretic results obtained. Different electrode geometries are described in the literature. The most common are spherical, cylindrical (Stoicheva *et al.*, 1985), and isomotive (Pohl *et al.*, 1981). The first two geometries give rise to a nonuniform electric field that is more intense than in the third configuration (Kaler *et al.*, 1984). Selection of the different configurations and geometries depends on the aims of the investigation.

3. EXPERIMENTAL METHODS

DEP was carried out in a pin–pin-type electrode arrangement (Fig. 1b). The platinum wire diameter was 65 μm and the electrodes were 1 mm apart. The pin-type tips were rounded (approximately spherical), the electrodes were mounted directly above the surface of a glass microscope slide, and the rounded ends were placed about 0.1 mm above the surface of the slide.

The orientation of microorganisms and spores in a uniform electric field was studied in a setup consisting of a microscope slide onto which two electrodes 0.75 mm apart had been painted with silver, as described by Teixeira-Pinto *et al.* (1960) (see Fig. 10a). A small drop (50 μl) of the suspension to be examined was placed between the tips.

The organisms used are described in Table 1. Details about the power supplies, culture media and growth conditions of cells and spores, killing methods, UV irradiation, toxic treatments, and so forth have been exhaustively described in previous works (Domínguez *et al.*, 1980; Iglesias *et al.*, 1984, 1985; López *et al.*, 1985). Cell permeability was determined by the methylene blue vital

Table 1. Microorganisms[a]

Strain	Genotype	Source
Escherichia coli K-12	ATCC 10798	ATCC
Saccharomyces cerevisiae X2180-1A	Mata,SUC2,mal,gal2,CUP1	Yeast Genetics Stock Center
Yarrowia lipolytica CX39-74C	Mat A, ural	Dr. J. Basel, University of California, Berkeley
Schizosaccharomyces pombe 972h$^-$		Dr. P. Nurse, Imperial Cancer Research Fund, London
Phycomyces blakesleeanus NRRL 1555		Dr. A. P. Eslava, our department
Penicillium expansum		Isolated in our laboratory
Verticillium chlamydosporium OILB		Dr. B. R. Kerry, Rothamsted Experimental Station, Harpenden, U.K.

[a]Abbreviations: ATCC, American Type Culture Collection; OILB, Organisation International pour la Lutte Biologique; NRRL, Northern Research Laboratory.

staining technique of Arnold (1972). Protoplast preparation was carried out as described by Sánchez *et al.* (1980).

The yield "Y," or dielectrophoretic collection rate, was defined as the average length of the chains of cells collected after a fixed time interval (Pohl, 1978). In this work, in order to render the measurements dimensionless, these values were divided by half the distance between the electrodes and have been termed Y_R. Of the various possible experimental errors, the largest was in measuring the yield. We used two types of measurement of the pearl chains formed: measuring directly and over microphotographs, at the times given. In both cases, the error was over 5% but never more than 10%.

4. PHYSICAL PARAMETERS

For the electrode configuration chosen and for particles with spherical symmetry, the length (average) of the radial chains adhered to the electrodes, or yield (Y), has been described by Crane and Pohl (1972):

$$Y = \frac{8\pi}{9} \frac{r_2}{r_1(r_2-r_1)} a^4 C\Phi \left(\frac{2t\chi}{\eta}\right)^{1/2} \tag{2}$$

where a is the radius of the particle considered as a sphere, C is the particle suspension concentration, Φ is the applied rms voltage, t is the elapsed time, η is the viscosity of the suspending medium, and r_1 and r_2 are the respective radii of the inner and outer spherical electrodes. χ is the excess polarizability and is defined as in Eq. (1).

Equation (2) predicts that the yield of the collected particles or cells is linear with respect to the cell concentration and with respect to the voltage (rms) applied. Also, Y is proportional to the square root of the elapsed time and inversely proportional to the square root of viscosity of the medium. The dependence of Y on the frequency of the electric field and on the electric conductivity is a complicated function and is represented in Eq. (2) by χ.

Equation (2) has been applied by Pohl (1978) to spherical yeast cells, both live and dead. However, many of the microorganisms in nature (bacteria, yeast, and spores) have other shapes—ellipsoidal (such as *Y. lipolytica* or *Phycomyces blakesleeanus* spores), rod-shaped (such as *Escherichia coli,* several bacilli, or *Schizosaccharomyces pombe*)—and different sizes. We have thus tried to extend the use of this equation to microorganisms with the geometrical forms described above. As examples, Fig. 3 shows the variation of Y_R with elapsed time for *S. pombe* and Fig. 4 shows the variation of Y_R with cell concentration for the three yeast strains. In both cases, a linear dependence is obtained which is in good agreement with Eq. (2). A higher voltage produces faster pearl chain formation and its length depends on the number of cells in suspension. Thus, for our experimental setup the best conditions (a linear relationship) were obtained with cell concentrations between 10^6 and 2×10^7 cells/ml and times up to 120 sec. Similar results were obtained with the fungal spores and *E. coli* (data not shown).

We also determined the relationship between the average length of the chains and the rms voltage (Iglesias, 1982; López, 1983; Iglesias *et al.*, 1984). In agreement with the results of Pohl (1978), we obtained a linear correspondence irrespective of the cell size or the geometric form of the microorganisms. However, in some cases, with lower concentrations of cells and higher voltages and at the usual collection times employed (100–120 sec), all the cells became attached to the electrodes. At high voltages the cells were either thrown away from the electrodes or phenomena of turbulence and stirring appeared (Pohl, 1978; Iglesias *et al.*, 1984). Voltages from 10 to 50 V rms were thus selected and in between this range the yield varied linearly, in accordance with theory. Moreover, this voltage did not damage the cells or the protoplasts (Förster and Emeis, 1985), as could be detected by measuring the number of cells forming colonies after the dielectrophoretic pulse (all the cells remained viable).

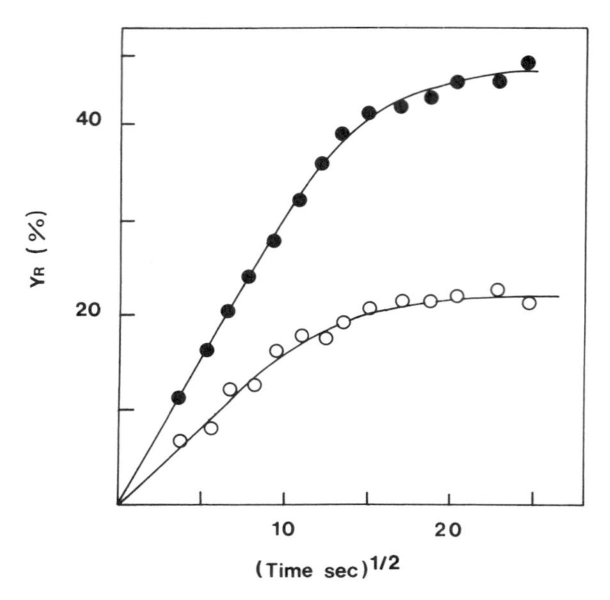

Figure 3. Variation of collection time with the square root of the length of time. ●, *Saccharomyces cerevisiae*, $\Phi = 40$ V rms, $C = 2 \times 10^7$ cells/ml; ○, *Schizosaccharomyces pombe*, $\Phi = 21$ V rms, $C = 7.2 \times 10^6$ cells/ml. Frequency 1 MHz. Reproduced from *Biochim. Biophys. Acta*, 1984, **804:**221–229, by copyright permission of Elsevier Science Publishers.

We also tested the influence of viscosity on the dielectrophoretic yield. Cell suspension was performed in sucrose, mannitol, or sorbitol instead of water. A decrease in the yield upon an increase in the viscosity of the medium was obtained. This is in good agreement with Eq. (2).

A more important point is the study of the variation of the dependence of yield on the frequency of the alternating field (yield spectrum) and on the electric conductivity of the medium which are closely correlated parameters. In Eq. (2), this dependence is due to the polarization factor, or excess polarizability, χ, which for the general case is a complex function of both of these parameters.

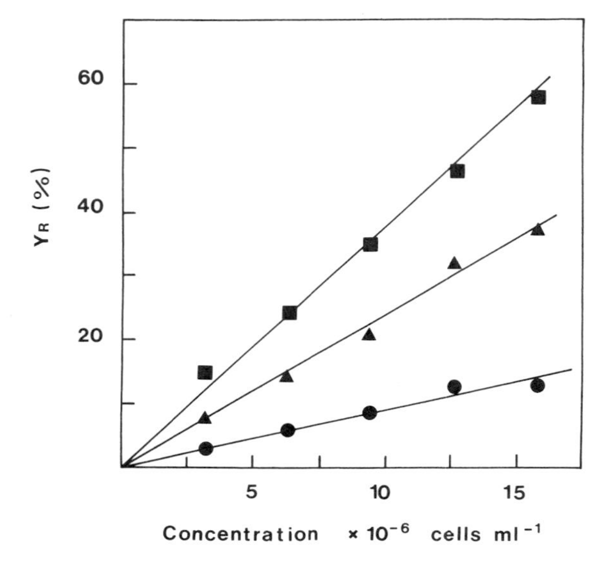

Figure 4. Dependence of yield on cell concentration in the suspension. ●, *Saccharomyces cerevisiae*, $\Phi = 21$ V rms; ▲, *Yarrowia lipolytica*, $\Phi = 40$ V rms; ■, *Schizosaccharomyces pombe*, $\Phi = 30$ V rms. Frequency 1 MHz. Conductivity 2×10^{-3} $(\Omega m)^{-1}$. Reproduced from *Biochim. Biophys. Acta*, 1984, **804:**221–229, by copyright permission of Elsevier Science Publishers.

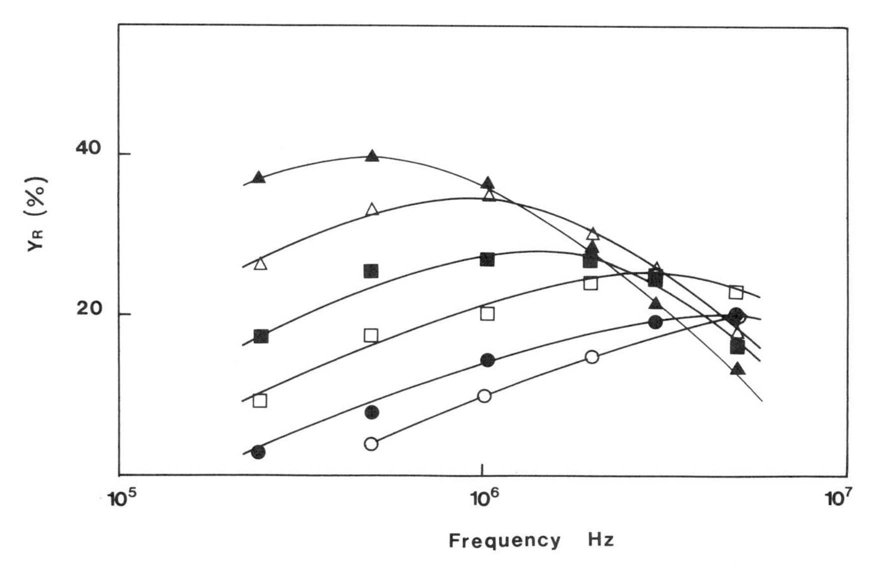

Figure 5. Dependence of yield on the frequency for aqueous suspensions of *Yarrowia lipolytica*. $\Phi = 21$ V rms; concentration 3×10^7 cells/ml; electric conductivity: ▲, 1×10^{-3}; △, 2×10^{-3}; ■, 4×10^{-3}; □, 8×10^{-3}; ●, 1.4×10^{-2}; ○, 2×10^{-2} $(\Omega m)^{-1}$.

Maintaining voltage, collection time, and cell concentration constant and fixing the conductivity, a complicated dependence of yield on the frequency (a spectrum) is obtained. For a given frequency, the curves showed a maximum yield. This maximum shifted toward higher frequencies as the electric conductivity of the medium increased. A decrease in the absolute value of the dielectrophoretic yield when conductivity increased was also observed (Fig. 5) (Pohl, 1978; López *et al.*, 1984). For an electric conductivity of 2×10^{-3} $(\Omega m)^{-1}$, the yield spectrum of all the spores and yeasts tested and that of yeast protoplasts shows a maximum at around 1 MHz (Iglesias *et al.*, 1984; López *et al.*, 1985). However, a slight displacement of the maximum toward lower frequencies was obtained for *E. coli* and PVC.

From the results obtained with yeast and spores, it was deduced that the dielectrophoretic behavior obtained is of the Maxwell–Wagner type (interfacial polarization). This mechanism depends largely on the difference in conductivity as well as on the dielectric constant of the medium and the particle. This bulk effect is due to the heterogeneity of the suspension. A similar argument was put forward by Pohl (1978) and Kaler and Pohl (1983), from their results with *Saccharomyces cerevisiae* and *Netrium digitus*. Recently, Mognaschi and Savini (1985) suggested that the separation of solid particles in liquid media can be based on differences in conductivities rather than differences in permittivities of the two media. Van Beek (1970) and Dukhin and Shilov (1974) revised the theory for this kind of suspension and found that for spherical particles the relaxation time which characterizes this dispersion is inversely proportional to the electric conductivity of the medium. This theoretical prediction is in good agreement with our experimental results (Fig. 5). However, the polarizing mechanism is not suited to adequately interpret the decrease in the yield (Fig. 6). Cations of different valencies were added to the cell suspensions, maintaining a constant value of electric conductivity. Similar results were obtained by our group with PVC (Santamaría *et al.*, 1985) and with silver bromide (Chen *et al.*, 1971). To integrate these data, it seems necessary to modify the Maxwell–Wagner mechanism to take into account an electric double layer surrounding the suspended particle which contributes a surface conduction. However, our efforts to produce a unified model to satisfy all the available data have been unsuccessful.

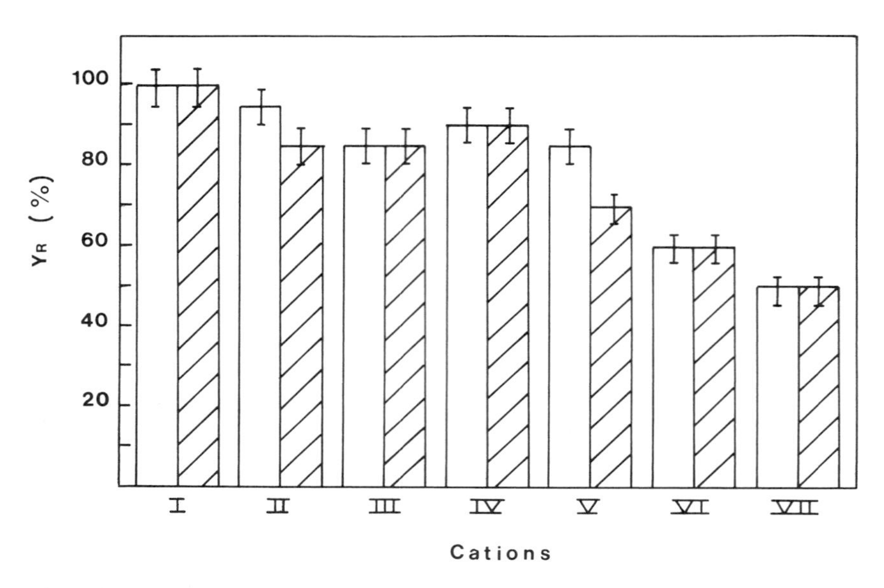

Figure 6. Influence of cations, anions, and growth medium on dielectrophoretic yield of *Saccharomyces cerevisiae:* □, minimal medium; ▨, complete medium; conductivity, 2×10^{-3} $(\Omega m)^{-1}$. Cation added: I, K^+ (Cl^-, NO_3^-, SO_4^{2-}); Na^+ (SO_4^{2-}); NH_4^+ (SO_4^{2-}); Cs^+ (Cl^-); II, Mg^{2+} (NO_3^-, SO_4^{2-}); Mn^{2+} (SO_4^{2-}); III, Ca^{2+}, (Cl^-); Ni^{2+} (SO_4^{2-}); Cd^{2+} (SO_4^{2-}); IV, Zn^{2+} (SO_4^{2-}); V, Cu^{2+} (SO_4^{2-}); VI, Fe^{3+}, (SO_4^{2-}); VII, Al^{3+} (NO_3^-). Measurements were taken 6 hr after addition of the cation.

5. BIOLOGICAL PARAMETERS

The physical parameters involved in DEP (Pohl, 1978) and in the cell fusion mediated by the electric field (Zimmermann, 1982; Sowers, 1987) are fairly well known. However, there is little information on the DEP of intracellular components.

A typical yeast cell is shown in Fig. 2. It is possible to subject the organism to different experimental procedures in order to evaluate the contribution of some of its parts to the dielectrophoretic phenomenon. First, using different yeast strains, the effect of the two main patterns of cell divisions was compared. The results (Fig. 7) show that the frequency spectra are similar in a yeast which divides by budding (*Saccharomyces cerevisiae*), another which divides by transversal fission (*Schizosaccharomyces pombe*), and in a spore (from *P. blakesleeanus*) which in our experimental conditions is unable to divide.

The results also show that differences in cell wall composition—β-glucan, mannan, and chitin in *Saccharomyces cerevisiae* (Cabib *et al.*, 1982) and α-glucan, β-glucan, and galactomannan in *Schizosaccharomyces pombe* (Bush *et al.*, 1974)—and in the organization of the cytoplasmic material have no effect on the frequency spectrum. Note that a great difference exists between a vegetative cell and a spore (Weber and Hess, 1976). On the same line, yeast protoplasts (López *et al.*, 1985), spinach chloroplasts (Pohl, 1978), and human lymphocytes (unpublished results) have similar yield spectra. From our results with protoplasts, we have estimated that the cell wall has a contribution to the yield of about 15%. Evaluation of the results with protoplasts is more complicated than with cells. This is due to difficulties in measuring the radii and to a certain degree of nonspecific aggregation due to the resuspension medium and membrane glycoproteins. However, our results (López *et al.*, 1985) were highly reproducible and suggest that the contribution of the cell

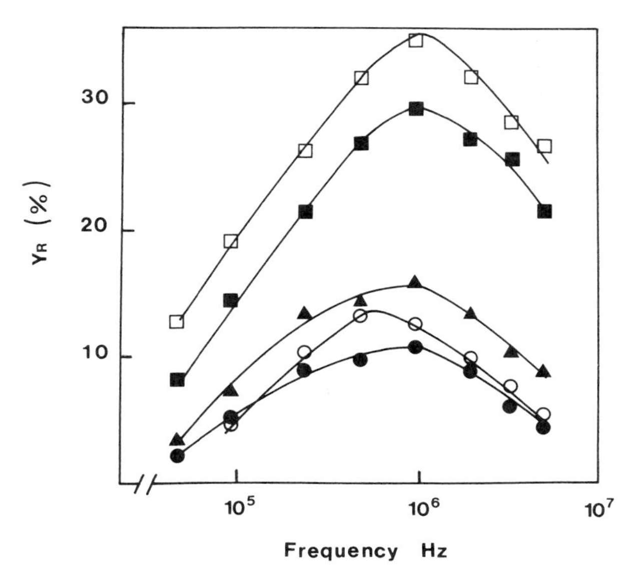

Figure 7. Frequency dependence of dielectrophoretic yield (Y_R). Conductivity 2×10^{-3} $(\Omega m)^{-1}$. $\Phi = 40$ V rms, except for *P. blakesleeanus* and *Schizosaccharomyces pombe* $\Phi = 20$ V rms. Collection time, 2 min. □, *P. blakesleeanus* 1.2×10^7 spores /ml; ■, *Schizosaccharomyces pombe* 1.5×10^7 cells/ml; ▲, *Saccharomyces cerevisiae* 1×10^7 cells/ml; ●, protoplasts of *Saccharomyces cerevisiae* 1×10^7 protoplasts/ml; ○, *E. coli* 1×10^9 cells/ml.

wall to yield is due to a net negative charge mainly attributable to the phosphate residues of the cementing polymer mannan (Cabib *et al.*, 1982). Interesting results on cell adhesion and measures of dielectrophoretic coefficients of pea protoplasts have recently been described by Stoicheva *et al.* (1985).

Dielectrophoretic behavior also depends on the metabolic state of the cells. Growth on nonfermentable carbon sources produces a dielectrophoretic response that decreases according to the amount of sugar present in the culture medium (Fig. 8). In the stationary phase of growth, a constant yield is obtained which corresponds to a minimum and is independent of the incubation time. The same yield is obtained if the yeast cells are resuspended in distilled water and placed at 4°C (for at least 5 hr) or if the carbon source is a nonfermentable one (i.e., ethanol or acetate instead of glucose or sucrose, Fig. 8). Our results suggest that old cells (in the stationary growth phase) or cells grown with nonfermentable carbon sources have deenergized plasma membranes. The results reported by de la Peña *et al.* (1982), Serrano (1983), Höfer and Künemund (1984), and Höfer *et al.* (1985) are in agreement with such a hypothesis. Experiments with tetraphenylphosphonium are under way. These considerations open the possibility that DEP could be used for measuring membrane potentials.

The dielectrophoretic yield, however, is independent of the nitrogen source. This can be demonstrated by changing an inorganic source, such as ammonium sulfate, by assimilable amino acids such as glutamine or asparagine in the culture medium regardless of the concentration used (López *et al.*, 1985). Our results disagree with those reported by Mason and Townsley (1971) who demonstrated that DEP separates yeast grown under different nutritional conditions and yeast of small differences in size. Although we cannot explain the discrepancy, it is difficult to understand how separation could be achieved with differences of about 10 μm^3 in the mean volume (40–45

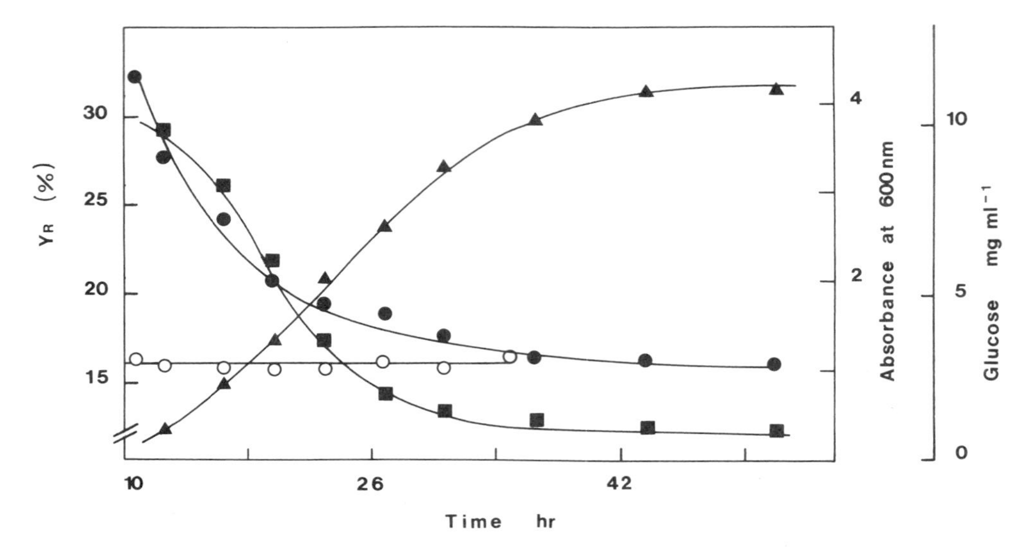

Figure 8. Dependence of dielectrophoretic yield (Y_R) on growth and carbon sources of a culture of *Saccharomyces cerevisiae* in minimal medium at 24°C. Cell concentration 1.9×10^7 cells/ml. Frequency 1 MHz. Conductivity 2×10^{-3} $(\Omega m)^{-1}$. ■, glucose concentration in the medium; ▲, absorbance at 600 nm; ●, yield on glucose (1 and 4%) or sucrose (2 and 4%), with or without shaking; ○, yield on ethanol (2%) or acetate (4%), with or without shaking. Adapted from *J. Bacteriol.*, 1985, **162:**790–793, by copyright permission of the American Society for Microbiology.

μm^3 for our strain of *S. cerevisiae* grown in minimal medium to 45–50 μm^3 grown in wort broth) with populations of 10^6–10^8 cells/ml and experimental errors of 5–10%.

While experiments to demonstrate and evaluate the contribution of the cell wall to the total dielectrophoretic behavior seem feasible, the evaluation of the contribution of internal structures in the range of frequencies studied in our work appears more problematic. If one considers eukaryotic cells (such as human lymphocytes, yeast protoplasts, and protoplasts from prokaryotic cells), several large differences are noticeable. First, prokaryotic cells have no internal compartmentalization. The spheroplasts of *E. coli* have a membrane more or less similar to the membrane of a eukaryotic cell with a homogeneous cytoplasm and DNA (as a packed structure). In eukaryotic cells, however, there is internal compartmentalization. There are organelles such as mitochondria, vacuoles, a Golgi apparatus, endoplasmic reticulum, and a nucleus surrounded by a double lipid bilayer with proteins. Cells without mitochondria, mutants ρ^- of *S. cerevisiae* (unpublished work), and spores with a much more compact cytoplasmic material (López *et al.*, 1985) show yield spectra similar to those reported for normal cells (Fig. 7). At present, it is not possible to evaluate the contribution of other organelles such as vacuoles or internal membranes.

The effect of the other most important eukaryotic cell component, the cell nucleus, has not been evaluated. Pohl (1978) presented an irregular yield spectrum, with a maximum around 3 MHz, for erythrocytes (cells without a nucleus). However, such results are difficult to evaluate because of the peculiar shape of the erythrocyte (biconcave disk) and because of the special medium required to stabilize these cells. To our knowledge, the yield spectrum of isolated nuclei has not been reported. Results obtained with prokaryotic cells, which lack a nucleus, will be discussed in the following section.

Permeabilized dead cells were not attached (or had a very low yield) in our experimental conditions. However, nonpermeabilized dead cells exhibited a more erratic behavior (see Sections 6, 8, and 9).

From the foregoing considerations and keeping in mind that size may be an important parameter (see Section 6), our hypothesis is that the biological membrane unit (the lipid bilayer with proteins and a width of 7.5 nm) is the main parameter responsible for the yield spectrum. This is obvious as long as it surrounds normal biological materials. In principle, this conclusion agrees with our hypothesis concerning the physical mechanism of dielectric dispersion observed at frequencies close to 1 MHz (see Section 4).

6. COMPARISON BETWEEN THE DIELECTROPHORETIC BEHAVIOR OF BACTERIA, YEAST, AND SPORES

The dielectrophoretic behavior of bacteria, spores, yeast, and human lymphocytes of different size, geometrical form, cell composition, mode of cell division, and evolutive group (prokaryotes and eukaryotes) is similar in a nonuniform AC field in the frequency range 0.1–5 MHz. For a given experimental setup, it is possible to measure values of cell concentration, voltage, collection times, conductivities, and viscosity, in which a proportional response occurs in the range of frequencies tested (Pohl, 1978; Iglesias *et al.*, 1984). Under identical experimental conditions, the same kind of spectra with a maximum at 1 MHz (Fig. 7) were obtained for the three yeast strains, yeast protoplasts, and spores. We also obtained another similar frequency spectrum, but with the maximum at 0.5 MHz for a bacterium (*E. coli*) and a particle (polyvinylchloride) (Santamaría *et al.*, 1985). In agreement with our results, Pohl (1978) also showed that spinach chloroplasts and *S.cerevisiae* have the same spectrum. However, those authors described different yield spectra for several bacteria: *Staphylococcus aureus, Bacillus cereus, B. subtilis, Pseudomonas fluorescens, P. aeruginosa, E. coli,* and others. These results contrast with our findings with yeast and deserve some comment. The two main differences between bacteria and yeast in analyzing dielectrophoretic behavior are the prokaryotic nature of bacteria and their small size. They do not have a nucleus and hence lack a shell. If one of these two parameters or the combination of both is important for dielectrophoretic behavior, one would expect to find a similar yield spectrum for all bacteria. This may or may not resemble the frequency spectrum obtained for yeast or spores. However, Chen (1972) and Pohl (1978) described a different frequency spectrum for each bacterium. The spectra reported cannot be grouped according to geometric form (*S. aureus* is spherical, *B. cereus* and *E. coli* have a bacillus form), type of cell wall (*S. aureus* and *B. cereus* are gram-positive and *E. coli* and *P. fluorescens* are gram-negative), or by a similar size (*S. aureus* is a very small bacterium and *B. subtilis* and *B. megaterium* are among the largest). It therefore seems that every bacterium, regardless of its biological and geometric differences, has a specific frequency spectrum. If this is so, the results are of enormous value from a taxonomic point of view. DEP could be used to classify bacteria by a cheaper and easier method than standard techniques. Unfortunately, our results show that all yeast and spores, although they are biologically diverse like bacteria, have an identical frequency spectrum. Thus, DEP is useless as a taxonomic criterion. A last point to consider is that the frequency spectra of bacteria reported by Pohl (1978) were obtained for a very broad frequency range, between 10^2 and 10^7 Hz.

From the comparison of the behavior of yeast, yeast protoplasts, and spores, we have examined whether a fourth power proportionality with respect to the radius exists in the dielectrophoretic yield and in the 1 MHz maximum of the frequency spectrum [Eq. (2)]. The relationship for three yeast strains, protoplasts of *S. cerevisiae,* and three different kinds of spores is shown in Table 2. For all of them, a good correlation between dielectrophoretic yield and the fourth power of the radius was obtained. This holds true for spherical and ellipsoidal volumes between 10 and 250 μm^3. The

Table 2. Relationship between the Dielectrophoretic Yield and the Fourth Power of the Radius in Yeast Cells and Spores[a]

Yeast	Y^*_R	a^4	Y^*_R/Y^*_{RSc}	a^4/a^4_{Sc}	δ (%)
Cells					
Saccharomyces cerevisiae	1.90	21.4	1	1	
Yarrowia lipolytica	2.90	36.0	1.53	1.68	11
Schizosaacharomyces pombe	10.60	61.5	5.57	2.87	
Protoplasts					
Saccharomyces cerevisiae	1.30	17.7	0.68	0.83	22
Spores					
Phycomyces blakesleeanus	14.50	168.0	7.63	7.84	5
Verticillium chlamydosporium	1.32	13.6	0.68	0.63	8
Penicillium expansum	0.57	5.8	0.30	0.27	9

[a]The radius (a) of the cells or spores was determined from the cell volumes calculated with a Coulter counter. To render the measurements comparable, normalized yield (Y^*_R) was calculated by dividing the real dielectrophoretic yield (Y_R) by voltage (V) and cell spore concentration (C): $Y^*_R = Y_R/VC$. Y^*_R/Y^*_{RSc} and a^4/a^4_{Sc} represent the dielectrophoretic yield and the fourth power of the radii compared with the values for *S. cerevisiae*. δ represents the deviation between the quotients of the normalized dielectrophoretic yield and those of the fourth power of the radii. All data deported here represent averages of four determinations which differed from one another by no more than 10%. Each experiment was carried out a minimum of three times with similar results. Reproduced from *J. Bacteriol.*, 1985, **162:**790–793, by copyright permission of the American Society for Microbiology.

results, together with those obtained with human lymphocytes (not shown), suggest that in the range of frequencies assayed by us, all eukaryotic microorganisms have a similar frequency spectrum and an equivalent yield. The only exceptions to the normal behavior were observed with *Schizosaccharomyces pombe*. This yeast, with the same frequency spectrum, has a yield higher than that corresponding to its equivalent volume. This could be explained by its almost cylindrical form (Iglesias *et al.*, 1984). Thus, a specific model must be designed for particles with this kind of geometry. We obtained a similar yield spectrum for *E. coli* (Fig. 7) cells. We are unable to attribute the differences in the maximum of the frequency spectrum (0.5 MHz instead of 1 MHz for yeast) to the lack of a nucleus or to their small volume (2 μm^3).

7. CELL ORIENTATION

7.1. Effect of Alternating Electric Fields

When unicellular organisms are subjected to an AC field, besides the translational movement, other kinds of responses, such as pearl chain formation, spinning or rapid rotation of cells, and cell orientation phenomena can occur. Orientational DEP is evidenced by the tendency of nonspherical cells to align with their longest axis parallel to the external field. However, changes in the frequency or in other parameters changed the cell alignment to other orientations (reviewed by Zimmermann, 1982).

Recently, the problems inherent to cell orientation have been the object of interest because of their applications to the study of the phenomena of electrofusion (Pilwat *et al.*, 1981) and electroporation (Sowers, 1987).

The stability of a predetermined orientation of a particle, in media with different dielectric properties in AC fields, was theoretically studied by Schwarz *et al.* (1965) and Saito *et al.* (1966). They showed that the direction of the stable orientation is determined by the geometry of the particle, by the dielectric properties of the material involved, and by the frequency of the field. As a special case, the same authors studied ellipsoidal particles with and without shells and showed that as the frequency of the electric field changes, there may be sudden jumps (turnover) in the stable direction. Some authors have found experimentally that the changes in cell orientation only occur in living nonspherical cells (Teixeira-Pinto *et al.*, 1960; Griffin and Stowell, 1966; Vienken *et al.*, 1984). Our results, however, show that this orientation phenomenon can also occur with spores and with nonliving yeast cells (Iglesias *et al.*, 1985).

7.2. Bacteria

With respect to cell orientation, we attempted to determine the behavior of several bacteria with bacillus shapes, such as *E. coli, B. subtilis,* and *B. megaterium.* Quantitative and reproducible results were hampered by the small size of the microorganisms and by the fact that at least 10% of the cultures are formed by short chains of cells (between two and seven). Dead bacterial cells are also able to orientate, although at the same conductivities lower frequency values are necessary (unpublished).

7.3. Yeast

In order to test the theoretical approach described in Section 7.1, living and dead cells of *Schizosaccharomyces pombe* (cylindrical form) were exposed to uniform and nonuniform electric fields (Iglesias *et al.*, 1985). In the frequency interval between 0.2 and 30 MHz at conductivities between 1 and 25×10^{-3} $(\Omega m)^{-1}$ and at a voltage of 10 V, living yeast always align with the longest axis parallel to the field lines. However, in suspensions of nonliving yeast, the orientation of the longest axis depended on both the frequency and the electric conductivity of the medium. The strength of the electric field seems to have no relevance in this phenomenon. Figures 9 and 10 show experimental results for cells exposed to uniform and nonuniform electric fields. Our results do not show the existence of orientation patterns other than parallel–perpendicular. However, Fig. 9 also shows an interval in which both orientations seem to coexist. In agreement with our results, Griffin

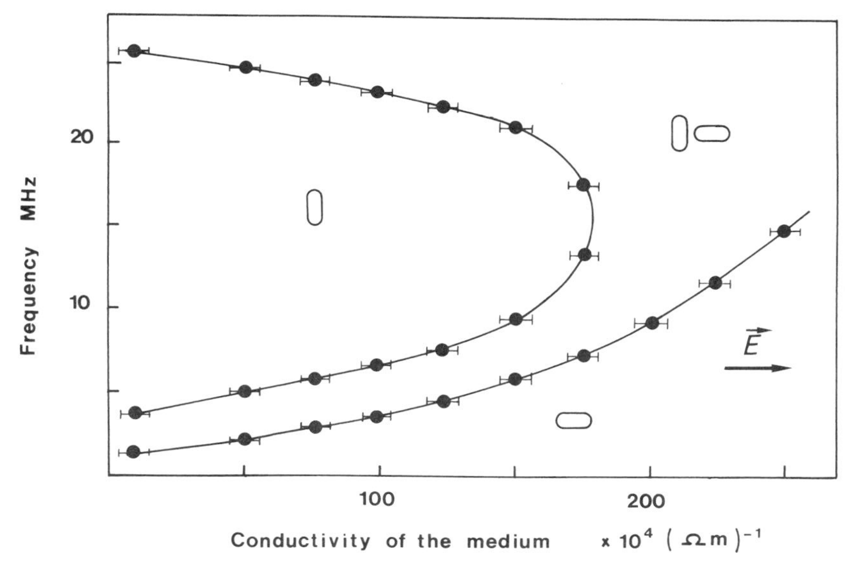

Figure 9. Orientation of nonliving cells of *Schizosaccharomyces pombe* in an alternating field as a function of frequency and of the conductivity of the medium. The values of each intersection point were calculated from a series of 10 micrographs by counting 100 cells in each, assuming that the cells were either parallel or perpendicular to the field lines when 90 ± 5% of them adopted one or the other disposition.

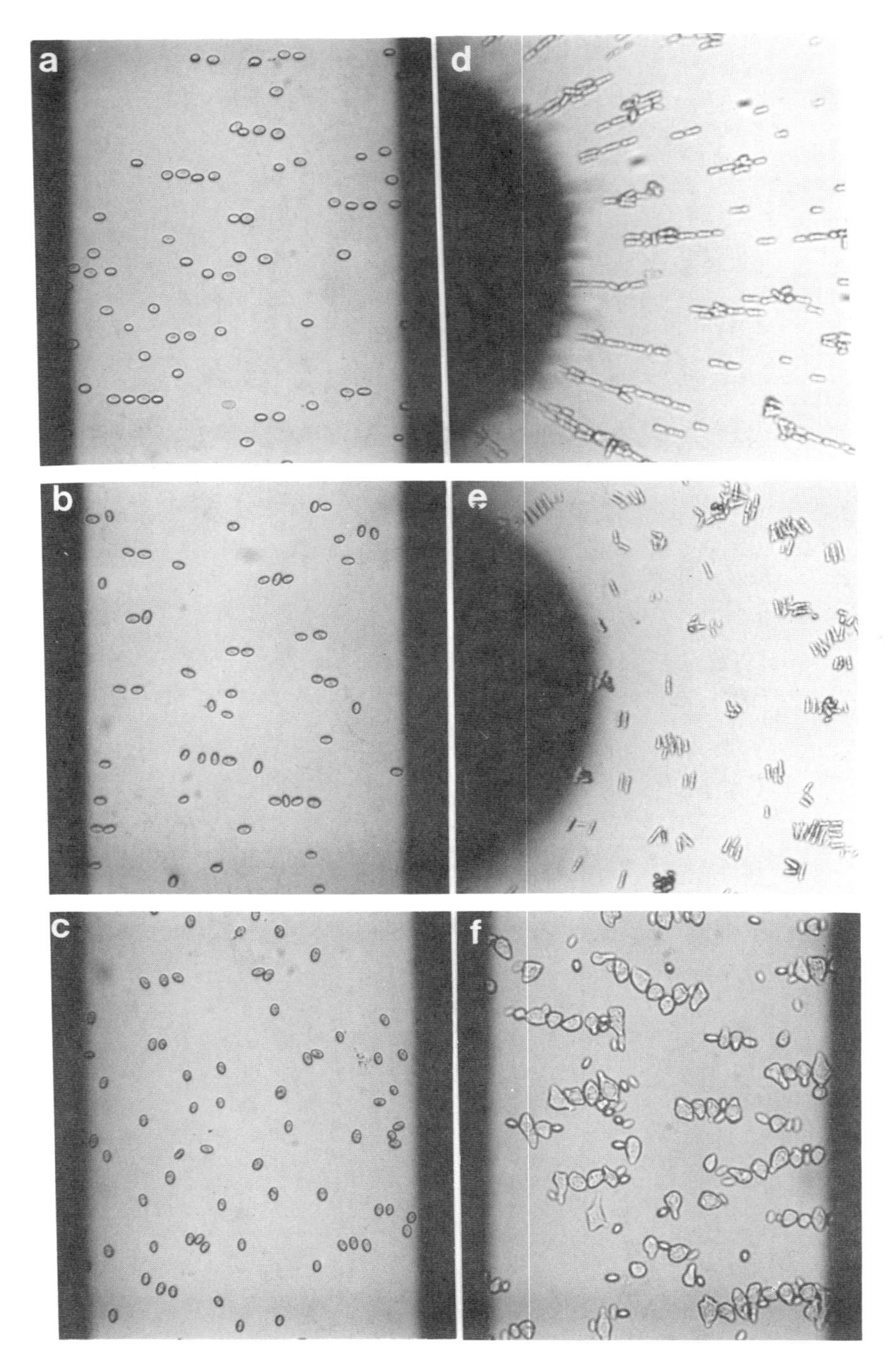

Figure 10. Sequential photomicrographs of the cellular orientation as a function of the frequency of the AC field. (a–c) *P. blakesleeanus* spores under a uniform electric field. (d, e) *Schizosaccharomyces pombe* nonliving cells under a nonuniform electric field. Frequency (a and d, parallel) 1 MHz, (b, mixed) 15 MHz, (c and e, perpendicular) 30 MHz. Conductivity $2 \times 10^{-3}\ (\Omega\mathrm{m})^{-1}$. (f) *P. blakesleeanus* germinated and UV-irradiated spores under a uniform electric field. Frequency 10 MHz; conductivity $6 \times 10^{-3}\ (\Omega\mathrm{m})^{-1}$.

(1970) showed a mixed orientation with functional human erythrocytes depending on the amount of glucose present in the suspending medium. These results cannot be fully interpreted by the theory of Saito *et al.* (1966) for ellipsoidal biological particles with only one shell. We attempted to verify the validity of our results, killing the cells by several different methods in order to obtain different degrees of preservation of cellular structures. By UV irradiation, toluene–ethanol, and detergent treatments, good preservation of cellular structure was achieved (Domínguez *et al.*, 1980). With autoclaving treatment, a complete disorganization of the cell cytoplasm and partial extraction of the cementing polymer of the cell walls occurred. Thus, by electron microscopy, cells subjected to the first type of treatment showed a similarity to viable cells; with the second treatment, cells adopted an empty-shell appearance. Both types of cells are permeable to methylene blue treatment. Identical results were obtained with all of these treatments regardless of whether the AC field was uniform or nonuniform (Fig. 10).

Pohl (1978) stated that changes in orientation only occur in living cells. This suggested that such changes are due to the polarizability of the lipid bilayer of the plasma membrane and to the ions bound to its surface. Thus, at high frequencies an alignment of the cells at right angles to the field takes place and the opposite occurs at low frequencies. Our experimental results disagree with the results in the previous paragraph. They also disagree with the hypothesis put forward by Saito *et al.* (1966) that for ellipsoidal particles with parameters of biological interest, only a sudden jump (90°) can occur.

The discrepancies reported could be due to heterogeneities in our working conditions that we were not able to detect, or to the fact that up to now the theoretical models elaborated are not supported by sufficient experimental data. We believe that for interpreting our results in future models, certain aspects should be taken into account. The first is that for dead yeast cells, the thickness of the ionic double layer associated with the cell surface must be reduced drastically in comparison with that attributed to living cells. In this way, the threshold frequency will take a lower value than that described for living cells—1 MHz for a conductivity of 2×10^{-3} (Ωm^{-1}). Second, the equalizing of the values of the electric conductivity of the medium and the cell by permeabilization of the plasma membrane (free diffusion of ions) should be considered. Finally, the previous results on cell orientation were obtained with organisms that lack a cell wall, i.e., *Euglena* and *Paramecium* (Teixeira-Pinto *et al.*, 1960; Griffin and Stowell, 1966). Thus, the contribution of the structural polymer of the cell wall to maintaining the shape of the yeast cells should be considered although at present we are unaware of how this can be done.

7.4. Spores

The latest results obtained by our team have indicated that the behavior of spores of *P. blakesleeanus* under AC electric fields is similar to that described for dead cells of *S. pombe*. For lower frequencies and at a conductivity of 2×10^{-3} $(\Omega m)^{-1}$, the spore is oriented with its longest axis parallel to the field lines. For frequencies of about 12 MHz, the orientation of the longest axis of the spores becomes perpendicular to the field lines. At 30 MHz, practically all the spores adopt a perpendicular orientation (Fig. 10c). Again, the change between the two orientations occurs in a gradual way. Qualitatively, we observed that normal spores, with a negligible metabolism, exhibit a behavior similar to dead yeast cells. As a common feature of both kinds of systems, we can only find that no exchange of ions occurs between the cell cytoplasm and the external medium. Rather, an identical flux in both directions is found in dead permeabilized cells. Again, we postulate a reduction in the ionic double layer anchored to the cell wall. The different values for the frequency spectrum of the parallel–perpendicular transition are obviously attributed to differences in shape (ellipsoidal versus cylindrical), in size (see Table 3), and in the dielectric properties of both systems. Spores of *P. blakesleeanus* must be subjected to heat-shock treatment before inoculation in liquid medium. Although the final causes of this behavior are unclear, failure to carry out such treatment produces very low levels of germination (Suarez *et al.*, 1985). Heating the spores in a water bath at 48°C for 15 min produces activation and high germination levels (about 90%) in liquid medium.

Spores treated in such a way were placed under a uniform electric field. In agreement with our working hypothesis, the spores were unable to undergo cell orientation under the previous frequencies and conductivities. Their behavior was identical to that of living cells. We also killed the heat-shock-treated spores by UV irradiation for 30 min at 4.5 J $(m^2\ sec)^{-1}$. The results were again identical to those obtained for dead yeast cells and for nonactivated spores. Germinated spores which were kept for 5 hr in liquid medium and treated with heat shock also revealed pearl chain formation and cell orientation in an AC field (Fig. 10). From all of these results, it seems obvious that the changes in the perpendicular–parallel orientation can be explained neither by the metabolic state of the cells nor by the equalization of the internal and external conductivities nor by a change in the ionic double layer surrounding cells and spores.

8. EFFECT OF CHEMICAL AND PHYSICAL AGENTS ON DIELECTROPHORESIS AND CELL ORIENTATION

Ionizing radiation, UV light from the sun, and a multitude of chemical agents (some man-made) cause alterations in the genetic material (DNA) of living organisms. These alterations are relevant in human health and have been clearly implicated in cancer and aging. We have tried to analyze by DEP the effect of some of these agents on yeast cells. The dielectric behavior of the yeast *S. pombe* was not affected by UV light or by chemical compounds that are able to inhibit cell metabolism, such as 2,4-dinitrophenol, sodium azide (de la Peña *et al.*, 1982), or detergents (López *et al.*, 1984), at concentrations affecting the proton flux through the cell membrane (Table 3). Neither could changes in cell orientation be obtained. Similar results were reported by Pohl (1978) using herbicides. However, if the treatment produces a permeabilization of the cell (Table 4), a decrease in the dielectrophoretic yield which corresponds to the amount of dead cells and changes in cell orientation occur (see Sections 6 and 9). Only short times of irradiation with UV light produce more irregular results. This could be due to the fact that some of all of the cells have lost their ability to grow and reproduce, whereas not all cells are completely permeabilized. These results open the possibility that extensive cell damage, detectable only by assays on cell reproduction, may be caused by DEP.

Table 3. Effect of Nonpermeabilizing Treatment on the Dielectrophoretic Behavior of *Schizosaccharomyces pombe*[a]

	Control (no treatment)	UV treatment (3 min)		2,4-DNF (5 mM)[b]		Na_3N (25 mM)[b]	
	1 or 5 MHz	1 MHz	5 MHz	1 MHz	5 MHz	1 MHz	5 MHz
	%	%	%	%	%	%	%
Dielectrophoretic yield	100	75 ± 5	75 ± 5	95 ± 5	96 ± 5	96 ± 5	95 ± 5
Methylene blue staining	0	32 ± 5	32 ± 5	1 ± 5	1 ± 5	1 ± 5	1 ± 5
Cell viability	100	76 ± 5	76 ± 5	99 ± 5	99 ± 5	99 ± 5	99 ± 5
Parallel	95 ± 5	95 ± 5	77 ± 5	95 ± 5	95 ± 5	95 ± 5	95 ± 5
Perpendicular	5 ± 5	5 ± 5	23 ± 5	5 ± 5	5 ± 5	5 ± 5	5 ± 5

[a]Aliquots of cells of a culture of *S. pombe* grown up on YED [2% yeast extract (Difco), 2% glucose] in the late exponential phase were centrifuged, washed, resuspended in distilled sterile water at a concentrate of 5×10^6 cells/ml, and treated under UV light (254 nm) or 1 hr with the reagent. After UV treatment the cells remained for 0.5 hr in the dark. The cells were centrifuged, washed, and resuspended at a conductivity of 2×10^{-3} $(\Omega m)^{-1}$, voltage 10 V rms. In each aliquot, dielectrophoretic yield, permeabilization to methylene blue, cell viability on plates, and the percentage of parallel and perpendicular cells to the field line forces at 1 and 5 MHz were determined. (The last percentage was determined by counting the cells in at least 10 photomicrographs of each experiment.) The results are the mean of at least three different experiments. Adapted from *Biophys. J.*, 1985, **48**:721–726, by copyright permission of the Biophysical Society.
[b]The same results were obtained with 2.5 mM 2,4-DNF or 5 mM Na_3N.

Table 4. Effect of Permeabilizing Treatment on the Dielectrophoretic Behavior of *Schizosaccharomyces pombe*[a]

	Control (no treatment)	Treatment in autoclave[b]		UV treatment (30 min)		CN^- (600 mM)	
	1 or 5 MHz	1 MHz	5 MHz	1 MHz	5 MHz	1 MHz	5 MHz
	%	%	%	%	%	%	%
Dielectrophoretic yield	100	8 ± 5	8 ± 5	12 ± 5	12 ± 5	5 ± 5	5 ± 5
Methylene blue staining	0	95 ± 5	95 ± 5	99 ± 5	99 ± 5	98 ± 5	98 ± 5
Cell viability	100	0	0	0	0	0	0
Parallel	95 ± 5	95 ± 5	5 ± 5	95 ± 5	0	95 ± 5	5 ± 5
Perpendicular	5 ± 5	5 ± 5	95 ± 5	5 ± 5	95 ± 5	5 ± 5	95 ± 5

[a]Aliquots of cells of a culture of *S. pombe* grown up on YED [2% yeast extract (Difco), 2% glucose] in the late exponential phase were centrifuged, washed, resuspended in distilled sterile water at a concentrate of 5×10^6 cells/ml, and treated under UV light (254 nm) or 1 hr with the reagent. After UV treatment the cells remained for 0.5 hr in the dark. The cells were centrifuged, washed, and resuspended at a conductivity of 2×10^{-3} $(\Omega m)^{-1}$, voltage 10 V rms. In each aliquot, dielectrophoretic yield, permeabilization to methylene blue, cell viability on plates, and the percentage of parallel and perpendicular cells to the field line forces at 1 and 5 MHz were determined. (The last percentage was determined by counting the cells in at least 10 photomicrographs of each experiment.) The results are the mean of at least three different experiments. Adapted from *Biophys. J.*, 1985, **48**:721–726, by copyright permission of the Biophysical Society.

[b]The same results were obtained with the toluene–ethanol treatment.

9. CELL VIABILITY

In Section 7 we have shown that dead and living cells differ in their behavior with respect to cell orientation. Pohl and Crane (1971) and Pohl (1978) studied the behavior of dead cells and reported contradictory results. When the cells were killed by heat at 60–70°C for 3 min and were then stained with crystal violet to determine cell viability, the unstained cells collected at the pin electrode, and the dead ones tended to remain uncollected in the suspension. Under these conditions, the dead cells were virtually unaffected by the nonuniform field. When dead cells were prepared by autoclaving at 121°C for 15 min, the dielectrophoretic yield for dead cells was similar to that for live cells. Mason and Townsley (1971) stated that a maximum yield for nonviable yeast cells exists at 10^4 Hz and for viable cells, at 10^5 Hz. Owing to the limitations in our experimental setup we were unable to assay the behavior of dead cells at frequencies lower than 5×10^5 Hz and thus cannot comment on these results. However, the rest of our results (Iglesias *et al.*, 1984, 1985) disagree completely with those of Mason and Townsley. We killed yeast cells by two treatments—autoclaving and treatment with toluene–ethanol. In both cases a complete permeabilization of the cells was obtained. This was demonstrated by methylene blue staining and by the incapacity to form colonies in a solid medium. Autoclaving is a drastic treatment which completely disorganizes the cell cytoplasm and solubilizes the cementing polymers of the cell wall. Treatment with toluene–ethanol is softer; the integrity of the cell cytoplasm is maintained and enzymatic activities can be assayed (Domínguez *et al.*, 1980). When we subjected to DEP dead yeast cells of three strains killed by both treatments, a linear response to voltage, time, or concentration was not obtained. A small deposit was formed at the tip of the electrode (Iglesias *et al.*, 1984), but the study of mixed living and dead cells clearly showed that the latter did not migrate to the electrode. However, our results fit in with the theoretical model (see Section 10) and with Pohl's (1978) calculations.

The above consideration is of practical value in cell separation. Our results show that at 10^6 Hz, living cells are collected at the electrodes while dead cells are not. It is thus possible to design (see Section 11) experimental devices to separate living from dead cells or to enrich a culture with either.

10. THEORETICAL MODEL

A theoretical model to understand the major features of the yield spectrum for both living and nonliving cells has been put forward by Crane and Pohl (1972) and Pohl (1978). In this model, the yeast cell is represented by a conductive dielectric sphere surrounded by two layers. The first layer is comprised of the cytoplasm, the nucleus, and the plasma membrane. The second layer is the surrounding ionic double layer coupled to the cell wall. The complete system is embedded in a moderately conductive medium. This model has two main simplifications. One is to assume that the ionic double layer surrounding the plasma membrane is the same as that surrounding the cell wall. The yeast cell wall is a complex structure (Cabib *et al.*, 1982) and the periplasmic space is not empty. However, as prokaryotic cells also have a cell wall, in order to compare dielectrophoretic behaviors one can assume that both cell walls have a similar function. The other simplification is to include the cytoplasm plus nucleus in only one shell. This assumption implies that in eukaryotic cells, the genetic material is in contact with the cytoplasm, as in prokaryotic cells. This also implies that the density of the nuclear material has the same value as that of the cytoplasm. The results described for spinach chloroplasts (Ting *et al.*, 1971) and for mitochondria (Pohl, 1978) indicate that biological material surrounded by a typical bilayer membrane has a yield spectrum. We have no knowledge of the dielectrophoretic behavior of the cell nucleus, but our guess is that its presence in a cell must produce some kind of contribution to the yield spectrum. Supporting this assumption is the fact that differences in density and dielectrophoretic behavior have been observed in yeast cells depending on their stage in the life cycle (Pohl, 1978). Assigning numerical values to most of the parameters, Crane and Pohl (1972) and Pohl (1978) found good agreement between the experimental and theoretical results. Using the same parameters and the real value for the radius of our yeast strain of *S. cerevisiae* (2 μm), we were unable to fit our experimental results for living cells with the proposed theoretical model. A radius of about 4 μm was necessary to obtain good agreement between the model and the experimental results. However, our results for dead cells can be interpreted fairly well by the Crane and Pohl model.

11. DIELECTROPHORESIS: APPLICATION FOR CELL SEPARATION

In the preceding sections we have discussed the possibility of using DEP in the separation of several kinds of cells, or organelles, or in the separation of living and dead yeast cells. DEP could also be used for collecting particles. An optimization in the process could be achieved through changes in the frequency of the applied field. Other factors should also be considered, such as avoiding local overheating, minimization of turbulence, and the like. Mason and Townsley (1971) were able to continuously separate *S. cerevisiae* using a separator model designed by Pohl (1958, 1978).

We have used a similar design: a 2- by 0.8-cm cylindrical column with central and cylindrical electrodes. The column was placed horizontally. Between 10^5 and 10^7 cells/ml were introduced through the poles with a peristaltic pump, at low fluxes. We detected the amount of collected cells by changes in absorbances and by cell plating. Experiments were also performed with mixtures of living and dead yeast cells. The results showed an enrichment in dead cells at the outlet of the columns. However, it is necessary to optimize parameters such as column size, flow rate, turbulence, and voltage rms.

Another advantage shown by continuous dielectrophoretic separators is the possibility of using negative DEP. Although positive DEP occurs over broad frequency ranges, there are also regions where the cells do not collect. The cells are repelled from the region of highest field intensity and therefore undergo negative DEP. In an electrode configuration called isomotive (Pohl, 1978; Pethig, 1979), the magnitude of the dielectrophoretic force is independent of the coordinate position. In this configuration, all the particles are subjected to identical forces regardless of their position in the space between the electrodes. In this way, either positive or negative DEP can be detected. Using a continuous separator with an isomotive electrode configuration, Pohl *et al.* (1981) were able to use both positive and negative DEP by changing the frequency of the electric field and the electric conductivity to study the behavior of *S. cerevisiae* and *Chlorella vulgaris*.

12. FUTURE OUTLOOK

The future of DEP is promising for the understanding and knowledge of the dielectric properties of inanimate particles and biological systems. Regarding the latter, beside its applications to the process of cell separation and cell orientation discussed in the preceding sections, DEP permits one to analyze other kinds of properties of living cells. One of the most interesting applications would be the study of the electric characteristics, specifically the polarizability of a single cell. By this technique, it would be possible to find subtle differences between the DEP behavior of single cells of several species of microorganisms or between the different kinds of human cells. Such findings could have important repercussions in the study of cellular degenerative processes, e.g., cancer. Moreover, using DEP it is possible to gain information about the polarizability of cells in the different phases of the cell division cycle.

DEP could also be an interesting approach for preparing oriented organized cell masses as Pohl suggested in 1978. It has been described that nonuniform electric fields produce translational and orientational effects. A combination of both could produce a controlled deposit over the electrodes. In order to maintain the orientation of the cell masses attached to the electrodes, it will be necessary to include the cells in a suspending medium of low melting point. In some cases, this technique could compete with the classical immobilization techniques for those enzymes and cells of great interest in biotechnology.

Zimmermann and Arnold (1983) interpreted the dielectrophoretic rotation of cells. They described for the first time a method for determining cell membrane capacitance and eliminating the use of intracellular electrodes. This procedure also opened an experimental line with great possibilities. Furthermore, a correct knowledge of the dielectrophoretic behavior of cells and protoplasts should improve the efficiency of electrofusion processes.

As a final remark, an improvement in experimental techniques, such as the use of quasi-elastic light scattering (Kaler *et al.,* 1984) for dielectrophoretic velocity measurements, permits one to reduce experimental errors which are inevitably present.

ACKNOWLEDGMENTS. We thank N. Skinner for revising the English version of this manuscript. M. C. L. is a recipient of a fellowship from Fondo de Investigaciones Sanitarias. This work was partially supported by a grant from the Fondo Nacional para el Desarrollo de la Investigación Científica y Técnica (0276-84).

REFERENCES

Arnold, W. M., 1972, The structure of the yeast cell wall: Solubilization of a marker enzyme, β-fructofuranoside, by the autolytic enzyme system, *J. Biol. Chem.* **247:**1161–1169.

Bates, G. W., Saunders, J. A., and Sowers, A. E., 1987, Electrofusion: Principles and applications, in: *Cell Fusion* (A. E. Sowers, ed.), Plenum Press, New York, pp. 367–395.

Bush, D. A., Horisberger, M., Horman, I., and Wursh, P., 1974, The wall structure of *Schizosaccharomyces pombe*, *J. Gen. Microbiol.* **81:**199–206.

Cabib, E., Roberts, R., and Bowers, B., 1982, Synthesis of the yeast cell wall and its regulation, *Annu. Rev. Biochem.* **51:**763–793.

Chen, C. S., 1972, On the nature and origin of biological dielectrophoresis, Ph.D. thesis, Oklahoma State University.

Chen, C. S., Pohl, H. A., Huebner, J. S., and Bruner, L. J., 1971, Dielectrophoretic precipitation of silver bromide suspensions. *J. Colloid Interface Sci.* **37:**354–362.

Crane, J. S., and Pohl, H. A., 1972, Theoretical models of cellular dielectrophoresis, *J. Theor. Biol.* **37:**15–41.

de la Peña, P., Barros, F., Gascón, S., Ramos, S., and Lazo, P. S., 1982, The electrochemical proton gradient of *Saccharomyces:* The role of potassium. *Eur. J. Biochem.* **123:**447–453.

Domínguez, A., Elorza, M. V., Villanueva, J. R., and Sentandreu, R., 1980, Regulation of chitin synthase activity in *Saccharomyces cerevisiae:* Effect of the inhibition of cell division and of synthesis of RNA and protein. *Curr. Microbiol.* **3:**263–266.

Domínguez, A., Varona, R. M., Villanueva, J. R., and Sentandreu, R., 1982, Mutants of *Saccharomyces cerevisiae* cell division cycle defective in cytokinesis: Biosynthesis of the cell wall and morphology. *Antonie van Leeuwenhoek J. Microbiol. Serol.* **48:**145–157.

Dukhin, S. S., and Shilov, V. N., 1974, *Dielectric Phenomena and the Double Layer in Disperse Systems and Polyelectrolytes,* Wiley, New York.

Fielding, G. H., Thompson, J. K., Borgardus, H. F., and Clark, R. C., 1975, Dielectrophoretic filtration of solid and liquid aerosol particulates, 68th Annual Meeting Air Pollution Control Association, Boston.

Förster, E., and Emesis, C. C., 1985, Quantitative studies on the viability of yeast protoplasts following dielectrophoresis, *FEMS Microbiol. Lett.* **26:**65–69.

Griffin, J. L., 1970, Orientation of human and avian erythrocytes in radio frequency fields, *Exp. Cell Res.* **61:** 113–120.

Griffin, J. L., and Stowell, R. E., 1966, Orientation of *Euglena* by radiofrequency fields, *Exp. Cell Res.* **44:** 684–688.

Höfer, M., and Künemund, A., 1984, Tetraphenylphosphonium ion is a true indicator of negative plasma-membrane potential in the yeast *Rhodotorula glutinis, Biochem. J.* **225:**815–819.

Höfer, M., Nicolay, K., and Robillard, G., 1985, The electrochemical H^+ gradient in the yeast *Rhodotorula glutinis, J. Bioenerg. Biomembr.* **17:**175–182.

Holzapfel, C., Vienken, J., and Zimmermann, U., 1982, Rotation of cells in an alternating electric field: Theory and experimental proof, *J. Membr. Biol.* **67:**13–26.

Iglesias, F. J., 1982, Separación dielectroforética en dispersiones de macromoléculas y levaduras: implicaciones biofísicas, Ph.D. thesis, Universidad de Salamanca.

Iglesias, F. J., López, M. C., Santamaría, C., and Domínguez, A., 1984, Dielectrophoretic properties of yeast cells dividing by budding and by transversal fission, *Biochim. Biophys. Acta* **804:**221–229.

Iglesias, F. J., López, M. C., Santamaría, C., and Domínguez, A., 1985, Orientation of *Schizosaccharomyces pombe* nonliving cells under alternating uniform and nonuniform electric fields, *Biophys. J.* **48:**721–726.

Kaler, K., and Pohl, H. A., 1983, Dynamic dielectrophoretic levitation of living individual cells, *IEEE Trans. Ind. Appl.* **IA-19:**1089–1093.

Kaler, K. V. I. S., Fritz, O. G., Jr., and Adamson, R. J., 1984, Quasi elastic scattering studies on yeast cells undergoing dielectrophoresis, *IEEE Trans. Ind. Appl.* **38A:**1110–1115.

López, M. C., 1983, Influencia de algunos parametios fisicos y biologicos sobre el comportamiento dielectroforetico de *Saccharomyces cerevisiae. Saccharomycopsis lipolytica* y *Schizosaccharomyces pombe,* M.S. thesis, Universidad de Salamanca.

López, M. C., Iglesias, F. J., Santamaría, C., and Domínguez, A., 1984, Dielectrophoretic behaviour of *Saccharomyces cerevisiae:* Effect of cations and detergents, *FEMS Microbiol. Lett.* **24:**149–152.

López, M. C., Iglesias, F. J., Santamaría, C., and Domínguez, A., 1985, Dielectrophoretic behavior of yeast cells: Effect of growth sources and cell wall and a comparison with fungal spores, *J. Bacteriol.* **162:**790–793.

Mason, B. D., and Townsley, D. M., 1971, Dielectrophoretic separation of living cells, *Can. J. Microbiol.* **17:** 879–888.

Mischel, M., and Lamprecht, I., 1980, Dielectrophoretic rotation in budding yeast cells, *Z. Naturforsch.* **35c:** 1111–1113.

Mognaschi, E. R., and Savini, A., 1985, Dielectrophoresis of lossy dielectrics, *IEEE Trans. Ind. Appl.* **IA-21:** 926–929.

Neumann, E., Gerisch, G., and Opatz, K., 1980, Cell fusion induced by high electric impulses applied to *Dictyostelium, Naturwissenschaften* **67:**414–415.

Odds, F. C., 1979, *Candida and Candidosis,* Leicester University Press, Leicester.

Pethig, R., 1979, *Dielectric and Electronic Properties of Biological Materials,* Wiley, New York.

Petkanchin, I., and Stoilov, S., 1975, Movement of spherical colloidal particles in an inhomogeneous electric field, *Res. Surf. Forces* **4:**138–141.

Pilwat, G., Richter, H. P., and Zimmermann, U., 1981, Giant culture cells by electric field-induced fusion, *FEBS Lett.* **133:**169–174.

Pohl, H. A., 1951, The motion and precipitation of suspensoids in divergent electric fields, *J. Appl. Phys.* **22:** 869–871.

Pohl, H. A., 1958, Some effects of nonuniform fields on dielectrics, *J. Appl. Phys.* **29:**1182–1189.

Pohl, H. A., 1978, *Dielectrophoresis,* Cambridge University Press, London.

Pohl, H. A., 1983, Natural oscillating fields of cells, in: *Coherent Excitations in Biological Systems* (H. Fröhlich and F. Kremer, eds.), Springer-Verlag, Berlin, pp. 199–210.

Pohl, H. A., and Crane, J. S., 1971, Dielectrophoresis of cells, *Biophys. J.* **11:**711–727.

Pohl, H. A., and Hawk, I., 1966, Separation of living and dead cells by dielectrophoresis, *Science* **152:**647–649.

Pohl, H. A., Kaler, K., and Pollock, K., 1981, Continuous positive and negative dielectrophoresis of micro-organisms, *J. Biol. Phys.* **9:**67–86.

Rodríguez, C., and Domínguez, A., 1984, The growth characteristics of *Saccharomycopsis lipolytica:* Morphology and induction of mycelium formation, *Can. J. Microbiol.* **30:**605–612.

Rose, A. H., and Harrison, J. S. (eds.), 1969, *The Yeast,* Volumes 1–3, Academic Press, New York.

Saito, M., Schwan, H. P., and Schwarz, G., 1966, Response of nonspherical biological particles to alternating electric fields, *Biophys. J.* **6:**313–327.

Sánchez, A., Larriba, G., Villanueva, J. R., and Villa, T. G., 1980, Glycosylation is not necessary for the secretion of exo-1, 3-β-D-glucanase by *Saccharomyces cerevisiae* protoplasts. *FEBS Lett.* **121:**283–286.

Santamaría, C., Iglesias, F. J., and Domínguez, A., 1985, Dielectrophoretic deposition in suspensions of macromolecules: Polyvinylchloride and Sephadex G-50, *J. Colloid Interface Sci.* **103:**508–515.

Schwarz, G., Saito, M., and Schwan, H. P., 1965, On the orientation of nonspherical particles in an alternating electrical field, *J. Chem. Phys.* **43:**3562–3569.

Serrano, R., 1983, In vivo glucose activation of the yeast plasma membrane ATPase, *FEBS Lett.* **156:**11–14.

Sowers, A. E. (ed.), 1987, *Cell Fusion,* Plenum Press, New York.

Stoicheva, N., Tsoneva, I., and Dimitrov, D. S., 1985, Protoplasts dielectrophoresis in axisymmetric fields, *Z. Naturforsch.* **40c:**735–739.

Suarez, T., Orejas, M., and Eslava, A. P., 1985, Isolation, regeneration and fusion of *Phycomyces blakesleeanus* spheroplasts, *Exp. Mycol.* **9:**203–211.

Teixeira-Pinto, A. A., Nejelski, L. L., Jr., Cutter, J. L., and Heller, J. H., 1960, The behavior of unicellular organisms in an electromagnetic field, *Exp. Cell Res.* **20:**548–564.

Ting, I. P., Jolley, K., Beasley, C. A., and Pohl, H. A., 1971, Dielectrophoresis of chloroplasts. *Biochim. Biophys. Acta* **234:**324–329.

Van Beek, L. K. H., 1970, Dielectric behaviour of heterogeneous systems, *Prog. Dielectr.* **7:**69–115.

Verschure, R. H., and Ijlst, L., 1966, Apparatus for continuous dielectric-medium separation of mineral grains, *Nature* **211:**619–620.

Vienken, J., Zimmermann, U., Alonso, A., and Chapman, D., 1984, Orientation of sickle red blood cells in an alternating electric field, *Naturwissenschaften* **71:**158–159.

Weber, D. J., and Hess, W. M. (eds.), 1976, *The Fungal Spore, Form and Function,* Wiley, New York.

Zimmermann, U., 1982, Electric field-mediated fusion and related electrical phenomena, *Biochim. Biophys. Acta* **694:**227–277.

Zimmermann, U., and Arnold, W. M., 1983, The interpretation and use of the rotation of biological cells, in: *Coherent Excitations in Biological Systems* (H. Fröhlich and F. Kremer, eds.), Springer-Verlag, Berlin, pp. 211–221.

Part II

Electroporation

Chapter 4

The Relaxation Hysteresis of Membrane Electroporation

Eberhard Neumann

1. INTRODUCTION

External electric fields have traditionally been applied in physical chemistry and biophysics to probe the ionic–electric properties and reactivities of molecules and molecular organizations such as biological membranes (Eigen and DeMaeyer, 1963; Neumann, 1986a). In recent years, electric field pulse techniques have also gained increasing importance in cellular and molecular biology, in gene technology, and in medicine. In particular, the methods of electroporation (Neumann *et al.*, 1982) and electrofusion (Senda *et al.*, 1979; Neumann *et al.*, 1980; Zimmermann and Scheurich, 1981; Weber *et al.*, 1981) have become powerful tools for cell manipulations (for reviews see Zimmermann, 1986; Berg, 1987; Sowers, 1987) and for the physical chemical study of electrically induced structural rearrangements in membranes (for review see Neumann, 1986a).

There is only indirect evidence that the applied electric pulses cause structural reorganizations in the cell membranes (for early references see Sale and Hamilton, 1968; Neumann and Rosenheck, 1972; Lindner *et al.*, 1977). Although there are a number of model approaches toward theories for electropermeabilization and electrofusion (e.g., see Abidor *et al.*, 1979; Benz *et al.*, 1979; Teissié and Tsong, 1981; Dimitrov and Jain, 1984; Sugar and Neumann, 1984; Weaver *et al.*, 1984; Powell *et al.*, 1986; Zimmermann, 1986; Sowers, 1987; Chernomordik *et al.*, 1987; Sugar *et al.*, 1987), it is fair to say that the detailed mechanisms of field-induced restructuring of membranes in electroporation and electrofusion processes are unknown.

In addition, in this new field there is still a need to classify the observations in terms of physical concepts and to establish an unequivocal terminology based on physical chemical principles.

The present account intends to provide a programmatic outline for the study of electroporation and of electrofusion in terms of field-induced structural rearrangements in the membranes. In particular, electroporation is viewed as a critical phenomenon and the concept of the relaxation hysteresis is introduced to elucidate the reversible and the irreversible aspects of electroporation and

EBERHARD NEUMANN • Faculty of Chemistry, Department of Physical and Biophysical Chemistry, University of Bielefeld, D-4800 Bielefeld 1, Federal Republic of Germany.

electrofusion. Since the early data of field effects on cell membranes provide fundamental knowledge, a brief digression on the very early findings is given. Because this contribution is not a review, some selection of references was necessary. Therefore, only the early references and some recent reviews are cited. Naturally, this essay is biased by the contributions of our own laboratory to this new field of biophysics.

2. EARLY ELECTROPORATION DATA

Electric gene transfer and electric cell fusion appear to be the most prominent applications of high electric field effects on cell membranes. Because the electric parameters are very similar for both techniques, electroporation and electrofusion can be related to one and the same primary field effect on membranes. This field effect permeabilizes the membranes and renders them at the same time fusiogenic.

Indeed, electropermeabilization was reflected in the early fundamental observation that short electric field pulses cause a transient increase in the membrane permeability for components of the cell interior and of the extracellular medium (Neumann and Rosenheck, 1972, 1973).

2.1. Lysis and Cell Death

There are a number of early reports on irreversible electric field effects on single cells which are of relevance in the discussion of membrane electroporation and electrofusion.

Pliquett (1967) found that AC electric fields (70 to 200 V/cm at 10 to 10,000 Hz) affect the envelope of Oxitrichide single cells (long diameter $\phi \approx 150$ μm). At $E \geq 130$ V/cm, the envelope vesiculates and releases small vacuoles (budding). If $E \geq 200$ V/cm, the envelope disintegrates and cytoplasm is released. The onset of the envelope processes is frequency dependent; at higher AC frequencies, higher field strengths are required.

Sale and Hamilton (1967, 1968) observed that short electric pulses (pulse duration $\Delta t \approx 20$ μs), above a critical field strength, cause lysis of bacterial protoplasts and spheroplasts and erythrocytes. Repetitive pulses (≈ 20), however, lead to cell death. Lysis and cell death were used as indicators to analyze dependencies on electric field strength and cell diameter. The results indicate that the electric field effect clearly is a cell membrane phenomenon; the authors suggested that the increased membrane potential may cause conformational changes in the membrane structure, resulting in lysis. The critical transmembrane voltage built up by the external field was found to be $V_{m,c} \approx 1$ V (for erythrocytes, bacterial protoplasts and spheroplasts). This value is much larger than the potential differences of 150–300 mV that are known from long-lasting voltage clamps which finally damage nerve membranes. The comparatively high critical voltage for cell lysis was suggested to be due to, among others, the short time scale of the electric pulse (20 μsec). The phenomenon was called *electric breakdown* by Sale and Hamilton (1968).

2.2. Transient Membrane Electropermeabilization

The cell death observed by Sale and Hamilton (1967) is apparently due to the high number of repetitive pulses. The concomitant massive material exchange and the dilution of the cell interior are certainly lethal whereas the cell membrane itself need not necessarily have experienced an irreversible electric breakdown (Neumann, 1986c).

It soon became apparent that the application of only a few pulses is not lethal. Therefore, the term *breakdown* appeared inadequate for describing a field-induced permeability increase being transient in nature. Indeed, dielectric breakdown, i.e., irreversible rupture of the membrane, was explicitly excluded in the case of the transient "Permeability Changes Induced by Electric Impulses in Vesicular Membranes" of the chromaffin granules (Neumann and Rosenheck, 1972). It was shown that above a threshold value of the initial field strength $E_0 = 18$ kV/cm (CD pulses), the granules release some of their content of catecholamines and ATP. The release gradually increases with increasing field strength. Below E_0 = 24 kV/cm, there is no release of intravesicular proteins. Since osmotic lysis leads to protein release, it was considered very unlikely that the short high-voltage pulses cause irreversible membrane damage (dielectric breakdown). It was later shown that above E_0 = 25 kV/cm there is protein release, too. In a theoretical study it was suggested that if a breakdown-like phenomenon is caused by the field pulse, a "local, transient and reversible breakdown" (with transient hole formation for material exchange) may not be excluded (Neumann and Rosenheck, 1973). It was shown by Rosenheck *et al.* (1975) that the electric field effect indeed only leads to a transient change of the chromaffin granules. In particular, the kinetics of the fluorescence change of the hydrophobic membrane probe diphenylhexatriene demonstrated that the change is in the membrane structure; it is *long-lived* (time constant ≈ 200 msec) compared with the short field decay time (10–150 μsec). The field effect is completely reversible if $E_0 \leq$ 25 kV/cm. The light scattering kinetics confirms the conclusion that the larger part of the material exchange is an *afterfield effect*.

Thus, the permeability change leading to release of catecholamines and ATP and intrusion of solvent components into the cell interior, is transient but long-lived compared with the field duration. The electric field-induced structural changes apparently *anneal* in multistep processes with time constants of 1–200 msec. The basic conclusions of reversibility, main material exchange after the field action, longevity of field-induced structural changes, exclusion of irreversible damage were confirmed by an elaborate electrooptic investigation by Lindner *et al.* (1977). Zimmermann *et al.* (1973) attributed resistance changes of *E. coli* in Coulter counter measurements to dielectric breakdown. The results of Coulter counter studies on bovine erythrocytes were analyzed in terms of a reversible dielectric breakdown (Zimmermann *et al.*, 1974).

2.3. Electric Gene Transfer and Electrofusion

An important aspect of field effects on cells is the artificial transfer of macromolecules or larger particles into the cell interior (Zimmermann *et al.*, 1976). The first application of the electric pulse technique to transfer genetic material was reported by Auer *et al.* (1976) who found uptake of SV 40 DNA and of mammalian cell RNA into human red blood cells.

The first electroporative gene transfer into living cells with the subsequent actual expression of the foreign gene was obtained by Neumann *et al.* (1982). This first actual cell electrotransformation was performed at the convenient laboratory temperature of 20°C (and not at the less favorable temperature of 30°C as claimed by Zimmermann, 1986). Beside a few "no data" statements, the first documented results on cell fusion by electric pulses were reported by Senda *et al.* (1979); the fused plant protoplasts were viable for at least several hours.

Figure 1 demonstrates the first example of viable giant cells, obtained by simple electropulsing of a suspension of cells of the eukaryotic microorganism *Dictyostelium discoideum* (Neumann *et al.*, 1980). In contrast to statements by Scheurich and Zimmermann (1981) and Zimmermann *et al.* (1981), the cell fusion induced by CD pulses does not require any chemical helper such as Ca ions or polyethylene glycol. The first yeast protoplasts were electrofused in the laboratory of H. Berg (Weber *et al.*, 1981). According to Zimmermann, the data obtained by Scheurich *et al.* (1980) did not reflect true fusion and are thus not comparable with the result of plant protoplasts (Zimmermann and Scheurich, 1981).

There are numerous reports on electroporative gene transfers and electrofusion, including systematic studies for the optimization of experimental conditions as well as attempts for a theoretical analysis of the field effects.

3. DIRECT FIELD EFFECTS AND STRUCTURAL RELAXATION

Electric fields act directly on charges: free ions, dipoles, ionic and dipolar as well as polarizable groups. When an electric field is strong enough, molecules and molecular organizations such as membranes undergo structural rearrangements. Usually the elementary reaction steps are coupled

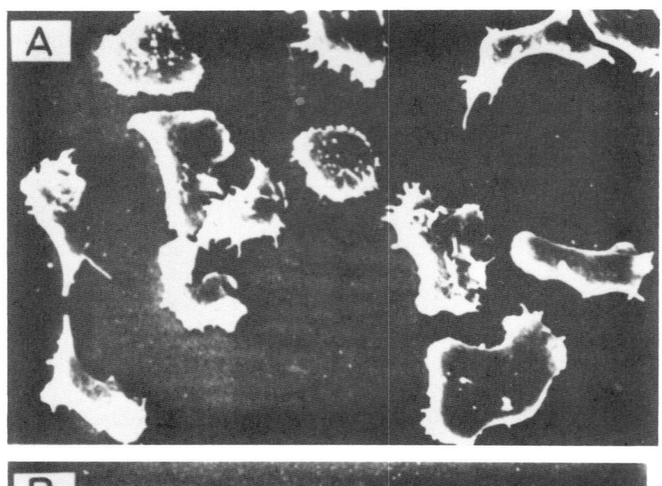

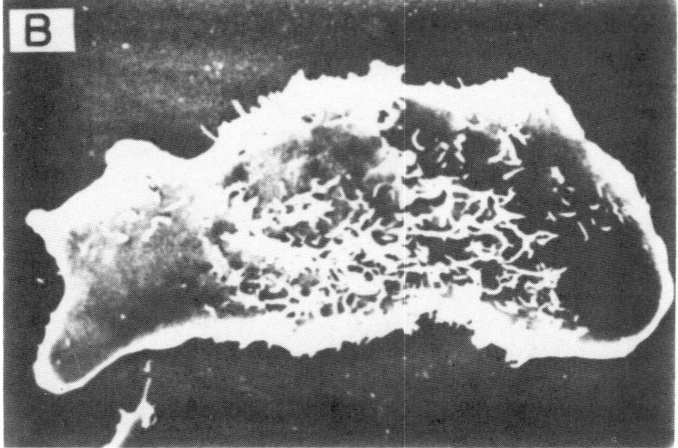

Figure 1. Electrofusion of living cells of the eukaryotic microorganism *Dictyostelium discoideum* (slime mold) at 23°C. (A) Single living cells of the strain Ax-2. (B) Giant living syncytia produced from nonagglutinated cells by electric CD impulses ($E_0 = 6$ kV/cm, $\tau_E = 40$ μsec) in the absence of $CaCl_2$. Cell fusion occurred within the first few seconds after exposure to pulses. The giant cells are mobile and chemotactically active (Neumann *et al.*, 1980; Neumann, 1984).

such that the measured response is a normal mode of the coupled system (Eigen and DeMaeyer, 1963).

Simple intramolecular transitions and intermolecular processes at small perturbations lead to exponential time (t) courses. Denoting by ξ the extent of a structural reorganization, the rate of the ith normal mode relaxation is given by

$$d\xi_i/dt = (\bar{\xi}_i - \xi_i)/\tau_i \tag{1}$$

where $\bar{\xi}_i$ is the amplitude and τ_i the relaxation time. Both, time constants and amplitudes are powerful parameters for the analysis of molecular mechanisms.

There are two major groups of experimental approaches to measure relaxation kinetic parameters: (1) impulse methods to measure the transient behavior as the response to impulse-forcing functions, and (2) stationary methods to measure the forced behavior as the response to oscillatory forcing functions.

The time course of ξ may be determined by optical and electrical methods (Eigen and De-Maeyer, 1963; Neumann, 1986a; Chernomordik *et al.*, 1987).

3.1. Rectangular Impulse Response

The response relaxation of rectangular pulses can be analyzed most straightforwardly, because the relaxation parameters refer to a constant electric field **E**.

Applying Eq. (A2) of the Appendix to Eq. (1), we obtain the result of Eq. (A3) in the form

$$\xi_i(t) - \xi_{i,0} = (\bar{\xi}_{i,\infty} - \xi_{i,0})(1 - e^{-t/\tau_i(E)}) \tag{2}$$

representing the exponential increase of $\xi_i(t)$ from ξ_0 at $t = 0$ to the stationary value $\bar{\xi}_\infty$ (at $t \rightarrow \infty$); see Fig. 2.

The off-field response, after a sudden switching off of the pulse, is given by

$$\xi_i(t) - \xi_{i,0} = (\bar{\xi}_{i,\infty} - \xi_{i,0})\, e^{-t/\tau(0)} \tag{3}$$

where $\bar{\xi}_{i,\infty}$ now refers to ξ_i at $t = 0$ for the off-field relaxation. Note that usually $\tau(0) \neq \tau(E)$; see the Appendix.

3.2. CD-Pulse Response

In the condenser discharge (CD) technique, the external force is an exponentially decaying field pulse:

$$E(t) = E_0 e^{-t/\tau_E} \tag{4}$$

where E_0 is the initial field strength and $\tau_E = RC$ is the time constant of the discharge circuit (with negligible inductance). R is practically given by the resistance of the sample chamber and C by the capacitance of the discharge capacitor.

If ξ_i (t) is directly responding to $E(t)$ and if the relaxation time of the system τ_i does practically not depend on E, the structural transition is described by

$$\xi_i(t) = \xi_{i,0} \frac{\tau_E}{\tau_E - \tau_i} (e^{-t/\tau_E} - e^{-t/\tau_i}) \tag{5}$$

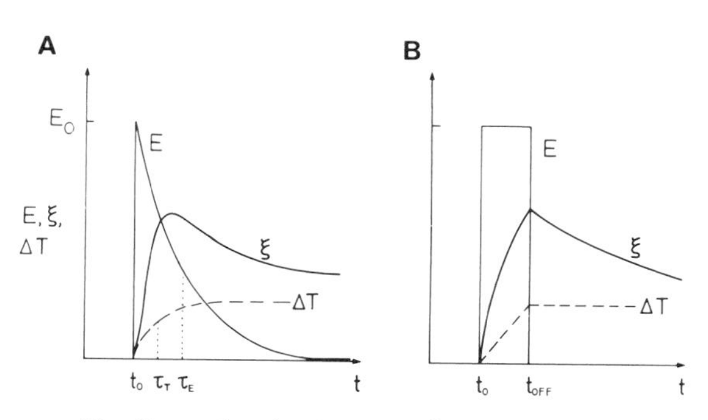

Figure 2. Structural rearrangement ξ (t) directly induced by electric field pulses; accompanying temperature changes ΔT. (A) Response to an exponentially decaying pulse (CD pulse) of field strength $E(t) = E_0 \exp(-t/\tau_E)$; $\tau_E = RC$ is the time constant of the capacitor discharge (CD) of the R,C circuit. If $\Delta T << T$ such that the sample resistance $R(T) \approx R(T + \Delta T)$, the temperature increase is given by $\Delta T \approx \Delta T_\infty$ $(1 - \exp[-t/0.5\ RC])$, where ΔT_∞ is the stationary value. (B) Response to a rectangular pulse, E = constant during $\Delta t = t_{OFF} - t_0$. The direct relaxational response ξ (t) is exponential, if a simple rearrangement process is measured. When $R(T) \approx R(T + \Delta T)$, the temperature increase ΔT is linear with time t within the pulse duration Δt.

Equation (5) is Eq. (A4). Often the response is rapid, i.e., $\tau_i << \tau_E$. For this case: $\xi_i(t) = \xi_{i,0}\, e^{-t/\tau_E}$. Equation (5) is represented in Fig. 2 for the case $\tau_i \approx \tau_E$.

The requirement of an E-independent τ seriously limits the application of CD pulses for the analysis of field-induced structural rearrangements.

3.3. Temperature Changes

Field pulse application to ionic solutions causes a temperature increase, mainly due to Joule heating. The differential increase dT of the temperature is given by

$$dT = \frac{U^2 dt}{R c_p \rho \cdot v} \tag{6}$$

where U is the applied voltage, R is the resistance and v/cm^3 is the volume of the sample chamber, respectively, c_p is the specific heat ($c_p \approx 4.18$ J/g K) and ρ is the specific mass of the solution ($\rho \approx 1$ g/cm^3), respectively.

3.3.1. Rectangular Pulses

For small temperature changes $\Delta T << T$, such that $R(T + \Delta T) \approx R(T)$, c_p and $m = \rho \cdot v$ remaining constant, the integration of Eq. (6) yields

$$\Delta T(t) = T(t) - T(0) = U^2 \Delta t/(R c_p \rho v) \tag{7}$$

describing the linear increase of T with time from $T(0)$ at $t_0 = 0$ up to $T(t)$, where $\Delta t = t - t_0$ is the pulse duration; see Fig. 2.

3.3.2. CD Pulses

For CD pulses, the voltage between the electrodes (distance b) decreases as $E = U/b$ or $U = U_0 e^{-t/\tau_E}$. Substitution into Eq. (6) and integration yields

$$\Delta T(t) = \Delta T_\infty (1 - e^{-t/\tau_T}) \tag{8}$$

where $\Delta T_\infty = T_\infty - T(0) = U_0^2 C/(2 c_p \rho v)$ is the maximum increase in T; $\tau_T = 0.5\, RC$, where C is the capacity of the discharge condenser; see Fig. 2.

It is stressed that the time courses $\xi(t)$ in Fig. 2 represent response functions of a direct field effect. The same curves are obtained if $\xi(t)$ is a secondary response (τ_i), induced by a rapidly equilibrating primary response (τ_p) to the external electric field, e.g., $\tau_E >> \tau_p$.

4. ELECTROPORATION AS A CRITICAL PHENOMENON

There are several experimental parallels between electroporation and electrofusion. Both field effect phenomena show threshold behavior. The numerical values of the threshold field strengths E_c are (almost) the same and are inversely proportional to the cell diameter. Both cell fusion and electroporative material exchange are clearly induced by the external field pulse. But, because of the longevity of the field-induced structural changes, the actual fusion events and the main part of the material exchange are, by and large, afterfield effects (Neumann, 1986c).

In summary, the data suggest that it is *one and the same primary field effect* on the membrane structure: leading to cell fusion if cell membranes are brought into contact before or after pulsing, or causing DNA uptake if the DNA is adsorbed to the cell surface before (or after) pulsing.

4.1. Threshold Parameters

The threshold field strength E_c for electroporation (electropermeabilization and electrofusion) is a kind of "point of no return" (Sugar and Neumann, 1984). If the electric field E ($\geq E_c$) is maintained, the electropores induced by the supercritical field increase in number and size (e.g., see Abidor *et al.*, 1979; Benz *et al.*, 1979; Chernomordik *et al.*, 1987) until, at a supercritical number density and pore size, the membrane ruptures (dielectric breakdown). If electric pulses of short duration Δt are applied, the field is already switched off before rupture can occur.

It is therefore pertinent to view membrane electroporation as a critical phenomenon, characterized by critical values for the extent ξ_c of structural rearrangement, for the field strength E_c, and for the pulse duration Δt_c. In our structural model, the primary requirement for the onset of electroporation is that the threshold ξ_c has to be reached. (The subcritical changes from ξ_0 to ξ_c represent reversible structural rearrangements such as the increase in number and size of hydrophobic defect sites and micropores in the bilayer.) The minimum field strength to attain the critical value ξ_c is the critical field E_c. Once the threshold ξ_c is reached ($E \geq E_c$), the actual electroporation starts and proceeds unidirectionally (no return) until the rupture threshold ξ_r is attained where the membrane disintegrates. If the field is reduced below E_c or switched off before ξ_r is reached, the electropores or electrocracks (Sugar *et al.*, 1987) reseal or anneal such that the original membrane state appears to be completely restored (reversible electroporation).

4.2. Strength–Duration Relationship

Similar to known electric membrane phenomena such as nerve excitation (Cole, 1968), the onset of electroporation is associated with a strength–duration relationship (e.g., see Lindner *et al.*, 1977; Zimmermann *et al.*, 1981). Since the threshold ξ_c is attained faster at a higher field strength (see Fig. 3), the minimum pulse duration Δt_c that is required for the onset of the electroporation process is the smaller the larger the applied external field. See Fig. 4.

If indeed the value E_c decreases with the independently chosen pulse length Δt (Zimmermann *et al.*, 1981), this feature may result from the stochastic nature of electropore formation (Sugar and Neumann, 1984; Chernomordik *et al.*, 1987). The larger Δt, the larger is the probability of nucleation of the electropores at a smaller field strength.

Due to the complexity of biological membrane organization (uneven surface charge distribution and membrane thickness; structural coupling to external matrix and to intracellular cytoskeletal elements), it is not possible to exactly calculate the strength–duration parameter set $E_c/\Delta t_c$ from first principles.

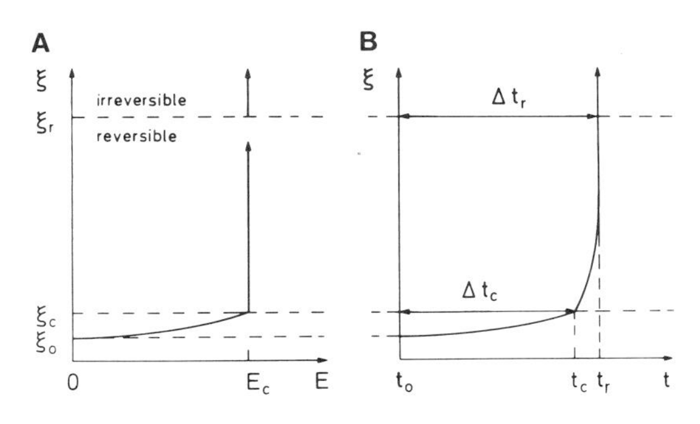

Figure 3. Electroporation viewed as a critical membrane phenomenon, associated with critical values for the field strength E_c, the pulse duration Δt_c, and the degree of structural rearrangement ξ_c. (A) Electroporation starts when the threshold ξ_c is attained; the minimum field strength to reach ξ_c is E_c. At $E \geqq E_c$ the electroporation process is unidirectional, i.e., irreversible, until the rupture threshold ξ_r is reached where the membrane breaks down irreversibly. If, however, the field is switched off before ξ_r is attained, the electroporated membrane reseals (reversible electroporation cycle). (B) The critical pulse duration Δt_c to trigger the electroporation process at ξ_c is the smaller the larger the field strength E ($> E_c$). The reversible electroporation cycle requires that the pulse length Δt is smaller than the rupture time Δt_r. The changes of ξ from ξ_0 to ξ_c represent reversible subcritical rearrangements such as the increase of size and number of hydrophobic defect sites.

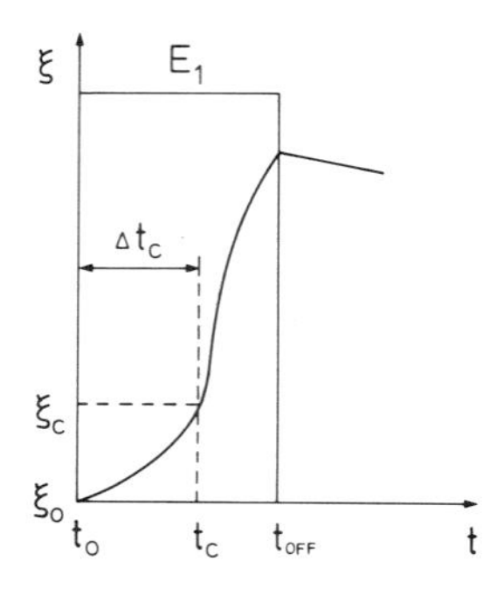

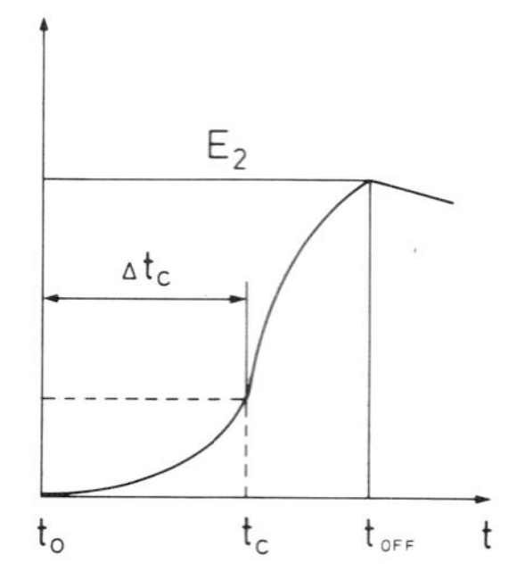

Figure 4. Strength/duration ($E/\Delta t_c$) relationship for the onset of the electroporation process at $\xi = \xi_c$. The critical pulse length Δt_c is the smaller the larger the field strength: $\Delta t_c(E_1) < \Delta t_c(E_2)$, if $E_1 > E_2$ (compare left and right).

The usual expression given for E_c in terms of the cell radius a and a critical transmembrane voltage is only an approximation. Nevertheless, the approximation for the maximum value at the pole cap regions (where $\cos \delta = +1, -1$, respectively):

$$E_c \approx 2V_{m,c}/3a \tag{9}$$

is useful, even for the estimate of the threshold of the initial field strength of exponentially decaying pulses (CD pulses). $V_{m,c} \approx 0.5$–1 V for short-duration pulses ($\Delta t \approx 10$ μsec) and $V_{m,c} \approx 0.2$–0.5 V for longer pulse duration ($\Delta t \geq 0.1$ msec).

The nature of some of the approximations inherent in the derivation of Eq. (9) is briefly elucidated in the treatment of the interfacial polarization.

5. ELECTROPORATION AS AN INDIRECT FIELD EFFECT

The magnitude of the applied field strength and the dependence of the threshold E_c on the cell radius a indicate that the field effect on the membrane structure is indirect. The data suggest that interfacial polarization precedes the structural transitions.

The actual membrane field affecting the lipids and proteins is strongly amplified by the interfacial polarization (e.g., see Schwan, 1985; Neumann, 1986b). The time constant (τ_p) of the buildup of the interfacial polarization $\Delta\varphi$ is dependent on a and the conductivities of the cell interior, the cell membrane, and the external medium (Schwan, 1957). In brief, the electroporation and electrofusion data indicate the sequence of events:

$$E \rightarrow \Delta\varphi \rightarrow \Delta\xi \tag{10}$$

where E causes the change $\Delta\varphi$ which in turn causes the change in extent $\Delta\xi$ of membrane rearrangements. In this sense the $\xi(t)$ function is delayed with respect to the application of the field pulse (Fig. 5).

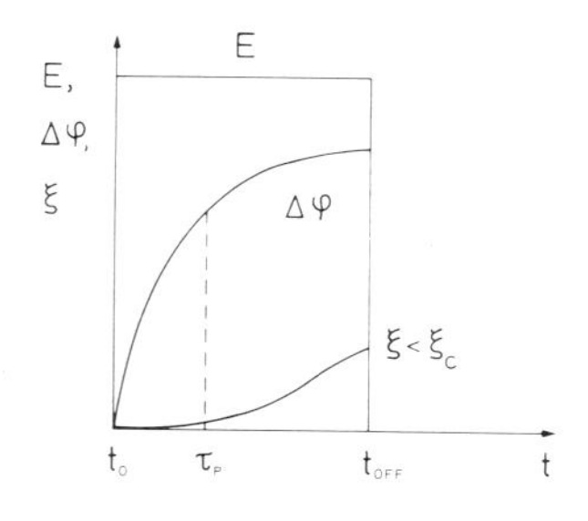

Figure 5. Amplified field effect. The external field pulse (E,Δt) causes the interfacial potential difference $\Delta\varphi$. The electric field, equivalent to $\Delta\varphi$, in turn induces structural rearrangements $\xi(t)$ in the membrane. Thus, $\xi(t)$ is delayed with respect to the application of E at t_0. The time course $\Delta\varphi(t)$ is represented as a simple exponential process (time constant τ_p). If the pulse duration at E is $\Delta t < \Delta t_c$, the time course $\xi(t) < \xi_c$ models a subcritical structural change.

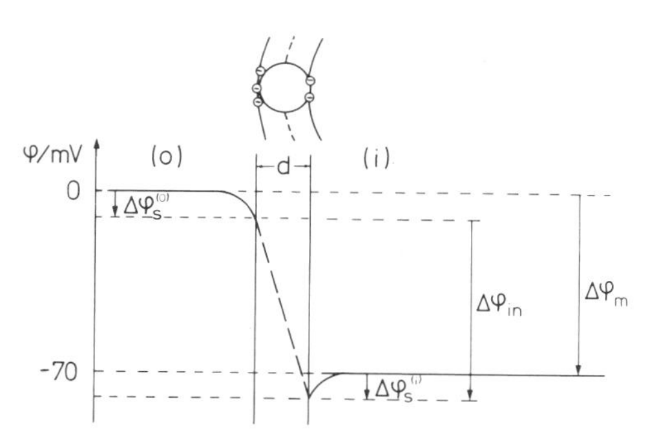

Figure 6. The profile of the electric potential φ across a cell membrane of thickness d. The measurable membrane potential $\Delta\varphi_m$ is the difference between the inside level $\varphi(i)$ and the potential $\varphi(o)$ of the outside medium, usually taken as the reference $\varphi(o) = 0$. In the usual case of negative surface charges, $\Delta\varphi_s^{(i)}$ and $\Delta\varphi_s^{(o)}$ represent the inner and the outer surface potentials, respectively. The intrinsic cross-membrane potential difference $\Delta\varphi_{in}$ is given by $\Delta\varphi_{in} = \Delta\varphi_m + \Delta\varphi_s^{(o)} - \Delta\varphi_s^{(i)}$. If $\Delta\varphi_s^{(o)} \approx \Delta\varphi_s^{(i)}$, then $\Delta\varphi_{in} \approx \Delta\varphi_m$. A typical value is $\Delta\varphi_m \approx -70$ mV.

5.1. Natural Transmembrane Voltage

It is well known that all cell membranes have a natural electric potential difference. This membrane potential $\Delta\varphi_m$ is defined as

$$\Delta\varphi_m = \varphi(i) - \varphi(o) \tag{11}$$

as the difference between inside (i) and outside (o); see Fig. 6. Classically, $\Delta\varphi_m$ is interpreted as diffusion potential arising from permselectivity and activity gradient of the potential-determining ion across the membrane; typically, $\Delta\varphi_m \approx -70$ mV, where $\varphi(o) = 0$ is taken as the reference (Fig. 6).

Biomembranes usually have an excess of negatively charged groups at the interface between membrane surface and solution. The contribution of these fixed charges and that of adsorbed small ions is denoted by $\Delta\varphi_s$; $|\Delta\varphi_s|$ decays with the distance from the surface and is dependent on the ionic strength I_c of the solution. The intrinsic potential difference $\Delta\varphi_{in}$ is given by

$$\Delta\varphi_{in} = \Delta\varphi_m - \Delta\varphi_s^{(o)} + \Delta\varphi_s^{(i)} \tag{12}$$

Even in the absence of a diffusion potential, $\Delta\varphi_{in}$ is finite, if $\Delta\varphi_s^{(i)} \neq \Delta\varphi_s^{(o)}$. When $\Delta\varphi_s^{(i)} \approx \Delta\varphi_s^{(o)}$, we may use $\Delta\varphi_{in} \approx \Delta\varphi_m$. Thus, the transmembrane voltage is given by the approximation

$$V_m \approx \Delta\varphi_m \tag{13}$$

In the presence of an applied field, the redistributions of ions in the interfacial regions lead to, probably asymmetric, changes in both $\Delta\varphi_m$ and $\Delta\varphi_s$. The displacement of the ionic atmosphere surrounding the membrane surfaces (e.g., see Schwan, 1985) changes the Debye–Hückel screening term asymmetrically relative to the field vector **E**.

In summary, the actual transmembrane voltage V_m, affecting directly the membrane structure, may be formally described by an asymmetric, field-dependent contribution from $\Delta\varphi_{in}$ and a symmetric contribution $\Delta\varphi$ from the field-induced interfacial polarization.

If the surface charge contributions are approximately equal such that $\Delta\varphi_{in} \approx \Delta\varphi_m$ applies, the transmembrane voltage V_m in external fields E is approximated by

$$V_m(E) \approx \Delta\varphi_m + \Delta\varphi(E,\delta) \tag{14}$$

where the interfacial potential term $\Delta\varphi$ is not only dependent on E but also on the angular position δ (see Fig. 7).

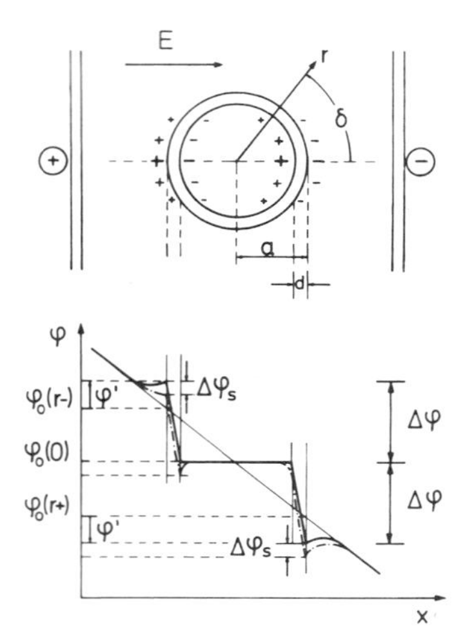

Figure 7. Interfacial polarization of a spherical nonconducting shell of thickness d and outer radius $r = a$ in a constant external field **E**. The stationary electric potentials are given in polar coordinates of the radius vector **r** and the angle δ, such that the conducting interior of the cell has the constant reference potential $\varphi_0 = 0$ for $0 \leqq r \leqq (a - d)$. For $r > a$, $\varphi_0 = -\mathbf{E} \cdot \mathbf{r} = -E \cdot r \cos \delta$. The total potential $\varphi(r)$, relative to $\varphi_0(0) = 0$, is given by $\varphi(r,\delta) = \varphi_0(r,\delta) + \varphi'(r,\delta)$, where $\varphi'(r,\delta)$ is the contribution of the interfacial polarization. The $\Delta\varphi$ terms are the interfacially induced, cross-membrane potential differences in the absence of fixed ionic groups and adsorbed ions, $\Delta\varphi_s = 0$. The dash/dot line models schematically the potential profile in the presence of fixed surface charges (here negative).

5.2. Interfacial Polarization

Since the main contribution to the actual transmembrane voltage V_m is due to interfacial polarization, the polarization term $\Delta\varphi$ is discussed separately for the case of zero fixed surface charge. According to the original treatment by Maxwell (1904), $\Delta\varphi$ is calculated from the solution of the Laplace equation $\nabla^2\varphi = 0$ for the electric potential $\varphi(r)$ in terms of the polar coordinates radius vector **r** and angular position δ relative to the direction of the applied field vector **E** (Fig. 7).

As usual, the cell membrane is modeled as a thin spherical shell of thickness d and outer radius a, and the definition $E(r) = -\partial\varphi/\partial r$ is applied.

In order to cover time dependencies and dielectric and conductive contributions, complex specific admittances are used (e.g., see Schwan, 1985). If, however, the conductivity terms λ are dominant, the stationary value ($\Delta\varphi$) and the time constant (τ_p) of interfacial polarization are expressed in terms of the specific conductivities of the membrane (λ_m), of the cell interior (λ_i), and of the external medium (λ_o).

In Fig. 7 the potential profiles $\varphi(r)$ and the interfacial transmembrane terms $\Delta\varphi$ are schematically represented. The total potential is expressed as a sum $\varphi(r,\delta) = \varphi_0(r,\delta) + \varphi'(r,\delta)$, where $\varphi_0(r,\delta) = -\mathbf{E} \cdot \mathbf{r} = -E \cdot r \cdot \cos \delta$ is the contribution due to the external field and $\varphi'(r,\delta)$ is due to the interfacial polarization. In this form the reference potential is $\varphi_0(0) = 0$ at $r = 0$; at the right pole cap where $\cos 0° = +1$, the reference potential is $\varphi_0(r) = -Er$ and at the left pole cap where $\cos 180° = -1$, $\varphi_0(r) = +Er$. It is mentioned that the dipole terms $\varphi'(r,\delta)$ are also expressed as functions of E, r, $\cos \varphi$ and, additionally, of the λ values. Finally, the transmembrane term is given by

$$\Delta\varphi = \varphi(a) - \varphi(a\text{-}d) \tag{15}$$

describing an *electric potential drop* for both hemispheres (left and right pole cap) in the direction of the external field, if $\lambda_m \leq \lambda_i,\lambda_o$, as realistically assumed in the graph of Fig. 7.

5.2.1. Impulse Polarization Responses

It is recalled from the relationship (10) that the interfacial polarization is the primary response to the external field.

5.2.1a. Rectangular Pulses. Applying Eq. (A3), the buildup of the interfacial polarization in rectangular pulses is given by

$$\Delta\varphi(t) = \Delta\varphi\ (1 - e^{-t/\tau_p}) \tag{16}$$

The stationary value $\Delta\varphi$ is obtained from Eq. (15):

$$\Delta\varphi = -(3/2)\ E\ a\ |\cos\ \delta|\ f(\lambda) \tag{17}$$

The conductivity term $f(\lambda)$ is given by

$$f(\lambda) = \frac{2\lambda_0[2\lambda_m + \lambda_i + (\lambda_m - \lambda_i)\ (\frac{a-d}{a})^3 - 3\lambda_m \frac{a-d}{a}\ ]}{(2\lambda_m + \lambda_i)(2\lambda_0 + \lambda_m) + 2\ (\frac{a-d}{a})^3\ (\lambda_i - \lambda_m)(\lambda_m - \lambda_0)} \tag{18}$$

Since for a typical case $a \approx 1$ µm compared with $d \approx 0.01$ µm, we introduce the approximation $d << a$ such that $[(a-d)/a]^3 \approx 1-3d/a$ and obtain:

$$f(\lambda) = \frac{\lambda_0\lambda_i(2d/a)}{(2\lambda_0 + \lambda_i)\lambda_m + (2d/a)(\lambda_0-\lambda_m)(\lambda_i-\lambda_m)} \tag{19}$$

Usually $\lambda_m << \lambda_0,\lambda_i$ (e.g., see Schwan, 1957, 1985) such that Eq. (19) takes the form:

$$f(\lambda) = \frac{1}{1 + \lambda_m\ (2 + \lambda_i/\lambda_0)/(2\lambda_i d/a)} \tag{20}$$

It is readily seen that an increase in the external ionic strength leading to an increase in λ_o will increase $\Delta\varphi$. This is consistent with the notion that the interfacial polarization is associated with ion accumulations at the interfaces of the membrane.

If the membrane can be considered nonconducting (before the onset of electroporation), we insert $\lambda_m = 0$ into Eq. (18) and obtain:

$$f(\lambda) = 1 \tag{21}$$

For this limiting case the time course of the interfacial membrane polarization is simply described by:

$$\Delta\varphi(t) = -(3/2)\ E\ a\ |\cos\ \delta|\ (1 - e^{-t/\tau_p}) \tag{22}$$

The time constant of the buildup of $\Delta\varphi$ is given by (Schwan, 1957):

$$\tau_p = a \cdot C_m \frac{\lambda_i + 2\lambda_0}{2\lambda_i\lambda_0 + a\ G_m(\lambda_i + 2\lambda_0)} \tag{23}$$

where $C_m \approx 1$ µF/cm² is the specific membrane capacitance and G_m is the specific membrane resistance.

For many cases of practical importance, the membrane conductance is small, $G_m \approx 10^{-3}$ S/cm², such that a $G_M << 2\lambda_i\lambda_o/(\lambda_i + 2\lambda_o)$ and the approximation

$$\tau_p \approx a\, C_m \frac{\lambda_i + 2\lambda_0}{2\lambda_i\lambda_0} \tag{24}$$

may be used.

It is remarked that in the case of high field strengths, the conductivity values refer to the field intensity E. In the absence of larger Wien effects (e.g., see Neumann, 1986a), we may use the approximations $\lambda(E) \approx \lambda(0)$, where $\lambda(0)$ are the conductivities at very low field.

5.2.1b. CD Pulses. The time course of the field-induced interfacial polarization is described by Eq. (A8) where the identities $\tau_F = \tau_E$ and $\tau_i = \tau_p$ have to be used:

$$\Delta\varphi(t) = \Delta\varphi \frac{\tau_E}{\tau_E - \tau_p} (e^{-t/\tau_E} - e^{-t/\tau_p}) \tag{25}$$

where $\Delta\varphi$ is given by Eq. (17). In the limit of $\tau_p << \tau_E$, Eq. (25) reduces to

$$\Delta\varphi(t) = -(3/2)\, E\, a\, |\cos\,\delta|\, e^{-t/\tau_E} \tag{26}$$

5.2.2. Forced Polarization Response

In AC fields the induced polarization is forced. If the field vector oscillates with $\mathbf{E}\,(t) = \hat{\mathbf{E}}\cos\omega t$ or, in complex notation,

$$\mathbf{E}(t) = \hat{\mathbf{E}} \cdot e^{j\omega t} \tag{27}$$

the induced potential difference is given by Eq. (A21); explicitly we obtain

$$\Delta\varphi(\omega,t) = \frac{\Delta\varphi}{(1 + \omega^2\,\tau_p^2)^{1/2}} \cos(\omega t - \alpha_p) \tag{28}$$

The forced amplitude $\Delta\varphi(\omega) = \Delta\varphi/(1 + \omega^2\tau_p^2)^{1/2}$ is dependent on frequency and τ_p:

$$\Delta\varphi(\omega) = \frac{-(3/2)\hat{E}\; a|\cos\delta| f(\lambda)}{(1 + \omega^2\tau_p^2)^{1/2}} \tag{29}$$

In the limit of small frequencies ($\omega << \tau_p^{-1}$), we see that $\Delta\varphi$ is independent of ω:

$$\Delta\varphi = -(3/2)\, \hat{E}\, a\, |\cos\,\delta|\, f(\lambda) \tag{30}$$

At higher frequencies ($\omega >> \tau_p^{-1}$), we obtain

$$\Delta\varphi(\omega) = \frac{-(3/2)\hat{E}\; a|\cos\delta| f(\lambda)}{\omega\tau_p} \tag{31}$$

For a nonconducting membrane ($G_m = 0$; $\lambda_m = 0$), Eqs. (21) and (24) are introduced into (31). In this limiting case, the high-frequency approximation is

$$\Delta\varphi(\omega) = -\frac{3\lambda_i\lambda_0\, \hat{E}\, |\cos\,\delta|}{\omega C_m\, (\lambda_i + 2\lambda_0)} \tag{32}$$

Thus, if $\omega >> \tau_p^{-1}$, the amplitude value of the induced interfacial membrane polarization is independent of the cell radius (Schwan, 1985; Zimmermann, 1986).

5.3. Transmembrane Voltage in Applied Fields

The explicit expressions of the transmembrane voltages V_m can now be specified for the various experimental conditions. We use Eq. (14) in the form

$$V_m\,(E,t) \approx \Delta\varphi_m + \Delta\varphi(E,\delta,t) \tag{33}$$

where $\Delta\varphi_m$ is assumed independent of E and of time. Furthermore, we only specify the maximum values of $|V_m|$ for the limiting case of a nonconducting membrane, i.e., $\lambda_m = 0$ such that $f(\lambda) = 1$.

For rectangular pulses the stationary value of V_m is derived by insertion of Eq. (17) into (33). With $f(\lambda) = 1$ we obtain for the transmembrane potential difference in the direction of the $\vec{E}$ vector:

$$V_m \approx -[(3/2)\,E\,a + \Delta\varphi_m/\cos\delta]\,|\cos\delta| \tag{34}$$

This notation correctly describes the signs and the angular position dependence of the transmembrane voltage relative to the direction of **E**. Note that $|\cos\delta|/\cos\delta = +1$ for the right hemisphere and $|\cos\delta|/\cos\delta = -1$ for the left hemisphere (Fig. 7). Normally, $|\Delta\varphi_m|$ is independent of position, but at the left pole cap $\Delta\varphi_m$ ($= -70$ mV) is in the same direction as $\Delta\varphi$ whereas at the right pole cap $\Delta\varphi_m$ is opposite to $\Delta\varphi$.

In CD pulses, $\Delta\varphi$ is transient and the complete term $\Delta\varphi(t)$ of Eq. (25) applies. In the limit of $\tau_p << \tau_E$ [see Eq. (26)], the initial (i.e., maximum) transmembrane voltage is given by

$$V_{m,0} \approx -[(3/2)\,E_0\,a + \Delta\varphi_m/\cos\delta]\,|\cos\delta| \tag{35}$$

For AC fields the amplitude of the transmembrane voltage is derived by insertion of Eq. (29) into (33). Hence,

$$V_m(\omega) \approx \frac{-[(3/2)\hat{E}\,a + \Delta\varphi/\cos\delta]|\cos\delta|}{(1 + \omega^2\tau_p^2)^{1/2}} \tag{36}$$

5.4. Structural Changes as Secondary Responses

In terms of Eq. (10), the structural changes induced by the amplified field resulting from interfacial polarization are secondary responses to the external field application.

For rectangular pulses the response relaxation $\xi_i\,(t)$ is derived from Eq. (A6) where τ_F is the interfacial polarization time constant τ_p specified in Eq. (23) or (24). The ith mode is given by

$$\xi_i(t) = \bar{\xi}_i(1 - \frac{\tau_i}{\tau_i - \tau_p}\,e^{-t/\tau_i} + \frac{\tau_p}{\tau_i - \tau_p}\,e^{-t/\tau_p}) \tag{37}$$

Note that both τ_i and τ_p refer to the values in the presence of the external field; at $t = 0$, $\xi_i\,(0) = 0$, otherwise instead of ξ_i (t) and $\bar{\xi}_i$ the pair $\xi_i\,(t) - \xi_i\,(0)$, $\bar{\xi}_i - \xi_i\,(0)$ has to be used.

The field-off relaxation (at $E = 0$) is described by Eq. (A15) in the form:

$$\xi_i(t) = \bar{\xi}_{i,0}\,\frac{1}{\tau_i - \tau_p}\,(\tau_i e^{-t/\tau_i} - \tau_p e^{-t/\tau_p}) \tag{38}$$

where $\bar{\xi}_{i,0}$ is the amplitude at $t = t_{OFF} = 0$; both τ_i and τ_p now refer to $E = 0$.

If the structural transitions are slow ($\tau_i >> \tau_p$), interfacial polarization can be considered as rapidly equilibrating to the field. In this case, Eq. (33) reduces to Eq. (2) and Eq. (38) reduces to Eq. (3) of the direct field effect.

The structural relaxations caused via interfacial polarization by a CD pulse are described by Eq. (A11), where $y_i(t)$, $y_{i,0}$, and τ_F are replaced by $\xi_i(t)$, $\xi_{i,0}$, and τ_E, respectively. The time course $\xi_i(t)$ is viewed relative to $\xi_i(0) = 0$ at $t = 0$. Note that Eq. (A11) only applies if both τ_i and τ_p can be considered to be field-independent.

The structural response to AC-field-induced oscillating interfacial potential differences $\Delta\varphi(\omega,t)$ is derived from Eqs. (A21) and (A25):

$$\xi_i(\omega,t) = \frac{\hat{\xi}_i}{[(1 + \omega^2\tau_p^2)(1 + \omega^2\tau_i^2)]^{1/2}} \cos(\omega t - \alpha_p - \alpha_i) \tag{39}$$

where the peak amplitude $\hat{\xi}_{i,\infty} = \hat{\xi}_i\,[(1 + \omega^2\tau_p^2)(1 + \omega^2\tau_i^2)]^{-1/2}$ is dependent on ω and on both τ_i and τ_p.

Here, too, slower structural relaxations ($\tau_i >> \tau_p$) may be analyzed as a direct response to $E(\omega,t)$, i.e., Eq. (39) reduces to the expression

$$\xi_i(\omega,t) = \frac{\hat{\xi}_i}{(1 + \omega^2\tau_i^2)^{1/2}} \cos(\omega t - \alpha_i) \tag{40}$$

The terms α_p and α_i are the phase angles, given by the arc tg functions of $\omega\tau_p$ and $\omega\tau_i$, respectively. See Eq. (A19).

6. THE ELECTROPORATION HYSTERESIS

It is emphasized that electroporation and electrofusion data indicate that reversible primary processes and irreversible secondary events are involved; see Table 1.

Ionic–dielectric polarization of the membrane/solution interfaces and structural rearrangements in the membrane are essentially reversible processes. Material exchange and the fusion processes are passive, unidirectionally occurring relaxation phenomena of irreversible nature. The data indicate that direct field effects on the membrane structure are of minor extent. Interfacial polarization leads to the strong (amplified) transmembrane fields which in turn induce the major structural rearrangements.

6.1. Electroporation Cycle and Crater Model

No doubt, a transient permeability increase (Neumann and Rosenheck, 1972) indicates transient membrane "openings": pores or cracks which reseal after pulsing. When the cycle of permeability increase and decrease is modeled on the level of electropores (Abidor *et al.*, 1979; Sugar and Neumann, 1984; Sugar *et al.*, 1987), the formation–resealing cycle of a pore is represented as a cyclic local change of the membrane structure. In addition, local deformations of the pore edges may lead to craterlike pore structures (Fig. 8).

During the transition to electropores, lipid molecules move rapidly away from the pore center; the remaining bilayer lipids have to change position relative to the planar configuration. Since global deformations such as elongations of the whole cell or vesicle have higher moments of inertia, local less inert deformations may occur more rapidly. Therefore, the pore edges may rapidly bend out relative to the membrane plane and form craterlike structures.

As discussed in Schwan's chapter of this book, the external field causes a net outward force on the outer membrane surface. Thus, the craterlike pore edges are predicted to have outward cur-

Table 1. Fundamental Processes of the Electroporation Hysteresis of Membranes

Physical chemical processes	Electric terms
Reversible primary processes	
Primary electric events	
Electric dipole induction and dipole orientation	Dielectric polarization
Redistribution of mobile ions at phase boundaries of membrane/solution, including (1) ionic atmosphere shifts and (2) local activity changes of effectors, e.g., H^+ (pH changes) or Ca^{2+} ions	Ionic–dielectric interfacial polarization (Maxwell–Wagner, β dispersion)
Structural rearrangements	Electrorestructuring
Conformational changes in protein and lipid molecules	
Phase transitions in lipid domains, resulting in pores, cracks (via pore coalescence), and percolation	Electroporation: electropores, electrocracks, electropercolation
Annealing and resealing processes	
Irreversible secondary processes	
Transient material exchange	Electropermeabilization
Release of internal compounds, e.g., hemolysis	Electrorelease
Uptake of external material, e.g., drugs, antibodies	Electroincorporation, electrosequestering
Transfer of genetic material (e.g., DNA, mRNA, viroids) with stable cell transformation	Electrotransfection: electrotransformation, electroporative gene transfer
Membrane reorganizations	
Cell fusion (if membrane contact)	Electrofusion
Vesicle formation (budding)	Electrovesiculation, electrobudding
Electromechanical rupture	Dielectric breakdown
Tertiary effects	
Temperature increase due to dissipative processes	Joule heating, dielectric losses
Metal ion release from metal electrodes	Electroinjection
Electrode surface H and O in statu nascendi	Electrolysis

vature. On the same speculative line, the crater edges sticking out from the membrane plane may provide the local contacts to initiate fusion between two membranes (e.g., see Sugar *et al.*, 1987).

6.2. Metastable States and Unidirectional Transitions

As outlined previously, the apparent dependence of the threshold field strength E_c on the pulse length (e.g., see Benz and Zimmermann, 1980) indicates a membrane specific critical threshold ξ_c at which the electroporation process is triggered. Once initiated, the process is unidirectional, i.e., irreversibly running at constant E ($> E_c$). Therefore, the membrane state just before the onset of the

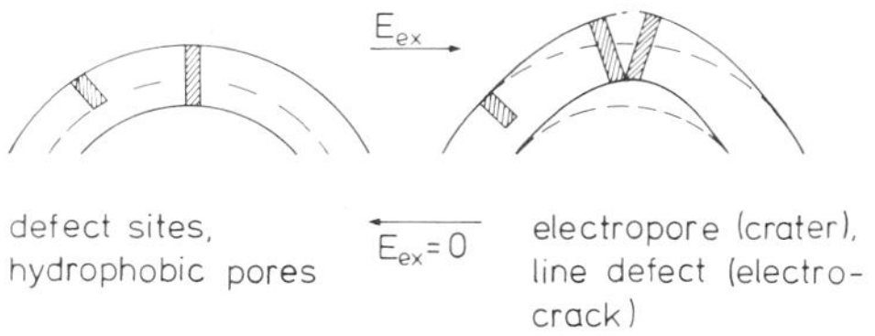

Figure 8. Crater model for the electropore. The cycle of membrane restructuring in pore regions is modeled as a cyclic transition from point defects (e.g., hydrophobic pores) to hydrophilic pores and cracks with craterlike edges. During electropore formation, lipid molecules move away from the pore space; the rest of the bilayer has to expand. To avoid energetically unfavorable global expansions (cell or vesicle elongations), the pore edges may rapidly bend out of the membrane plane and form craterlike structures.

E_{ex}

defect sites, hydrophobic pores

$E_{ex} = 0$

electropore (crater), line defect (electrocrack)

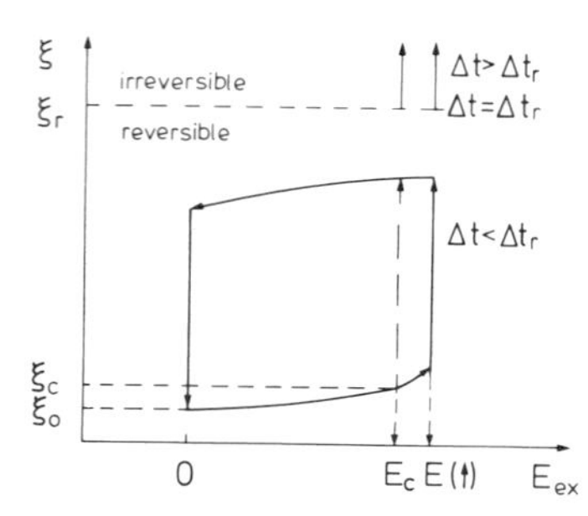

Figure 9. The relaxation hysteresis of the membrane electroporation. The cyclic change in the extent of structural rearrangements ξ [at $E > E_c$ and $\Delta t(E) < \Delta t_r$] are displayed as a function of the external field (E_{ex}). The cycle comprises reversible as well as irreversible elements. The subcritical changes of ξ between ξ_0 and ξ_c ($E < E_c$) are reversible. At the supercritical field strength $E(\uparrow)$ ($> E_c$), the structural rearrangements associated with the electroporation process are unidirectional, i.e., irreversible. If the field duration is larger than the rupture threshold (Δt_r), the membrane ruptures. If, however, the field pulse is switched off at $\Delta t < \Delta t_r$, we remain in the reversible electroporation domain. The (slow) return at $E = 0$ from the upper branch to $\xi = \xi_0$ is unidirectional; the resealing or annealing processes are irreversibly occurring relaxations to the initial state ξ_0.

electroporation process must be metastable. The minor structural rearrangements before electroporation ($\xi_0 < \xi < \xi_c$) are apparently reversible but occur on a metastable level (Neumann, 1973).

At short pulse durations ($\Delta t < \Delta t_r$) such that the rupture threshold ξ_r was not reached, the electroporated (and fusiogenic) membrane is again metastable with respect to less porous structures. The pore resealing process (at $0 < E < E_c$) is also unidirectional, i.e., irreversibly running until intact bilayer structures are restored.

In this sense, membrane electroporation represents a cycle of structural rearrangements (Fig. 9), where the intermediate states of the annealing process are probably different from those of the electroporation process in the presence of the external field.

The physical conception that comprises both (reversible) metastable states and (irreversible) unidirectional transitions in a cyclic manner is called hysteresis (e.g., see Neumann, 1973). Therefore, the electroporation/resealing cycle may be analyzed and understood in terms of a structural relaxation hysteresis.

6.3. Energetics of the Electroporation Hysteresis

The energetics of the field-induced structural rearrangements in the electroporation/resealing cycle can be thermodynamically treated independent of any special mechanism.

The characteristic reaction free enethalpy $\Delta_r G(E)$ for the transition from the intact bilayer state to pore configurations is given by

$$\Delta_r G(E) = \Delta_r G - \int_0^E \Delta_r M \, dE_m \tag{41}$$

where E_m is given by $|V_m|/d$ with $|V_m|$ from Eq. (33) or (34) and $\Delta_r G$ is the reaction free enthalpy in the absence of E (e.g., see Schwarz, 1967; Neumann, 1986a). Obviously, pore formation is energetically unfavorable at $E = 0$; hence, $\Delta_r G > 0$.

The reaction dipole moment $\Delta_r M$ is related to the difference between the moments M_w of the water-filled pore and M_m of the bilayer having the same size as the aqueous pore.

Applying the dielectric continuum model, the polarization moments (relative to vacuum) are given by

$$M = \epsilon_0 (\epsilon - 1) v E_m \tag{42}$$

where ϵ_0 is the dielectric permittivity of the vacuum, ϵ_m is the dielectric constant of the medium, and v is the volume of the electroporated membrane.

The polarization of the water ($\epsilon_w \approx 80$) near the pore edge in the pore wall (Sugar and Neumann, 1984) is energetically more favorable (by a factor of about 40) than the polarization of an equal volume of lipid bilayer ($\epsilon_m \approx 2$) (e.g., see Abidor *et al.*, 1979). With $\Delta_r M = N_A (M_w - M_m)$, where N_A is Avogadro's constant, we can specify

$$\Delta_r M = N_A \epsilon_0 (\epsilon_w - \epsilon_m) \cdot v(\delta) \cdot E_m \tag{43}$$

If $d << a$, the electroporated membrane volume $v(\delta)$ is given by $v(\delta) \approx 4\pi a^2 d\,(1 - |\cos \delta|)$ where the fraction of spherical membrane shell affected by a supercritical field is $(1 - |\cos \delta|) = 1 - E_c/E$ (Fig. 7).

It is seen from Eq. (43) that, because of $\epsilon_w > \epsilon_m$, we obtain $\Delta_r M > 0$. Therefore, the unidirectional electroporation process is associated with $\Delta_r G(E) < 0$; obviously because $\int \Delta_r M \, dE_m > \Delta_r G$ [see Eq. (41)]. On the same line, the initial driving force for the unidirectional pore resealing process is given by $\Delta_r G_{res.} = -|\Delta_r G(E)|$.

In the case of unidirectional transitions, starting from metastable states, the rate equation for the extent ξ of structural rearrangement, generally given by

$$d\xi_i/dt = k_{ij} \cdot \xi_i - k'_{ij}(1 - \xi_i) \tag{44}$$

is reduced to

$$d\xi_i/dt = k_{ij} \cdot \xi_i \tag{45}$$

Because of the irreversibility the reserve process can be neglected. The general rate coefficient k_{ij} represents a combination of the rate constants k_j of all elementary steps j contributing to the mode i. On the same line, the relaxation time τ_i is a combination of the k_{ij} and k'_{ij}.

Finally, the field dependence of the rate constant k_j is described by

$$k_j(E) = k_j(0) \cdot e^{\int \Delta_r M_j^* dE/RT} \tag{46}$$

where $\Delta_r M_j^*$ is the transition dipole moment of the activated state, R the gas constant, T the absolute temperature, and $k_j(0)$ the value at $E = 0$.

In summary, the explicit expressions derived in this chapter provide a general framework for the thermodynamic and kinetic analysis of electroporation and electrofusion processes in terms of the relaxation hysteresis formalism.

7. APPENDIX

7.1. Responses to Impulse Forcing Functions

If $y(t)$ is the response function caused by the forcing function $\bar{y}(t)$, the (linear) relaxation equation is:

$$dy(t)/dt = [\bar{y}(t) - y(t)]/\tau \tag{A1}$$

The forcing function $\bar{y}(t)$ is the maximum value (or amplitude) of $y(t)$ and represents the limiting case of an infinitely rapid response to the actual forcing process.

Equation (A1) is an inhomogeneous differential equation with the general solution:

$$y(t) = \frac{e^{-t/\tau}}{\tau} \int_0^t \bar{y}(t) e^{t/\tau} dt \tag{A2}$$

where it is implicit that at $t = 0$, $y(0) = 0$.

7.1.1. Rectangular Forcing Function

If a constant external force such as an electric field pulse is suddenly applied (step function) at $t = 0$, the forcing function $\bar{y}_i(t)$ is constant. The response function is derived from Eq. (A2):

$$y_i(t) = y_{i,\infty}\,(1 - e^{-t/\tau_i(F)}) \tag{A3}$$

describing the exponential increase of the ith normal mode from $y_i(0) = 0$ to the stationary value $y_{i,\infty} = \bar{y}$ (at $t \rightarrow \infty$). It is explicitly expressed that the relaxation time $\tau_i(F)$ refers to the presence of the external force. It is readily seen that the response to a sudden removel of the force, e.g., switching off the field pulse, is given by

$$y_i(t) = y_i(0) \cdot e^{-t/\tau(0)} \tag{A4}$$

where $y_i(0) = \bar{y}$ is the value of $y_i(t)$ at t_{OFF}; $y_i(0)$ is the initial value (at $t = 0$) for the field-off relaxation; usually $\tau(0) \neq \tau(F)$.

7.1.2. Exponentially Increasing Forcing Function

The response to the forcing function

$$\bar{y}(t) = \bar{y}_\infty(1 - e^{-t/\tau_F}) \tag{A5}$$

where τ_F is the time constant and $\bar{y}_\infty$ is the value of $\bar{y}(t)$ at $t \rightarrow \infty$, is also derived from Eq. (A2). We obtain for the ith mode:

$$y_i(t) = y_{i,\infty}(1 - \frac{\tau_i}{\tau_i - \tau_F}\, e^{-t/\tau_i} + \frac{\tau_F}{\tau_i - \tau_F}\, e^{-t/\tau_F}) \tag{A6}$$

where the amplitude $y_{i,\infty} = \bar{y}_\infty$ is the stationary value of $y_i(t)$ at $t \rightarrow \infty$.

If $\tau_i \gtrsim \tau_F$, the response is delayed with respect to $\bar{y}(t)$. If the response is very rapid ($\tau_i << \tau_F$), the measured relaxation is rate-determined by the forcing function, i.e., $y_i(t) = y_{i,\infty}(1 - e^{-t/\tau_F})$.

7.1.3. Exponentially Decreasing Forcing Function

The forcing function of a CD pulse is

$$\bar{y}(t) = \bar{y}_0\, e^{-t/\tau_F} \tag{A7}$$

From Eq. (A2) we obtain the response function

$$y_i(t) = y_{i,0}\, \frac{\tau_F}{\tau_F - \tau_i}\, (e^{-t/\tau_F} - e^{-t/\tau_i}) \tag{A8}$$

where $y_{i,0} = \bar{y}_0$ is the amplitude term, representing the limit of infinitely rapid response to the external force F (at $t = 0$).

If the response is delayed ($\tau_i \gtrsim \tau_F$), the response function starting with $y_i(0)$ at $t = 0$ increases with time, proceeds through a maximum at

$$t_{max} = \frac{\tau_i \tau_F}{\tau_F - \tau_i}\, \ln(\tau_F/\tau_i) \tag{A9}$$

and then approaches the stationary value $y_i\,(\infty)$ at $t \rightarrow \infty$.

Note that the relaxation time of the system τ_i must be independent of the force F in order to justify the analysis leading to Eqs. (A8) and (A9). In addition, the value of $y_{i,0}$ must be evaluated by curve fitting.

If the response is very rapid ($\tau_i << \tau_F$), the response is determined by the decay of the forcing function: $y_i(t) = y_{i,0}\, e^{-t/\tau_F}$.

7.1.4. Secondary Exponential Responses

The forcing function $\bar{y}(t)$ itself can result from a primary forcing function, decaying exponentially with the time constant τ_F. Inserting Eq. (A8) in the form

$$\bar{y}_i(t) = \bar{y}_{i,0} \frac{\tau_F}{\tau_F - \tau_p} (e^{-t/\tau_F} - e^{-t/\tau_p}) \tag{A10}$$

where τ_p is the time constant of the primary response to the force F.

The response function to this forcing function, Eq. (A10), is given by

$$y_i(t) = \frac{y_{i,0}\tau_F}{\tau_F - \tau_p} \left[\frac{\tau_F(e^{-t/\tau_F} - e^{-t/\tau_i})}{\tau_F - \tau_i} - \frac{\tau_p\,(e^{-t/\tau_p} - e^{-t/\tau_i})}{\tau_p - \tau_i}\right] \tag{A11}$$

where τ_i is the time constant of the system responding to the force P that itself is caused by F.

If the primary force is suddenly switched off, the forcing function for secondary processes is

$$\bar{y}(t) = \bar{y}_0 \cdot e^{-t/\tau_F} \tag{A12}$$

Insertion in Eq. (A2) yields the response function of the ith mode:

$$y_i(t) = y_{i,0} \cdot \frac{1}{\tau_i - \tau_F} (\tau_i e^{-t/\tau_i} - \tau_p e^{-t/\tau_p}) \tag{A13}$$

where $y_{i,0} = \bar{y}_0$ is the amplitude at $t = t_{OFF} = 0$.

7.2. Responses to Oscillating Forcing Functions

In the simplest practical case, the external force represents a harmonic oscillatory forcing function, e.g.,

$$\bar{y}(t) = \hat{y} \cos \omega t \tag{A14}$$

where $\hat{y}$ is the peak amplitude, $\omega = 2\pi\upsilon$ is the angular frequency and υ the frequency. In complex notation, with $j^2 = -1$,

$$\bar{y}(t) = \hat{y}\, e^{j\omega t} = \hat{y}(\cos \omega t + j \sin \omega t) \tag{A15}$$

7.2.1. Primary Forced Response

The forced response is stationary and oscillates, too, generally with a retardation (phase shift) but with the same frequency:

$$y(t) = B e^{j\omega t} \tag{A16}$$

Differentiation of Eq. (A14) yields

$$d\,y(t)/dt = j\omega \cdot y(t) \tag{A17}$$

Comparison with the general relaxation equation (A1) yields the solution

$$y(t) = \frac{1}{1 + j\omega\tau} \cdot \bar{y}(t) = \sigma\bar{y}(t) \tag{A18}$$

Applying complex number calculus, the transfer function σ of the relaxation mode i is given by

$$\sigma_i = \frac{1}{1+j\omega\tau_i} = \frac{e^{-j\alpha_i}}{(1+\omega^2\tau_i^2)^{1/2}} = \frac{1-j\omega\tau_i}{1+\omega^2\tau_i^2} \tag{A19}$$

where α_i = arc tg$(\omega\tau_i)$ is the phase angle; it describes the lagging behind of the response mode i.

The combination of Eqs. (A13), (A16), and (A17) leads to

$$y_i(t) = \frac{\hat{y}_i e^{j(\omega t - \alpha_i)}}{(1+\omega^2\tau_i^2)^{1/2}} \tag{A20}$$

Inserting the real parts of $e^{j(\omega t - \alpha_i)}$ and of $\sigma_i e^{j\omega t}$, we obtain

$$y_i(t) = \frac{\hat{y}_i \cos(\omega t - \alpha_i)}{(1+\omega^2\tau_i^2)^{1/2}} = \frac{\hat{y}_i \cos\omega t + \omega\tau_i \sin\omega t}{1 + \omega_i^2\tau_i^2} \tag{A21}$$

In the low-frequency limit $\omega << \tau_i^{-1}$, the response $y_i(t)$ rapidly equilibrates, i.e., $\sigma_i = 1$ and hence $y_i = \hat{y}_i\, e^{j\omega t}$. In the high-frequency limit $(\omega >> \tau_i^{-1})$, we have $(1 + \omega^2\tau_i^2)^{1/2} \approx \omega\tau_i$. Hence,

$$y_i(t) = \frac{\hat{y}_i e^{j\omega t - \alpha_i}}{\omega\tau_i} = \frac{\hat{y}}{\omega\tau_i} \cos(\omega t - \alpha_i) \tag{A22}$$

It is readily seen that an increase in the frequency decreases the amplitude.

7.2.2. Secondary Forced Response

If the forcing function $\bar{y}(t)$ itself is caused by an oscillatory force and has the form

$$\bar{y}_p(t) = \hat{y}_p \cdot e^{j\omega t} \cdot \sigma_p \tag{A23}$$

the response function for the *i*th normal mode is given by

$$y_i\,(t) = \sigma_i\sigma_p \cdot \hat{y}_i \cdot e^{j\omega t} \tag{A24}$$

where

$$\sigma_i\sigma_p = \frac{1}{1 + j\omega\tau_i} \cdot \frac{1}{1 + j\omega\tau_p}$$

The final equation for the forced response is:

$$y_i(t) = \frac{\hat{y}_i e^{j(\omega t - \alpha_p - \alpha_i)}}{(1+\omega^2\tau_i^2)^{1/2} \cdot (1+\omega^2\tau_p^2)^{1/2}} \tag{A25}$$

ACKNOWLEDGMENTS. The technical help of Mrs. A. Tiemann and Mrs. M. Pohlmann in the preparation of the figures and the typing of the manuscript is gratefully acknowledged. I thank the Deutsche Forschungsgemeinschaft for Grant DFG 227/4.

REFERENCES

Abidor, I. G., Arakelyan, V. B., Chernomordik, L. V., Chizmadzhev, Y. A., Pastushenko, V. F., and Tarasevich, M. R., 1979, Electric breakdown of bilayer lipid membranes. I. The main experimental facts and their qualitative discussion, *Bioelectrochem. Bioenerg.* **6:**37–52.

Auer, D., Brandner, G., and Bodemer, W., 1976, Dielectric breakdown of the red blood cell membrane and uptake of SV 40 DNA and mammalian cell RNA, *Naturwissenschaften* **63:**391.

Benz, R., and Zimmermann, U., 1980, Relaxation studies on cell membranes and lipid bilayers in the high electric field range, *Bioelectrochem. Bioenerg.* **7:**723–739.

Benz, R., Beckers, F., and Zimmermann, U., 1979, Reversible electrical breakdown of lipid bilayer membranes: A charge-pulse relaxation study, *J. Membr. Biol.* **48:**181–204.

Berg, H., 1987, Electrotransfection and electrofusion of cells and electrostimulation of their metabolism, *Stud. Biophys.* **119:**17–29.

Chernomordik, L. V., Sukharev, S. I., Popov, S. V., Pastushenko, V. F., Sokirko, A. V., Abidor, I. G., and Chizmadzhev, Y. A., 1987, The electrical breakdown of cell and lipid membranes: The similarity of phenomenologies, *Biochim. Biophys. Acta* **902:**360–373.

Cole, K. W., 1968, *Membranes, Ions and Impulses,* University of California Press, Berkeley.

Dimitrov, D. S., and Jain, R. K., 1984, Membrane stability, *Biochim. Biophys. Acta* **779:**437–468.

Eigen, M., and DeMaeyer, L., 1963, Relaxation methods, in: *Techniques of Organic Chemistry,* Volume 8(2) (S. L. Friess, E. S. Lewis, and A. Weissberger, eds.), Wiley, New York, pp. 895–1054.

Lindner, P., Neumann, E., and Rosenheck, K., 1977, Kinetics of permeability changes induced by electric impulses in chromaffin granules, *J. Membr. Biol.* **32:**231–254.

Maxwell, J. C., 1904, *A Treatise on Electricity and Magnetism,* 3rd ed., Oxford University Pres, London, pp. 435–441.

Neumann, E., 1973, Molecular hysteresis and its cybernetic significance, *Angew. Chem. Int. Ed. Engl.* **12:**356–369.

Neumann, E., 1984, Electric gene transfer into culture cells, *Bioelectrochem. Bioenerg.* **13:**219–223.

Neumann, E., 1986a, Chemical electric field effects in biological macromolecules, *Prog. Biophys. Mol. Biol.* **47:**197–231.

Neumann, E., 1986b, Digression on biochemical membrane reactivity in weak electromagnetic fields, *Bioelectrochem. Bioenerg.* **16:**565–567.

Neumann, E., 1986c, Mechanisms of membrane processes, electric gene transfer and cell fusion, ZiF Proc., University of Bielefeld Press, pp. 6–14.

Neumann, E., and Rosenheck, K., 1972, Permeability changes induced by electric impulses in vesicular membranes, *J. Membr. Biol.* **10:**279–290.

Neumann, E., and Rosenheck, K., 1973, Potential difference across vesicular membranes, *J. Membr. Biol.* **14:**194–196.

Neumann, E., Gerisch, G., and Opatz, K., 1980, Cell fusion induced by high electric impulses applied to Dictyostelium, *Naturwissenschaften* **67:**414–415.

Neumann, E., Schaefer-Ridder, M., Wang, Y., and Hofschneider, P. H., 1982, Gene transfer into mouse lyoma cells by electroporation in high electric fields, *EMBO J.* **1:**841–845.

Pliquett, F., 1967, Das Verhalten von Oxitrichiden unter dem Einfluß des elektrischen Feldes, *Z. Biol.* **116:**9–22.

Powell, K. T., Derrick, E. G., and Weaver, J. C., 1986, A quantitative theory of reversible electrical breakdown in bilayer membranes, *Bioelectrochem. Bioenerg.* **15:**243–255.

Rosenheck, K., Lindner, P., and Pecht, I., 1975, Effect of electric fields on light-scattering and fluorescence of chromaffin granules, *J. Membr. Biol.* **20:**1–12.

Sale, A. J. H., and Hamilton, W. A., 1967, Effects of high electric fields on microorganisms. I. Killing of bacteria and yeasts, *Biochim. Biophys. Acta* **148:**781–788.

Sale, A. J. H., and Hamilton, W. A., 1968, Effects of high electric fields on microorganisms. III. Lysis of erythrocytes and protoplasts, *Biochim. Biophys. Acta* **163:**37–43.

Scheurich, P., and Zimmermann, U., 1981, Giant human erythrocytes by electric-field-induced cell-to-cell fusion, *Naturwissenschaften* **68:**45.

Scheurich, P., Zimmermann, U., Mischel, M., and Lamprecht, I., 1980, Membrane fusion and deformation of red blood cells by electric fields, *Z. Naturforsch.* **35c:**1081–1085.

Schwan, H. P., 1957, Electrical properties of tissue and cell suspensions, *Adv. Biol. Med. Phys.* **5:**147–209.

Schwan, H.-P., 1985, Biophysical principles of the interaction of ELF fields with living matter, in: *Biological Effects and Dosimetry of Static and ELF Electromagnetic Fields* (M. Grandolfo, S. M. Michaelson, and A. Rindi, eds.), Plenum Press, New York, pp. 221–271.

Schwarz, G., 1967, On dielectric relaxation due to chemical rate processes, *J. Phys. Chem.* **71:**4021–4030.

Senda, M., Takeda, J., Shunnosuke, A., and Nakamura, T., 1979, Induction of cell fusion of plant protoplasts by electrical stimulation, *Plant Cell Physiol.* **20**(7)**:**1441–1443.

Sowers, A. E. (ed.), 1987, *Cell Fusion,* Plenum Press, New York.

Sugar, I. P., and Neumann, E., 1984, Stochastic model for electric field-induced membrane pores—electroporation, *Biophys. Chem.* **19:**211–225.

Sugar, I. P., Förster, W., and Neumann, E., 1987, Model of cell electrofusion—Membrane electroporation, pore coalescence and percolation, *Biophys. Chem.* **26:**321–335.

Teissié, J., and Tsong, T. Y., 1981, Electric field induced transient pores in phospholipid bilayer vesicles, *Biochemistry* **20:**1548–1554.

Weaver, J. C., Powell, K. T., Mintzer, R. A., Ling, H., and Sloan, S. R., 1984, The electrical capacitance of bilayer membranes: The contribution of transient aqueous pores, *Bioelectrochem. Bioenerg.* **12:**393–412.

Weber, H., Förster, W., Jacob, H.-E., and Berg, H., 1981, Microbiological implications of electric field effects, *Z. Allg. Mikrobiol.* **21:**555–562.

Zimmermann, U., 1986, Electrical breakdown, electropermeabilization and electrofusion, *Rev. Physiol. Biochem. Pharmacol.* **105:**175–255.

Zimmermann, U., and Scheurich, P., 1981, High frequency fusion of plant protoplasts by electric fields, *Planta* **151:**26–32.

Zimmermann, U., Schulz, J., and Pilwat, G., 1973, Transcellular ion flow in Escherichia coli B and electrical sizing of bacterias, *Biophys, J.* **13:**1005–1013.

Zimmermann, U., Pilwat, G., and Riemann, F., 1974, Reversibler dielektrischer Durchbruch von Zellmembranen in elektrostatischen Feldern, *Z. Naturforsch.* **29c:**304–305.

Zimmermann, U., Riemann, F., and Pilwat, G., 1976, Enzyme loading of electrically homogeneous human red blood cell ghosts prepared by dielectric breakdown, *Biochim. Biophys. Acta* **436:**460–474.

Zimmermann, U., Scheurich, P., Pilwat, G., and Benz, R., 1981, Zellen mit manipulierten Funktionen: Neue Perspektiven für Zellbiologie, Medizin und Technik, *Angew. Chem.* **93:**332–351.

Chapter 5

Electrical Breakdown of Lipid Bilayer Membranes

Phenomenology and Mechanism

Leonid V. Chernomordik and Yuri A. Chizmadzhev

1. INTRODUCTION

It is known that in the case of a sufficiently strong polarization of cell membranes by an external electric field, processes develop in a membrane which lead to a very significant increase in conductance and permeability. When the field is switched off, the membrane can return from such a high-conducting state to the initial one. This phenomenon is called reversible electrical breakdown (Stampfli, 1958; Zimmermann, 1982). If the amplitude or duration of a pulse is sufficiently large, irreversible damage of the cell membranes occurs. Interest in the study of this phenomenon is based on the existence of important biotechnological applications, many of which have been reflected in this collective book.

The transient or irreversible loss of one of the most important functions of cell membranes—the barrier function—which occurs in the course of electrical breakdown, has been investigated by different methods and for different objects. It is established that the phenomenon is universal in character. Electrical breakdown of membranes of microorganisms, algae, plant protoplasts, erythrocytes, lymphocytes, nerve cells, and myoblasts, as well as of membrane organelles is characterized by a variety of general regularities (Zimmermann, 1982; Benz and Conti, 1981; Tsong, 1983). Therefore, it is natural to assume that electrical breakdown is associated with processes developing in the lipid matrix of cell membranes. The lipid matrix is the universal basis of all membrane structures. The lipid phase determines the isolating "barrier" properties of biomembranes. Planar lipid membranes, which are formed on the opening in the Teflon septum of the experimental chamber when a drop of lipid solution in liquid hydrocarbon (e.g., decane) is deposited on it, turned out to be a convenient experimental model for studying the effect of a high electric field on the properties of the bimolecular layer of lipid molecules (Mueller *et al.*, 1962). In this system, both irreversible and reversible electrical breakdown of bilayers were discovered and investigated. It may be considered to be established that the appearance and development of pores underlie these

Leonid V. Chernomordik and Yuri A. Chizmadzhev • The A. N. Frumkin Institute of Electrochemistry, Academy of Sciences of the USSR, 117071 Moscow, USSR.

phenomena. However, there is as yet no common view on their structure and the mechanism of their appearance. In this chapter we summarize our investigations into the theoretical and experimental nature of electrical breakdown. In outlining the theory, emphasis is placed only on those results which lead to the predictions verified experimentally. The details of calculations and of the experimental procedure can be found in the papers cited. We believe that it is already possible to draw a unified picture of the phenomenon under discussion, which includes both irreversible and reversible breakdown.

2. EXPERIMENTAL METHODS

For studying the electroporation of lipid membranes, use is made of the methods of charge relaxation (Benz *et al.*, 1979; Benz and Zimmermann, 1981) or voltage clamp (Abidor *et al.*, 1979; Chernomordik *et al.*, 1983, 1987). In the former case, the kinetics of the voltage decrease across the membrane after the application of a pulse of short duration (20 nsec to 10 μsec) have been investigated. It is important to stress that the actual breakdown occurs even before the start of recording of the membrane discharge. This method enables one to elucidate how quickly a state of breakdown can occur, but does not allow one to follow the evolution of pores in time.

The time resolution of the voltage-clamp method is 5–10 μsec. However, it enables one to continuously monitor changes in the specific conductance of membranes from 10^{-8} to 10^{-1} Ω^{-1} cm^{-2} and thereby to observe the accumulation of pores and changes in their size. Thus, the charge relaxation and voltage-clamp methods well complement each other.

3. IRREVERSIBLE BREAKDOWN OF MEMBRANES IN AN ELECTRIC FIELD

With a voltage step applied to membranes, the stage of rapid current buildup associated with mechanical rupture of the membrane is preceded by the latent stage where the processes occurring in the membrane have a relatively weak effect on the conductance (or leads to reversible changes). The duration of the latent phase, i.e., the lifetime of the intact membrane in the field, is a random quantity. Therefore, the current oscillograms of irreversible breakdown of each membrane in the field are individual in character (Abidor *et al.*, 1979). The phenomenon of irreversible breakdown has to be characterized by stochastic quantities, the most important of which is the mean lifetime of membranes t_l, which abruptly decreases with increasing voltage (Fig. 1).

The mechanism of irreversible breakdown has been explored comprehensively (Abidor *et al.*, 1979; Pastushenko *et al.*, 1979; Chernomordik and Abidor, 1980). It is established that the basis of

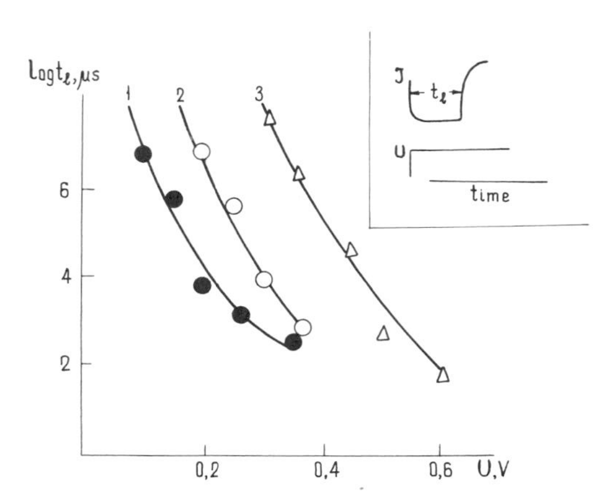

Figure 1. Effect of the voltage (U) on the mean lifetimes (t_l) of lipid bilayer membranes of different compositions. The experimental points were obtained for bilayers of phosphatidylcholine (○); phosphatidylcholine in the presence of the electrolyte solution of lysophosphatidylcholine (4×10^{-4} g/liter) (●); and phosphatidylethanolamine (△). 0.1 M KCl. Theoretical curves 1, 2, and 3 were obtained from Eq. (2) with the use of the corresponding values of γ and D: (1) $\gamma = 3.3 \times 10^{-12}$ N, $D = 2.5 \times 10^{-5}$; (2) $\gamma = 8.6 \times 10^{-12}$ N, $D = 8 \times 10^{-8}$; (3) $\gamma = 1.66 \times 10^{-11}$ N, $D = 3.1 \times 10^{-7}$. (Inset) A typical oscillogram of irreversible breakdown current I with the application of a step of voltage U to the membrane.

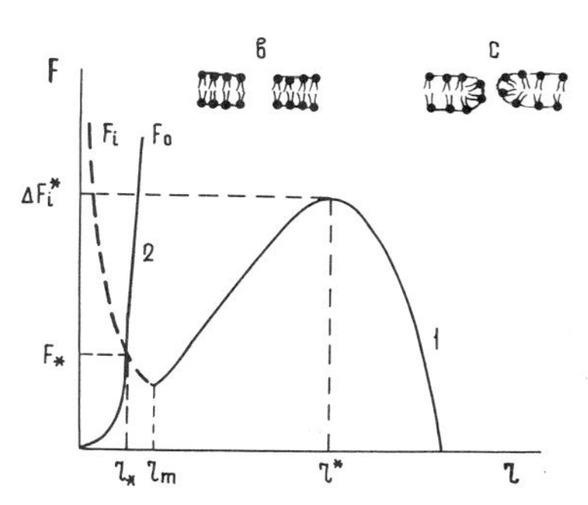

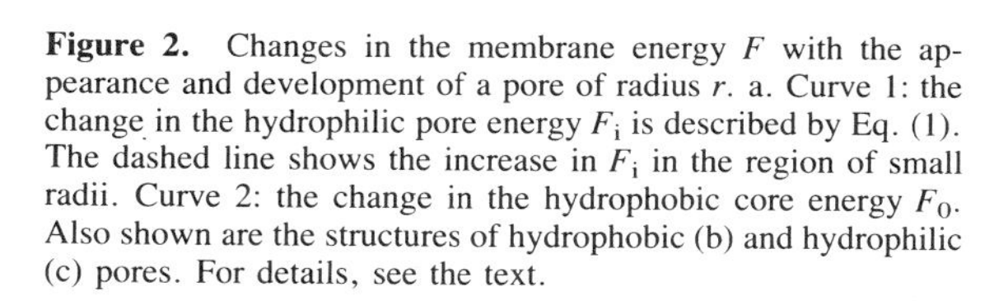

Figure 2. Changes in the membrane energy F with the appearance and development of a pore of radius r. a. Curve 1: the change in the hydrophilic pore energy F_i is described by Eq. (1). The dashed line shows the increase in F_i in the region of small radii. Curve 2: the change in the hydrophobic core energy F_0. Also shown are the structures of hydrophobic (b) and hydrophilic (c) pores. For details, see the text.

this phenomenon is the appearance and the development of hydrophilic pores in a membrane, i.e., pores of which the inner surface is covered with polar lipid head groups. The voltage dependence of the lifetime of bilayers is attributed to water (having a greater dielectric permittivity than the membrane) drawn into the region of the high electric field, i.e., into a pore. We do not, however, discuss another possibility associated with the contribution to the electrostatic energy of the system: the dipole moments of the polar heads of the phospholipid molecules which form a pore (Derzhanski *et al.*, 1979; Neumann *et al.*, 1982).

The work of formation of a cylindrical pore of radius r is given by (Albidor *et al.*, 1979):

$$\Delta F = 2\pi r\gamma - \pi r^2\sigma - \pi r^2 C_m \left(\frac{\epsilon_w}{\epsilon_m} - 1\right)\frac{U^2}{2} \qquad (1)$$

Here F is the free energy, γ is the linear tension in the pore, σ and C_m are the tension and specific capacitance of the membranes, ϵ_w and ϵ_m are the dielectric constants of water and the membrane, and U is the voltage across the membrane. The value of γ for small pores is a function of r both due to the presence of elastic energy of membrane bending and as a result of the action of hydration repulsion forces. Assuming that γ = constant, we can represent the dependence of the energy of the system on the pore radius as a curve with a maximum (Fig. 2a) (Abidor *et al.*, 1979). The critical pore radius r^*, corresponding to its maximum point, and the height of the energy barrier, decrease with increasing voltage applied. If, as a result of thermal fluctuations, the radius of any pore exceeds its critical value, then such a pore will tend to spontaneously expand because this process is accompanied by a decrease in the free energy of the system. Rupture of the membrane in the electric field occurs if one of the pores overcomes the energy barrier $\Delta F_i^*(U)$. In theory (Abidor *et al.*, 1979), the mean lifetime, t_l, can be expressed as:

$$t_l = D \exp\left\{\frac{\pi\gamma^2}{kT\left[\sigma + C_m\left(\frac{\epsilon_w}{\epsilon_m} - 1\right)\frac{U^2}{2}\right]}\right\} \qquad (2)$$

where k is the Boltzmann constant, T is the temperature in kelvin, and D is the preexponential factor dependent on the number of pores in the membrane and on the rate of their diffusion in the plane of the membrane. In the more general case where the rate of formation of pores is low, t_l is determined by the kinetics of pore appearance and disappearance. The stochastic theory of such a process has been developed in the papers by Chizmadzhev *et al.* (1979), Pastushenko *et al.* (1979), and Arakelyan *et al.* (1979).

Figure 1 presents the theoretical curves log $t_l(U)$ obtained from Eq. (2) for membranes of certain compositions. For the measurement of σ and C_m, Chernomordik *et al.* (1985) determined γ and D, which gave the minimum deviation of the theoretical curve from the experimental points. We see that, in general, at certain values of γ and D, the theoretical curves represent the voltage dependences of the lifetimes of bilayers of different composition. It should be noted that the γ values thus obtained (0.3 to 1.7×10^{-11} N) are relatively close to the value given by Harbich and Helfrich (1979) and to those estimated from the radius of pores induced in giant lecithin liposomes by a high electric field. Since γ is related to the bending energy of the pore edge, the stability of bilayers should depend on the molecular geometry of lipid molecules (Petrov *et al.*, 1980). Experiments have shown (Fig. 1) that the presence of lysolecithin in membranes (a cone promoting the formation of pores) decreases t_l by several orders of magnitude (Chernomordik *et al.*, 1985).

In conclusion we emphasize that during irreversible breakdown, the appearance of pores in a membrane is revealed only after the radius of the first induced pore reaches the critical radius. The subsequent growth of this pore inevitably leads to irreversible breakdown of the membrane. The early phase of hydrophilic pore formation remains, however, unexplained.

4. REVERSIBLE ELECTRICAL BREAKDOWN: PHENOMENOLOGY

In 1979, Benz *et al.*, using the charge relaxation method, found that when membranes of oxidized cholesterol are rapidly (within ~ 500 nsec) charged to approximately 1 V, their resistance is reversibly decreased by nearly nine orders of magnitude. Thus, for the first time it became possible to observe reversible breakdown for planar lipid membranes. Later this phenomenon was also investigated by using azolectin bilayers modified with UO_2^{2+} ions (Abidor *et al.*, 1982; Chernomordik *et al.*, 1982) as well as membranes of lecithin and cholesterol in the presence of alkaloid holoturin A (Sukharev *et al.*, 1983). In spite of marked quantitative differences between these experimental systems, in all cases the same qualitative regularities were observed.

Figure 3 presents a set of current oscillograms obtained by Chernomordik *et al.* (1983) from the successive application of voltage pulses of different amplitude U to a membrane of oxidized cholesterol. The oscillograms illustrate some of these regularities. Dominating at the start of a 20-μsec pulse, however, is a charging current which rapidly falls off. Simultaneously, a conduction current develops, and just after 5 μsec this current becomes dominating. For this reason the level of background current is not recorded on the oscillogram. At the end of a pulse of $U = 0.7$ V (curve 1), the conduction current is about 5×10^{-5} A, which corresponds to a conductance value of 10^{-5} to 10^{-6} Ω^{-1}. Thus, the conductance increases against the background conductance (10^{-9} to 10^{-10} Ω^{-1}) by a factor of 10^4 to 10^5. If the pulse duration is not too long, the increase in conductance is reversible and, in contrast to the irreversible breakdown oscillograms which are of random character, it is fully reproducible after repeated application of the same pulse (curve 2). A pulse with a long duration or a train of pulses separated by small intervals between them will cause the reversible increase in conductance to become irreversible (Chernomordik *et al.*, 1983). The increase in the pulse amplitude leads to a rapid increase in the rate of current development (curves 3–5). The strong voltage dependence of the time of development of reversible breakdown is also revealed by the

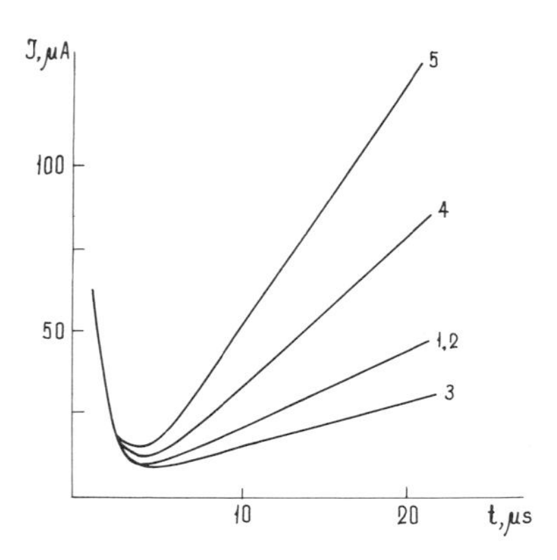

Figure 3. Current oscillograms of the bilayer of oxidized cholesterol with the successive application to the membrane of voltage pulses of 20 μsec duration and of 0.68 (curve 3), 0.72 (curve 4), and 0.73 (curve 5) V amplitude. The pulse of 0.7 V amplitude was applied twice (curves 1 and 2). The intervals between the pulses were more than 1 min.

charge relaxation method (Benz and Zimmermann, 1980a,b). Since the voltage causing a certain increase in conductance depends on the duration of exposure to that voltage, the notion of the potential of reversible breakdown and the notion of the potential of irreversible breakdown are physically meaningless.

4.1. Reversible Breakdown—Accumulation of Pores

The mechanism of reversible electrical breakdown of a lipid bilayer also involves large-scale deformations of the membrane such as changes in the thickness or in the area or in the dielectric permittivity. This leads to a change in capacitance during the buildup of charge. In the experiments by Chernomordik *et al.* (1982), the capacitance of UO_2^{2+}-modified membranes was measured in that phase of breakdown where the conduction current is still small compared with the capacitance-based current. In this case, it was found that the capacitance of membranes remains unchanged to within 2%. It was unambiguously concluded that this behavior arises from local defects (i.e., the induced or preexisting pores). Based on the current value at a resistance of about 10^4 Ω, the total area of the pores is approximately 10^{-8} cm^2. This is much less than the area of the whole membrane (about 10^{-3} cm^2). The reproducibility of the current oscillograms of reversible breakdown at successive applications of the same pulses (Fig. 3, curves 1 and 2) as well as the absence of visible fluctuations at even relatively large currents, indicate that, in this system, the law of large numbers holds true—a host of pores arise in the membrane.

4.2. Characteristic Times of Development and Resealing of Pores

Important information on the properties of pores developed during breakdown was obtained from the kinetics of transmembrane resistance increase after application of multistep voltage pulses. Figure 4 shows the current responses of UO_2^{2+}-modified membranes after the application of two identical voltage pulses with a certain interval between them at a relatively lower voltage. At the instant of decrease in the voltage after the first pulse, there is a very rapid (< 5 μsec) decrease in the membrane conductance (Chernomordik *et al.,* 1982). A similar result was obtained by Benz and Zimmermann (1981) in analyzing the characteristic time of restoration of transmembrane resistance of oxidized cholesterol bilayers. It has been suggested that such a decrease in conductance reflects the rapid disappearance of pores (Benz and Zimmermann, 1981; Zimmermann, 1982). However, special experiments have shown that this is not the case (Chernomordik *et al.,* 1987). The response to the repeated pulses coincident in amplitude with the first one, but applied after 4 msec, is a buildup of the conduction current. In the second pulse, the current is the same as that at the end of the first pulse (Fig. 4). Thus, in this experiment, the duration of an interval at a relatively low voltage turned out to be insufficiently long to ensure a significant decrease in the number and the

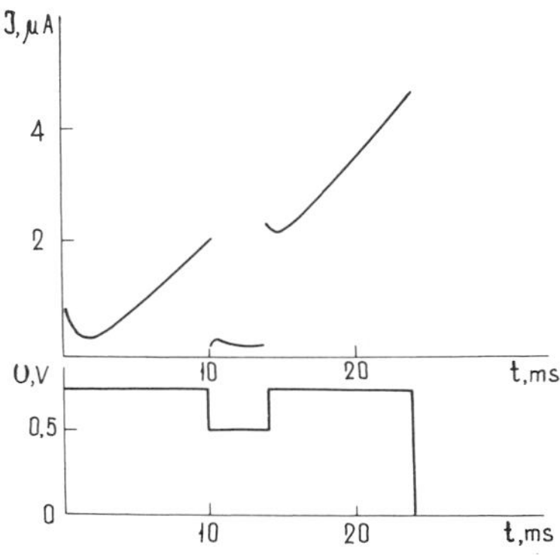

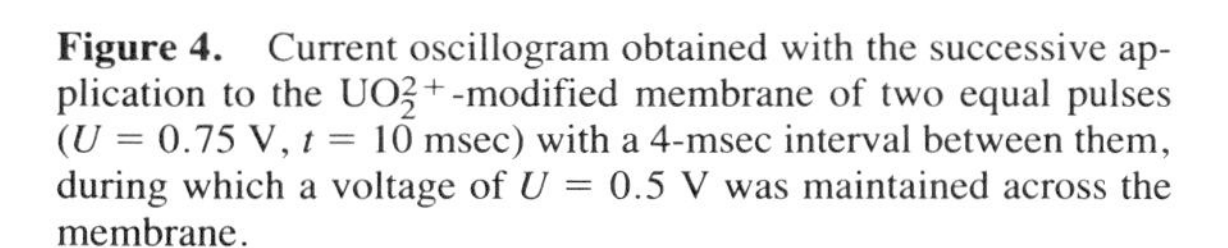

Figure 4. Current oscillogram obtained with the successive application to the UO_2^{2+}-modified membrane of two equal pulses (U = 0.75 V, t = 10 msec) with a 4-msec interval between them, during which a voltage of U = 0.5 V was maintained across the membrane.

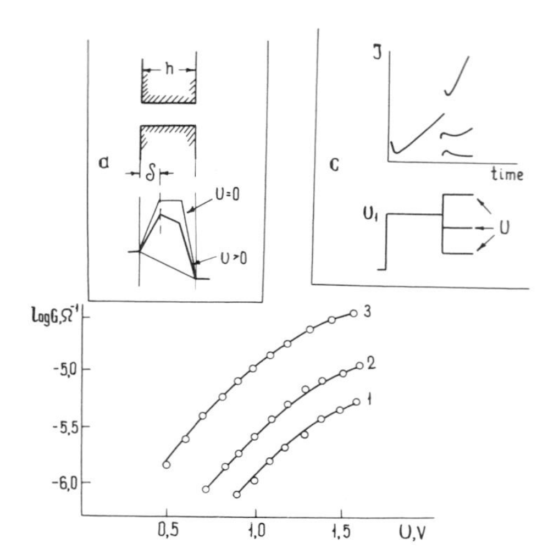

Figure 5. Effect of the voltage U on the conductance G of the UO_2^{2+}-modified bilayer in a state of reversible breakdown. (a) The energy profile of a narrow pore in the membrane. The external electric field lowers the barrier and thereby facilitates the passage of ions through the pore. (b) The dependences $\log G(U)$ obtained immediately after the application to the membrane of voltage pulses of different amplitude and duration: 1.6 V, 0.1 msec (1); 1.6 V, 0.2 msec (2); 1.6 V, 0.5 msec (3). (c) The experiments performed to obtain the dependences $\log G(U)$. Changes of current through the bilayer were investigated with changes in the voltage from the same level U_1 at the first step of pulse to different levels of U at the second step.

radius of pores developed in the membrane during the first pulse. With a gradual increase in the duration of the interval between pulses, Chernomordik *et al.* (1987) determined the interval after which the current at the start of the second pulse was the same as that at the start of the first one. In this way, the times of resealing of the membrane after reversible breakdown were determined; they varied from 1 to 100 sec depending upon the electrical treatment. The same characteristic times of resealing (disappearance of pores) were obtained after reversible breakdown caused by a strong but very short (20 nsec) polarization of the membrane (Chernomordik *et al.*, 1987).

After the pulse terminates, the conductance G abruptly decreases. However, the number and the mean radius of the pores are preserved. Thus, G depends nonlinearly on the voltage. Chernomordik *et al.* (1987) have shown that this dependence is caused by the electrostatic expulsion of an ion from a narrow pore (Parsegian, 1969). According to the model developed by Chernomordik *et al.* (1987), the ions passing through a narrow pore overcome a certain energy barrier that is lowered if a voltage is applied to the membrane.

For simplicity, we assume that the membrane conductance G is determined by pores of equal size and shape and that the energy profile of the ion in the pore, $W(x)$, has the shape of a trapezoid (Fig. 5a). At a voltage of $U >> 25$ mV and at a mean pore radius $\bar{r}$, which is much less than the membrane thickness h, we obtain:

$$\log G = \log \frac{\pi h \kappa N}{4} - \log \left[\left(1 + \frac{n\beta U}{W_0 - n\beta U} \right) \exp\,(W_0 - n\beta U) - \frac{n\beta U}{W_0 - n\beta U} \right] \quad (3)$$

Here N is the number of pores; κ is the specific conductivity of the electrolyte; $\beta = e/kT$, e is the proton charge, k is the Boltzmann constant, and T is the absolute temperature; $n = \delta/h$ is the dimensionless length of the entrance region of the pore (see Fig. 5a). This length is mainly determined by the geometry of the edge. It is weakly dependent on r. W_0 is the energy of the ion at the center of the pore, which is unambiguously related to $\bar{r}$ (Parsegian, 1969). Comparing the experimental dependence of $\log G$ on U with Eq. (3), it is possible to estimate the values of $\bar{r}$ and N, and thus obtain information on the dynamics of the changes in the size and the number of pores in the membrane during and after reversible breakdown. It appears from Fig. 5b that the theoretical curves fit well the experimental points obtained from experiments investigating the voltage dependence of the conductance of a bilayer in a state of reversible breakdown (Fig. 5c).

The changes in mean radius and number of pores during reversible breakdown were considered in terms of our theoretical model (Glaser *et al.*, unpublished). By using pulses of equal amplitude but different duration, it was possible to determine that the increase in duration of the *effect* of a

Table 1. Changes in the Mean Radius and in the Number of Pores, Developed in the UO_2^{2+}-Modified Membrane, as Functions of the Duration (t) of the Voltage Pulse U and of the Time of Resealing t_r of the Membrane after the Voltage Pulse[a]

	U = 1.6 V, t_r = 0		U = 1.35 V, t_r = 0		U = 1.45 V, t_r = 2 msec	
t (msec)	r (Å)	$N \times 10^5$	r (Å)	$N \times 10^5$	r (Å)	$N \times 10^5$
0.1	6.5	1				
0.2	6.7	2				
0.5	8	4				
2			6.7	0.2		
5			6.9	0.4		
10			7.2	0.8		
0					9.4	2
1					7.6	2
20					6.8	2
10^4 (10 sec)					6.0	2

[a]The experiments at U = 1.6 V, 1.35 V, and 1.45 V were carried out on different membranes.

pulse was linearly dependent on the time and thus on the number of pores (Table 1). Simultaneously, the mean pore radius increases the rates of both processes (pore number and pore radius) since they are dependent on the voltage across the membrane. Similar experiments have also demonstrated that the same level of conductance resulting from different electrical treatments corresponds to populations which are different in the number and the size of the pores. This important conclusion is also confirmed by results of experiments on the influence of blocking agents (sucrose and hemoglobin) on the conductance of a lipid bilayer at different stages of reversible breakdown (Chernomordik *et al.*, 1983).

The process of pore resealing after reversible breakdown was also studied (Glaser *et al.*, unpublished). The analysis of log $G(U)$, obtained at different intervals after the application of the same voltage pulses which produce a certain population of pores in the bilayer, showed that the decrease in conductance in the characteristic time range of milliseconds to tens of milliseconds is determined by a change in the size of the pores (Table 1). Only at times of over 10 sec does the number of pores start to decrease. Thus, the characteristic time of change in the size of pores (about 1–10 msec) turns out to be significantly shorter than the characteristic time of decrease in the number of pores (resealing time).

5. PORES: STRUCTURE AND MECHANISM OF FORMATION

The pores in a lipid bilayer can in principle be hydrophobic or hydrophilic (Abidor *et al.*, 1979). In the former case, the walls of the pores are formed by hydrocarbon lipid tails (Fig. 2b), and in the latter case, the inner surface of the pores is covered by polar heads (Fig. 2c). Estimates show that the hydrophobic pores filled with water are energetically unfavorable (Chernomordik *et al.*, 1983) and thus should be short-lived. Since the defects developed in the membrane during reversible breakdown turned out to be long-lived (lifetimes of 10 to 100 sec), it may be assumed that these are hydrophilic pores. Such an assumption is also supported by an increase in the rate of accumulation of pores during reversible breakdown in a membrane containing lysolecithin (Glaser *et al.*, unpublished). This effect can be attributed to a decrease in the linear tension of hydrophilic pores in the presence of lysolecithin (Chernomordik *et al.*, 1985) which also contributes to their formation.

Thus, the evolution of hydrophilic pores causes the phenomena of reversible and irreversible electrical breakdown of lipid membranes, which are observed in a broad range of duration and intensity of the electrical treatment. The mechanism of pore formation has been repeatedly discussed in the literature (Zimmermann, 1982; Petrov *et al.*, 1980; Pastushenko and Petrov, 1984; Dimitrov, 1984; Sugar and Neumann, 1984; Weaver *et al.*, 1986).

Leikin *et al.* (1986) argued that hydrophilic pores may arise (Fig. 2b) as a result of lateral thermal fluctuations of lipid molecules in a bilayer. The energy of such a pore, F_0, depends on the radius as shown in Fig. 2a. The hydrophobic interaction, which leads to a decrease in the surface tension at the interface of the hydrophobic surface with water, significantly decreases the energy in the case of small r ($\lesssim$ 10 Å). Thereby, it increases the probability of pore induction. On the contrary, hydrophilic pores (Fig. 2c) with the same range of radii ($\lesssim$ 10 Å) should lead to increased pore energy, F_i, with decreasing radius (Fig. 2a) due to an increase in elastic edge energy and hydration repulsive forces. It should be stressed that it is very difficult to calculate F_i for such small r within the framework of macroscopic approaches because nontrivial model assumptions are needed (Petrov *et al.*, 1980). The above reasoning concerning the increase in F_i at $r \lesssim 10$ Å makes one doubt the validity of the arguments encountered in the literature about the presence of hydrophilic pores of zero radius (Weaver *et al.*, 1986; Powell *et al.*, 1986).

Comparison of the dependences of the energies of hydrophilic and hydrophobic pores on the radius suggests that at small radii the existence of hydrophobic pores in a bilayer is more favorable (Fig. 2a). If the radius of the pore exceeds the critical value r_*, where $F_0(r_*) = F_i(r_*)$, then the transformation of the hydrophobic pore to a hydrophilic one becomes favorable. The further growth of the hydrophilic pore depends on the energy profile $F_i\ (r)$.

Consider first the case where the value $r = r_m$, corresponding to the local minimum of $F_i(r)$, exceeds r_*. Second, assume the presence of an overcritical pore and that the diffusion barrier to growth of the hydrophilic pore is sufficiently high, then the irreversible breakdown of the bilayer is unlikely. Although the diffusion barrier at very high voltages U becomes low [see Eq. (1)], the increase in the pore radius in the regions $r^*(U) < r < r^*(U = 0)$ will lead (at $r \sim h$) to a decrease in the voltage across the pore. The energy diagram of the system now looks like that of a membrane with a pore in the absence of a field (Pastushenko and Chizmadzhev, 1983). In this case, the irreversible breakdown of the membrane will begin only after one of the pores has overcome the diffusion barrier $F^*(U = 0)$. When $r_m > r_*$, and diffusion barrier conditions are high, metastable hydrophilic pores may be formed in the membrane in the region $r \sim r_m$. Assuming that the accumulation of such pores in the electric field is the cause of reversible breakdown of lipid bilayers, the kinetics of this process may be characterized by the decrease in the energy in the electric field as water is drawn into the region of a high field, i.e., into a pore.

Leikin *et al.* (1986) calculated the rate, K_p, of spontaneous formation of metastable pores as follows:

$$\ln K_p\ (U) = \ln\left(\frac{\nu S}{a_0}\right) - \frac{\Delta W_p(U = 0)}{kT} + \frac{\pi r_*^2\ (\epsilon_w - \epsilon_m) F_0 U^2}{2hkT} \tag{4}$$

where ν is the characteristic frequency which is equal in the order of magnitude to the frequency of lateral thermal fluctuations of lipid molecules in the bilayer; a_0 is the area per molecule; and s is the bilayer area. Leikin *et al.* (1986) have shown that r_* is independent of U. This circumstance enables one to verify experimentally Eq. (4) and the following relationships.

Three-step voltage pulses, consisting of two measuring parts with $U = U_m$ and one testing part with $U = U_t$ (Fig. 6a), were applied to a UO_2^{2+}-modified membrane. By a change in the current $\Delta I(U_t)$ between the end of the first and the start of the second measuring part, the dependence of K_p on the amplitude U_t at a constant U_m was investigated. Voltage pulses were selected so as to ensure a linear increase in current through the membrane in the testing part. In this case, the rate of appearance of pores is much higher than that of their disappearance, and the latter process can be

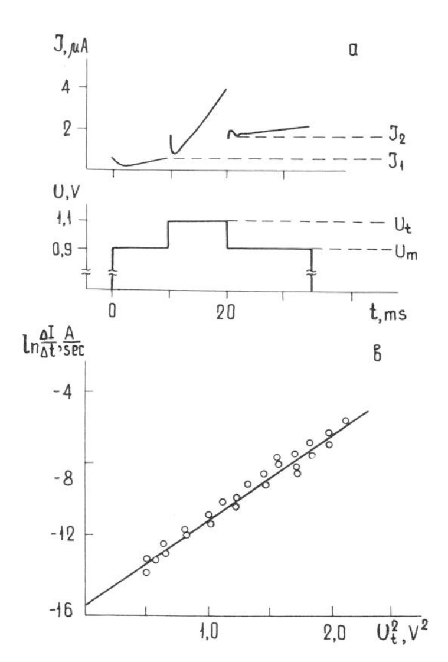

Figure 6. Kinetics of the appearance of pores in the course of reversible breakdown of the UO_2^{2+}-modified membrane. (a) Scheme of the experiments performed. The oscillogram of current response (at the top) and the shape of voltage pulse (at the bottom) at $U_m = 900$ mV and $U_t = 1.1$ V. (b) The dependence of ln $(\Delta I/\Delta t)$ on U_t^2 at $U_m = 0.9$ V.

neglected. It was assumed that $\Delta I(U_t) = \Delta N(U_t) \times U_m g(U_m)$, where $\Delta N(U_t)$ is the number of hydrophilic pores formed in the bilayer in the testing part of the pulse, and $g(U_m)$ is the average conductance of these pores at $U = U_m$. By comparing the experimental data with the theoretical predictions, and since $K_p(U_t) = \Delta N(U_t)/\Delta t$, Eq. (4) was reduced to

$$\ln\left(\frac{\Delta I}{\Delta t}\right) = A + BU_t^2 \tag{5}$$

where

$$A = \ln\left(\frac{U_m g(U_m)\nu s}{a_c}\right) - \frac{\Delta W_p(U = 0)}{kT} \tag{6}$$

$$B = \frac{\pi r_*^2 \epsilon_w \cdot \epsilon_0}{2hkT} \tag{7}$$

Figure 6b shows the dependence of ln $(\Delta I/\Delta t)$ on U_t^2. It is evident from Fig. 6 that for $0.65 < U_t < 1.5$ V, the value of ln $(\Delta I/\Delta t)$ linearly increases with increasing U_t^2 in accordance with Eqs. (4) and (5). We note that from the coefficient B one can estimate the value of r_* as ≈ 3 Å (Leikin *et al.*, 1986). The value of the coefficient A determines the height of the energy barrier for appearance of hydrophilic pores using the familiar expression for $F_0(r)$ (Leikin *et al.*, 1986). Such estimations give r_* values of ≈ 4 Å. We emphasize that A and B are independent and reflect quite different features of the process of formation of pores. Therefore, the similarity of the results of our estimates favors the proposed model for the appearance of pores. The mechanism proposed is equally applicable, also, to irreversible breakdown. The membranes which are unable to display a reversible increase in conductance seem to be characterized by a relationship between the energy branches F_0 and F_i in which the intermediate minimum $r = r_m$ either is located high or is altogether absent. The capability of membranes for the reversible or irreversible type of breakdown is probably determined primarily by the effective geometry of lipid molecules. This appealing hypothesis certainly deserves a special experimental test.

6. SIMILARITY OF THE BASIC REGULARITIES OF ELECTRICAL BREAKDOWN OF CELL AND LIPID MEMBRANES

According to modern ideas, the cell membrane can be regarded as a sea of lipids surrounding protein molecules. The proteins interact in some way with each other and with lipids. It is just the low permeability of the lipid bilayer of membranes for hydrophilic particles that is the basis for the barrier function of membranes. The qualitative similarity of the electric field effect on the conductance and permeability of membranes of a great variety of cells and organelles indicates that the decisive role in the phenomenon of electrical breakdown of membranes is played by a lipid bilayer. The results obtained from the study of electrical breakdown of lipid bilayer membranes and those obtained by equivalent methods, such as voltage clamp and charge relaxation for cell membranes, support the lipid bilayer as the site of the electric field effect.

Certainly the breakdown of cell membranes has specific features which distinguish it from that of lipid bilayers. Specifically, under certain conditions the increase in the permeability of cell membranes during breakdown leads to secondary processes such as colloid-osmotic lysis (Kinosita and Tsong, 1977a,b; Zimmermann *et al.*, 1981). The absence of a meniscus in the closed vesicular membranes of cells and organelles may also lead to changes in the character of the breakdown especially at later stages (Pastushenko and Chizmadzhev, 1983). Finally, for cell membranes one can observe in some cases an interesting effect of "accommodation" of membranes (or cells as a whole) due to a high electric field. After the application of a train of equal-voltage pulses with sufficiently long intervals between them, the conductance value attained by the end of a pulse decreases with each following pulse (Zimmermann, 1982; Chernomordik *et al.*, 1987). The mechanism of this phenomenon has not been elucidated.

In spite of these differences, the basic regularities describing the behavior of cell and model membranes in a high electric field are remarkably alike. We primarily emphasize the similarity of oscillograms of breakdown current with that of clamping of voltage across cell membranes (Conti *et al.*, 1973; Chernomordik *et al.*, 1987) and across planar bilayers (Chernomordik *et al.*, 1983, 1987). In both cases, the irreversible damage to the membrane is preceded by a reversible decrease in the resistance. The lifetime (t_l) of lipid bilayers and cell membranes in an electric field is a random quantity. The dependences of the mean lifetime on the voltage applied in both cases have a similar slope. When the voltage is increased by 0.1 V, there is a decrease in t_l by approximately one order of magnitude (Chernomordik *et al.*, 1987). At the reversible stage of breakdown, one can also observe common regularities. The cell membranes and lipid bilayers possess qualitatively the same dependences of reversible breakdown current on the time and amplitude of the voltage applied. The oscillograms of reversible breakdown current, presented in the paper by Conti *et al.* (1973) for the giant axon of a squid, and those obtained for the membranes of L-cells and erythrocytes (Chernomordik *et al.*, 1987), very closely resemble those of reversible breakdown of lipid bilayer membranes. The dependences of the reversible breakdown voltage on the duration of a charging pulse are similar for membranes of the alga *Valonia* and for planar lipid membranes (Benz and Zimmermann, 1980a,b). It was established that the breakdown potential of cell and lipid membranes depends in a similar manner on the concentration and composition of the electrolyte as well as on temperature. As in the case of lipid bilayer membranes, an increase in intensity and duration of electrical treatment leads to a simultaneous increase in both the number and mean radius of pores which develop in cell membranes during reversible breakdown (Kinosita and Tsong, 1977a,b). The decrease in the voltage across cell and model membranes, which are in the state of reversible breakdown, leads to a very rapid (within a few microseconds) decrease in conductance (Benz and Conti, 1981). The time of resealing of pores in lipid bilayers also turned out to be very close to that characteristic of biomembranes (seconds or minutes) (Zimmermann, 1982; Dressler *et al.*, 1983; Sowers and Lieber, 1986). We note, however, that in some cell membranes the times of resealing after breakdown turn out to be markedly longer (tens of minutes or hours) (Zimmermann, 1982; Kinosita and Tsong, 1977a,b; Glaser *et al.*, 1986). Most likely, as long as the pores developed during breakdown are located in the lipid region of the membrane, the resealing times correspond to those for bilayer membranes. However, with a very great number of pores or a sufficient increase in the radius of a pore, protein particles may find themselves at the edge of the pore. This may lead to a peculiar stabilization of the edge of the pore and to a significant increase in its lifetime.

The similarity of the phenomenologies of electrical breakdown of cell and lipid membranes suggests that the pores which develop in the course of electrical breakdown of cell membranes arise in the lipid matrix. Thus, the question of the mechanism of the breakdown of lipid bilayer membranes is also applicable for describing the electroporation of biomembranes.

7. CONCLUSION

It appears that the early stages of electroporation of biomembranes are primarily associated with the appearance of pores in the lipid matrix, most likely in the protein-free domains of membranes. The simplicity of planar lipid bilayers as an experimental system (i.e., the absence of proteins and a controllable lipid composition) and sensitive and easily interpretable electrical measurements enables one to use them for studying the fairly delicate details of these early stages. We refer, for example, to the mechanism of formation of hydrophilic pores, to the properies of these pores, and to the regularities of their development and resealing. The results from lipid bilayers may be regarded as an approximation to the molecular mechanism of the more complicated phenomenon of reversible loss of the barrier function of cell membranes in the case of electrical treatment of short duration.

ACKNOWLEDGMENTS. We wish to thank Drs. S. L. Leikin, M. M. Kozlov, and V. F. Pastushenko for helpful discussions. We are also grateful to Mrs. G. F. Voronina for assistance in preparing the manuscript.

REFERENCES

Abidor, I. G., Arakelyan, V. B., Chernomordik, L. V., Chizmadzhev, Y. A., Pastushenko, V. F., and Tarasevich, M. R., 1979, Electric breakdown of bilayer lipid membrane. I. The main experimental facts and their qualitative discussions, *Bioelectrochem. Bioenerg.* **6**:37–52.

Abidor, I. G., Sukharev, S. I., Chernomordik, L. V., and Chizmadzhev, Y. A., 1982, The reversible electrical breakdown of bilayer lipid membranes modified by UO_2^{2+} ions, *Bioelectrochem. Bioenerg.* **9**:141–148.

Arakelyan, V. B., Chizmadzhev, Y. A., and Pastushenko, V. F., 1979, Electric breakdown of bilayer lipid membrane. V. Consideration of the kinetic stage in the case of the membrane containing arbitrary number of defects, *Bioelectrochem. Bioenerg.* **6**:81–88.

Benz, R., and Conti, F., 1981, Reversible electrical breakdown of squid giant axon membrane, *Biochim. Biophys. Acta* **645**:115–123.

Benz, R., and Zimmermann, U., 1980a, Pulse-length dependence of the electrical breakdown in lipid bilayer membranes, *Biochim. Biophys. Acta* **597**:637–642.

Benz, R., and Zimmermann, U., 1980b, Relaxation studies on cell membranes and lipid bilayers in the high electric field range, *Bioelectrochem. Bioenerg.* **7**:723–739.

Benz, R., and Zimmermann, U., 1981, The resealing process of lipid bilayers after reversible electrical breakdown, *Biochim. Biophys. Acta* **640**:169–178.

Benz, R., Beckers, F., and Zimmermann, U., 1979, Reversible electrical breakdown of lipid bilayer membranes: A charge-pulse relaxation study, *J. Membr. Biol.* **48**:181–204.

Chernomordik, L. V., and Abidor, I. G., 1980, The voltage-induced local defects in unmodified bilayer lipid membrane, *Bioelectrochem. Bioenerg.* **7**:617–623.

Chernomordik, L. V., Sukharev, S. I., Abidor, I. G., and Chizmadzhev, Y. A., 1982, The study of the BLM reversible electrical breakdown mechanism in the presence of UO_2^{2+} ions, *Bioelectrochem. Bioenerg.* **9**: 149–155.

Chernomordik, L. V., Sukharev, S. I., Abidor, I. G., and Chizmadzhev, Y. A., 1983, Breakdown of lipid bilayer membranes in an electric field, *Biochim. Biophys. Acta* **736**:203–213.

Chernomordik, L. V., Kozlov, M. M., Melikyan, G. B., Abidor, I. G., Markin, V. S., and Chizmadzhev, Y. A., 1985, The shape of lipid molecules and monolayer membrane fusion, *Biochim. Biophys. Acta* **812:** 643–655.
Chernomordik, L. V., Sukharev, S. I., Popov, S. V., Pastushenko, V. F., Sokirko, A. V., Abidor, I. G., and Chizmadzhev, Y. A., 1987, The electrical breakdown of cell and lipid membranes: The similarity of phenomenologies, *Biochim. Biophys. Acta* **902:**360–373.
Chizmadzhev, Y. A., Arakelyan, V. B., and Pastushenko, V. F., 1979, Electric breakdown of bilayer lipid membranes. III. Analysis of possible mechanisms of defect origination, *Bioelectrochem. Bioenerg.* **6:**63–71.
Conti, F., Fioravanti, R., and Wanke, E., 1973, Breakdown dielettrico della membrana dell assone gigante di calamaro, in: *Atti della Prima Riunione Scientifica Plenaria della Societa di Biofisica Pura e Applicata,* Camogli, pp. 401–412.
Derzhanski, A., Petrov, A. G., and Mitov, M. D., 1979, Electric field induced pores in erythrocyte membrane—a discussion, in: *Proceedings of the Fifth School on Biophysics of Membrane Transport,* Wroclaw, Poland, pp. 285–286.
Dimitrov, D. S., 1984, Electric field-induced breakdown of lipid bilayers and cell membranes: A thin viscoelastic film model, *J. Membr. Biol.* **78:**53–60.
Dressler, V., Shwister, K., Haest, C. W. M., and Deuticke, B., 1983, Dielectric breakdown of the erythrocyte membrane enhances transbilayer mobility of phospholipids, *Biochim. Biophys. Acta* **732:**304–307.
Glaser, R. W., Wagner, A., and Donath, E., 1986, Volume and ionic composition changes in erythrocytes after electric breakdown: Simulation and experiment, *Bioelectrochem. Bioenerg.* **16:**455–470.
Harbich, W., and Helfrich, W., 1979, Alignment and opening of giant lecithin vesicles by electric field, *Z. Naturforsch.* **34a:**1063–1065.
Kinosita, K., and Tsong, T. Y., 1977a, Voltage-induced pore formation and haemolysis of human erythrocytes, *Biochim. Biophys. Acta* **471:**227–242.
Kinosita, K., and Tsong, T. Y., 1977b, Formation and resealing of pores of controlled sizes in human erythrocyte membrane, *Nature* **268:**438–441.
Leikin, S. L., Glaser, R. W., and Chernomordik, L. V., 1986, Mechanism of pore formation under electrical breakdown of membranes, *Biol. Membr.* **3:**944–951.
Mueller, P., Rudin, D. O., Tien, H. T., and Wescott, W. C., 1962, Reconstitution of cell membrane structure in vitro and its transformation into an excitable system, *Nature* **194:**979–980.
Neumann, E., Shaefer-Ridder, M., Wang, Y., and Hofschneider, P. H., 1982, Gene transfer into mouse lyoma cells by electroporation in high electric fields, *EMBO J.* **1:**841–845.
Parsegian, V. A., 1969, Energy of an ion crossing a low dielectric membrane: Solution to four relevant electrostatic problems, *Nature* **221:**844–846.
Pastushenko, V. F., and Chizmadzhev, Y. A., 1983, Electrical breakdown of lipid vesicles, *Biofizika* **28:**1036–1039.
Pastushenko, V. F., and Petrov, A. G., 1984, Electro-mechanical mechanism of pore formation in bilayer lipid membranes, in: *Proceedings, Seventh School on Biophysics of Membrane Transport,* Poland, pp. 69–81.
Pastushenko, V. F., Chizmadzhev, Y. A., and Arakelyan, V. B., 1979, Electric breakdown of bilayer lipid membranes. II. Calculation of the membrane lifetime in the steady-state diffusion approximation, *Bioelectrochem. Bioenerg.* **6:**53–63.
Petrov, A. G., Mitov, M. D., and Derzhanski, A., 1980, Edge energy and pore stability in bilayer lipid membranes, in: *Advances of Liquid Crystal Research Applications,* Volume 1 (L. Bata, ed.), Pergamon Press, New York, pp. 605–626.
Powell, K. T., Derrick, E. G., and Weaver, J. C., 1986, A quantitative theory of reversible electrical breakdown in bilayer membranes, *Bioelectrochem. Bioenerg.* **15:**243–255.
Sowers, A. E., and Lieber, M. R., 1986, Electropore diameters, lifetimes, numbers and locations in individual erythrocyte ghosts, *FEBS Lett.* **205:**179–184.
Stampfli, R., 1958, Reversible electrical breakdown of the excitable membrane of a Ranvier node, *Ann. Acad. Bras. Cien.* **30:**57–63.
Sugar, I. P., and Neumann, E., 1984, Stochastic model for electric field-induced membrane pores—Electroporation, *Biophys. Chem.* **19:**211–225.
Sukharev, S. I., Chernomordik, L. V., and Abidor, I. G., 1983, Reversible electrical breakdown of holoturin modified bilayer lipid membranes, *Biofizika* **28:**423–426.
Tsong, T. Y., 1983, Voltage modulation of membrane permeability and energy utilization in cells, *Biosci. Rep.* **3:**487–505.

Weaver, J. C., Mintzer, R. A., Ling, H., and Sloan, S. R., 1986, Conduction onset criteria for transient aqueous pores and reversible electrical breakdown in bilayer membranes, *Bioelectrochem. Bioenerg.* **15:** 229–241.

Zimmermann, U., 1982, Electric field-mediated fusion and related electrical phenomena, *Biochim. Biophys. Acta* **694:**227–277.

Zimmermann, U., Scheurich, P., Pilwat, G., and Benz, R., 1981, Cells with manipulated functions: New perspectives for cell biology, medicine and technology, *Angew. Chem. Int. Ed. Engl.* **20:**325–344.

Chapter 6

Stochastic Model of Electric Field-Induced Membrane Pores

Istvan P. Sugar

1. INTRODUCTION

Two basic mechanisms have been suggested to describe the experimentally observed properties of electroporation: the electromechanical model and the statistical model of pore expansion. These models have been reviewed by Dimitrov and Jain (1984). This chapter considers the statistical model of electroporation for a one-component planar lipid bilayer membrane. At zero electric field, the membrane is populated with microscopic pores by the fluctuation clustering of vacancies (i.e., molecule-free sites) in the bilayer. Under the effect of a transmembrane electric field, the average pore size increases. The driving force of the electric field-mediated pore opening is associated with the enhancement of the electric polarization of the solvent molecules during their transfer from the bulk solvent space to the region of the larger electric field spreading from the pore wall into the solution of the pore interior (Sugar and Neumann, 1984; Powell *et al.*, 1986).

The characteristic features of the present model are the following: (1) the model results are valid independently from the molecular details of the pore structure; (2) the phenomena of electroporation are described quantitatively in the case of both stable and metastable planar bilayer membranes when the transmembrane voltage is changed stepwise; (3) the uniform description of reversible and irreversible electroporation and mechanical breakdown is presented; (4) exact solutions of the stochastic equations of the electroporation are determined; (5) the pores are considered to be independent of each other. The consequences of the pore–pore interaction, pore coalescence, and integral proteins are discussed elsewhere (Sugar *et al.*, 1987). The results of the model are compared with the available experimental data, such as: membrane lifetime, critical transmembrane voltage, and membrane conductance during the resealing process.

ISTVAN P. SUGAR • Departments of Biomathematical Sciences and Physiology and Biophysics, Mt. Sinai Medical Center, New York, New York 10029, and Institute of Biophysics, Semmelweis Medical University, 1444 Budapest, Hungary.

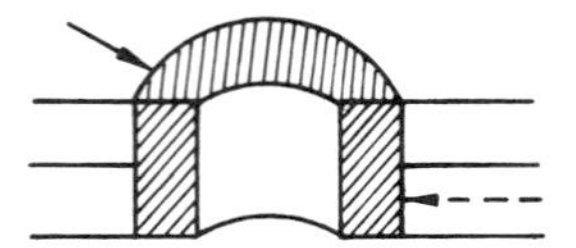

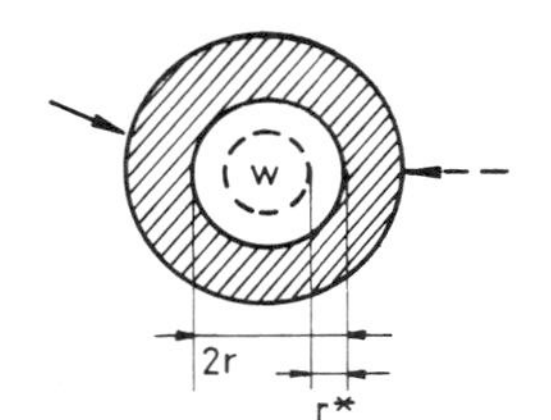

Figure 1. Side and top views of the general pore structure. The water-filled pore interior (w) is surrounded by the pore wall (shaded area), which in turn is surrounded by the bulk bilayer lipid membrane. r is the pore radius; r^* is the thickness of the region of the larger electric field; solid and dashed arrows show possible sites of lipid transfer from the upper and lower monolayer of the bilayer to the pore wall, respectively.

2. THE MODEL

On the basis of energetic considerations, different pore structures have been proposed: hydrophobic pore and inverted pore model (Abidor *et al.*, 1979), partially hydrophobic pore (Pastushenko and Petrov, 1984), and periodic block model (Sugar and Neumann, 1984). The description of the present electroporation model does not require a detailed knowledge of the pore structure.

2.1. General Pore Structure

The following general pore properties are assumed in the model (Fig. 1): (1) A common feature of every pore model is that the water-filled pore interior is surrounded by the pore wall, and the pore wall is surrounded by the bulk bilayer lipid membrane. The structure of the lipid molecules within the pore wall is different from the bilayer structure of the bulk membrane. Consequently, the specific energy of the pore wall is different from that of the bulk membrane. (2) During the pore opening and closing process, the number of lipid molecules within the pore wall increases and decreases, respectively. In the case of planar membranes, lipid transfer into the pore wall takes place simultaneously from both monolayers of the bilayer membrane. However, it is assumed that the sites of these two transfer processes along the pore circumference are independent of each other.

2.2. The Stochastic Variable

Assuming circular pore geometry, both the state of the pore and the process of electroporation can be described by one stochastic variable, a. This variable is proportional to the number of lipid molecules in the pore wall. The stochastic variable a and the pore radius r are related by:

$$a = \mathrm{ENT}\ (2r\pi/l) \tag{1}$$

where ENT is the integer of a numerical value p.qrs, e.g., ENT(p.qrs) = p, and l is the characteristic length along the pore circumference. The value of the characteristic length depends on the molecular structure of the pore wall. For example, in the case of the "periodic block model" (Sugar and Neumann, 1984), l is twice the cross-sectional diameter of a lipid molecule.

The opening and closing process of a single membrane pore may be treated as a stochastic process in terms of a Markov chain (Goel and Richter-Dyn, 1974). The general stochastic state transitions are given by the scheme:

$$\underset{0}{\bullet} \underset{\overleftarrow{w}_1}{\overset{\overrightarrow{w}_0}{\rightleftarrows}} \underset{1}{\bullet} \underset{\overleftarrow{w}_2}{\overset{\overrightarrow{w}_1}{\rightleftarrows}} \underset{2}{\bullet} \rightleftarrows \ldots \rightleftarrows \underset{a-1}{\bullet} \underset{\overleftarrow{w}_a}{\overset{\overrightarrow{w}_{a-1}}{\rightleftarrows}} \underset{a}{\bullet} \underset{w_{a+1}}{\overset{\overrightarrow{w}_a}{\rightleftarrows}} \underset{a+1}{\bullet}$$

where $\overrightarrow{w}_a$ and $\overleftarrow{w}_a$ represent the "rate constants" or transition probabilities per unit time for the $a \rightarrow (a+1)$ pore opening step and $a \rightarrow (a-1)$ pore closing step, respectively. Within the framework of this general scheme, the physics of electroporation is concentrated in the functions of the transition probabilities. In the next three sections, the transition probabilities in the function of the external electric field strength are determined and then the stochastic equations of the electroporation are constructed.

2.3. Transition Probabilities

Denoting by Δt a small time interval within which one state transition $a \rightarrow (a+1)$ occurs, the transition probability of this state change is given by

$$\overrightarrow{w}_a \cdot \Delta t = (\Delta t/\tau) \exp - [(\Delta \bar{G}^*(a + 0.5) - \Delta \bar{G}(a))/kT] \tag{2}$$

and the transition probability of $a \leftarrow (a+1)$ process is

$$\overleftarrow{w}_{a+1} \cdot \Delta t = (\Delta t/\tau) \exp - [(\Delta \bar{G}^* (a + 0.5) - \Delta \bar{G}(a + 1))/kT] \tag{3}$$

where τ is the characteristic transition time, $1/\tau$ is the transition frequency as the number of trials per unit time, T is the absolute temperature, k is the Boltzmann constant, $\Delta \bar{G}(a) = \bar{G}(a) - \bar{G}(0)$ is the Gibbs free energy change of the membrane/solution system in the presence of an electric field when a single pore of state a forms in the bilayer, and $\Delta \bar{G}^*(a + 0.5)$ is the free energy change when an activated pore structure forms between state a and $a + 1$ in the bilayer.

The activation Gibbs free energies are given by

$$\Delta \bar{G}^*(a + 0.5) - \Delta \bar{G}(a) = \alpha - kT \ln (\beta a)^2 \tag{4}$$

$$\Delta \bar{G}^*(0.5) - \Delta \bar{G}(0) = \alpha \tag{5}$$

where α is the activation energy. The second term in Eq. (4) contains the activation entropy. For the sake of simplicity, the activation energy is independent of the pore state. The transfer of lipid molecules from one monolayer to the pore wall can take place at different sites. The number of possible sites, βa, is proportional to the pore circumference, and also to the stochastic variable a [Eq. (1)]; β is the proportionality constant. Since the sites of transfer from the two monolayers to the pore wall are independent (see Section 2.1), the square of βa gives the thermodynamic probability of the activated pore between a and $a + 1$ states.

2.4. Free Energy Function at Zero Electric Field

According to the nucleation theory of bilayer stability (Kashchiev and Exerowa, 1983; Kashchiev, 1987) at zero electric field, the free energy change of the membrane/solution system when a single pore of radius r forms is

$$\Delta G(r) = -r^2\pi(\mu_b - \mu_s)/A_0 + 2r\gamma = -r^2\pi \cdot \Delta g + 2r\gamma \tag{6}$$

where μ_s and μ_b are the chemical potential of the monomer lipid molecule in the solution (s) and in the bilayer membrane (b), respectively; A_0 is the cross-sectional area of a lipid molecule. The second term is the energy expended on the creation of the pore wall; γ is the energy of the pore wall per unit length. When $\Delta\mu = (\mu_b - \mu_s) < 0$, the free energy change $\Delta G(r)$ increases with increasing pore radius. In this case, pores of any size tend to shrink and the bilayer is stable with respect to rupture. However, in the case of $\Delta\mu > 0$, the $\Delta G(r)$ function decreases from a certain pore radius r_c; if $r < r_c$, the pore tends to shrink and the bilayer is metastable with respect to rupture; if $r > r_c$, the pore can grow spontaneously and the bilayer is unstable.

The chemical potential difference, and consequently the bilayer stability, depends on the monomer lipid concentration C in the solution. Denoting C_e as the concentration where $\Delta\mu = 0$ and denoting CMC as the critical micelle concentration (Kashchiev and Exerowa, 1983), then the bilayer is stable when $C_e < C < CMC$. The bilayer is metastable when $C < C_e < CMC$ or $C < CMC < C_e$.

Although experimental data are not available on stable planar bilayers, the model calculations were performed for stable systems. These calculations have biological relevance because the electroporation of the stable planar bilayer should be analogous to the electroporation of large lipid vesicles or to the electroporation of the protein-free domains of cell membranes (see Section 3.1).

2.5. Free Energy Function in the Presence of an Electric Field

In the presence of a transmembrane electric field, one has to introduce additional energy terms into the free energy function [Eq. (6)] describing the change in the electric polarization energy of the system $\Delta G_{el}(r)$ when a single pore of radius r forms in the bilayer (Sugar and Neumann, 1984). If $r < r^*$

$$\Delta G_{el}(r) = 0.5\ \epsilon_0(\epsilon_m - \epsilon_w)\ \pi r^2 d E_m^2 \qquad (7)$$

If $r > r^*$

$$\Delta G_{el}(r) = 0.5\ \epsilon_0 E_m^2\ \pi d[(\epsilon_m - 1)r^2 - (\epsilon_w - 1)\ (r^2 - \{r - r^*\}^2)] \qquad (8)$$

In Eqs. (7) and (8), d is the bilayer thickness, E_m is the electric field strength in the bilayer, and ϵ_0, ϵ_m, and ϵ_w are the vacuum dielectric permittivity, relative dielectric permittivity of the membrane, and relative dielectric permittivity of water, respectively. During the derivation of Eqs. (7) and (8), the following assumptions have been made.

(1) At higher ionic strengths (> 0.1 M), the electric conductivity of the aqueous solution is so much larger than that of the planar bilayer that the applied voltage U only drops across the bilayer. In terms of the constant field approximation, the average field in the bilayer is given by $E_m = U/d$; the field strength in the bulk electrolyte may be approximated by $E_s \simeq 0$.

(2) The electric field within the solvent-filled pore is inhomogeneous, decreasing from the value E_m $(= U/d)$ at the pore wall/solvent interface toward the pore center. In the layer of solvent molecules adjacent to the inner cylindrical part of the pore wall of thickness r^*, the field intensity E_p is approximated by $E_p = E_m$ (Fig. 1). For larger pores where $r > r^*$, the electric field strength in the central region of radius $r - r^*$ is considered to be $E_s \simeq 0$. For small pores where $r < r^*$, the homogeneous field approximation $E_p = E_m$ holds. According to Jordan (1982), at 1 M ionic strength the electric field becomes highly inhomogeneous if $r > d/5$. For these conditions, the relation $r = r^* \simeq d/5$ specifies the largest pore size to which the small-pore field approximation $E_p = E_m$ may be applied. In the case of oxidized cholesterol membranes, $d = 3.3$ nm (Benz *et al.*, 1979). Thus, $r^* = 0.86$ nm at 0.1 M ionic strength (I); r^* increases with decreasing I (Jordan, 1982).

Introducing the pore state a, defined in Eq. (1), into Eqs. (6)–(8), the Gibbs free energy function is given by $\Delta\bar{G}(a) = \Delta G(a) + \Delta G_{el}(a)$; see the transition probability functions in Eqs. (2)–(5). Equations (2)–(8) permit the calculation of the transition probabilities as a function of the pore state and transmembrane voltage. For this purpose, the numerical values of the oxidized cholesterol bilayer system were used (Benz *et al.*, 1979): $\gamma = 1.25 \times 10^{-11}$ N (Abidor *et al.*, 1979), $l = 1.8$ nm (Sugar and Neumann, 1984), $T = 313$ K, $\epsilon_m = 2.1$, $\epsilon_w = 80$. The calculations were performed with three values of $\Delta\mu/A_0$ ($= \Delta g$): -0.001, 0, and 0.001 N/m. These values are typical for stable and metastable bilayers (Tien, 1974), respectively.

2.6. Master Equation of Electroporation

The probability $P_{a,a_0}(t, t + \Delta t)$ of occurrence of a pore state a at time $t + \Delta t$ if the state is a_0 at time $t = 0$ is given by

$$P_{a,a_0}(t + \Delta t) = \vec{w}_{a-1}\Delta t P_{a-1,a_0}(t) + \overleftarrow{w}_{a+1}\Delta t P_{a+1,a_0}(t) + (1-[\vec{w}_a + \overleftarrow{w}_a]\Delta t)P_{a,a_0}(t) \tag{9}$$

The "difference" equation (9) may be transformed into the "forward master" equation (Goel and Richter-Dyn, 1974):

$$dP_{a,a_0}(t)/dt = \vec{w}_{a-1}P_{a-1,a_0}(t) + \overleftarrow{w}_{a+1}P_{a+1,a_0}(t) - [\vec{w}_a + \overleftarrow{w}_a]P_{a,a_0}(t) \tag{10}$$

3. RESULTS AND DISCUSSIONS

Since the transition probabilities $\overleftarrow{w}_a$ and $\vec{w}_a$ in the master equation of the electroporation [Eq. (10)] are nonlinear functions of the pore state a, no general solution is possible. However, one can determine the exact stationary and quasi-stationary solutions, the lifetime of the metastable membrane, the kinetics of pore opening and closing as well as other parameters.

3.1. Stationary Solutions

If the pore state a is confined in the $(0,a^*)$ interval, the stationary pore size distribution $P_{a,a_0}(t = \infty)$ can be determined by means of Eq. (11) in Table 1. The different solutions of the master equations [Eq. (10)] in Table 1 were taken from Goel and Richter-Dyn (1974).

Physically, the upper limit of the pore size is comparable to the membrane size itself, i.e., $a^* = \infty$ is a good approximation. In the case of *stable membranes,* the infinite sum in Eq. (11) is convergent. The logarithm of the pore state distributions are shown in Fig. 2a at different transmembrane voltages. At subcritical electric fields ($U^s_{cr} < 0.4$ V), the pore size distribution is homogeneous with a cusplike maximum at $a = 0$. This maximum of the distribution curve defines a phase termed the poreless phase of the stable bilayer. At supercritical electric fields ($U^s_{cr} > 0.4$ V), the pore size distribution becomes inhomogeneous with two maxima: one at $a = 0$ represents the poreless phase and one at $a > 0$ defines the so-called porous phase of the membrane. When the maximum belonging to the porous phase exceeds the cusplike maximum at the poreless phase, the membrane lipids form pore walls almost exclusively. In the frame of our independent pore model, this loose net structure of the membrane *is still stable.* This is a transmembrane voltage-induced phase transition of the membrane from the poreless phase into the porous phase (Sugar, 1987). In Fig. 3a and b, characteristic values of the pore state distribution curves are plotted as a function of transmembrane voltage. The solid lines represent the average pore state calculated by

$$\langle a \rangle = \sum_{a=0}^{a^*} aP_{a,a_0}(t=\infty) \tag{15}$$

The average pore state sharply changes at the transition voltage U_m defined by the voltage where $d\langle a \rangle/dU$ is maximal. The transition voltage decreases with increasing Δg (see Fig. 3a,b), while

Table 1. Expressions for the Equilibrium Distribution $P_{a,a_0}(\infty)$ and for the "First Passage Time" $T^{(1)}_{a,a_0}$

Condition	Expression	Eq.
$0 \leq a_0 \leq a^*$	$P_{a,a_0}(\infty) = \dfrac{\vec{w}_0\, \vec{w}_1 \ldots \vec{w}_{a-1}}{\overleftarrow{w}_1\, \overleftarrow{w}_2 \ldots \overleftarrow{w}_a} \Big/ \sum_{i=1}^{a^*} \dfrac{\vec{w}_0\, \vec{w}_1 \ldots \vec{w}_{i-1}}{\overleftarrow{w}_1\, \overleftarrow{w}_2 \ldots \overleftarrow{w}_i}$	(11)
$0 \leq a_0 \leq a < \infty$	$T^{(1)}_{a,a_0} = \sum_{i=a_0}^{a-1} \sum_{n=0}^{i} \vec{w}_n^{-1}\, \Pi_{n+1,i}$	(12)
$0 \leq a < a_0 < a^*$	$T^{(1)}_{a_0,a} = \sum_{i=a_0}^{a^*-1} \sum_{n=a+1}^{i} \vec{w}_n^{-1} \Pi_{n+1,i}\, R_{a,n} - R_{a,a_0} \sum_{i=a+1}^{a^*-1} \sum_{n=a+1}^{i} \vec{w}_n^{-1}\, \Pi_{n+1,i}\, R_{a,n}$	(13)
$0 \leq a < a_0 < \infty$	$T^{(1)}_{a_0,a} = \sum_{i=a+1}^{a_0} M_{i-1,i}$	(14)

$$\Pi_{i,j} = \frac{\overleftarrow{w}_i \overleftarrow{w}_{i+1} \ldots \overleftarrow{w}_j}{\vec{w}_i \vec{w}_{i+1} \ldots \vec{w}_j} \text{ if } i \leq j \text{ and } \Pi_{i,i-1} = 1$$

$$R_{a,a_0} = \sum_{i=a_0}^{a^*-1} \Pi_{a+1,i} \Big/ \sum_{i=a}^{a^*-1} \Pi_{a+1,i}$$

$$M_{i-1,i} = \frac{1}{\overleftarrow{w}_i} + \frac{\vec{w}_i}{\overleftarrow{w}_i \overleftarrow{w}_{i+1}} + \frac{\vec{w}_i \vec{w}_{i+1}}{\overleftarrow{w}_i \overleftarrow{w}_{i+1} \overleftarrow{w}_{i+2}} + \ldots$$

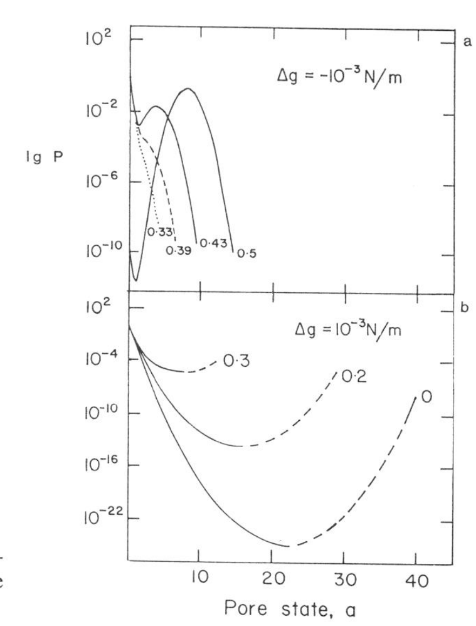

Figure 2. Stationary and quasi-stationary pore state distributions at different transmembrane voltages. (a) Stable membrane; (b) metastable membrane.

the width of the phase transition remains about 0.05 V. The broken lines show the deviation of the pore states from the average value, as calculated by

$$\langle a\rangle + 3\langle\Delta a\rangle = \langle a\rangle + 3\ [\sum_{a=0}^{a^*} (a - \langle a\rangle)^2 P_{a,a_0}(\infty)]^{1/2} \tag{16}$$

During the phase transition, the relative pore state fluctuation ($\langle\Delta a\rangle/\langle a\rangle$) drops by two orders of magnitude! The dotted lines show the most probable state in the porous phase. The porous phase appears at the critical voltage, U^{s}_{cr}, and the size of the pores increases with increasing trans-

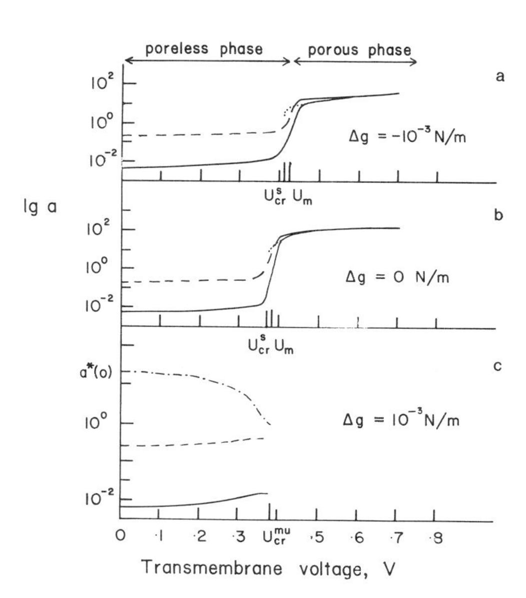

Figure 3. Characteristic quantities of electroporation as a function of transmembrane voltage. (a, b) Stable membranes; (c) metastable membrane. (——) Average pore state calculated by Eq. (15); (– – –) pore state fluctuation calculated by Eq. (16); (·····) the most probable state of the porous phase.

membrane voltage. In this voltage range, the *membrane acts as a filter,* where the pore size is in the nanometer range and can be regulated by the transmembrane voltage.

Reversible Electroporation in Stable Membranes

The poreless–porous phase transition is reversible. Upon application of a supercritical voltage, opening of the pores takes place, resulting in the equilibrium porous phase. Upon switching off the field, coherent resealing of the pores starts immediately (Section 3.4).

Electroporation measurements have not been performed on stable planar lipid bilayer membranes. However, phenomenologically a large unilamellar vesicle or the protein-free domain of a cell membrane may be analogous to a stable planar bilayer. This is the case because the free energy change of a large vesicle $\Delta G_v(r) = \gamma 2r\pi$ increases with increasing pore radius (Sugar and Neumann, 1984) which is the criterion of stability (see Section 2.4). Comparing $\Delta G_v(r)$ with Eq. (6), we get $\Delta g = 0$ for large vesicles. In the case of large unilamellar DPPC vesicles, the critical transmembrane voltage was found to be 0.25 V (Teissie and Tsong, 1981) while according to the model $U^{s}_{cr} = 0.36$ V (Fig. 3b).

3.2. Quasi-Stationary Solutions

In the case of *metastable membranes,* the infinite sum in Eq. (11) is divergent, i.e., there is *no stationary state* of a metastable system. However, one can define the *quasi-stationary pore size distribution* if the free energy barrier $\Delta\bar{G}(r_c)$ is much higher than the thermal energy unit kT. Within the lifetime of the metastable membrane, the upper limit of the pore state a^* is defined by the minimum of the $P_{a,a_0}(t = \infty)$ function.

In Fig. 2b the quasi-stationary pore state distributions are shown at different transmembrane voltages. The limit state of the membrane a^* decreases with increasing transmembrane voltage (see the beginning of the dashed lines in Fig. 2b). When a^* is close to zero, i.e., $\Delta\bar{G}(r_c) \simeq kT$, the metastable state of the membrane ceases to exist. The transmembrane voltage at which the metastable state of the membrane disappears is U^{mu}_{cr}.

Using Eqs. (11), (15), and (16) for the voltage-dependent limit state a^*, the average pore state $\langle a \rangle$ and the deviation of the pore state from the average are determined within the voltage range $0 < U < U^{mu}_{cr}$; see solid line and broken line in Fig. 3c.

Mechanical Breakdown, Reversible and Irreversible Electroporation

The average pore size is essentially zero while the membrane is in its metastable phase. Because of the thermal fluctuations, pores are opening temporarily, but their state exceeds $a = 1$ very rarely (see Fig. 3c). In spite of this, there is a finite probability, $P_{a^*,a_0}(t = \infty)$ that the state of a pore exceeds the limit state, $a^*(U)$. As a consequence of this, the membrane becomes unstable and pore opening proceeds until the mechanical rupture of the membrane (*mechanical breakdown*) takes place. Since $P_{a,a_0}(t = \infty)$ increases with increasing transmembrane voltage, the probability of the mechanical breakdown increases or the *lifetime* of the metastable membrane decreases.

Applying supercritical transmembrane voltage ($U > U^{mu}_{cr}$), the membrane becomes unstable immediately and coherent unlimited opening of micropores takes place. This results in rupture of the membrane (*irreversible electroporation*). If, however, the voltage is switched off before none of the pore states have attained the limit state $a^*(0 \text{ V})$, the membrane jumps back into the metastable state and resealing of the pores takes place. This is the mechanism of *reversible electroporation* in the case of metastable membranes.

In the voltage interval $U^{mu}_{cr} < U < U^{us}_{cr}$, there is neither a stationary nor a quasi-stationary solution of Eq. (11) and therefore the membrane is unstable. Strong electric fields ($U > U^{us}_{cr}$) stabilize the membrane. Qualitatively, this is equivalent to the porous phase of stable membranes (Section 3.1). According to the calculations for oxidized cholesterol membranes, $U^{us}_{cr} \simeq 1.2$ V. At

$U > U_{cr}^{us}$, the most probable pore state decreases with increasing transmembrane voltage, while it levels off after 1.7 V at $a \simeq 1000$ (not shown in Fig. 3c). Since these states are larger than $a^*(0\ \text{V})$, if the field is switched off, the membrane becomes unstable and ruptures.

The phenomena of mechanical breakdown and electroporation (reversible and irreversible) have been demonstrated on oxidized cholesterol membranes (Abidor *et al.*, 1979; Benz *et al.*, 1979). It is interesting that bilayers made from phosphatidylcholine, phosphatidylethanolamine, and monoolein were fragile and always ruptured after the application of the supercritical field (Benz *et al.*, 1979). In spite of this, the voltage relaxation curves of these membranes just after the application of charge pulses of 400 μsec duration were similar to those of the oxidized cholesterol membranes. This is presumably because the threshold pore state of the conductivity, a_{th} (where the pore size is equal to the size of the hydrated ions), and the threshold pore state of metastability, $a^*(0\ \text{V})$, are comparable for the fragile membranes while $a^*(0\ \text{V}) > a_{th}$ in the case of oxidized cholesterol membranes. Therefore, when a "fragile" membrane has attained its highly conducting state, some of its pores have exceeded $a^*(0\ \text{V})$, too. After switching off the field, the majority of the pores start annealing. This leads to a characteristic voltage relaxation. The largest pores continue opening until rupture of the membrane.

3.3. Nonstationary Solutions

If the pore state at $t = 0$ is a_0, the function $F_{a,a_0}^{(1)}(t)dt$ is the probability that the pore reaches state a for the first time within the time interval t and $t + dt$. The "first passage time" is the average time of the $a_0 \rightarrow a$ process:

$$T_{a,a_0}^{(1)} = \int_0^{\infty} tF_{a,a_0}^{(1)}(t)dt \tag{17}$$

where $F_{a,a_0}^{(1)}(t)$ is related to $P_{a,a_0}(t)$ through the relation (Goel and Richter-Dyn, 1974):

$$P_{n,a_0}(t) = \int_0^{t} F_{a,a_0}^{(1)}(t - \tau)\, P_{n,a}(\tau)\, d\tau \tag{18}$$

The "first passage time" is an exact function of the transition probabilities in Eq. (10). The form of the function depends on the actual restrictions of the stochastic process. The stochastic variable of the electroporation a is restricted; it is never a negative value. The lowest permitted state, $a = 0$, is called the reflecting state of the process.

In the case of metastable membranes, the upper limit state, a^*, is called the *adsorbing state* of the process. After reaching this state, rupture of the membranes takes place and the stochastic variable never returns to the $(0, a^*)$ interval.

In Table 1, expressions for "first passage time" are shown in the case of different restrictions on the stochastic process. Using these exact formulas, we can calculate: the time delay of mechanical breakdown; reversible and irreversible electroporation; and the time course of the coherent pore opening and closing process, when stepwise change of the transmembrane voltage takes place.

3.3.1. Time Delay of Mechanical Breakdown—Membrane Lifetime

In the case of a metastable membrane, sooner or later the system becomes unstable. This takes place when one of the pores exceeds the limit state of metastability. Considering only a *single pore*, one can determine the "first passage time" from the closed pore state ($a_0 = 0$) to the limit state (a^*) by means of Eq. (13). The obtained "first passage time", $T_{a^*,0}^{(1)}$, as a function of transmembrane voltage ($0 \leq U \leq U_{cr}^{mu}$) is shown in Fig. 4.

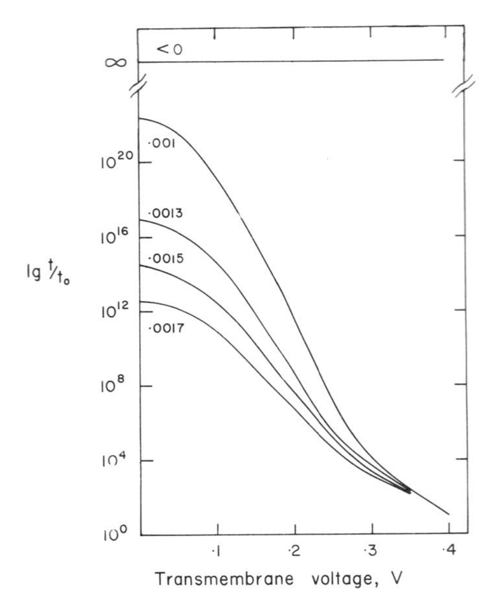

Figure 4. Lifetime of planar membranes as a function of transmembrane voltage at different values of Δg ($= \Delta\mu/A_0$) shown at each curve in N/m. $t = T^{(1)}_{a^*,0}$ and $t_0 = (\beta^2/\tau)\exp - (\alpha/kT)$.

If the average number of pores in the membrane is larger than one, the membrane lifetime is smaller. An approximate relation between the membrane lifetime $T^{(N)}_{a^*,0}$ and the average number of the pores N has been derived by Arakelyan *et al.* (1979) in the case of a large number of pores:

$$T^{(N)}_{a^*,0} \simeq T^{(1)}_{a^*,0} N^{-0.5} \tag{19}$$

Thus, apart from a constant term, semilogarithmic plots of $T^{(1)}_{a^*,0}(U)$ and $T^{(N)}_{a^*,0}(U)$ functions are the same. In accordance with the experimental lifetime data (Abidor *et al.*, 1979; Green and Andersen, personal communication), the logarithmic plot of $T^{(1)}_{a^*,0}(U)$ is a straight line within the 0.1–0.2 V interval and then the curve starts to level off.

According to the calculations, the membrane lifetime is very sensitive to Δg especially at zero electric field (Fig. 4). Since $\Delta g = \Delta\mu/A_0$ is strongly related to the lipid concentration in the aqueous solution (Section 2.4), reliable lifetime data cannot be obtained without a controlled lipid concentration in the aqueous medium.

3.3.2. Time Delay of Electroporation

After applying a supercritical voltage $U > U^{\mathrm{mu}}_{\mathrm{cr}}$, a more or less coherent opening of the pores takes place. When the average pore size a becomes comparable to the size of the hydrated ions in the aqueous medium a_{th}, the electric conductivity of the membrane increases by about 10 orders of magnitude (Benz *et al.*, 1979). The *delay time* between the application of the high electric field and the increase of the membrane conductivity has been estimated by determining $T^{(1)}_{a_{\mathrm{th}},0}$ [Eqs. (12, 13)] at different transmembrane voltages [Fig. 5a,b at (U_{cr}, 1 V) interval].

The calculated $T^{(1)}_{3,0}$ is the upper limit of the delay time because, for example, pores with $a_0 = 1$ reach $a_{\mathrm{th}} = 3$ much faster than pores with $a_0 = 0$. In the case of Fig. 5b, the vertical line at $U^{\mathrm{mu}}_{\mathrm{cr}}$ separates the metastable and unstable nonconducting states of the membrane, while in Fig. 5a the vertical line at $U^{\mathrm{s}}_{\mathrm{cr}}$ separates the poreless and porous conducting states of the stable membrane. At a subcritical field, the metastable membrane becomes conductive after a very long delay time and then the mechanical breakdown takes place [Fig. 5b, solid line in (0, $U^{\mathrm{mu}}_{\mathrm{cr}}$) interval]. In the case of *metastable membranes*, the measured and calculated delay time data of oxidized cholesterol mem-

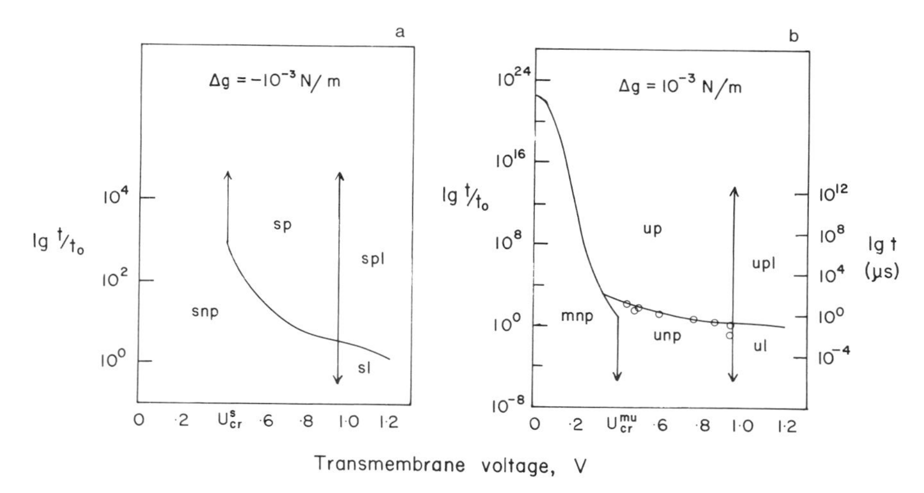

Figure 5. Time delay of electroporation as a function of transmembrane voltage. (a) Stable membrane; (b) metastable membrane. Circles: experimental data from Benz and Zimmermann (1980), right side ordinate. m, metastable membrane; u, unstable membrane; s, stable membrane; np, nonconductive (small) pores; p, conductive pores; l, conductive lipid membrane. The threshold pore state of the conductive pores $a_{th} = 3$ (Benz and Zimmermann, 1981), $t = T^{(1)}_{3,0}$, and $t_0 = (\beta^2/\tau)\exp - (\alpha/kT)$. In the transmembrane voltage interval $(0, U^{mu}_{cr})$ the membrane lifetime is plotted; in this interval $t = T^{(1)}_{a^*,0}$.

branes are in good agreement if t_0 is chosen properly (see circles in Fig. 5b). In spite of the encouraging agreement, it is important to note that the calculations assume a stepwise change of the transmembrane voltage and this is not the case in the experiments of Benz and Zimmermann (1980). Powell *et al.* (1986) developed a model of the electroporation which simulates the conditions of these experiments properly. The experimental data show a drastic change in the tendency of the delay time–voltage curve at 0.95 V. According to Benz and Zimmermann (1980), at this voltage not only the aqueous membrane pores but also the bulk lipid part of the membrane becomes conductive. At this transmembrane voltage, the electric field energy becomes comparable with the Born energy of the ion in the membrane.

3.4. Transition Kinetics

Up to this point, the "first passage time" was calculated as a function of transmembrane voltage at given starting and final state. Let us now determine the "first passage time" as a function of the final state both at a given starting state and at a given transmembrane voltage. Physically, this function is analogous to the time dependence of the average pore state.

The exact form of the "first passage time" function is known (Table 1) while in the case of the $\langle a(t) \rangle$ function only approximate solutions are available (Sugar and Neumann, 1984).

3.4.1. Poreless⇌Porous Transition

In the case of *stable membranes*, a poreless→porous first-order phase transition (Sugar, 1987) can be induced by supercritical fields. The starting state is $a_0 = 0$ and the most probable stationary porous state (Fig. 3a) is the final state of the transition process. The "first passage time" can be determined by Eq. (12). In Fig. 6a the kinetics of this transition are shown at different transmembrane voltages. In every case the transition process begins with a small increase in the pore state. During a latency period, the pore state increases very slowly. The higher the transmembrane voltage, the shorter is the latency period. Finally, the pore state reaches the stationary value with a very fast process. Upon switching off the field, the porous–poreless process takes place. By means

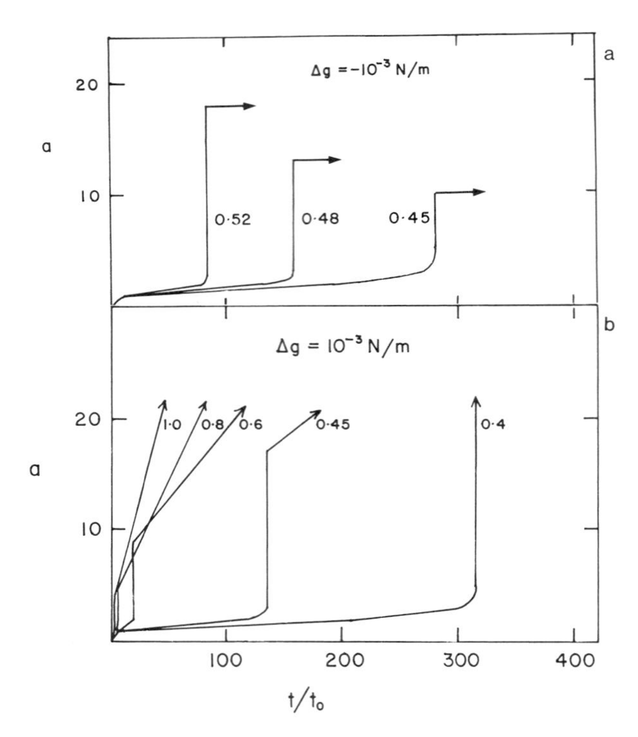

Figure 6. Transition kinetics of electroporation. (a) Poreless→porous transition in a stable membrane; (b) rupture process in an unstable membrane. Transmembrane voltage is shown for each curve in volts. $t = T^{(1)}_{a,0}$ and $t_0 = (\beta^2/\tau)\exp - (\alpha/kT)$.

of Eq. (14) the time course of the resealing process has been calculated in the case of two different starting states. The semilogarithmic plot of the relaxation process is shown in Fig. 7a.

3.4.2. Membrane Rupture

After applying a supercritical field, the *metastable membrane* becomes unstable immediately. The beginning state is $a_0 = 0$. The time course of the process can be determined by Eq. (12). In Fig. 6b the rupture process is shown at different transmembrane voltages. The process is similar to the poreless→porous transition, but the pore state further increases after the very fast increase in pore state, resulting in the rupture of the membrane.

3.4.3. Resealing Process

Upon switching off the supercritical field before the limit state of metastability, $a^*(0)$, has been attained, the membrane becomes metastable again. By means of Eq. (13) the time course of the relaxation process has been calculated both at different starting states and at different Δg (see Fig. 7b). The dot–dash line is the transformation of $a(t/t_0)$ function to $a^2(t/t_0)-a^2_{th}$ function in the case of $\Delta g = 0.001$ N/m, $a_0 = 10$, $a_{th} = 3$. For $a<a_{th}$, no ions are conducted. The shape of this dot–dash curve is the same as that of the measured conductivity versus time curves during the resealing of oxidized cholesterol membranes (Benz and Zimmermann, 1981). Therefore, as is physically plausible, the ion conductance is proportional to a^2 (i.e., the surface of the pore mouth). Using the relaxation time ($\tau_r = 0.55$ μsec) of the experimental resealing process at 40°C (Benz and Zimmer-

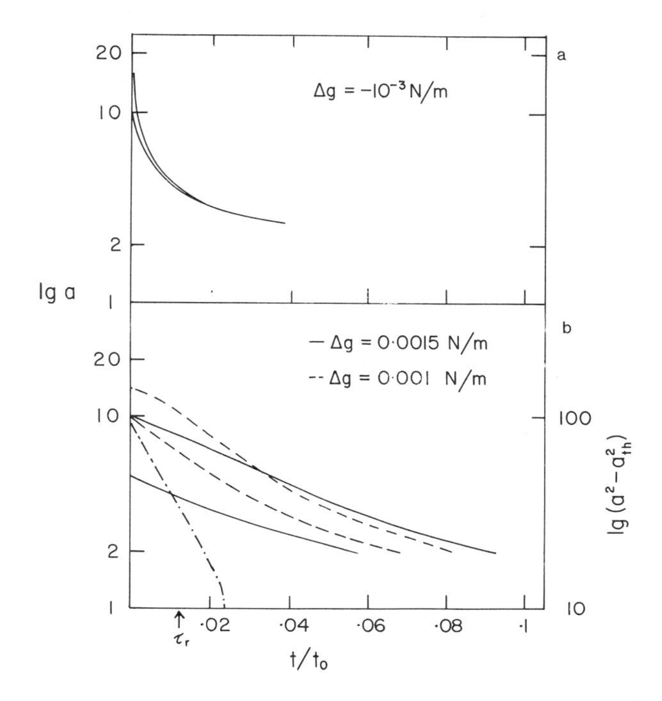

Figure 7. Resealing process. (a) Porous→poreless transition in stable membrane at different starting states, a_s. Upper curve: transmembrane voltage is 0.5 V at $t < 0$ and 0.0 V at $t > 0$; lower curve: transmembrane voltage is 0.45 V at $t < 0$ and 0.0 V at $t > 0$. (b) Resealing processes of metastable membrane both at different starting states $a_s[< a^*\ (0)]$ and Δg. $t = T^{(1)}_{a,a_s}$ and $t_0 = (\beta^2/\tau) \exp - (\alpha/kT)$. The dot–dash line represents the transformation a to $a^2 - a^2_{th}$ with $a_{th} = 3$ (Benz and Zimmermann, 1981); $\Delta g = 0.001$ N/m, $a_s = 10$ and the respective ordinate at the right-hand side; τ_r is the relaxation time of the annealing process.

mann, 1981) and the calculated relaxation curve in Fig. 7b, one can determine the absolute time scale: $t_0 = 0.55\ \mu\text{sec}/0.0125 = 44\ \mu\text{sec}$.

Acknowledgments. I thank Professor T. E. Thompson and W. van Osdol for critical evaluation of the manuscript before its submission. This work was supported by NIH Grants GM-14628, GM-23573, and GM-26894.

REFERENCES

Abidor, I. G., Arakelyan, V. B., Chernomordik, L. V., Chizmadzhev, Y. A., Pastushenko, V. F., and Tarasevich, M. R., 1979, Electric breakdown of bilayer lipid membranes I, *Bioelectrochem. Bioenerg.* **6:** 37–52.

Arakelyan, V. B., Chizmadzhev, Y. A., and Pastushenko, V. F., 1979, Electric breakdown of bilayer lipid membranes V, *Bioelectrochem. Bioenerg.* **6:**81–87.

Benz, R., and Zimmermann, U., 1980, Relaxation studies on cell membranes and lipid bilayers in the high electric field range, *Bioelectrochem. Bioenerg.* **7:**723–739.

Benz, R., and Zimmermann, U., 1981, The resealing process of lipid bilayers after reversible electrical breakdown, *Biochim. Biophys. Acta* **640:**169–178.

Benz, R., Beckers, F., and Zimmermann, U., 1979, Reversible electrical breakdown of lipid bilayer membranes: A charge-pulse relaxation study, *J. Membr. Biol.* **48:**181–204.

Dimitrov, D. S., and Jain, R. K., 1984, Membrane stability, *Biochim. Biophys. Acta* **779:**437–468.

Goel, N. S., and Richter-Dyn, N., 1974, *Stochastic Models in Biology,* Academic Press, New York.

Jordan, P. C., 1982, Electrostatic modeling of ion pores: Energy barriers and electric field profiles, *Biophys. J.* **39:**157–164.

Kashchiev, D., 1987, On the stability of membrane, foam and emulsion with respect to rupture by hole nucleation, *Colloid Polymer Sci.* **265:**436–441.

Kashchiev, D., and Exerowa, D., 1983, Bilayer lipid membrane permeation and rupture due to hole formation, *Biochim. Biophys. Acta* **732:**133–145.

Pastushenko, V. F., and Petrov, A. G., 1984, Electro-mechanical mechanism of pore formation in bilayer lipid membranes, in: Biophysics of Membrane Transport, School Proceedings, Poland, pp. 70–91.
Powell, K. T., Derrick, E. G., and Weaver, J. C., 1986, A quantitative theory of reversible electrical breakdown in bilayer membranes, *Bioelectrochem. Bioenerg.* **15:**243–255.
Sugar, I. P., 1987, Cooperativity and classification of phase transitions: Application to one- and two-component phospholipid membranes, *J. Phys. Chem.* **91:**95–101.
Sugar, I. P., and Neumann, E., 1984, Stochastic model for electric field-induced membrane pores: Electroporation, *Biophys. Chem.* **19:**211–225.
Sugar, I. P., Forster, W., and Neumann, E., 1987, Model of cell electrofusion: Membrane electroporation, pore coalescence and percolation, *Biophys, Chem.* **26:**321–337.
Teissie, J., and Tsong, T. Y., 1981, Electric field induced transient pores in phospholipid bilayer vesicles, *Biochemistry* **20:**1548–1554.
Tien, H. T., 1974, *Bilayer Lipid Membranes,* Dekker, New York.

Chapter 7

Theory of Electroporation

James C. Weaver and Kevin T. Powell

1. INTRODUCTION

The electroporation process leads to a membrane state which is characterized by pores and is associated with large transmembrane potentials. The dramatic phenomena attributed to electroporation include rupture, reversible electrical breakdown, the transient high-permeability state, and electrical fusion. All are associated with transmembrane potentials significantly larger than cellular resting potentials. Generally, a theory of electroporation should provide a unified, quantitative description of many experimental observations and be capable of predicting new behavior. In addition to describing the behavior of pore size and number, a theory should also include specific mechanisms for membrane charging by external sources, and specific internal mechanisms for membrane discharging, such as ionic conduction through pores under particular conditions. Further, a theory should yield specific quantitative predictions of experimentally measured behavior, so that comparisons between theory and experiment can be readily made. We first outline essential features of the dramatic phenomena, and then describe basic features of a quantitative theory based on transient aqueous pores. Throughout we use the notation $U(t)$ for the instantaneous transmembrane potential, $R(t)$ for the instantaneous membrane resistance, and $P_{d,m}(t)$ for the instantaneous diffusive permeability of the membrane.

2. ELECTRICAL RUPTURE

Sudden, nonthermal rupture (also "irreversible mechanical breakdown") occurs in planar bilayer membranes exposed to a transmembrane potential, U, in the approximate range $200 \lesssim U \lesssim 500$ mV for a relatively long time (i.e., $\Delta t \gtrsim 10^{-4}$ sec) (Crowley, 1973; Abidor *et al.*, 1979; Benz, *et al.*, 1979; Weaver and Mintzer, 1981; Sugar, 1981; Powell and Weaver, 1986). However, prompt rupture does not occur in vesicles or cells at the same U (Chizmadzhev, 1983, personal communication; Sugar and Neumann, 1984). A delayed lysis of cells which sometimes follows the

James C. Weaver and Kevin T. Powell • Harvard–MIT Division of Health Sciences and Technology, Massachusetts Institute of Technology, Cambridge, Massachusetts 02139.

transient high-permeability state appears caused by chemical or osmotic imbalances rather than a direct electrical effect (Zimmermann *et al.*, 1976b; Kinosita and Tsong, 1977a,b; Schwister and Deuticke, 1985). Heating appears to be negligible and just before rupture the membrane capacitance, C, has not changed significantly, which rules out significant electrocompression as the rupture mechanism (Crowley, 1973; Chernomordik and Abidor, 1980; Chizmadzhev and Abidor, 1980; Sugar, 1981). Strikingly, membranes often avoid rupture when exposed to the shorter, but larger transmembrane potentials which cause reversible electrical breakdown (Benz *et al.*, 1979).

3. REVERSIBLE ELECTRICAL BREAKDOWN

Instead of causing rupture, larger but shorter U results in a nondamaging, more rapid discharge of the membrane, i.e., reversible electrical breakdown (''REB''; sometimes ''dielectric breakdown''). REB appears to immediately precede the longer-lasting high-permeability state. During REB the membrane resistance, $R(t)$, of planar, vesicular, and cellular membranes is believed to rapidly drop to low values, and then to recover over times ranging from seconds to minutes or even hours. Typical square pulse characteristics which cause REB are a pulse width in the range $10^{-7} \lesssim \Delta t \lesssim 10^{-4}$ sec, and amplitudes, at the membrane, in the range $1500 \gtrsim U \gtrsim 500$ mV. Overall, the resistance recovery process is rapid in artificial bilayers, but much slower and strongly temperature dependent in cells. REB has several characteristic attributes, including an increasingly rapid discharge as the applied pulse amplitude, U_i, is increased for fixed pulse width, Δt, or as Δt is increased at fixed pulse amplitude, and a breakdown potential, $U_{0,c}$, which diminishes as Δt increases (Benz *et al.*, 1979). Of major importance is the fact that either rupture or REB can be caused to occur simply by varying the magnitude and duration of U. This interdependence places significant constraints on possible explanations of rupture and REB (Crowley, 1973; Coster and Zimmermann, 1975; Abidor *et al.*, 1979; Chizmadzhev and Abidor, 1980; Sugar, 1981). A closely related feature is that there is a low, but nonzero, probability of experiencing rupture during or following REB (Neumann and Rosenheck, 1972; Zimmermann *et al.*, 1973; Kinosita and Tsong, 1977a,c; Benz *et al.*, 1979; Benz and Zimmermann, 1980a,b, 1981; Sukharev *et al.*, 1982; Abidor *et al.*, 1982; Chernomordik *et al.*, 1982, 1983; Schwister and Deuticke, 1985).

4. TRANSIENT HIGH-PERMEABILITY STATE

Immediately following reversible electrical breakdown, a transient high-permeability state generally occurs (Neumann and Rosenheck, 1972; Zimmermann *et al.*, 1975, 1976a; Auer *et al.*, 1976; Kinosita and Tsong, 1977a,c; Sowers and Lieber, 1986). Although this phenomenon itself is sometimes termed ''electroporation'' (Neumann *et al.*, 1982), the high-permeability state appears to be only one consequence of such pores, so that it is better to use ''electroporation'' to refer to all pore phenomena caused by exposure of membranes to large U. Although electrical measurements are not readily made for cells wherein significant molecular transport occurs, it is believed that first REB occurs, followed by a much longer interval wherein ''electrically silent'' transport of both charged and neutral molecules occurs. Compared to artificial planar bilayers, cell membranes exhibit a much slower, temperature-dependent recovery or ''resealing,'' during which the significant transport of even large macromolecules may occur. The high-permeability state can also lead to a delayed, nonthermal cell lysis, which seems not to be the direct result of electric effects on the membrane, but to resulting cellular imbalances such as short-circuiting of proton pumps or an osmotic pressure difference (Zimmermann *et al.*, 1976b; Kinosita and Tsong, 1977a,b; Schwister and Deuticke, 1985). However, a significant fraction of cells can take up macromolecules, survive, and grow (Neumann *et al.*, 1982; Potter *et al.*, 1984; Hashimoto *et al.*, 1985; Fromm *et al.*, 1986; Riggs and Bates, 1986; Toneguzzo and Keating, 1986; Tur-Kaspa *et al.*, 1986; Chu *et al.*, 1987; Stopper *et al.*, 1987; MacNeil, 1987; Weaver *et al.*, 1987, 1988; Bliss *et al.*, 1988).

Present understanding of the transient high-permeability state derives partially from studies which utilize optical or radioactive methods to determine release or uptake from an entire population of cells (Neumann and Rosenheck, 1972; Zimmermann *et al.*, 1974, 1975, 1976a; Kinosita and Tsong, 1977a–c; Lindner *et al.*, 1977; Teissie and Tsong, 1981; El-Mashak and Tsong, 1985; Schwister and Deuticke, 1985; Serpersu *et al.*, 1985), partially from studies which make kinetic measurements on individual cells (Mehrle *et al.*, 1985; Sowers, 1984, 1986; Sowers and Lieber, 1986), and partially from studies which introduce DNA into cells, wherein success is determined by scoring a biological endpoint, viz. the number of viable transformants. However, transformation depends on the sequential occurrence of several processes: electroporative uptake, cell survival, cytoplasmic transport of DNA in the presence of degradative enzymes, entry into the nucleus, and incorporation and expression with the host DNA. It is generally not known whether the cells that transform are representative, or are in some sense special. Typical overall electroporative transformation probabilities are on the order of 10^{-4} to 10^{-6} (Neumann *et al.*, 1982; Potter *et al.*, 1984; Hashimoto *et al.*, 1985; Fromm *et al.*, 1986; Riggs and Bates, 1986; Toneguzzo and Keating, 1986; Tur-Kaspa *et al.*, 1986; Chu *et al.*, 1987; Stopper *et al.*, 1987; MacNeil, 1987).

In contrast, the use of two different fluorescent molecules and single cell measurements by flow cytometry show that more than 10% of the cells can exhibit electroporation (Weaver *et al.*, 1987). This further supports the view that electroporation is a universal phenomenon of cell membranes. In such investigations, fluorescent-labeled test macromolecules (e.g., "green") are provided in solution before or after exposure to electric fields, and a second fluorescent molecule (e.g., "red") can be provided at a later time, such that both electroporative uptake and the degree of subsequent resealing can be determined on a single cell basis. Flow cytometry allows quantitative measurements on large numbers (e.g., $> 10^4$) of individual cells exposed to the same applied field. Subsequent computation allows determination of the population distributions of macromolecular uptake, resealing, percentage of cells with uptake, average uptake per cell, and indication of relative size or shape changes at a particular time, and also allow the temporal changes of distributions over times of minutes or longer.

5. ELECTROFUSION

Cells or vesicles can fuse with high probability upon application of short duration (e.g., $10^{-5} \lesssim \Delta t \lesssim 10^{-4}$ sec; U reaches $500 < U < 1500$ mV) electric fields before or during contact. Although transient aqueous pore involvement seems likely (Sowers, 1984; Sowers and Lieber, 1986; Sugar *et al.*, 1987), there also is some evidence against their participation (Sowers, 1986; Stenger and Hui, 1986).

6. PRESENT STATUS OF A TRANSIENT AQUEOUS PORE THEORY

Most of the above phenomena have been observed in both artificial bilayers and biological membranes, from which it follows that the fundamental mechanism must involve the bilayer portion of a membrane. Here we adopt the hypothesis of transient aqueous pores, wherein pores are regarded as fundamental, dynamic structures of the bilayer portion of membranes. By assuming the presence of many such pores, it has been possible to quantitatively describe electrical rupture (and the associated critical potential, U_c) (Abidor *et al.*, 1979; Weaver and Mintzer, 1981; Powell and Weaver, 1986), reversible electrical breakdown (and the associated breakdown potential $U_{0,c}$), and other measurable quantities (Weaver *et al.*, 1986; Powell *et al.*, 1986). Further, some features of the transient high-permeability state can be quantitatively described with a modification of the theory (Weaver *et al.*, unpublished), and the presence of many pores can be consistent with known values and the voltage dependence of the membrane capacitance (Weaver *et al.*, 1984a), the diffusive permeability to small solutes (Weaver *et al.*, 1984b), and the high electrical resistance of artificial bilayer membranes at low transmembrane potentials (Weaver *et al.*, 1986).

7. BASIC PROPERTIES OF TRANSIENT AQUEOUS PORES

The cell membrane is generally described in terms of the fluid mosaic membrane model (Singer and Nicolson, 1972). Our basic hypothesis is that transient aqueous pores are to be included in the bilayer portion (Fig. 1), where they are regarded as fundamental structures, which are not static and permanent, but instead dynamic and transient (Litster, 1975; Taupin *et al.*, 1975; Abidor *et al.*, 1979; Petrov *et al.*, 1979; Weaver and Mintzer, 1981; Sugar, 1981; Kashchiev and Exerowa, 1983; Sugar and Neumann, 1984; Powell and Weaver, 1986). Therefore, a pore population at any particular time is described by a probability density function, $n(r,t)$. Individual pores are regarded as

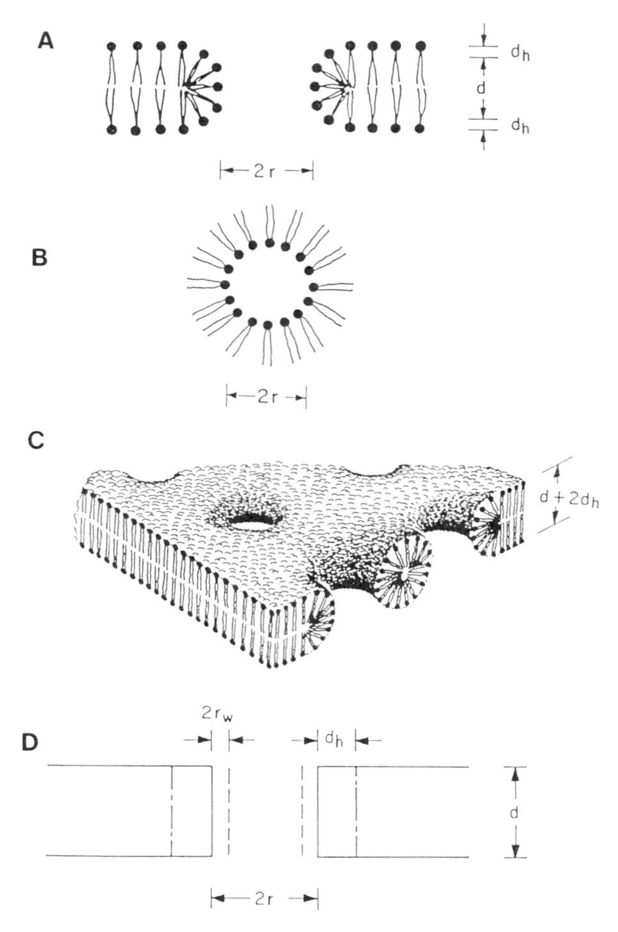

Figure 1. (A) Cross section of a hypothetical transient aqueous pore. The radius is r, the thickness of the insulating hydrocarbon is d, and the thickness of the head group region is d_h. (B) Top view of the pore, at the center of the membrane. (C) View of several pores in an artificial bilayer membrane; inspection suggests that at the highest pore density, the smallest pore separation is approximately $d + 2d_h$. (D) Simple cylindrical pore used in the present theory. An assumed bound monolayer of water of thickness r_w is shown at the pore edge. From Weaver *et al.* (1984b).

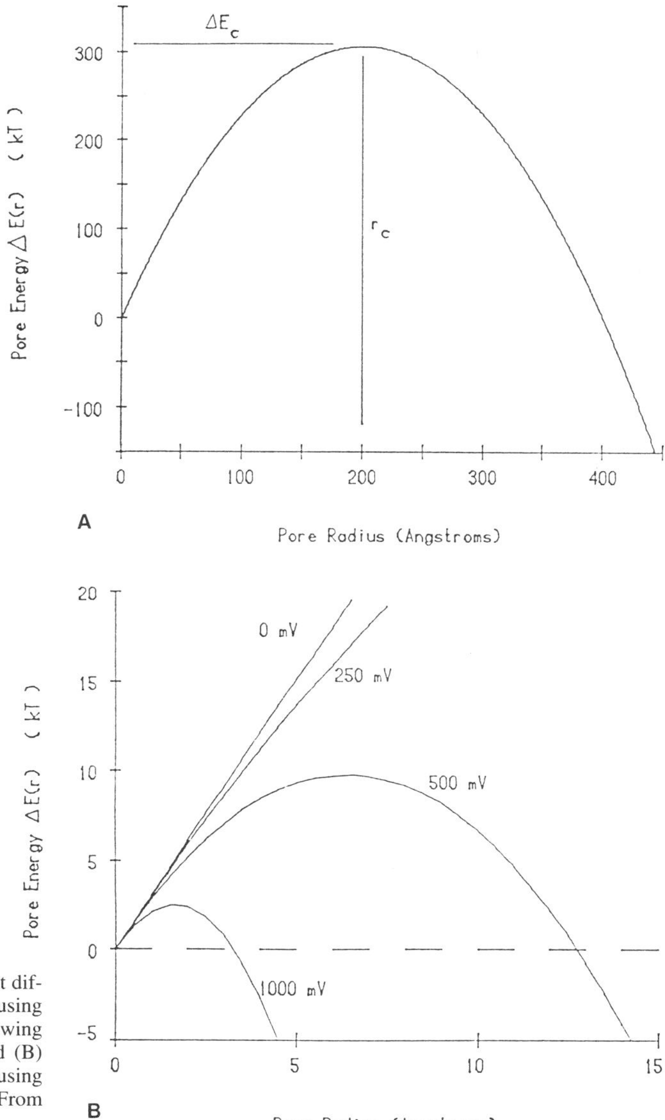

Figure 2. Pore energy, $\Delta E(r)$, at different transmembrane potentials using Eq. (4): (A) ΔE at $U = 0$, showing $r_c(U = 0)$ and $\Delta E_c(U = 0)$, and (B) ΔE at 0, 250, 500, and 1000 mV using different scales for both axes. From Powell and Weaver (1986).

resulting from a varying, microscopic balance between thermal fluctuations and restraining pressures associated with a pore's energy, ΔE (Fig. 2). Although different pore shapes can be considered (Abidor *et al.*, 1979; Petrov *et al.*, 1980; Miller, 1981; Sugar and Neumann, 1984; Pastushenko and Petrov, 1984), we presently use a simple cylindrical geometry (Fig. 1d). Born energy considerations dictate that small pores cannot readily admit ions (Parsegian, 1969; Levitt, 1978; Jordan, 1981; Pastushenko and Chizmadzhev, 1982; Levitt, 1985). Thus, we approximate "switch-on" criteria for determining which pores have significant conduction at large U (Weaver *et al.*, 1986). Pores with low conductance are approximated as electrical capacitors with infinite resistance, whereas conducting pores are regarded as having both resistance, with a steric hindrance factor, and capacitance. In short, a pore is treated as a liquid capacitor which converts the electrical force associated with U into an expanding pressure within the aqueous pore interior (Abidor *et al.*, 1979; Weaver and Mintzer, 1981; Sugar, 1981; Powell and Weaver, 1986). An equivalent, thermodynamic derivation of pore free energies has also been presented (Sugar and Neumann, 1984).

A pressure balance is used to calculate pore creation energy, ΔE, by regarding pore formation as the removal of planar area πr^2 and creation of a cylindrical pore edge of length $2\pi r$, giving

$$\Delta E = 2\pi\gamma r - \pi r^2 \Gamma \tag{1}$$

Typical surface energies are approximately $\Gamma = 1 \times 10^{-3}$ J/m^2, and the edge energies, γ, are believed to be in the range $\gamma = 1$ to 6×10^{-11} J/m (Litster, 1975; Petrov *et al.*, 1980; Chernomordik *et al.*, 1983). Here γ is treated as a constant, even though it is likely a function of r (Petrov *et al.*, 1980; Miller, 1981; Weaver and Mintzer, 1981). The ΔE function for vesicles and cells differs from that for planar bilayers, in that the $\pi r^2 \Gamma$ term is absent (Chizmadzhev, personal communication; Sugar and Neumann, 1984)). The $\pi r^2 \Gamma$ term in ΔE causes metastability in planar membranes, so occurrence of even one critical pore with a radius

$$r > r_c = \frac{\gamma}{\Gamma} = r_c(U=0) \tag{2}$$

can lead to unrestrained expansion and membrane rupture. No significant temperature rise occurs, and the term "nonthermal rupture" is therefore appropriate. The corresponding critical energy is $\Delta E_c = \pi\gamma^2/\Gamma$. Qualitatively, pores would contract to zero radius if there were no expansive pressure. However, random bombardment of the membrane by water molecules generates a fluctuating, microscopic pressure, leading to a distribution of pore radii (Powell, 1984). If a series of fluctuations causes a pore to achieve the critical size, further, unlimited expansion leading to rupture is likely. The rupture probability can be calculated by using a mean membrane lifetime (Abidor *et al.*, 1979). or by estimating an absolute appearance rate for pores with $\Delta E \geq \Delta E_c$ (Weaver and Mintzer, 1981).

8. PORES IN THE PRESENCE OF A TRANSMEMBRANE POTENTIAL

Additional pore energy and an expansive pressure occur due to electrical polarization if $U > 0$. In the present version this is due to the difference in polarization of pore water compared to membrane lipid (Abidor *et al.*, 1979; Weaver and Mintzer, 1981; Sugar and Neumann, 1984; Weaver *et al.*, 1984a; Powell and Weaver, 1986). Permanent dipoles of membrane constituents may also contribute (Petrov *et al.*, 1979; Neumann *et al.*, 1982), but are not yet included. Here we first consider the capacitance, C_1, of a lipid membrane region, which to a good approximation is $C_1 = \epsilon_0 K_l A_1/d$, where the total bilayer thickness is $h = d + 2d_h$ (Fig. 1), $\epsilon_0 = 8.85 \times 10^{-12}$ farad/m, $K_1 = 2.1$ is the dielectric constant of the lipid, and A_1 is the lipid area. The electrical energy is $E_e = CU^2/2$, and the corresponding electrocompressive pressure is

$$P_e = \frac{\partial}{\partial z}\left[\frac{CU^2}{2}\right] = \frac{\epsilon_0 K_l U^2}{2d^2} \qquad z = \text{coordinate normal membrane} \tag{3}$$

The capacitance of an aqueous pore is similar in form, but with area πr^2 and the dielectric constant of water ($K_w = 80 >> K_1$). Aqueous pores store more charge per area, and experience a larger compressive pressure, than lipid regions. This pressure difference results in expansion of the pore in the radial direction, so that as U is increased, larger pore radii are favored.

Quantitatively, ΔE is obtained by calculating the $(\Delta p)dV$ work as a pore expands from zero radius to r (Abidor *et al.*, 1979; Powell and Weaver, 1986):

$$\Delta E(r) = 2\pi\lambda r - \pi r^2[\Gamma + aU^2] \qquad \text{with } a = \frac{\epsilon_0 K_1}{2d}\left[\frac{K_w}{K_l} - 1\right] \tag{4}$$

A thermodynamic approach yields equivalent results (Sugar and Neumann, 1984). The present expression for a differs from an earlier one, wherein constant charge rather than constant transmembrane potential was assumed (Weaver and Mintzer, 1981). Here we assume that U is constant in the vicinity of the pore during size fluctuations driven by thermal fluctuations, and that the same local U is experienced over the entire pore (but see Sugar and Neumann, 1984). Various activation energies, which can be represented by an additional energy barrier at very small r, may be included for the creation of pores (Abidor *et al.*, 1979; Petrov *et al.*, 1980; Neumann *et al.*, 1982; Sugar and Neumann, 1984; Pastushenko and Petrov, 1984). In the simplest version of the theory, presented here, we omit these additional features, and assume the presence of sufficient pores to saturate the membrane. This is essentially a method of providing a normalization factor without explicitly introducing other membrane excitations.

9. KEY FUNCTIONS AND EQUATIONS

The present version of the theory utilizes two differential equations and several functions which describe either observable or nonobservable quantities:

1. $\Delta E(r,U)$, the pore energy of Eq. (1), which is a function of r and the local transmembrane potential, U_p, given below (Powell and Weaver, 1986).
2. $n(r,t)$, the pore probability density function, such that the instantaneous number of pores with radii between r and $r + dr$ is $n(r,t)dr$ (Powell and Weaver, 1986).
3. $U(t)$, the instantaneous transmembrane potential (an observable quantity), which through Eq. (5) is a function of $R(t)$, C, and R_i, a fixed resistance representing the contribution of the bulk bathing electrolyte, electrodes, and pulse generator through which the membrane is charged (Benz *et al.*, 1979; Weaver *et al.*, 1986; Powell *et al.*, 1986).
4. $r^*_{on}(U)$, an approximate but quantitative conduction criterion which sharply divides pores with negligible conduction from those with significant conduction. Pores with $r < r^*_{on}(U)$ are regarded as insulating, those with $r \geq r^*_{on}(U)$ as fully conducting, but with a hindrance factor appropriate to small channels. This allows computation of individual pore conductance, $G_p(a',r)$ (Weaver *et al.*, 1986).
5. $R(r)_{spread}$, the spreading resistance associated with potential drops within the bathing electrolyte near the entrance to each pore, is approximated by $1/\pi r\sigma$, where σ is the bulk electrolyte conductivity, and the total resistance associated with each pore is taken to be $R(r)_{spread} + 1/G_p(a',r)$ (Neuman, 1966;Pastushenko and Chizmadzhev, 1982; Powell *et al.*, 1986).
6. $U_p(t)$, the local value of U experienced by each pore. For nonconducting pores $U_p = U$, but conducting pores have $U_p < U$ because of a voltage divider effect associated with $R(r)_{spread}$ and $1/G_p(a',r)$ (Powell *et al.*, 1986).
7. $R(t)$, the instantaneous membrane resistance (an observable quantity), is obtained by determining its inverse, the pore conductance, $G(t)$, and integrating $n(r,t)[R(r)_{spread} + 1/G_p(a',r)]^{-1}$ from $r^*_{on}(U)$ to infinity (Powell *et al.*, 1986).
8. $P_{d,m}(t)$, the instantaneous diffusive permeability, is calculated using $n(r,t)$ and an estimate of the hindered diffusion through each pore. $P_{dm}(t)$ is computed using Eq. (9), with implicit time dependence through the use of $n(r,t)$ instead of $n(r)$.
9. Q, the "injected charge" (Benz *et al.*, 1979) (an observable quantity), is obtained by integrating the total current through the membrane over the duration of the applied pulse.
10. $U_{0,c}$, the "breakdown potential" (Benz *et al.*, 1979) (an observable quantity), is obtained

by iterative computation of $U(t)$ in order to find U_0, the value of U at the end of the pulse. The maximum attainable value of U_0 is $U_{0,c}$. As in experiments, U_0 first rises and then falls as U_i or the pulse width is increased (Powell *et al.*, 1986).

11. U_c, the critical potential for rupture, is given approximately by Eq. (8).

In addition to these functions, two differential equations are used to provide a self-consistent description of the membrane pores interacting with the external charging source (i.e., pulse generator). The first describes the relation between an applied potential and $U(t)$. Assuming that C is a constant (Chernomordik and Abidor, 1980; Chizmadzhev and Abidor, 1980), and that a square voltage pulse is applied, the differential equation which describes the relation between the applied pulse of amplitude U_i, and $U(t)$ is (Powell *et al.*, 1986)

$$\frac{dU(t)}{dt} + \left[\frac{1}{R(t)C} + \frac{1}{R_i C}\right] U(t) = \frac{U_i}{R_i C} \tag{5}$$

The pulse generator is regarded as an ideal potential source in series with an output resistance, R_i (Benz *et al.*, 1979; Benz and Zimmermann, 1980a; Powell *et al.*, 1986). Thus, $U(t)$ is the result of charging the membrane capacitance, C, through a fixed resistance R_i, and discharging the membrane through the parallel membrane resistance, $R(t)$, which varies dramatically as a function of $n(r,t)$ and $U(t)$. Equation (5) is applicable, with modification, to other pulse shapes, such as the exponential pulse often used in electroporation protocols for introduction of DNA into cells.

The second differential equation describes both the quasi-steady state and dynamic behavior of $n(r,t)$ as $U(t)$, and hence $\Delta E(r,U)$, is changed. This equation is obtained by modifying the equation of Deryagin and Gutop (1962) (Pastushenko *et al.*, 1979; Powell, 1984; Powell and Weaver, 1986; Powell *et al.*, 1986) to yield

$$\frac{\partial n}{\partial t} = D_p \frac{\partial^2 n}{\partial r^2} + u_p \left(\frac{\partial \Delta E}{\partial r}\right) \frac{\partial n}{\partial r} + u_p \left(\frac{\partial^2 \Delta E}{\partial r^2}\right) n \tag{6}$$

The parameters D_p and u_p are the effective diffusivity and mobility, respectively, of a pore's radius, and are related by the Einstein relation $u_p = D_p/kT$.

Equations (5) and (6) can be solved simultaneously by numerical methods. By using the approximate but quantitative descriptions provided by the conduction criterion, r^*_{on}, $R(r)_{spread}$, and $G_p(a',r)$, quantitative descriptions of experimentally observable quantities such as $U(t)$ (Fig. 3), the breakdown potential, $U_{o,c}$ (Figs. 3 and 6), the membrane resistance, $R(t)$ (Fig. 5), the post-pulse decay time constant, RC (Table 1), and the "injected charge," Q (Table 1), can be computed under REB conditions (Powell *et al.*, 1986). Further, by using Eq. (6) in the quasi-steady-state approximation, Eq. (8) is derived, which estimates the critical potential for rupture, U_c. Together these provide a unified, quantitative description of rupture and REB.

Nonobservable quantities which allow additional insight into mechanism can also be calculated. Figure 4 gives $n(r,t)$ during and after REB, and shows that under some conditions a large number of 10- to 20-Å pores are transiently present. These pores readily conduct ions and rapidly discharge the membrane, causing REB, before any pore can expand to $r_c(U = 0)$ and thereby rupture the membrane.

A recent extension of the present version of the theory investigates the effects of various energy barriers associated with pore formation kinetics or with interactions of pores with other membrane structures. This extension yields solutions for $n(r,t)$ which exhibit key features of both REB and the high-permeability state, i.e., a rapid electrical discharge, followed by a thermally activated, slow resealing of the large pores. A single large pore with an extended lifetime appears capable of providing the significant diffusive transport of macromolecules observed during the high-permeability state associated with electroporation.

One approximate treatment of the transient high-permeability state also uses Eq. (9), but with

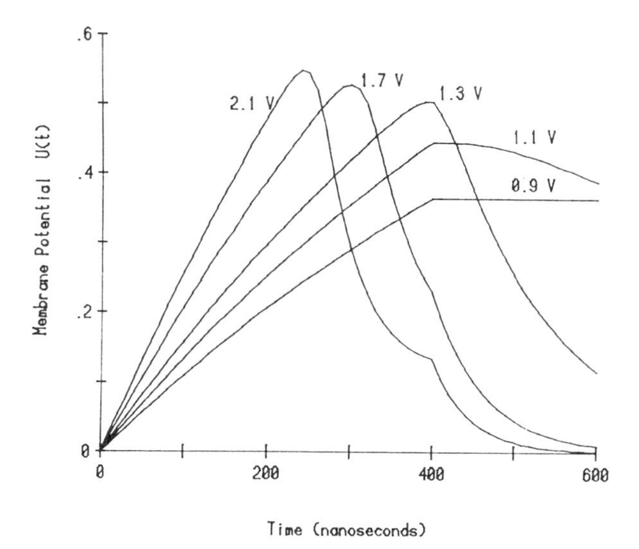

Figure 3. Theoretical behavior of $U(t)$, in volts, versus time, in nanoseconds, for applied 400-nsec pulses with $U_i = 0.9, 1.1, 1.3, 1.7$, and 2.1 V. A planar membrane has area $A_m = 2 \times 10^{-6}$ m²; other parameters are in the text. The theoretical behavior of the $U(t)$ curves is in good agreement with experiments on planar bilayer and cell membranes (Benz *et al.*, 1979; Benz and Zimmermann, 1980a). The 0.9-V pulse is subbreakdown, even though R has dropped from an initial 5×10^9 ohms to about 10^4 ohms at 600 nsec, because the RC time constant for post-pulse decay is too long. Thus, on the time scale of the experiment, $U(t)$ following the 0.9-V pulse is a horizontal line. The potential at the end of the pulse, $U_{o,c}$, is largest for $U_{ij} = 1.3$ *v*. *This maximum value has been defined to be the "Breakdown potential,"* $U_{o,c}$, for REB (Benz *et al.*, 1979). Here the 400-nsec pulse has $U_{o,c} \approx 0.5$ V. Consistent with experiments, U_o decreases for still larger U_i, and R drops even lower, as revealed by the progressively smaller RC decay time constant for post-pulse U. As shown in Fig. 4, there is a high probability that REB occurs without rupture. From Powell *et al.* (1986).

$\Delta E(r)$ modified to include the possibility of proteins or other membrane macromolecules, or macromolecular aggregates, interacting with large transient pores caused by large U. This amounts to the assumption of an additional, hypothetical barrier, over which pores are driven at high U. In this version, REB still occurs, but some large pores can be metastably "trapped" behind the barrier, such that thermal activation provides their slow return. This hypothetical mechanism is consistent with the temperature-dependent resealing time of cell membranes, which can be much longer than that observed in pure artificial bilayers. It is also consistent with the realization that significant diffusive transport of macromolecules can occur through one large pore, i.e., slightly larger than the macromolecular size, if the single pore is held open for seconds to minutes (Weaver *et al.*, unpublished).

For smaller transmembrane potentials with slow time variation, the time dependence of Eq. (5) can be neglected. This yields an approximate quasi-steady-state solution for the pore population given by the following equation, and shown in Fig. 7:

$$n(r) = n(0)e^{-\beta\Delta E(r)} \qquad \text{with } \beta = 1/kT \tag{7}$$

A Boltzmann factor, $e^{-\beta\Delta E}$, governs $n(r)$ to a good approximation provided $r << r_c$. Equation (7) is not a result of using the simple cylindrical pore energy, so that any well-behaved $\Delta E(r)$ should be acceptable.

A more complicated, exact quasi-steady-state solution (not shown) for $n(r)$ can be found and used to determine the rate at which critical pores ($r > r_c$) appear and then continue their expansion to rupture the membrane (Pastushenko *et al.*, 1979; Powell, 1984; Powell and Weaver, 1986). This

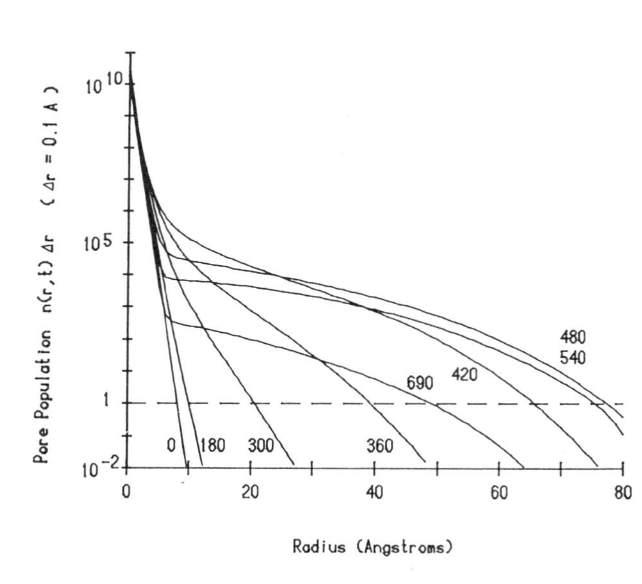

Figure 4. Theoretical behavior of $n(r,t)\Delta r$, the pore population, for different times during and after the 400-nsec pulse of amplitude 1.3 V of Fig. 3. The pore probability density, $n(r,t)$, is multiplied by $\Delta r = 0.1$ Å, and plotted on a logarithmic scale. The number near each curve is the time in nanoseconds. At $t = 0$ the pulse generator instantaneously reaches U_i, but $U(t)$ increases from 0 only as the membrane charges through R_i. Thus, at $t = 0$ the preexisting quasi-steady-state distribution, $n(r)$, occurs. As $U(t)$ then rapidly rises, $n(r,t)$ shifts dramatically toward large pore radii, with particularly rapid changes for pores with $r > r_c(U)$. The expansion of very large pores is slowed by the onset of conduction through such pores, which lowers the value of U at such pores because of the spreading resistance. By the end of the pulse there are many ($\sim 10^5$) pores with $r > 40$ Å. As $U(t)$ rapidly decays after the pulse, some pores continue to expand, with many reaching radii of ~80 Å before $n(r,t)$ begins to collapse (e.g., 690 nsec) to the initial state. The dashed line corresponds to the average presence of one pore within $\Delta r = 0.1$ Å. There is a high probability that REB occurs without rupture, because there is a low probability that a single pore will reach r_c $(U = 0) = 200$ Å. In this way, REB can occur reversibly, without rupture. From Powell *et al.* (1986).

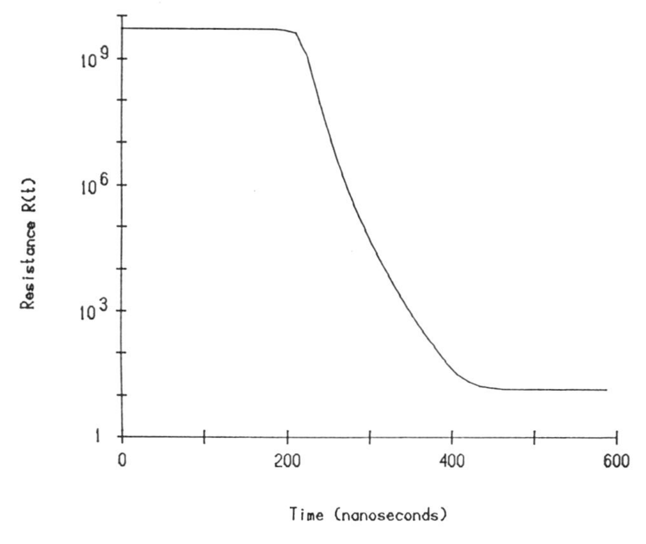

Figure 5. Theoretical behavior of $R(t)$, in ohms, for REB caused by a 400-nsec pulse of amplitude $U_i = 1.3$ V as shown in Fig. 3. Initially, $R = 5 \times 10^9$ ohms, so that REB results in an eight order of magnitude drop within about 200 nsec. and the final value is only 13 ohms. Still larger pulse amplitudes result in still lower final values of R (see Table 1), in agreement with experiments. The present version does not, however, describe the recovery of $R(t)$ for $t \gtrsim 450$ nsec, due to the simplifying computational assumption that R can only decrease (Powell *et al.*, 1986).

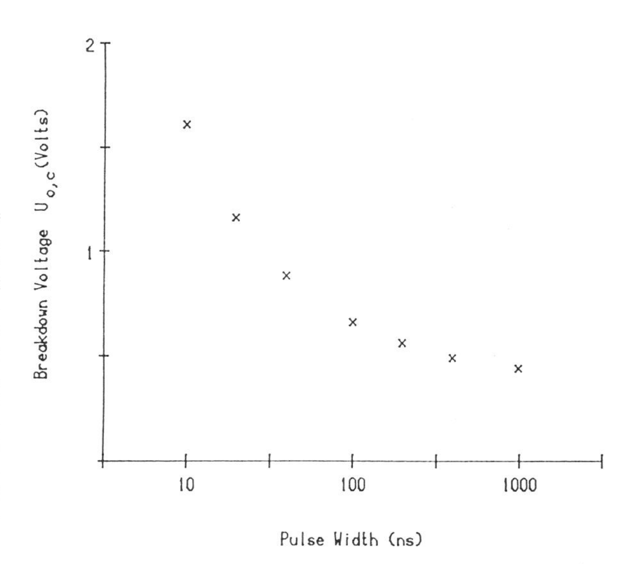

Figure 6. Theoretical behavior of the breakdown potential, $U_{o,c}$ (Benz *et al.*, 1979), in volts, as a function of the pulse width, in nanoseconds. For very short pulses (e.g., < 100 nsec), $R(t)$ can continue to decrease significantly after the pulse has ended, due to futher changes in $n(r,t)$ as shown in Fig. 4. For this reason, simple exponential decay extrapolation of experimental potentials (Benz *et al.*, 1979) can yield values of U_o which are too low, and which differ from theoretical values. In general, however, the theory is in good agreement with experiment (Powell *et al.*, 1986).

in turn allows estimation of the critical rupture potential, U_c, at which membrane bursting has a high probability of occurring. Alternatively, a simpler closed form expression for U_c is obtained by using the absolute rate approximation (Weaver and Mintzer, 1981; Powell and Weaver, 1986):

$$U_c \approx a^{-1/2} \left[\frac{\pi \beta \gamma^2}{\ln \left[\frac{\nu_0' V_B}{\nu_c} \right]} - \Gamma \right]^{1/2} \tag{8}$$

where a is given by Eq. (4), V_B is the volume of the membrane, ν_0' is an attempt rate, and ν_c is the rate of appearance and expansion of critical pores which is associated with exhibiting rupture (e.g., $\nu_c = 1\ \text{sec}^{-1}$).

10. LOW-POTENTIAL BEHAVIOR

The existence of transient aqueous pores can be consistent with the known behavior of bilayer membranes at low U. Specifically, the membrane electrical capacitance, $C(U)$, has been shown to

Table 1. Theoretical Values of Postpulse R, the Discharge Time Constant, $\tau = RC$, using $C = 1.3 \times 10^{-8}$ Farad and $R = R(600\ \text{nsec})$, and Q, the "Injected Charge," for Several U_i[a]

U_i (volts)	R(400 nsec) (ohms)	R(600 nsec) (ohms)	$\tau = RC$ (nsec)	Q (coulomb)
0.9	3×10^6	8×10^4	4.0×10^7	3.5×10^{-9}
1.1	3×10^3	83	4.0×10^4	4.3×10^{-9}
1.3	38	13	510	5.1×10^{-9}
1.7	6.4	6.4	85	6.9×10^{-9}
2.1	4.7	4.7	63	9.0×10^{-9}

[a]Note that once REB occurs ($U_i = 1.1$ volt), τ decreases as U_i is further increased.

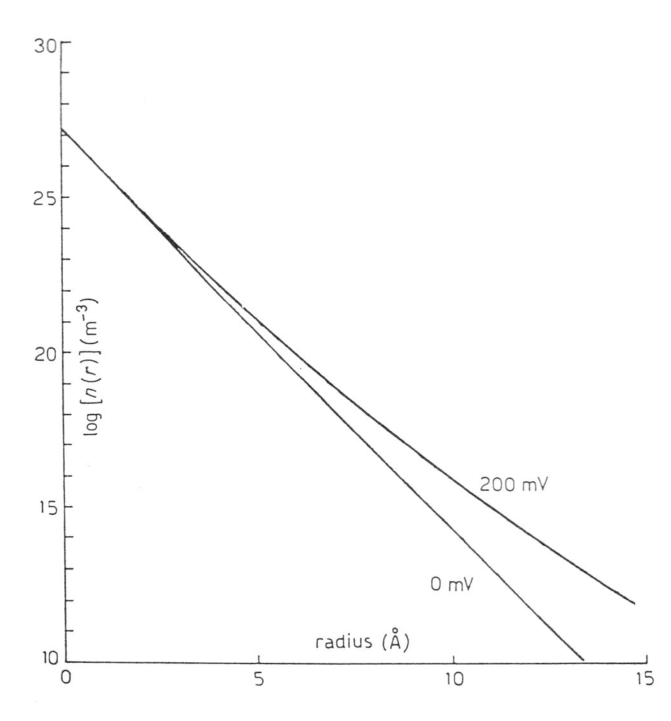

Figure 7. The quasi-steady-state pore probability density, $n(r)$, obtained from Eq. (7) at $U = 0$ and 200 mV. The average number of pores with radii between r and $r + dr$ is $n(r)$ dr. Note that $\log[n(r)]$ is plotted, because of the rapid decrease in $n(r)$ with increasing r. The coupling of electric fields through changes in ΔE (r, U) produces the greatest fractional change in $n(r)$ for the largest pores (Weaver *et al.*, 1984a; Powell, 1984; Powell and Weaver, 1986). Over the approximate range $0 \lesssim U \lesssim 200$ mV, a significant increase in $P_{d,m}$ is predicted, because pore lifetimes are long enough to use the approximation of steady-state diffusion. The dominant contribution is by the larger pores, which, as shown here, exhibit the largest fractional change (Weaver *et al.*, 1984b).

have the functional dependence $C(U) = C(0)[1 + \alpha U^2]$ to a good approximation. Here $C(0)$ is the membrane capacitance at $U = 0$, α is on the order of 10^{-2} volt^{-2}, and the characteristic response time following changes in U is on the order of 10^{-8} sec (Weaver *et al.*, 1984a). This agrees with solvent-free bilayer membrane experiments wherein $\alpha \approx 2 \times 10^{-2}$ V^{-2}, and a response within 10^{-6} sec is observed (Alvarez and Latorre, 1978). A pore-saturated membrane can also be consistent with known diffusive permeabilities at $U = 0$. The quasi-steady-state distribution, $n(r)$, of Eq. (7) predicts that most pores are too small to allow transport of even small solutes. Calculation of the potential-dependent pore-mediated diffusive permeability, $P_{d,m}(U)$, is more complicated than that of $C(U)$, i.e.,

$$P_{d,m}(U) = \frac{D_s}{A_m h} \int_{(2r_w + r_s)}^{r_c} \pi(r - 2r_w)^2 H(r) n(r) dr \tag{9}$$

Here the quasi-steady-state pore probability density, $n(r)$, is implicitly a function of U, D_s is the solute diffusion constant in water, r_w is the effective radius of a bound water molecule, r_s is the effective radius of a solute molecule, and $H(r) = H_F\,(r - 2r_{w,}r_s)$ is the steric factor for hindered diffusion (Bungay and Brenner, 1973). The integration involves the function, $n(r)$, which describes a large range of pore sizes (Fig. 6), not one size as often assumed by others. However, the rapid decrease in $n(r)$ with increasing r results in most of the predicted pore-mediated diffusive flux occurring through pores with radii just larger than $r = 2r_w + r_s$. Computation of $P_{d,m}(0)$ yields values less than experimental values, and shows that the presence of many pores does not necessarily render the membrane a "leaky sieve." The computation of $P_{d,m}(U)$ predicts a significant fractional increase in permeability as U is increased, but the experiments which would test this prediction have apparently not been conducted (Weaver *et al.*, 1984b).

Transport of charged molecules is more complicated, because both hindered transport (Bungay and Brenner, 1973), and Born energy changes must be considered (Parsegian, 1969; Jordan, 1981; Levitt, 1985). Presently we estimate Born energies governing transport of the ubiquitous small ions, and obtain an estimate of $r^*_{on}(U)$, the size pore that will conduct or "switch on." Using the quasi-steady-state $n(r)$ distribution, it is found that the pore population conducts few ions at low U. Using "worst case" conditions, we estimate that 1 mm^2 bilayer membrane does not conduct significantly until U exceeds either ~270 mV ("low" case) or ~510 mV ("high" case) (Weaver *et al.*, 1986). This is consistent with the known high resistance of artificial bilayers at low U, and the low permeability of cell membranes to charged solutes. The numerical solutions describing REB reveal that the large difference in these conduction thresholds has an insignificant effect on $U(t)$, $R(t)$, and other quantities. However, a more accurate quantitative treatment of pore conduction at lower U is needed.

11. CONCLUSIONS

Many features of electroporation can be quantitatively described by a unified theory which invokes transient aqueous pores that preferentially expand in response to increased U. These phenomena include electrical rupture, reversible electrical breakdown, and basic features of the high-permeability state. More specifically, the present theory gives successful descriptions of: (1) U_c, the critical potential for rupture of planar bilayer membranes, (2) $U_{o,c}$, the breakdown potential of REB, (3) the crossing of $U(t)$ curves following REB, (4) decay time constants following REB, (5) the magnitude of the injected charge associated with REB, (6) avoidance of rupture during REB with high, but not unity, probability, (7) a rise and fall of U_o as the pulse amplitude U_i is increased for fixed pulse width, (8) a rise and fall of U_o as the pulse width is increased for fixed pulse amplitude, (9) high membrane resistance at low U, and (10) the potential dependence of the membrane capacitance.

However, it is not yet possible to use experimental results to distinguish between the present hypothesis that a large population of pores is always present, expanding rapidly in response to large potentials, and an alternative hypothesis that pores are rapidly created by large potentials, followed immediately by rapid pore expansion. Further, despite the success of the present theory, several significant deficiencies are notable, given here in the form of needed improvements:

1. A quantitative, time-dependent treatment of pore creation is needed, so that $n(r,t)$ can be determined without the present assumption of a "pore-saturated" membrane. A successful theory of pore formation would quantitatively describe the energetics and kinetics of pore evolution in both planar and cell membranes, beginning from an initial, perhaps pore-free, state.
2. There is also a critical need for a basic treatment of pore-membrane molecule interactions during the recovery or resealing phase which follows REB, and which may govern the slow, highly temperature-dependent recovery from the high-permeability state.
3. A "continuous conduction" version which avoids the use of the sharp conduction criteria (r^*_{on}), and involves appropriate Born energies and hindered transport to account for conduction of all pores.
4. An explicit recognition and treatment of other coexisting membrane excitations is needed. Other excitations presumably occupy a major area of the membrane, which is presently allocated to very small pores. In this case, normalization of the $n(r,t)$ function would no longer be based on the assumption of a pore-saturated membrane. Instead, excitations such as the "breathing modes" suggested by Bach and Miller would occupy much of the bilayer portion of the membrane (Bach and Miller, 1980). The present assumption that the many very small pores are real is a major limitation; pores predicted to have radii smaller than atomic dimensions cannot be taken seriously. Instead, another form of normalization of the function $n(r,t)$ should be sought, in which it is explicitly recognized that other excitations can occupy the area of the membrane which is allocated by the present theory to very small pores. Initial computations wherein the number of total pores is reduced by a factor of 10^1 to 10^3 show that the basic features of the theory are preserved.

Presently, however, the theory quantitatively describes several key features of rupture and reversible electrical breakdown, and also appears capable, with modification, of describing the transient high-permeability state. Thus, while several important modifications of the theory appear both warranted and possible, the success of the theory suggests that transient aqueous pores should be seriously considered as fundamental entities in bilayer membranes, and as key participants in electrical phenomena, particularly electroporation.

NOTE ADDED IN PROOF

A significantly improved version of the present theory has been recently completed (Barnett and Weaver, unpublished data).

REFERENCES

Abidor, I. G., Arakelyan, V. B., Chernomordik, L. V., Chizmadzhev, Y. A., Pastushenko, V. F., and Tarasevich, M. R., 1979, Electric breakdown of bilayer membranes. I. The main experimental facts and their qualitative discussion, *Bioelectrochem. Bioenerg.* **6:**37–52.

Abidor, I. G., Chernomordik, L. V., Sukharev, S. I., and Chizmadzhev, Y. A., 1982, The reversible electrical breakdown of bilayer lipid membranes modified by uranyl ions, *Bioelectrochem. Bioenerg.* **9:**141–148.

Alvarez, O., and Latorre, R., 1978, Voltage-dependent capacitance in lipid bilayers made from monolayers, *Biophys. J.* **21:**1–17.

Auer, D., Brandner, G., and Bodemer, W., 1976, Dielectric breakdown of the red blood cell membrane and uptake of SV 40 DNA and mammalian cell RNA, *Naturwissenschaften* **63:**391.

Bach, D., and Miller, I. R., 1980, Glyceryl monooleate black lipid membranes obtained from squalene solutions, *Biophys. J.* **29:**183–188.

Benz, R., and Zimmermann, U., 1980a, Relaxation studies on cell membranes and lipid bilayers in the high electric field range, *Bioelectrochem. Bioenerg.* **7:**723–739.

Benz, R., and Zimmermann, U., 1980b, Pulse-length dependence of the electrical breakdown in lipid bilayer membranes, *Biochim. Biophys. Acta* **597:**637–642.

Benz, R., and Zimmermann, U., 1981, The resealing process of lipid bilayers after reversible electrical breakdown, *Biochim. Biophys. Acta* **640:**169–178.

Benz, R., Beckers, F., and Zimmermann, U., 1979, Reversible electrical breakdown of lipid bilayer membranes: A charge-pulse relaxation study, *J. Membr. Biol.* **48:**181–204.

Bliss, J. G., Harrison, G. I., Mourant, J. R., Powell, K. T., and Weaver, J. C., 1988, Electroporation: The population distribution of macromolecular uptake and shape changes in red blood cells following a single 50 μs square wave pulse, *Bioelectrochem. Bioenerg.* **20:**57–71.

Bungay, P. M., and Brenner, H., 1973, The motion of a closely fitting sphere in a fluid-filled tube, *Int. J. Multiphase Flow* **1:**25.

Chernomordik, L. V., and Abidor, I. G., 1980, The voltage-induced local defects in unmodified BLM, *Bioelectrochem. Bioenerg.* **7:**617–623.

Chernomordik, L. V., Sukharev, S. I., Abidor, I. G., and Chizmadzhev, Y. A., 1982, The study of the BLM reversible electrical breakdown mechanism in the presence of $UO^{2+}{}_2$, *Bioelectrochem. Bioenerg.* **9:**149–155.

Chernomordik, L. V., Sukharev, S. I., Abidor, I. G., and Chizmadzhev, Y. A., 1983, Breakdown of lipid bilayer membranes in an electric field, *Biochim. Biophys. Acta* **736:**203–213.

Chizmadzhev, Y. A., and Abidor, I. G., 1980, Bilayer lipid membranes in strong electric fields, *Bioelectrochem. Bioenerg.* **7:**83–100.

Chu, G., Hayakawa, H., and Berg, P., 1987, Electroporation for the efficient transfection of mammalian cells with DNA, *Nucleic Acids Res.* **15:**1311–1326.

Coster, H. G. L., and Zimmermann, U., 1975, The mechanism of electrical breakdown in the membranes of Valonia utricularis, *J. Membr. Biol.* **22:**73–90.

Crowley, J. M., 1973, Electrical breakdown of biomolecular lipid membranes as an electromechanical instability, *Biophys. J.* **13:**711–724.

Deryagin, B. V., and Gutop, Y. V., 1962, Theory of the breakdown (rupture) of free films, *Kolloidn. Zh.* **24:** 370–374.

El-Mashak, E. M., and Tsong, T. Y., 1985, Ion selectivity of temperature-induced and electric field induced pores in dipalmitoylphosphatidylcholine vesicles, *Biochemistry* **24:**2884–2888.

Fromm, M. E., Taylor, L. P., and Walbot, V., 1986, Stable transformation of maize after gene transfer by electroporation, *Nature* **319:**791–793.

Hashimoto, H., Morikawa, H., Yamada, H., and Kimura, A., 1985, A novel method for transformation of intact yeast cells by electroinjection of plasmid DNA, *Appl. Microbiol. Biotechnol.* **21:**336–339.

Jordan, P. C., 1981, Energy barriers for passage of ions through channels: Exact solution of two electrostatic problems, *Biophys. Chem.* **13:**203–212.

Kashchiev, D., and Exerowa, D., 1983, Bilayer lipid membrane permeation and rupture due to hole formation, *Biochim. Biophys. Acta* **732:**133–145.

Kinosita, K., and Tsong, T. Y., 1977a, Voltage-induced pore formation and hemolysis of human erythrocytes, *Biochim. Biophys. Acta* **471:**227–242.

Kinosita, K., Jr., and Tsong, T. Y., 1977b, Formation and resealing of pores of controlled sizes in human erythrocyte membrane, *Nature* **268:**438–441.

Kinosita, K., Jr., and Tsong, T. Y., 1977c, Hemolysis of human erythrocytes by a transient electric field, *Proc. Natl. Acad. Sci. USA* **74:**1923–1927.

Levitt, D. G., 1978, Electrostatic calculations for an ion channel. I. Energy and potential profiles and interactions between ions, *Biophys. J.* **22:**209–219.

Levitt, D. G., 1985, Strong electrolyte continuum theory solution for equilibrium profiles, diffusion limitation, and conductance in charged ion channels, *Biophys. J.* **48:**19–31.

Lindner, P., Neumann, E., and Rosenheck, K., 1977, Kinetics of permeability changes induced by electric impulses in chromaffin granules, *J. Membr. Biol.* **32:**231–254.

Litster, J. D., 1975, Stability of lipid bilayers and red blood cell membranes, *Phys. Lett.* **53A:**193–194.

MacNeil, D. J., 1987, Introduction of plasmid DNA into *Streptomyces lividans* by electroporation, *FEMS Microbiol. Lett.* **42:**239–244.

Mehrle, W., Zimmermann, U., and Hampp, R., 1985, Evidence for asymmetrical uptake of fluorescent dyes through electro-permeabilized membranes of Avena mesophyll protoplasts, *FEBS Lett.* **185:**89–94.

Miller, I. R., 1981, Structural and energetic aspects of charge transport in lipid layers and biological membranes, in: *Topics in Bioelectrochemistry and Bioenergetics* (G. Milazzo, ed.), Wiley, New York, pp. 161–224.

Neumann, E., and Rosenheck, K., 1972, Permeability changes induced by electric impulses in vesicular membranes, *J. Membr. Biol.* **10:**279–290.

Neumann, E., Schaefer-Ridder, M., Wang, Y., and Hofschneider, P. H., 1982, Gene transfer into mouse lyoma cells by electroporation in high electric fields, *EMBO J.* **1:**841–845.

Newman, J., 1966, Resistance for flow of current to a disk, *J. Electrochem. Soc.* **113:**501–502.

Parsegian, V. A., 1969, Energy of an ion crossing a low dielectric membrane: Solutions to four relevant electrostatic problems, *Nature* **221:**844–846.

Pastushenko, V. F., and Chizmadzhev, Y. A., 1982, Stabilization of conducting pores in BLM by electric current, *Gen. Physiol. Biophys.* **1:**43–52.

Pastushenko, V. F., and Petrov, A. G., 1984, Electro-mechanical mechanism of pore formation in bilayer lipid membranes, in: *Proceedings of the Seventh School on Biophysics of Membrane Transport,* Poland, pp. 70–91.

Pastushenko, V. F., Chizmadzhev, Y. A., and Arakelyan, V. B., 1979, Electric breakdown of bilayer membranes. II. Calculation of the membrane lifetime in the steady-state diffusion approximation, *Bioelectrochem. Bioenerg.* **6:**53–62.

Petrov, A. G., Seleznev, S. A., and Derzhanski, A., 1979, Principles and methods of liquid crystal physics applied to the structure and functions of biological membranes, *Acta Phys. Pol.* **A55:**385–405.

Petrov, A. G., Mitov, M. D., and Derzhanski, A. I., 1980, Edge energy and pore stability in bilayer lipid membranes, in: *Advances in Liquid Crystal Research and Applications* (L. Bata, ed.), Pergamon Press, Elmsford, NY, pp. 695–737.

Potter, H., Weir, L., and Leder, P., 1984, Enhancer-dependent expression of human κ immunoglobulin genes introduced into mouse pre-B lymphocytes by electroporation, *Proc. Natl. Acad. Sci. USA* **81:**7161–7165.

Powell, K. T., 1984, A statistical model of transient aqueous pores in bilayer lipid membranes, M.S. thesis, Massachusetts Institute of Technology.

Powell, K. T., and Weaver, J. C., 1986, Transient aqueous pores in bilayer membranes: A statistical theory, *Bioelectrochem. Bioenerg.* **15:**211–227.

Powell, K. T., Derrick, E. G., and Weaver, J. C., 1986, A quantitative theory of reversible electrical breakdown, *Bioelectrochem. Bioenerg.* **15:**243–255.

Riggs, C. D., and Bates, G. W., 1986, Stable transformation of tobacco by electroporation: Evidence for plasmid concatenation, *Proc. Natl. Acad. Sci. USA* **83:**5602–5606.

Schwister, K., and Deuticke, B., 1985, Formation and properties of aqueous leaks induced in human erythrocytes by electrical breakdown, *Biochim. Biophys. Acta* **816:**332–348.

Serpersu, E. H., Kinosita, K., and Tsong, T. Y., 1985, Reversible and irreversible modification of erythrocyte membrane permeability by electric field, *Biochim. Biophys. Acta* **812:**779–785.

Singer, S. J., and Nicolson, G. L., 1972, The fluid mosaic model of the structure of cell membranes, *Science* **175:**720–731.

Sowers, A. E., 1984, Characterization of electric field-induced fusion in erythrocyte ghost membranes, *J. Cell Biol.* **99:**1989–1996.

Sowers, A. E., 1986, A long-lived fusogenic state is induced in erythrocyte ghosts by electric pulses, *J. Cell Biol.* **102:**1358–1362.

Sowers, A. E., and Lieber, M. R., 1986, Electropore diameters, lifetimes, numbers, and locations in individual erythrocyte ghosts, *FEBS Lett.* **205:**179–184.

Stenger, D. A., and Hui, S. W., 1986, Kinetics of ultrastructure changes during electrically-induced fusion of human erythrocytes, *J. Membr. Biol.* **93:**43–53.

Stopper, H., Jones, H., and Zimmermann, U., 1987, Large scale transfection of mouse L-cells by electropermeabilization, *Biochim. Biophys. Acta* **900:**38–44.

Sugar, I. P., 1981, The effects of external fields on the structure of lipid bilayers, *J. Physiol. Paris* **77:**1035–1042.

Sugar, I. P., and Neumann, E., 1984, Stochastic model for electric field-induced membrane pores: Electroporation, *Biophys. Chem.* **19:**211–225.

Sugar, I. P., Forster, W., and Neumann, E., 1987, Model of cell electrofusion: Membrane electroporation, pore coalescence and percolation, *Biophys. Chem.* **26:**321–335.

Sukharev, S. I., Chernomordik, L. V., Abidor, I. G., and Chizmadzhev, Y. A., 1982, Effects of UO^{2+}_{2} ions on the properties of bilayer membranes, *Bioelectrochem. Bioenerg.* **9:**133–140.

Taupin, C., Dvolaitzky, M., and Sauterey, C., 1975, Osmotic pressure induced pores in phospholipid vesicles, *Biochemistry* **14:**4771–4775.

Teissie, J., and Tsong, T. Y., 1981, Electric field induced transient pores in phospholipid bilayer vesicles, *Biochemistry* **20:**1548–1554.

Toneguzzo, F., and Keating, A., 1986, Stable expression of selectable genes introduced into human hematopoietic stem cells by electric field-mediated DNA transfer, *Proc. Natl. Acad. Sci. USA* **83:**3496–3499.

Tur-Kaspa, R., Teicher, L., Levine, B. J., Skoultchi, A. I., and Sharfitz, D. A., 1986, Use of electroporation to introduce biologically active foreign genes into primary rat hepatocytes, *Mol. Cell Biol.* **6:**716–718.

Weaver, J. C., and Mintzer, R. A., 1981, Decreased bilayer stability due to transmembrane potentials, *Phys. Lett.* **86A:**57–59.

Weaver, J. C., Powell, K. T., Mintzer, R. A., Ling, H., and Sloan, S. R., 1984a, The electrical capacitance of bilayer membranes: The contribution of transient aqueous pores, *Biolectrochem. Bioenerg.* **12:**393–404.

Weaver, J. C., Powell, K. T., Mintzer, R. A., Sloan, S. R., and Ling, H., 1984b, The diffusive permeability of bilayer membranes: The contribution of transient aqueous pores, *Bioelectrochem. Bioenerg.* **12:**405–412.

Weaver, J. C., Mintzer, R. A., Ling, H., and Sloan, S. R., 1986, Conduction onset criteria for transient aqueous pores and reversible electrical breakdown in bilayer membranes, *Bioelectrochem. Bioenerg.* **15:** 229–241.

Weaver, J. C., Bliss, J. G., Harrison, G. I., Mourant, J. R., and Powell, K. T., 1987, Electroporation in individual cells: Measurements using light scattering and fluorescence by flow cytometry, in: *Proceedings, 9th Annual IEEE Engineering in Medicine and Biology Society,* Boston, pp. 708–709.

Weaver, J. C., Harrison, G. I., Bliss, J. G., Mourant, J. R., and Powell, K. T., 1988, Electroporation: High frequency of occurrence of a transient high-permeability state in erythrocytes and intact yeast, **229:**30–34.

Zimmermann, U., Schultz, J., and Pilwat, G., 1973, Transcellular ion flow in Escherichia coli B and electrical sizing of bacterias, *Biophys. J.* **13:**1005–1013.

Zimmermann, U., Pilwat, G., and Riemann, F., 1974, Dielectric breakdown of cell membranes, *Biophys. J.* **14:** 881–899.

Zimmermann, U., Pilwat, G., and Riemann, F., 1975, Preparation of erythrocyte ghosts by dielectric breakdown of the cell membrane, *Biochim. Biophys. Acta* **375:**209–219.

Zimmermann, U., Riemann, F., and Pilwat, G., 1976a, Enzyme loading of electrically homogeneous human red blood cell ghosts prepared by dielectric breakdown, *Biochim. Biophys. Acta* **436:**460–474.

Zimmermann, U., Pilwat, G., Holzapfel, C., and Rosenheck, K., 1976b, Electrical hemolysis of human and bovine red blood cells, *J. Membr. Biol.* **30:**135–152.

Chapter 8

Leaks Induced by Electrical Breakdown in the Erythrocyte Membrane

Bernhard Deuticke and Karl Schwister

1. INTRODUCTION

Among animal cells, the erythrocyte has played a very important role in the study of the *electrical breakdown** of biological membranes. Much of the original data on this process stem from erythrocytes. This is due to the ready availability and the easy handling of this model system as well as the easy means to detect the formation of membrane leaks by virtue of their final consequence, the release of hemoglobin due to colloid-osmotic lysis. Moreover, the erythrocyte is the simplest system that can be envisaged for studying electrical breakdown of animal cell membranes, since only one membrane surrounding a macroscopically homogeneous cytoplasmic space is involved. Finally, there is probably no other eukaryotic cell for which so many details concerning the membrane constituents, their three-dimensional organization, their biological function, and their overall dynamics and transport properties are already known.

There are numerous studies on permeability changes induced in erythrocytes by electrical breakdown (Sale and Hamilton, 1968; Zimmermann *et al.*, 1974, 1976; Riemann *et al.*, 1975; Kinosita and Tsong, 1977a,b,c; see Zimmermann, 1982, for a review of the earlier literature). However, these did not provide the details required for a molecular description of the events occurring when erythrocytes are exposed to a high-voltage pulse. Moreover, the properties of the electrical leak causing colloid-osmotic lysis were not always unambiguously distinguished from those of the leaks caused by colloid-osmotic lysis. Recent results from the authors' laboratory (Schwister, 1985; Schwister and Deuticke, 1985; Dressler *et al.*, 1983) have provided information that may be helpful in this respect. These results and the methods used to obtain them are reviewed in what follows.

*This term is used by us synonymously with the term ''electroporation'' preferred by other contributors to this volume.

Bernhard Deuticke and Karl Schwister • Department of Physiology, Medical Faculty, RWTH Aachen, D-5100 Aachen, Federal Republic of Germany.

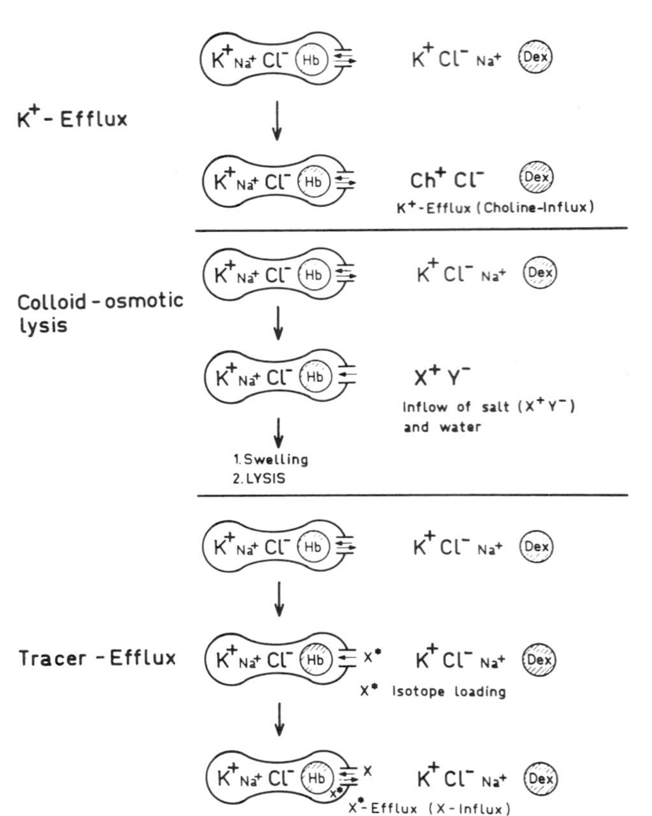

Figure 1. Approaches for quantifying leak permeabilities in electroporated erythrocytes. For details see text.

2. METHODS

2.1. Formation of Leaks and Assessment of Leak Permeability

Basically, the extent of leakiness induced by a field pulse was assessed by three approaches (Fig. 1). In all three approaches, washed erythrocytes are first suspended at a hematocrit of 30% in a medium containing (mM) KCl (115), NaCl (25), Dextran 4 (30), imidazole (1), pH 7.4. Ten milliliters of this suspension is cooled to 0°C, filled into a discharge chamber originally developed by Riemann *et al.* (1975), and exposed to a brief exponentially decaying pulse (τ = 1–40 μsec) of high voltage (3–20 kV/cm) by means of a discharge equipment with a spark gap as a switch. The time constant τ of the pulse is derived from RC, where R is the ohmic resistance of the cell suspension and C is the capacitance of the storage capacitors. After the discharge, the cells are kept strictly at 0–2°C until further use. Under these conditions, resealing of the cells is essentially prevented. Swelling and lysis are suppressed by the dextran added as a colloid-osmotic protectant, which balances the osmotic drag of hemoglobin and other impermeant cellular constitutents (e.g., organic phosphates, GSH).

2.1.1. Approach I: Measurement of K^+ Efflux by an Ion-Sensitive Electrode

The perforated cells are packed in a refrigerated centrifuge (12,000 *g*) and then injected, by a device based on a commercial fluid dispenser, into 50 volumes of ice-cold imidazole-buffered medium containing a 1:1 electrolyte not interfering with potentiometric K^+ measurement, e.g., choline chloride or LiCl, and Dextran 4 (30 mM). K^+ release into the medium is followed continuously by a combined K^+-sensitive glass electrode. K^+ contamination of the suspension medium by loss of K^+ from the reference electrode should be prevented by using an electrode with an interposed bridge compartment containing tris sulfate. The measured voltage and its time-dependent changes are recorded by a pen-writer but may also be stored on a computer. Changes of voltage (= increase of extracellular K^+) are followed to about 90–95% completion. The equilibrium level of K^+ required for evaluation of the data in terms of pseudo-first-order kinetics is obtained by lysing the cells with a small amount (100 μl) of Triton X-100. At the low hematocrit (2–3%) and in view of the response of all the cells in the range of field strengths and pulse lengths applied, total K^+ can be set equal to extracellular K^+ at equilibrium. The voltages are converted into K^+ concentrations using appropriate calibration curves. After suitable correction for background, rate coefficients are obtained by evaluation of the data in terms of two-compartment kinetics by the equation

$$\ln (1 - K_t^+ / K_\infty^+) = -kt$$

2.1.2. Approach II: Measurement of Rates of Hemolysis

2.1.2a. Discontinuous Procedure. The permeability of leaky erythrocytes to ions or salts can be derived from the time course of colloid-osmotic hemolysis occurring upon suspending cells in electrolyte solutions not containing impermeable protectants (Deuticke *et al.*, 1984). One hundred microliters of packed, electroporated cells obtained as described above is squirted rapidly into 6 ml of isotonic (300 mosmol/liter) salt solutions at 0°C. After various time intervals, the extent of lysis is determined by measuring the hemoglobin content in the supernatant and relating this value to the total hemoglobin content of the suspension.

2.1.2b. Continuous Procedure. Rates of rapid hemolysis can be be measured conveniently by following the changes of light-scattering or absorbance of a very dilute cell suspension during hemolysis. Fifty microliters of a cell suspension subjected to an electric pulse as described above is mixed rapidly into a photometric cuvette containing 2.5 ml of isotonic solution at 0°C. The suspensions are stirred magnetically and changes of absorbance at 700 nm registered. The change of the signal is proportional to the extent of hemolysis.

Both techniques provide a half time of hemolysis. The reciprocal of this value can serve as a rate coefficient. Provided that additional information is available on normal volume and surface area, as well as the critical hemolytic volume of the cell, a salt permeability may also be calculated using a derivation introduced by Jacobs (1952).

2.1.3. Approach III: Measurement of Tracer Fluxes

Leak permeabilities on an absolute scale can be obtained by measuring tracer fluxes. Perforated cells are loaded with labeled test-solutes in a medium containing colloid-osmotic protectant (0°C, 2 hr). The loaded cells are washed once very rapidly to remove extracellular radioactivity, and resuspended in tracer-free medium at 0°C. The increase of supernatant radioactivity is followed and rate coefficients are calculated (following the principles outlined for Approach I).

2.2. Measurement of Transbilayer Mobility of Phospholipid Analogues

In order to establish local perturbations of the membrane lipid domain in chemically perturbed erythrocytes, measurements of the transbilayer mobility of phospholipids and of suitable analogues of these compounds have proven useful (Bergmann *et al.*, 1984a,b). This technique can also be applied to cells subjected to electric breakdown (Dressler *et al.*, 1983).

Briefly, a labeled amphiphile (e.g., a lysophospholipid) is first incorporated into the outer leaf of the membrane lipid bilayer. Its slow reorientation to the inner leaf is then quantified by following the time-dependent decrease of extractability of the incorporated radioactivity by bovine serum albumin. Serum albumin selectively removes such amphiphiles from the outer layer of the erythrocyte membrane, provided the transbilayer mobility is not too high. This technique can be applied to nonresealed or partly resealed cells when all steps are carried out at 0°C, or to fully resealed cells at any temperature desired.

2.3. Resealing Experiments

To establish rates of resealing, electroporated cells are brought to an elevated temperature for strictly defined periods of time. This can be achieved by using temperature jump equipment allowing for equilibration within about 5 sec. Heat exchangers of appropriate surface area-to-volume ratios (20–40 cm^2/ml) can easily be constructed.

Cells are suspended in a protectant-containing medium (see above) and subjected to electric breakdown at 0°C. The time course of resealing is determined by exposing aliquots at the suspension to elevated temperatures in a heat exchanger for increasing, strictly defined periods of time. Subsequently, they are rapidly brought back to 0°C by a second heat exchanger (Schwister, 1985). Residual leak permeabilities are determined by Approach I or II as described above. The time resolution of this resealing procedure is about 20 sec.

3. RESULTS AND DISCUSSION

3.1. Voltage Dependence of the Induced Leaks

Short electric field pulses of appropriate amplitude are known to induce, in pure lipid membranes, very short-lived, reversible states of high conductance (Benz and Zimmermann, 1981; Chernomordik *et al.*, 1987). In most (see Knight and Scrutton, 1986, for a review), although not all (e.g., see Lindner *et al.*, 1977; Benz and Conti, 1981) biological membranes, electrical breakdown goes along with the occurrence of much more stable states of increased permeability, particularly when cells are kept at low temperature. The reasons for this high stability are unknown. It allows, however, the characterization of the induced leaks in a situation of approximate steady state for the leak and for the cell in toto, since colloid-osmotic cell swelling and lysis due to the influx of water and salt through the induced leaks can be prevented by impermeable macromolecules acting as colloid-osmotic protectants (Wilbrandt, 1941; Pooler, 1985).

Taking advantage of these possibilities to keep cells and their leaks stable after electrical breakdown, the relationship between field strength (and duration) of the leak-inducing pulse and the induced leak permeability can be established. The leak permeability of electroporated human erythrocytes, as characterized by K^+ efflux (Fig. 2A), rates of hemolysis in isotonic NaCl (Fig. 2B) or isotonic erythritol (Fig. 2C), and tracer fluxes of erythritol (Fig. 2C), increases nonlinearly with increasing field strength above the field strength at which all cells are affected by the pulse.

This latter field strength is not identical to the minimal external field strength required to cause the first leaks in suspended erythrocytes. The "critical" field strength has been a major object of earlier studies on electrical breakdown of erythrocyte membranes. Using various methods, it could be demonstrated that breakdown occurs at a threshold transmembrane potential of around 1–1.5 V, when long pulses (> 20 μsec) are applied (Riemann *et al.*, 1975; Kinosita and Tsong, 1977b). Similar values have been reported for other membranes (see Zimmermann, 1982, for a review). In comparing data, a marked dependence of the threshold potential on temperature has to be considered. As shown by Coster and Zimmermann (1975) in algal cells, the critical field strength increases by a factor of about 2 when the temperature is decreased from 30°C to 0°C.

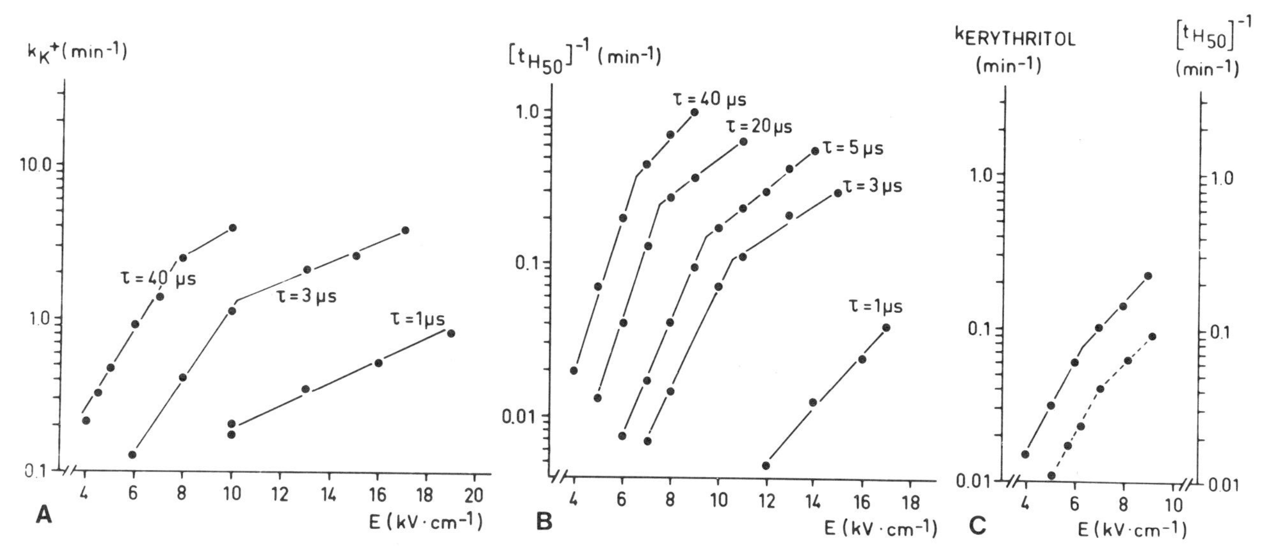

Figure 2. Field strength dependence of electrically induced leakiness in human erythrocytes. (A) Rates of K^+ net efflux into choline chloride media, containing Dextran 4 (30 mM) as a stabilizer, 0°C. (B) Reciprocal half times of hemolysis in isotonic NaCl at 0°C. (C) Leak permeability to erythritol, characterized by tracer fluxes at chemical equilibrium (left-hand ordinate, continuous line) in the presence of Dextran 4 or by reciprocal half times of hemolysis of erythrocytes in isotonic erythritol solution (right-hand ordinate, dashed line). Measurements carried out at 0°C. Mediated fluxes of erythritol blocked by addition of cytochalasin B, 10 μM.

The relationship between induced transmembrane potential and external field strength in the discharge chamber can be described by the Laplace equation

$$V_m = E_{ext}\, r\, f \cos \alpha \tag{1}$$

where r is the spherical cell radius, f a shape factor (1.5 for a sphere), and α the angle between the field vector and the radial vector for a given point on the cell surface. At a given E_{ext}, V_m is highest for $\cos \alpha = 1$. Therefore, with increasing field strength, the critical transmembrane voltage is first reached at $\alpha = 0°$, i.e., at the poles of the cell facing the electrodes. Depending on the model for electrical breakdown of membranes, the critical transmembrane potential is the voltage at which electrical compression of the bilayer exceeds the opposing elastic forces of the lipids (e.g., see Zimmermann, 1982; Dimitrov, 1984) and pores are formed (e.g., see Sugar and Neumann, 1984). Alternatively, small preexisting nonconducting aqueous pores may become sufficiently enlarged by the external field to become conducting and permeable to solute (Chernomordik *et al.*, 1983; Weaver and Powell, this volume).

At the critical field strength, only a few erythrocytes are electroporated. This is probably due to various kinds of heterogeneity of the system. The range between the lowest field strength just inducing electroporation and the field strength at which all cells are affected comprises about 1.5 kV/cm in human erythrocytes (Riemann *et al.*, 1975; Kinosita and Tsong, 1977b) but also in other human blood cells (Smith and Cleary, 1983). In view of this heterogeneity, all permeabilities obtained for electroporated erythrocytes are mean values comprising a broad range of permeabilities of individual cells.

In order to obtain characteristic parameters for the leak/field strength relationship, a log-linear presentation can be used (Fig. 2). This—purely descriptive—evaluation results in a biphasic linear relationship with a break where the slope decreases considerably. The curves are shifted to higher field strengths and the break points displaced to lower leak permeabilities when shorter pulses are used. The absolute values for the induced leak permeabilities differ depending on the test permeant studied, but also as a function of the driving forces which are high in Fig. 2A and C due to the steep chemical gradients of the permeants, but low in Fig. 2B since the colloid-osmotic pressure difference between cells and medium is rather small. The time resolution of the lysis approach is thus much higher than that of the K^+ efflux approach, while the latter has the advantage of being more straightforward and less prone to artifacts.

The break points obtained at a given field strength and pulse length coincide for the various approaches. Below and above the break points the log-linear relationships follow a function of the general type $\log y = x \log b$ (the logarithmic form of $y = b^x$). For the special situation of changes of the leak flux rate k with increasing voltage E_D in a range above the critical field strength E_{crit}, the appropriate equation is

$$\log (k - k_{crit}) = (E_D - E_{crit}) \log b \tag{2}$$

This type of relationship may suggest that leak formation involves an avalanche type or cooperative process. The parameter b in Eq. (2) (the antilogarithm of the slope in Fig. 2) is thus something like an amplification factor. Its values, which can easily be obtained from log-linear regression analysis, are on the order of 1.5–3 below the break and decrease to values of 1.2–1.3 above the break. It is interesting to note that in pure lipid membranes the conductance of the leak induced by electrical breakdown increases much more steeply with breakdown voltage than in the erythrocyte membrane (Chernomordik *et al.*, 1983).

As a simple, geometric explanation for the relationship in Fig. 2, one might consider the nonlinear increase—with field strength—of the fractional cell surface area experiencing transmembrane voltages that exceed the critical membrane potential required for leak development. This interpretation, which is essentially based on the relationship $V_m = E_{ext}\, r f \cos \alpha$, can be refuted on the basis of quantitative considerations (Schwister and Deuticke, 1985).

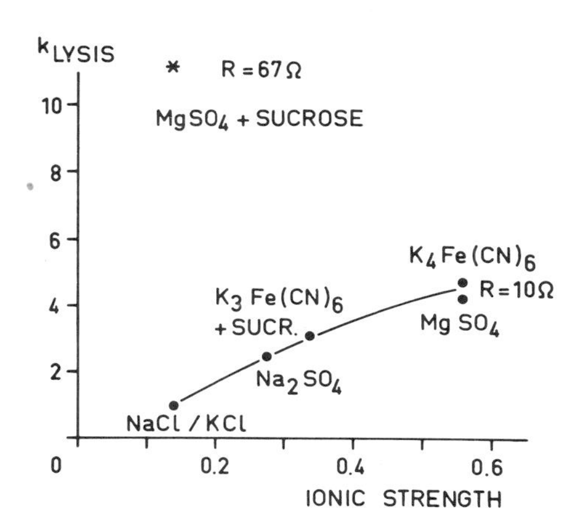

Figure 3. Influence of the ionic strength of the suspension medium during electrical breakdown on field-induced leak permeability. Cells (hematocrit 30%) subjected to 6 kV/cm in media (R = 10 Ω) containing the constituents indicated at each data point, plus Dextran 4 (30 mM) (for details see Schwister and Deuticke, 1985). Subsequently, rates of colloid-osmotic lysis (Approach II) were determined in KCl/NaCl media as usual. The asterisk demonstrates the influence of changes of the ohmic resistance (R) in the suspension during breakdown, at constant ionic strength. k values normalized to values in NaCl/KCl medium.

The slopes in Fig. 2 and in particular the position of the lines in the coordinate system vary with the duration of the electric pulse. By replotting the experimental data, a hyperbolic relationship can be demonstrated between E_D and τ (Schwister and Deuticke, 1985). To obtain a certain leak permeability, a constant value of the product τE_D is thus required. A similar relationship was already shown to hold for the critical field strength (Riemann *et al.*, 1975; Kinosita and Tsong, 1977b). It is obviously also valid in the supercritical range.

Besides field strength and pulse length, the ionic strength of the suspending medium influences the extent of the induced leak. In an earlier study it was claimed (Kinosita and Tsong, 1977b) that leak permeability *increases* with *decreasing* ionic strength, e.g., when electrolyte is replaced by nonelectrolyte in the suspension medium used for electroporation. This increase is, however, in fact due to the increase of the ohmic resistance of the medium and consecutive changes of the time constant $\tau = R\ C$ of the system. When special incubation media are devised which provide for increased ionic strengths (up to fourfold) during the breakdown, at constant ohmic resistance (Schwister and Deuticke, 1985), leak permeability (measured at normal ionic strength) *increases* as a function of the ionic strength prevailing during the electric pulse (Fig. 3). The relevance of this finding will be considered in Section 4.

3.2. Properties of the Induced Leak in Human Erythrocytes

The high permeability of the induced leaks to small inorganic ions and small polar nonelectrolytes clearly suggests the formation of an aqueous pathway in erythrocytes subjected to electrical breakdown. In line with this view, K^+ leak fluxes in cells electroporated at 6 kV/cm exhibit an apparent Arrhenius activation energy of 29 ± 5 kJ, as would be expected (Stein, 1986, p. 97) for diffusion in a fluid aqueous environment.

3.2.1. Apparent Pore Size

An aqueous pathway in a membrane may be expected to discriminate penetrating solutes according to their size. If the leak is assumed to have the structure of a cylindrical pore, the dependence of the induced permeabilities on the permeants' radius can be fitted to the Renkin equation (Renkin, 1955), which yields an apparent pore radius. If this fitting procedure cannot be used or provides unsatisfactory results, semiquantitative information may still be helpful. Earlier attempts to establish the size of the leaks induced by electrical breakdown had already shown that the membrane becomes permeable to molecules of the size of the tetrasaccharide, stachyose, indicating pores with a radius of at least 0.6 nm (Kinosita and Tsong, 1977c). Other investigators postulated larger sizes (Benz and Zimmermann, 1981).

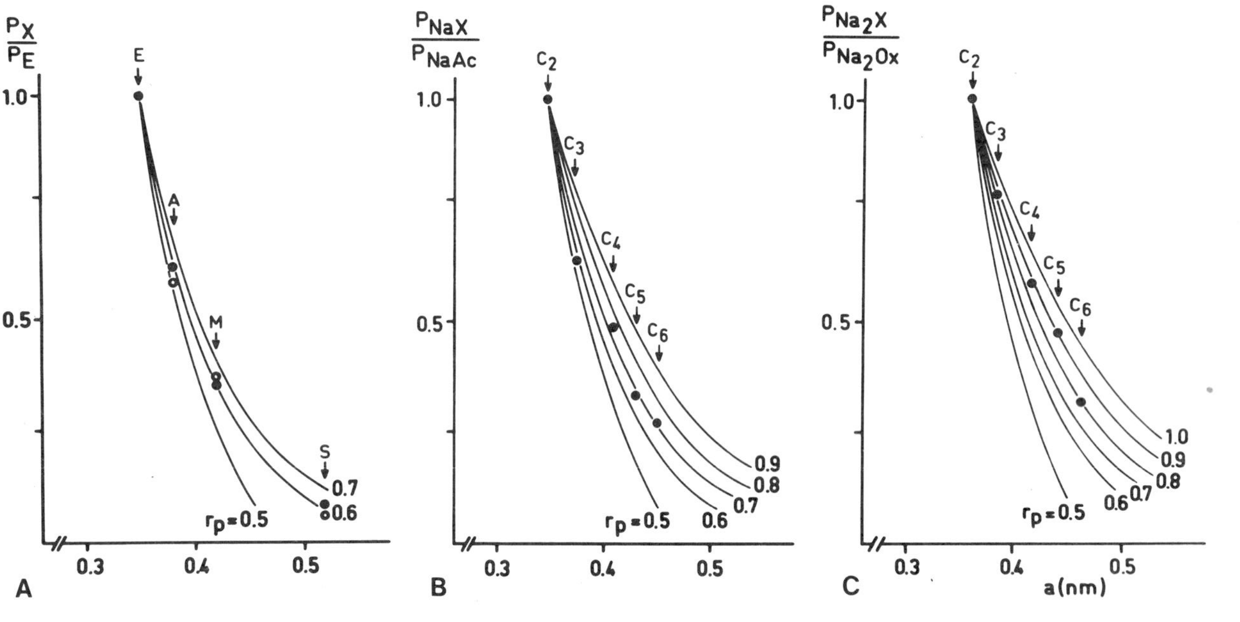

Figure 4. Equivalent pore radii, derived from normalized permeabilities of polyols (A), monocarboxylates (B), and dicarboxylates (C), plotted against the radii of these solutes. The continuous lines are obtained from an equation (Renkin, 1955) relating the diffusion of a solute through cylindrical pores of radius r_p to the solute radius a under consideration of steric hindrance and of frictional resistances. Values were obtained for cells electroporated at 6 or 8 kV/cm, τ = 40 μsec. (A) Data from tracer flux measurements (0°C) (E, erythritol; A, arabinose; M, mannitol; S, sucrose), normalized to those for erythritol. (B) Data from rates of hemolysis (t_{H50}^{-1}) in isotonic solutions of the Na salts of aliphatic monocarboxylic acids (C_2 = acetate, C_3 = propionate, etc.), normalized to data for acetate. (C) Data from rates of hemolysis (t_{H50}^{-1}) in isotonic solutions of the Na salts of aliphatic dicarboxylic acids (C_2 = oxalate, C_3 = malonate, etc.), normalized to data for oxalate. Radii for polyols are hydrodynamic radii (Deuticke *et al.*, 1984), radii of mono- and dicarboxylates are derived from molecular models. From Schwister and Deuticke (1985), by permission.

Using tracer flux measurements (Approach III) and rates of hemolysis (Approach II), we have collected systematic data on pore size. Leak permeabilities to polyols, when normalized to values for erythritol and plotted against the radius of the permeants, can be fitted to an apparent radius of about 0.6 nm (Fig. 4A). Somewhat larger radii (0.7–0.8 nm) emerge from an evaluation of the rates of hemolysis of electroporated erythrocytes in isotonic solutions of the Na salts of small (C_2–C_6) aliphatic mono- and dicarboxylic acids (Fig. 4B,C). In such systems, differences in the rate of uptake of salt, which is the prerequisite for prelytic cell swelling, are determined by the anions since the cation is the same in all media (Deuticke *et al.*, 1984). Application of the Renkin equation seems permitted.

The calculation of such numbers is, of course, based on simplifying assumptions including that of a cylindrical pore, in which the solute experiences frictional drag. In a more direct approach, the approximate size of leaks is derived from the size of nonelectrolytes that protect leaky cells from colloid-osmotic lysis by establishing an osmotic pressure counterbalancing the osmotic drag of intracellular impermeable constituents. Only such added nonelectrolytes that do not permeate at all through the induced pores will be protective for an unlimited time. Slowly permeating protectants will permit swelling or lysis after finite periods.

This approach can be realized by incubating the leaky cells for 20 hr at 0°C in saline media containing oligosaccharides, polyethyleneglycols, or dextrans at 30 mM. The results obtained can only be regarded as approximate due to some resealing of the cells at 0°C and to the molecular weight dispersity of the polymer used. However, it can be demonstrated (Fig. 5) that the size of the induced leaks increases slightly with increasing breakdown voltage at low field strengths and more pronouncedly at higher values. At 10 kV/cm, protectants with mean radii of 1.9 nm are required to protect 95% of the cells, while at 6 kV/cm, radii of only 0.8 nm are required for this purpose. Moreover, at high field strengths the radii seem to exhibit a much broader dispersion over the cell population than at lower ones.

In summary, the size selectivity of the leaks induced by electrical breakdown indicates the presence of porelike structures. This conclusion was recently confirmed by Ginsburg and Stein (1987). They reevaluated the experimental data of Schwister and Deuticke (1985) using radii of the

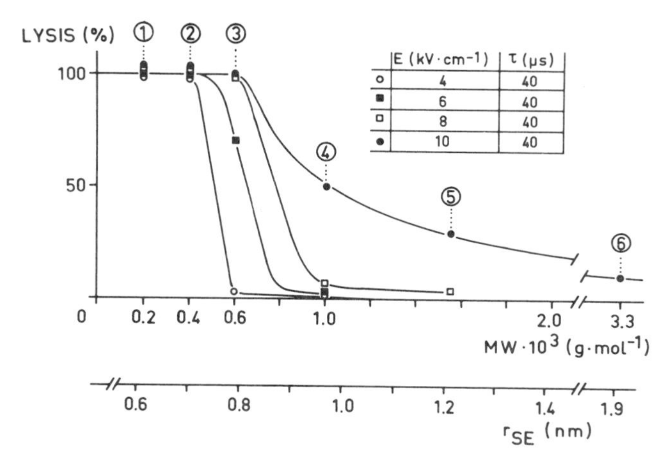

Figure 5. Protective action of nonelectrolytes of increasing molecular size against colloid-osmotic lysis. Cells were subjected to electric pulses (τ = 40 μsec) at varying field strengths—(○) 4 kV/cm, (■) 6 kV/cm, (□) 8 kV/cm, (●) 10 kV/cm—in the usual medium containing 30 mM Dextran 4 as colloid-osmotic protectant. Subsequently, they were incubated at 0°C for 20 hr in media containing the various protectants at 30 mM. The lysis observed after this incubation is plotted against the radii (Scherer and Gerhard, 1971) of the test molecules: (1) PEG 200, r_{SE} = 0.56 nm; (2) PEG 400, r_{SE} = 0.68 nm; (3) PEG 600, r_{SE} = 0.80 nm; (4) PEG 1000, r_{SE} = 1.00 nm; (5) PEG 1550, r_{SE} = 1.20 nm; (6) PEG 3300, r_{SE} = 1.9 nm. PEG, poly(ethylene glycol). From Schwister and Deuticke (1985), by permission.

Table 1. Characteristics of the Postulated Pores Induced in the Erythrocyte Membrane by Electrical Breakdown[a]

	Field strength (kV/cm)	Leak permeability (10^{-7} cm/sec)	Equivalent pore radius (nm)	Apparent number of pores/cell	Pore density (10^5 cm^{-2})
Tracer fluxes (erythritol)	4	0.111	0.55	0.2	1.4
	6	0.431	0.60	0.7	5.0
	7	0.724	0.65	1.0	7.1
	8	1.044	1.00	0.6	4.3
	9	1.723	1.50	0.4	2.8
K^+ net flux	4	0.79	0.55	1.0	7.1
	6	2.67	0.60	2.9	20.6
	7	4.72	0.65	4.3	30.5

[a]Leak permeabilities calculated for erythritol from tracer fluxes and for K^+ from net fluxes in Fig. 2, using equivalent pore radii derived from Figs. 4 and 5.

permeants as derived from the van der Waals volume (Bondi, 1964) and arrived at an equivalent pore radius of 0.4–0.5 nm for the lower range of breakdown voltages. Surprisingly, however, the data could also be fitted by an alternative model assuming non-Stokesian diffusion (Stein, 1986, Chapter 2) through a network of intertwined hydrocarbon chains comparable to a soft polymer (Ginsburg and Stein, 1987). This ambiguity requires further investigation.

3.2.2. Apparent Pore Density

Based on total leak permeabilities and apparent pore radii, apparent numbers of pores per cell can also be approximated. For this purpose, assumptions are required concerning the diffusion coefficient within the pore, D_p, and the length of the pore, l_p (Deuticke *et al.*, 1983). Assuming l_p = 5 nm and $D_p = D_w$, the diffusion coefficient of the test permeant in bulk solution, one arrives at the numbers compiled in Table 1 for a range of supercritical field strengths. The data seem to permit the conclusion that the number of static pores necessary to account for the induced leak permeabilities is very low, even lower than the number of about 10^7 cm^{-2} reported for an artificial lipid membrane (Benz and Zimmermann, 1981). The pore density increases with increasing field strengths, but may rediminish at high intensities. The differences between numbers obtained for erythritol and K^+ may not be real, since various assumptions were made in calculating K^+ permeabilities from net fluxes. The formation of, on the average, less than one membrane defect per cell is unlikely under conditions where at least a large part of the total cell surface area of 140 μm^2 has experienced a supercritical transmembrane voltage. Thus, one has to consider the involvement of dynamic, fluctuating structural defects, appearing and disappearing at varying sites. Such fluctuating defects would be comparable to defects observed in artificial lipid/protein membranes (Gerritsen *et al.*, 1980; Van der Steen *et al.*, 1982) but also in chemically modified erythrocytes (Deuticke *et al.*, 1983, 1987; Heller *et al.*, 1984).

3.2.3. Chemical Selectivity of the Pores

Electrically induced leaks also exhibit some chemical selectivity. This follows from comparing rates of colloid-osmotic lysis of leaky cells in isotonic salt solutions sharing one ion, but differing in the counterion (see also Deuticke *et al.*, 1984). In general, monovalent ions are preferred over divalent ones, the ratio being larger for cations (about 10) than for anions (2–3). Alkali and earth alkali cations are only discriminating among themselves to the extent expected from the differences in their bulk diffusion coefficients. In contrast, halides are selected to a higher extent in the sequence $F^- : Cl^- : Br^- : I^- = 0.5:1:1.5:2.7$, in accordance with selectivity patterns observed for leaks induced by chemical modification of the erythrocyte membrane (Deuticke *et al.*, 1984, 1986). This corresponds to sequence I of Wright and Diamond (1977). Sequence I is to be expected when weakly charged groups inside the "pore" determine the selectivity. More intricate selectivity patterns among organic anions of comparable size but different substituents have also been observed (Schwister and Deuticke, 1985). The cation/anion discrimination of the leak induced by electrical breakdown has not been studied. In view of the close resemblance of this leak to that induced by chemical modification (Deuticke *et al.*, 1984), some cation selectivity (P_K:$P_{Cl} \approx 5-10$) may be expected, which is comparable to the cation selectivity of the single-length pore produced by insertion of the polyene antibiotic, amphotericin B, into the erythrocyte membrane (Deuticke *et al.*, 1984).

3.3. Enhancement of Transbilayer Mobility of Long-Chain Amphiphiles in Electroporated Erythrocytes

Structural defects induced by chemical perturbation of the erythrocyte membrane (e.g., by oxidants or SH reagents, by insertion of channel-forming peptide analogues, or by certain drugs) manifest themselves not only in leak formation but also in an acceleration of the transbilayer reorientation (= flip-flop) of long-chain amphiphiles intercalated among the membrane lipids (Bergmann *et al.* 1984a; Schneider *et al.*, 1986; Deuticke *et al.*, 1987). The same is true in the case of electrical breakdown. From Fig. 6 it becomes evident that the flip rates of oleoyl-lysolecithin, which at 0°C reorients with a half time of 20 hr in native erythrocytes (Bergmann *et al.*, 1984b), are enhanced by at least three orders of magnitude ($t_{1/2} \approx 15$ min). The slope of the field strength–response curve is the same for leak permeability and flip rates, suggestive of similar underlying events. Concomitantly, the preferential orientation of phosphatidylethanolamine (the major membrane aminophospholipid) to the inner membrane surface is essentially lost, while the other outer leaflet phospholipid, phosphatidylserine, and the choline phospholipids maintain their orientation (Dressler *et al.*, 1983). This points to a somewhat selective alteration of the lipid domain which will have to be considered in molecular models of electroporation. Moreover, these observations may bear a direct relationship to the membrane fusion of erythrocytes subjected to electrical breakdown.

3.4. Species Differences in Leak Formation

Erythrocytes from different mammalian species differ in their membrane phospholipid (Deuticke, 1977) and protein (Hamaguchi and Cleve, 1972) patterns. For these reasons, variations in

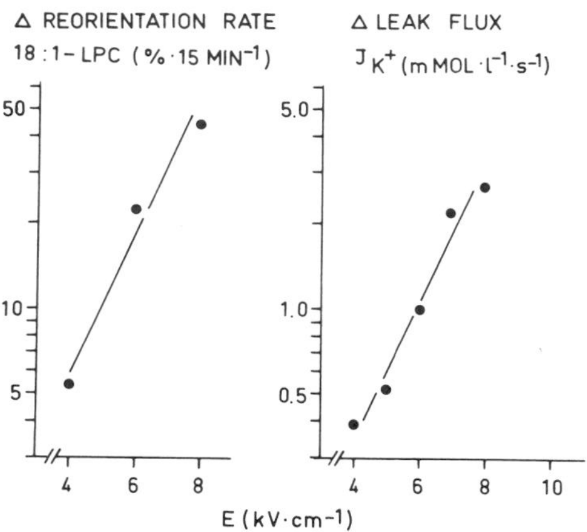

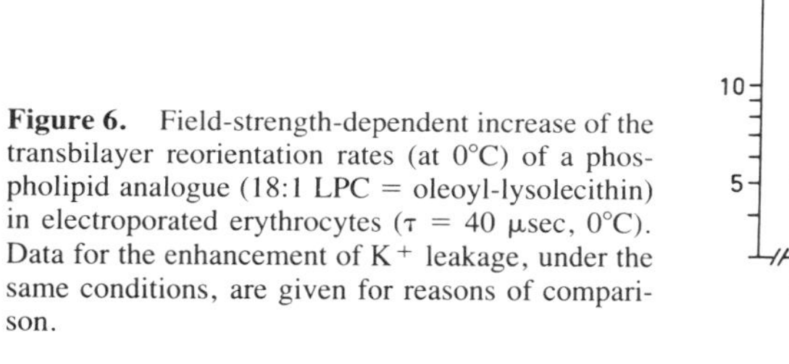

Figure 6. Field-strength-dependent increase of the transbilayer reorientation rates (at 0°C) of a phospholipid analogue (18:1 LPC = oleoyl-lysolecithin) in electroporated erythrocytes (τ = 40 μsec, 0°C). Data for the enhancement of K^+ leakage, under the same conditions, are given for reasons of comparison.

their response to electric field pulses might be expected. Such differences can in fact be demonstrated (Schwister, 1985). Most conspicuously, bovine erythrocytes exhibit a response qualitatively different from that of human erythrocytes.

1. Although having a diameter somewhat smaller than human erythrocytes (Gruber and Deuticke, 1973), bovine cells require a field strength for breakdown that is about twice that required for human erythrocytes. This difference is evident from the earlier data of Zimmermann *et al.* (1976). From the relationship $V_m = E_{ext} r f \cos \alpha$, one would expect the opposite. Obviously, the bovine red cell membrane requires intrinsically a higher field strength for breakdown than does the human erythrocyte membrane.

2. On the other hand, the parameter b, in the leak permeability/field strength relationship as defined in Eq. (2), is somewhat higher in bovine than in human erythrocytes at the same pulse length. Furthermore, bovine erythrocytes do not exhibit the break in this relationship observed in human erythrocytes (Fig. 7A). Consequently, much higher leak permeabilities are reached at low field strength which cannot be quantified by K^+ fluxes. Since the apparent pore radii as determined by the protection technique are only about 20% larger than those in human erythrocytes (Schwister, 1985), pore densities (or opening frequencies) are probably much higher in bovine than in human erythrocytes. The lack of a break in bovine (and other, see below) erythrocytes also provides evidence that the break in human cells is not an artifact.

3. Most conspicuously, however, the leak permeabilities of bovine erythrocytes increase with time when a certain characteristic value of initially induced leak permeability is exceeded. As evident from Fig. 7B, leak permeabilities in cells exposed to a field pulse of $\tau = 5$ μsec remain constant over 4 hr for a field strength of 8 or 10 kV/cm. This corresponds to the behavior of human erythrocytes. For field strengths of 12 or 14 kV/cm, which induce lysis with rate constants > 0.2 min^{-1} when tested immediately after breakdown, the rate constant increases about tenfold when the cells are kept at 0°C for 4 hr after the breakdown. Obviously, an unstable state is produced in the bovine but not in the human erythrocyte membrane under conditions producing a leak (marked by the dashed line in Fig. 7A). Estimates of pore size at these high field strengths (Schwister, 1985) suggest that the spontaneous increase of permeability at 0°C is largely due to an increase of the apparent density (or frequency of opening) of the induced pores.

4. Finally, bovine differ from human erythrocytes in the ion selectivity of the field-induced leak. In contrast to human erythrocytes, the leak in bovine erythrocytes discriminates very little not only between alkali chlorides but also between sodium halides (Schwister, 1985).

Leak formation in bovine erythrocyte membranes thus exhibits quite a number of peculiar features. To check whether these peculiarities are common properties of ruminant erythrocytes, ovine and camel erythrocytes were also studied (Schwister, 1985). These species have erythrocyte membrane phospholipid patterns rather similar to bovine erythrocytes but different from human erythrocytes. Their content of sphingomyelin is much higher and their lecithin content much lower than those of human erythrocytes (Deuticke, 1977). The patterns of leak permeabilities proved to be inconclusive. Like bovine cells, the two other species lack the break in the leak permeability/field strength relationship. Their leak permeabilities do not increase, however, with time at any permeability. Leaks in sheep erythrocytes have the type of ion selectivity that is typical for human cells; camel cells resemble bovine erythrocytes. Pore sizes are not conspicuously different from those in human or bovine cells. The features of the leak permeabilities induced by electrical breakdown are thus certainly not simply governed by the phospholipid patterns but by more refined structural properties and their alterations.

3.5. Resealing of Electrically Induced Membrane Leaks

Leaks induced in artificial lipid membranes by electrical breakdown reseal very rapidly (Benz and Zimmermann, 1981; Chernomordik *et al.*, 1987). In most biomembranes, resealing of the leaks also occurs, at least after breakdown at moderate field strengths. The time constants seem to vary considerably among membrane systems (see Zimmermann, 1982, and Knight and Scrutton 1986,

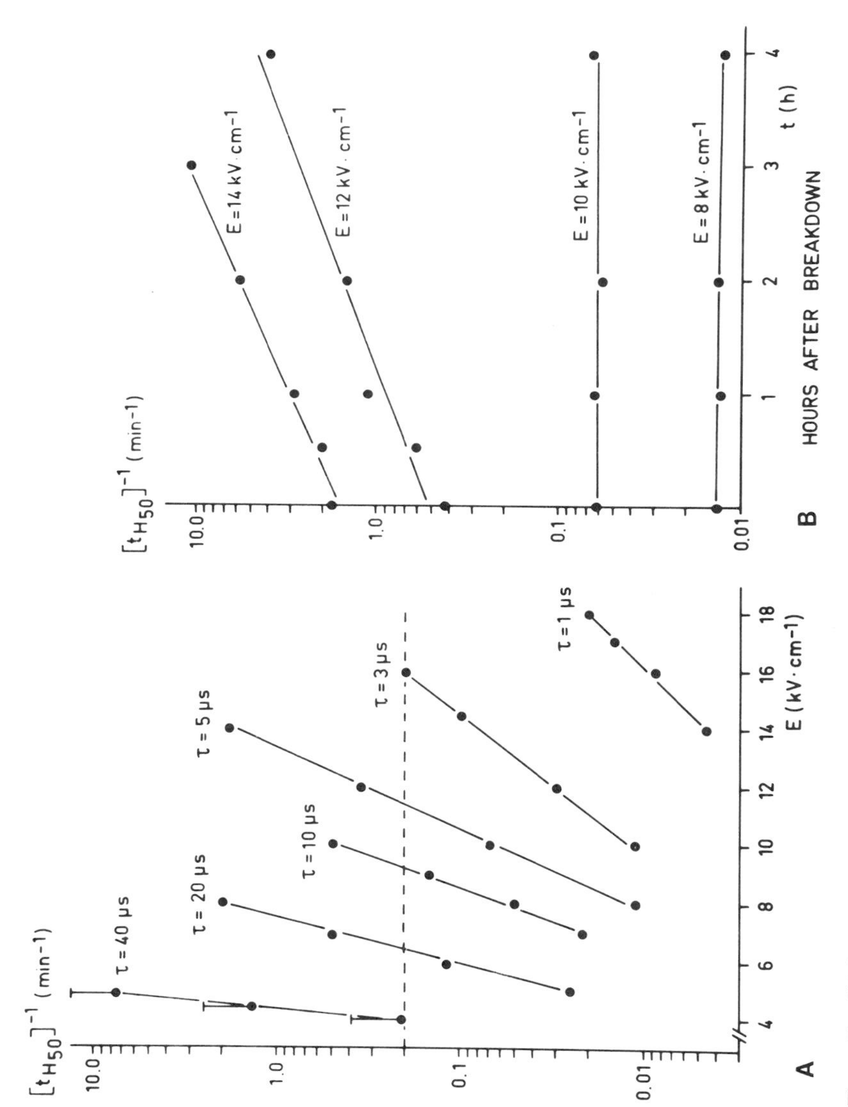

Figure 7. (A) Log-linear relationship between electric field strength and induced leak permeability in bovine erythrocytes, as derived from rates of colloid-osmotic lysis at 0°C. Note the absence of a break. Dashed line: critical permeability above which leak permeabilities increase with time (see panel B and text). (B) Instability of the leak induced by high field strengths ($\tau = 5$ μsec) in bovine erythrocytes. Cells were subjected to the field pulse at 0°C, kept for 0–4 hr at 0°C in protecting medium, and assayed for rates of colloid-osmotic lysis (Approach II) after various time intervals.

for reviews). Early studies on erythrocytes demonstrated that in these cells the leaks are highly stable at 0°C but reseal rapidly when the temperature is increased (Kinosita and Tsong, 1977c). Studies employing temperature jump equipment for heating and recooling electroporated erythrocytes have recently provided more quantitative data (Fig. 8). The time course for the decrease of permeability upon warming leaky cells does not obey first-order kinetics, as was recently also reported (Glaser *et al.*, 1986). The data can be fitted to a second-order process which allows determination of the rate constant k_R (Fig. 8B). The values of k_R obtained at various temperatures for a fixed initial leak permeability can be fitted by Arrhenius relationship $\ln k_R = (E_a/R)(1/T)$ (Fig. 8C). From the slope, an apparent activation energy for resealing was calculated which proved to be independent of the field strength originally applied and thus of the initial leak permeability. A mean value of 117.3 ± 7.7 kJ/mole (28.1 ± 1.8 kcal/mole) indicates that a considerable energy barrier has to be overcome to reconstitute the normal permeability. This energy barrier is independent of the size of the leaks and their number. On the other hand, the rate constant of resealing, k_R, at a given temperature is an inversely linear function of the field strength applied, thus being higher for small and few leaks than for larger and more numerous leaks. In terms of the Eyring Absolute Rate Theory, an entropy term will have to account for these differences in rates at constant activation enthalpy.

Simultaneous measurements of permeability and pore size suggest that resealing in erythrocytes is based to a major extent on a closing (or less frequent opening) of pores while a decrease in diameter contributes only to a minor extent. For example, a 95% resealing goes along with a 35% decrease in pore diameter but a tenfold diminution of the apparent number of pores (Schwister, 1985). This pattern would seem to suggest that resealing of the pores is almost an all-or-none process in which the individual defect whenever it appears maintains its size for most of the resealing period. Resealing also pertains to the enhancement of transbilayer reorientation rates, although a complete normalization of flip rates cannot be obtained by procedures essentially restoring the permeability barrier. The enhanced flip rates still observed after "resealing" have a markedly lower activation energy than the flip in native cells (Dressler *et al.*, 1983), suggesting that the induced flip sites are different from the native ones.

Other mammalian species exhibit a resealing pattern of the leaks not very different from humans. A few observations deserve to be mentioned:

1. Bovine cells reseal normally below the "characteristic permeability" (see above). Above this characteristic permeability, i.e., in the range in which leak permeabilities increase with time (see Fig. 7B), warming of the electroporated cells goes along with either resealing or enhancement of leak permeability, depending on the experimental conditions (Schwister, 1985).

2. Sheep cells reseal about six times faster than human cells at the same induced leak permeability. However, the Arrhenius activation energy for resealing in sheep cells is about twice that for human erythrocytes (205 versus 117 kJ/mole). This finding again emphasizes that the rate of resealing is dominated by an entropy term.

3. The apparent activation energies of resealing when studied comparatively in five species (sheep, ox, camel, human, dog) correlate well with the differences in the sphingomyelin (and lecithin) contents (Fig. 9). These two phospholipids constitute about 90% of the outer lipid leaflet in the erythrocyte membrane (Deuticke, 1977). This correlation points to an involvement of membrane lipids in the resealing process.

4. CONCLUDING COMMENTS: HOW TO ENVISAGE LEAK FORMATION AND RESEALING?

According to the data reviewed in the foregoing chapters, electric field pulses induce in erythrocytes aqueous leaks which increase in both size and number (or frequency of opening) with increasing field strengths and pulse lengths. Combined measurements of total leak permeabilities and apparent pore sizes suggest that in the lower range of induced leak permeabilities, the increase

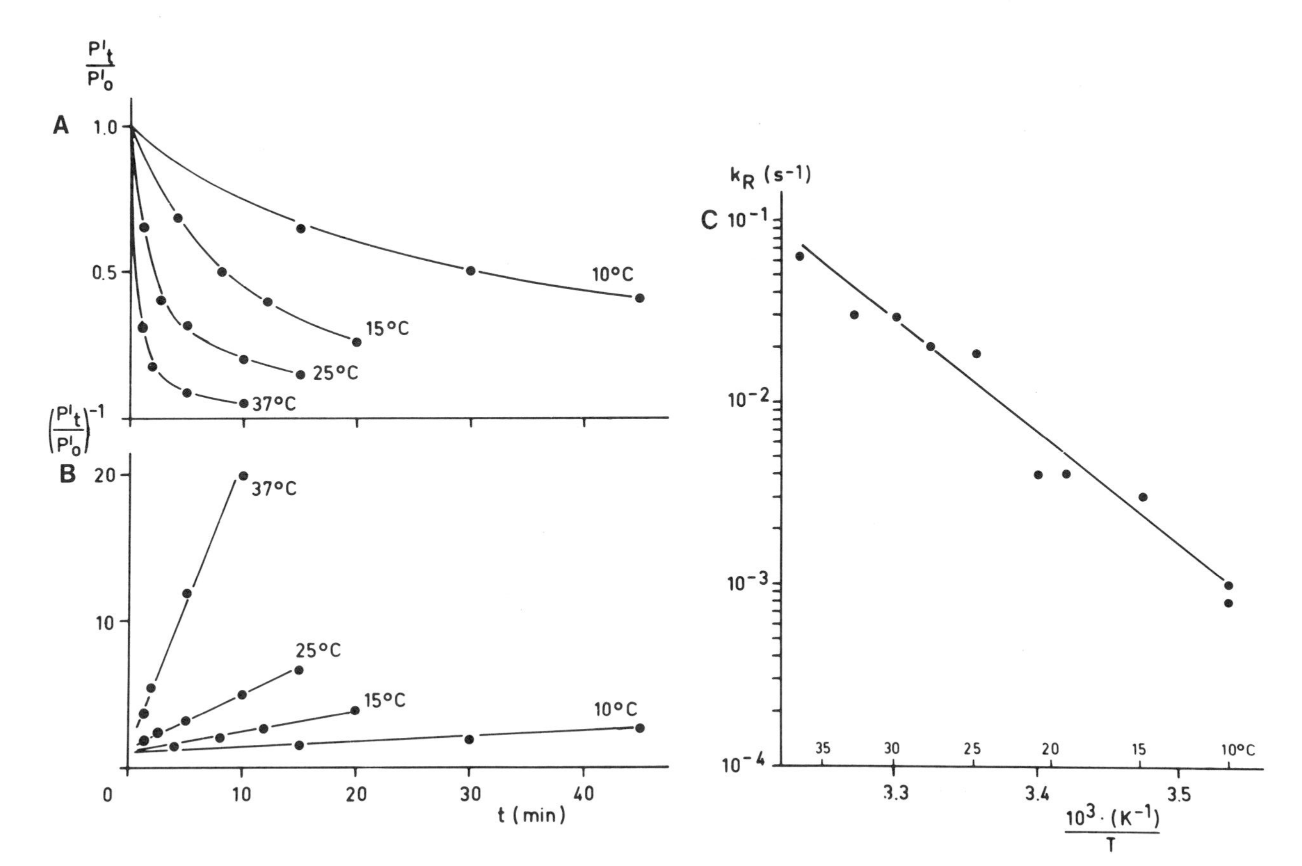

Figure 8. Time course of the resealing of human erythrocytes after electrical breakdown (E_D = 10 kV/cm, τ = 40 μsec). (A) Primary data, obtained as described in Section 2. Leak permeabilities derived from colloid-osmotic hemolysis. (B) Data from panel A evaluated according to the equation $P'\text{t}/P'_0 - 1 = k_R\, t$, where P_0 and P_t are leak permeabilities measured immediately after breakdown and after resealing at the temperature indicated for a time period t. Leak permeabilities assayed by measuring rates of colloid-osmotic lysis. (C) Arrhenius plot for the resealing of electrically induced leaks. Primary data (obtained for E_D = 6 kV/cm, τ = 40 μsec) obtained as shown in panel A.

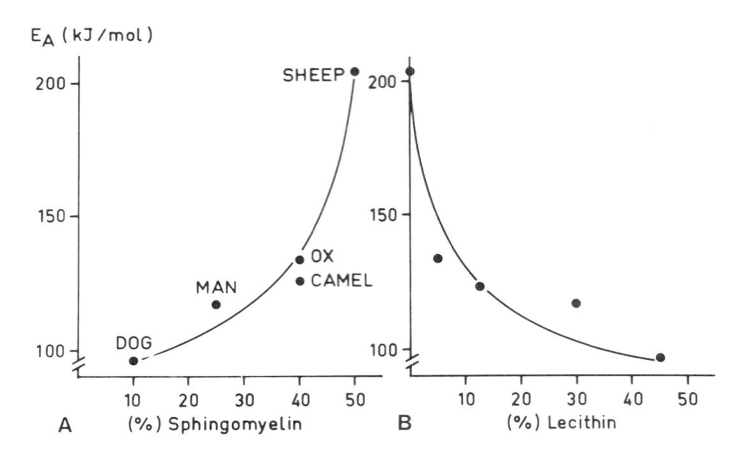

Figure 9. Relationship between the activation energy for the resealing of electric leaks in erythrocytes from various species and their membrane contents (Deuticke, 1977) of the choline phospholipids lecithin and sphingomyelin.

in number (or frequency of opening) may be the dominating parameter. In the case of extensive leakiness, the increase in pore size becomes dominating. It is not known whether the number of pores actually decreases at higher field strengths. The leaks discriminate among anions but not among cations and distinguish divalent from monovalent ions, suggesting that weakly charged groups are present in the pores in a low density.

With respect to the structure of the pores, a comparison with apparent pore radii reported for pure lipid membranes may be of interest. In view of the short lifetime of the pores in lipid membranes, their sieving properties have not yet been established in detail. From indirect estimates, equivalent radii between 0.4 and 4 nm (Benz and Zimmerman, 1981; Chernomordik *et al.*, 1983) have been deduced. The short-lived defects in electroporated lipid membranes thus seem to be larger than the long-lived ones in biological membranes. This discrepancy is to some extent, however, an apparent one and not the relevant difference between artificial lipid and natural membranes in the response to electrical breakdown. Recent investigations by Sowers and Lieber (1986), Chernomordik *et al.* (1987), Weaver *et al.* (1988), and Kinosita *et al.* (1988) have provided evidence that in erythrocytes, as in artificial membranes, large defects are formed during the application of the field pulse. These defects, however, which may have radii up to 8 nm (Sowers and Lieber, 1986), reseal within less than 10 sec, as shown by high-resolution techniques of permeability and conductance measurements. Short-lived membrane perturbations can also be demonstrated in erythrocyte membranes by electron microscopic strategies (Stenger and Hui, 1986). Unlike in artificial membranes, however, resealing of the initial leaks in biological membranes does not lead to full recovery of the normal low-permeability state. Instead the cells maintain long-lived smaller leaks accessible to analytical procedures used in the work presented here. This dual response to a field pulse (short-lived high leakiness followed by longlived smaller leakiness) is obviously the major feature that distinguishes biological lipid–protein membranes from artificial lipid membranes in the response to electric field pulses.

Concerning their apparent size and density, ion selectivity, and parallelism with flip site formation, the electrically induced long-lived pores in the erythrocyte membrane bear a close relationship to leaks that are induced in this membrane by chemical modification, e.g., oxidative cross-linking of spectrin via disulfide bridges (Deuticke *et al.*, 1983, 1984), or protein damage combined with lipid peroxidation induced by oxidants (Heller *et al.*, 1984) and radical-forming agents and treatments (Deuticke *et al.*, 1986, 1987). Due to their chemical origin, the resealing properties of these leaks cannot be directly compared with those arising from a merely physical perturbation.

Physical leaks of other origin, on the other hand, seem to be less well comparable to the electrically induced pores. The membrane defect detectable after osmotic lysis of erythrocytes, i.e., after exposure to an isotropic tensile stress, is much larger, ranging from 0.7 to about 30 nm in radius (Lieber and Steck, 1982a,b). A single defect per cell has been claimed to be present. Its size is very dependent on the ionic milieu. Its resealing occurs in time domains comparable to those observed for electrically induced leaks (Lieber and Steck, 1982b; Lee *et al.*, 1985). Interestingly, the activation energies of resealing for osmotic (Lee *et al.*, 1985) and electrical (Fig. 8) leaks are almost identical (between 10 and 37°C). However, in electrical pores, resealing is mainly a matter of opening frequencies and not of leak size, which is the only parameter presumably involved in the resealing of the single osmotic hole. These activation energies are about twice as high as those reported for the resealing of artificial lipid membranes after electrical breakdown (50 kJ/mole; Benz and Zimmermann, 1981). Not unexpectedly, this rather low activation energy is of the same order as that for the lateral diffusion of phospholipids in artificial and cellular membranes (Lindblom *et al.*, 1981; Golan *et al.*, 1984).

Leaks induced by high pressure in erythrocyte membranes (Pequeux *et al.*, 1980; Onuchi and Lacaz-Vieira, 1985) have not been characterized in great detail. Nothing seems to be known concerning their size and density. In contrast to electrical and osmotic leaks, these defects seem not to outlast the phase of exposure to the high pressure. The same is true for leaks induced by exposure of erythrocytes to viscometric shear, i.e., an anisotropic deformation in the plane of the membrane (P. Barth and B. Deuticke, 1987, unpublished results).

Various structural alterations have been discussed as a basis for the chemical leaks (Deuticke *et al.*, 1983, 1986). Similar phenomena may in principle also underlie the electrically induced leaks (Fig. 10). In general, one has to consider the question of proteinaceous versus lipidic structures. Structural defects in the lipid domain might well account for the dynamic nature of the leaks, their rather variable size, the parallelism between the leak and flip site formation, and the species differences in the activation energies for resealing. On the other hand, the remarkable stability of the leaks and the high activation energy of the resealing process seem to be hard to reconcile with mere lipid structures. However, for the large leak induced by osmotic lysis, which is also very stable, Lieber and Steck (1982a,b) have proposed this type of structure. In our view, some involvement of proteins in the formation of electrically induced pores in biomembranes is very likely, although an important contribution of the lipid domain is almost certain. Evidence for this view stems from the fusogenic action of electrical breakdown. Membrane fusion quite generally seems to require a local rearrangement in the lipic domain (Verkleij, 1984; Rand and Parsegian, 1986), e.g., a transition from a lamellar to a nonlamellar arrangement. The presence of such nonlamellar structures has been shown to be paralleled by the occurrence of leaks and flip sites in artificial (Noordam *et al.*, 1980) and even in erythrocyte membranes (Claßen *et al.*, 1987; Tournois *et al.*, 1987).

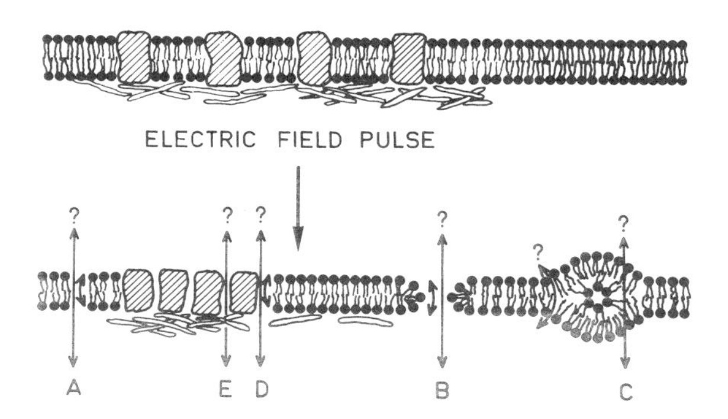

Figure 10. Schematic view of the various types of membrane alterations and defects that might cause leakiness in erythrocytes after electrical breakdown. The scheme illustrates the concept that the field pulse might perturb integral and/or peripheral (membrane skeletal) proteins and thereby affect the organization of the lipid domain. Leaks (↕) and flip sites (⟲) could be located in (A) hydrophobic pores, (B) hydrophilic pores, (C) nonlamellar phases, or (D) areas of a mismatch between lipid and integral proteins. Leaks might also be formed between aggregated proteins (E).

A lasting perturbation of the membrane lipid domain after electrical breakdown is also indicated by the reversible occurrence of flip sites for long-chain amphiphiles. Although the structure of such flip sites and the mechanism of "flip-flop" of amphiphiles are not understood, it seems reasonable to assume that defects in the ordinary lamellar structure are required to allow for the passage of the polar head group through the hydrophobic core of the membrane. Long-lived (at least 3 hr at 0°C) alterations of the cell membrane lipid domain after electrical breakdown have also become evident in NMR analyses (Lopez *et al.*, 1988). The relationship between these phenomena, pore formation, and electrofusion is not clear.

If mere lipid structures are unlikely to form the leaks, how then could proteins be involved? On the one hand, perturbations of the membrane skeletal proteins (e.g., spectrin, actin, band 4.1 protein, and ankyrin), which somehow contribute to the stabilization of the erythrocyte lipid domain (Haest, 1982; Dressler *et al.*, 1984; Chasis and Shohet, 1987), might play a role. On the other hand, structural rearrangements of integral membrane proteins might also disturb their contacts with surrounding lipids and induce a destabilization (Israelachvili, 1977) allowing for the formation of locally perturbed lipid structures. Evidence is available that indeed membrane-spanning peptide structures may stabilize or destabilize the bilayer configuration of lipids (de Kruijff *et al.*, 1985). The open question is whether independent evidence can be adduced for long-lasting rearrangements of proteins in erythrocytes subjected to a field pulse and whether these rearrangements can be localized.

According to the equation $V_m = E_{ext}\, r f \cos\alpha$, the induced damage will spread over increasing segments of the cell surface when E_{ext} exceeds E_{crit}. Therefore, the primary damage will have to be studied in the range of E_{crit}. As already mentioned, evidence has recently become available by patch-clamp studies (see Chernomordik *et al.*, 1987) that a breakdown event comparable to that in pure lipid membranes may actually occur in the erythrocyte membrane. Therefore, the critical field strength should induce a first leak in the lipid domain of the erythrocyte at those faces of the cell where $\cos\alpha = 1$. Recent data of Mehrle *et al.* (1985) for a plant protoplast and of Sowers and Lieber (1986) and Kinosita *et al.* (1988) for erythrocytes have provided direct evidence that the defects responsible for initial high leak fluxes in spherical cells indeed seem to be localized in this way. Further inferential evidence for such localization may come from fusion studies if one accepts that formation of leaks, flip sites, and "fusion sites" are corresponding events. Fusion seems indeed to become possible first at the poles of the cell facing the electrodes (Teissie and Blangero, 1984; Sowers, 1986).

The crucial open question that remains to be answered is that of the link between the "primary" short-lived (≤ 100 sec) breakdown of the erythrocyte membrane and the formation of the "secondary," much more stable leaks leading to the events described here. Unless one claims that long-lived instabilities, merely triggered by the field pulse, are possible in the lipid domain of biomembranes, e.g., due to its marked heterogeneity, one has to consider proteins. It could be a reversible change of conformation (Tsuji and Neumann, 1983) or lateral reorganization of membrane skeletal proteins induced by the current flow through the primary defect in the lipid domain. Alternatively, a vertical displacement of intrinsic proteins, perpendicular to the plane of the membrane may occur (Blumenthal *et al.*, 1983).

As long as no further information is available, an additional or alternative view of electroporation cannot be completely discarded. The field pulse might induce a lateral displacement of membrane constituents (in particular proteins), i.e., an electrophoretic process. This effect would have its maximum at the two poles of the cell where $\sin\alpha = 1$. The principal possibility of such lateral displacements of proteins by external electric fields is well established (Poo, 1981; Sowers and Hackenbrock, 1981). The crucial question is whether they can be induced by the very short-lived but extremely strong field pulses applied in electroporation. The available evidence is somewhat equivocal. On the one hand, the increase of leak formation with increasing ionic strength of the breakdown medium (Fig. 3) may result from a reduction of repulsive forces between membrane proteins due to charge screening which would facilitate lateral displacement by a field vector parallel to the plane of the membrane. On the other hand, the freeze-fracture faces of electroporated erythrocytes studied by Stenger and Hui (1986) do not reveal long-lasting irregularities in the lateral distribution of the membrane-intercalated protein particles.

All these hypotheses presently still fail to explain the high but also highly temperature-dependent stability of the leaky state. The intriguing problem of the molecular basis of the long-lived barrier defects created by electric pulses in erythrocytes and other biological membranes is therefore still unresolved and remains a challenge for future joint work of biophysicists, biochemists, and morphologists.

ACKNOWLEDGMENT. Work from the authors' laboratory was supported by the Deutsche Forschungsgemeinschaft (Sonderforschungsbereich 160/C3).

REFERENCES

Benz, R., and Conti, F., 1981, Reversible electrical breakdown of squid giant axon membrane, *Biochim. Biophys. Acta* **645:**115–123.

Benz, R., and Zimmermann, U., 1981, The resealing process of lipid bilayers after reversible electrical breakdown, *Biochim. Biophys. Acta* **640:**169–178.

Bergmann, W. L., Dressler, V., Haest, C. W. M., and Deuticke, B., 1984a, Crosslinking of SH-groups in the erythrocyte membrane enhances transbilayer reorientation of phospholipids, *Biochim. Biophys. Acta* **769:** 390–398.

Bergmann, W. L., Dressler, V., Haest, C. W. M., and Deuticke, B., 1984b, Reorientation rates and asymmetry of distribution of lysophospholipids between the inner and outer leaflet of the erythrocyte membrane, *Biochim. Biophys. Acta* **772:**328–336.

Blumenthal, R., Kempf, C., Van Renswoude, J., Weinstein, J. N., and Klausner, R. D., 1983, Voltage-dependent orientation of membrane proteins, *J. Cell. Biochem.* **22:**55–67.

Bondi, A., 1964, Van der Waals volumes and radii, *J. Phys. Chem.* **68:**441–451.

Chasis, J. A., and Shohet, S. B., 1987, Red cell biochemical anatomy and membrane properties, *Annu. Rev. Physiol.* **49:**237–248.

Chernomordik, L. V., Sukharev, S. I., Abidor, I. G., and Chizmadzhev, Y., 1983, Breakdown of lipid bilayer membranes in an electric field, *Biochim. Biophys. Acta* **736:**203–213.

Chernomordik, L. V., Sukharev, S. I., Popov, S. V., Pastushenko, V. F., Sokirko, A. V., Abidor, I. G., and Chizmadzhev, Y. A., 1987, The electrical breakdown of cell and lipid membranes: The similarity of phenomenologies, *Biochim. Biophys. Acta* **902:**360–373.

Claßen, J., Haest, C. W. M., Tournois, H., and Deuticke, B., 1987, Gramicidin-induced enhancement of transbilayer reorientation of lipids in the erythrocyte membrane, *Biochemistry* **26:**6604–6612.

Coster, H. G. L., and Zimmermann, U., 1975, The mechanism of electrical breakdown in the membranes of Valonia utricularis, *J. Membr. Biol.* **22:**73–90.

de Kruijff, B., Cullis, P. R., Verkleij, A. J., Hope, M. J., Van Echteld, C. J. A., and Taraschi, T. F., 1985, Lipid polymorphism and membrane function, in: *The Enzymes of Biological Membrane* (2nd ed.), Volume 1 (A. N. Martonosi, ed.), Plenum Press, New York, pp. 131–204.

Deuticke, B., 1977, Properties and structural basis of simple diffusion pathways in the erythrocyte membrane, *Rev. Physiol. Biochem. Pharmacol.* **78:**1–97.

Deuticke, B., Poser, B., Lütkemeier, P., and Haest, C. W. M., 1983, Formation of aqueous pores in the human erythrocyte membrane after oxidative cross-linking of spectrin by diamide, *Biochim. Biophys. Acta* **731:** 196–210.

Deuticke, B., Lütkemeier, P., and Sistemich, M., 1984, Ion selectivity of aqueous leaks induced in the erythrocyte membrane by crosslinking of membrane proteins, *Biochim. Biophys. Acta* **775:**150–160.

Deuticke, B., Heller, K. B., and Haest, C. W. M., 1986, Leak formation in human erythrocyte by the radical-forming oxidant t-butylhydroperoxide, *Biochim. Biophys. Acta* **854:**169–183.

Deuticke, B., Heller, K. B., and Haest, C. W. M., 1987, Progressive oxidative membrane damage in erythrocytes after pulse treatment with t-butylhydroperoxide, *Biochim. Biophys. Acta* **899:**113–124.

Dimitrov, D. S., 1984, Electric field-induced breakdown of lipid bilayers and cell membranes: a thin viscoelastic film model, *J. Membr. Biol.* **78:**53–60.

Dressler, V., Schwister, K., Haest, C. W. M., and Deuticke, B., 1983, Dielectric breakdown of the erythrocyte membrane enhances transbilayer mobility of phospholipids, *Biochim. Biophys. Acta* **732:**304–307.

Dressler, V., Haest, C. W. M., Plasa, G., Deuticke, B., and Erusalimsky, J. D., 1984, Stabilizing factors of phospholipid asymmetry in the erythrocyte membrane, *Biochim. Biophys. Acta* **775:**189–196.

Gerritsen, W. J., Henricks, W. J., De Kruijff, P. A. J., and Van Deenen, L. L. M., 1980, The transbilayer movement of phosphatidylcholine vesicles reconstituted with intrinsic proteins from the human erythrocyte membrane, *Biochim. Biophys. Acta* **600:**607–619.

Ginsburg, H., and Stein, W. D., 1987, Biophysical analysis of novel transport pathways induced in red blood cell membranes, *J. Membr. Biol.* **96:**1–10.

Glaser, R. W., Wagner, A., and Donath, E., 1986, Volume and ionic composition changes in erythrocytes after electric breakdown, *Bioelectrochem. Bioenerg.* **16:**455–467.

Golan, D. E., Alecio, M. R., Veatch, W. R., and Rando, R. R., 1984, Lateral mobility of phospholipid and cholesterol in the human erythrocyte membrane: Effects of protein–lipid interactions, *Biochemistry* **23:** 332–339.

Gruber, W., and Deuticke, B., 1973,Comparative aspects of phosphate transfer across mammalian erythrocyte membranes, *J. Membr. Biol.* **13:**19–36.

Haest, C. W. M., 1982, Interactions between membrane skeleton proteins and the intrinsic domain of the erythrocyte membrane, *Biochim. Biophys. Acta* **694:**331–352.

Hamaguchi, H., and Cleve, H., 1972, Solubilization and comparative analysis of mammalian erythrocyte membrane glycoproteins, *Biochem. Biophys. Res. Commun.* **47:**459–464.

Heller, K. B., Poser, B., Haest, C. W. M., and Deuticke, B., 1984, Oxidative stress of human erythrocytes by iodate and periodate: Reversible formation of aqueous membrane pores due to SH-group oxidation, *Biochim. Biophys. Acta* **777:**107–116.

Israelachvili, J. N., 1977, Refinement of the fluid-mosaic model of membrane structure, *Biochim. Biophys. Acta* **469:**221–225.

Jacobs, M. H., 1952, The measurement of cell permeability with particular reference to the erythrocyte, in: *Modern Trends in Physiology and Biochemistry* (E. S. Guzman Barron, ed.), Academic Press, New York, pp. 149–171.

Kinosita, K., Jr., and Tsong, T. Y., 1977a, Hemolysis of human erythrocytes by a transient electric field, *Proc. Natl. Acad. Sci. USA* **74:**1923–1927.

Kinosita, K., Jr., and Tsong, T. Y., 1977b, Voltage-induced pore formation and hemolysis of human erythrocytes, *Biochim. Biophys. Acta* **471:**227–242.

Kinosita, K., Jr., and Tsong, T. Y., 1977c, Formation and resealing of pores of controlled sizes in human erythrocyte membrane, *Nature* **268:**438–441.

Kinosita, K., Jr., Ashikawa, I., Saita, N., Yoshimura, H., Itoh, H., Nagayama, K., and Ikegami, A., 1988, Electroporation of cell membrane visualized under a pulsed-laser fluorescence microscope, *Biophys. J.* **53:** 1015–1019.

Knight, D. E., and Scrutton, M. C., 1986, Gaining access to the cytosol: The technique and some applications of electropermeabilization, *Biochem. J.* **234:**497–506.

Lee, B., McKenna, K., and Bramhall, J., 1985, Kinetic studies of human erythrocyte membrane resealing, *Biochim. Biophys. Acta* **815:**128–134.

Lieber, M. R., and Steck, T. L., 1982a, A description of the holes in human erythrocyte membrane ghosts, *J. Biol. Chem.* **257:**11651–11659.

Lieber, M. R., and Steck, T. L., 1982b, Dynamics of the holes in human erythrocyte membrane ghosts, *J. Biol. Chem.* **257:**11660–11666.

Lindblom, G., Johansson, L. B. A., and Arvidson, G., 1981, Effect of cholesterol in membranes: Pulsed nuclear magnetic resonance measurements of lipid lateral diffusion, *Biochemistry* **20:**2204–2207.

Lindner, P., Neumann, E., and Rosenheck, K., 1977, Kinetics of permeability changes induced by electric impulses in chromaffine granules, *J. Membr. Biol.* **32:**231–254.

Lopez, A., Rols, M. P., and Teissie, J., 1988, ^{31}P NMR analysis of membrane phospholipid organization in viable, reversibly electropermeabilized Chinese hamster ovary cells, *Biochemistry* **27:**1222–1228.

Mehrle, W., Zimmermann, U., and Hampp, R., 1985, Evidence for asymmetrical uptake of fluorescent dyes through electro-permeabilized membranes of Avena mesophyll protoplasts, *FEBS Lett.* **185:**89–94.

Noordam, P. C., Van Echteld, C. J. A., de Kruijff, B., Verkleij, A. J., and de Gier, J., 1980, Barrier characteristics of membrane model systems containing unsaturated phosphatidylethanolamines, *Chem. Phys. Lipids* **27:**221–232.

Onuchi, J. F., and Lacaz-Vieira, F., 1985, Glycerol-induced baroprotection in erythrocyte membranes, *Cryobiology* **22:**438–445.

Pequeux, A., Gilles, R., Pilwat, G., and Zimmermann, U., 1980, Pressure-induced variations of K^+-permeability as related to a possible reversible electrical breakdown in human erythrocytes, *Experientia* **36:** 565–566.

Poo, M.-m., 1981, In situ electrophoresis of membrane components, *Annu. Rev. Biophys. Bioeng.* **10:**245–276.

Pooler, J. P., 1985, The kinetics of colloid osmotic hemolysis. I. Nystatin-induced lysis, *Biochim. Biophys. Acta* **812:**193–198.

Rand, R. P., and Parsegian, V. A., 1986, Mimicry and mechanism in phospholipid models of membrane fusion, *Annu. Rev. Physiol.* **48:**201–212.

Renkin, E. M., 1955, Filtration, diffusion, and molecular sieving through porous cellulose membranes, *J. Gen. Physiol.* **38:**225–243.

Riemann, F., Zimmermann, U., and Pilwat, G., 1975, Release and uptake of haemoglobin and ions in red blood cells induced by dielectric breakdown, *Biochim. Biophys. Acta* **394:**449–462.

Sale, A. J. H., and Hamilton, W. A., 1968, Effects of high electric fields on micro-organisms, *Biochim. Biophys. Acta* **163:**37–43.

Scherer, R., and Gerhard, P., 1971, Molecular sieving by the Bacillus megaterium cell wall and protoplast, *J. Bacteriol.* **107:**718–735.

Schneider, E., Haest, C. W. M., Plasa, G., and Deuticke, B., 1986, Bacterial cytotoxins, amphotericin B and local anesthetics enhance transbilayer mobility of phospholipids in erythrocyte membranes: Consequences for phospholipid asymmetry, *Biochim. Biophys. Acta* **855:**325–336.

Schwister, K., 1985, Bildung, Charakteristika aund Ausheilverhalten elektrisch induzierter Poren in der Erythrozyten-Membran, Ph.D. thesis, RWTH Aachen.

Schwister, K., and Deuticke, B., 1985, Formation and properties of aqueous leaks induced in human erythrocytes by electrical breakdown, *Biochim. Biophys. Acta* **816:**332–348.

Smith, G. K., and Cleary, St. F., 1983, Effects of pulsed electric fields on mouse spleen lymphocytes in vitro, *Biochim. Biophys. Acta* **763:**325–331.

Sowers, A. E., 1986, A long-lived fusogenic state is induced in erythrocyte ghosts by electric pulses, *J. Cell Biol.* **102:**1–5.

Sowers, A. E., and Hackenbrock, C. R., 1981, Rate of lateral diffusion of intramembrane particles: Measurement by electrophoretic displacement and rerandomization, *Proc. Natl. Acad. Sci. USA* **78:**6246–6250.

Sowers, A. E., and Lieber, M. R., 1986, Electropore diameters, lifetimes, numbers, and locations in individual erythrocyte ghosts, *FEBS Lett.* **205:**179–184.

Stein, W. D., 1986, *Transport and Diffusion across Cell Membranes,* Academic Press/Harcourt Brace Janovich, New York.

Stenger, D. A., and Hui, S. W., 1986, Kinetics of ultrastructural changes during electrically induced fusion of human erythrocytes, *J. Membr. Biol.* **93:**43–53.

Sugar, I. P., and Neumann, E., 1984, Stochastic model for electric field-induced membrane pores, *Biophys. Chem.* **19:**211–225.

Teissie, J., and Blangero, C., 1984, Direct experimental evidence of the vectorial character of the interaction between electric pulses and cells in cell electrofusion, *Biochim. Biophys. Acta* **775:**446–448.

Tournois, H., Leunissen-Bijvelt, J., Haest, C. W. M., de Gier, J., and de Kruijff, B., 1987, Gramicidin induced hexagonal H_{II} phase formation in erythrocyte membranes, *Biochemistry* **26:**6604–6612.

Tsuji, K., and Neumann, E., 1983, Conformational flexibility of membrane proteins in electric fields. I. Ultraviolet absorbance and light scattering of bacteriorhodopsin, *Biophys. Chem.* **17:**153–163.

Van der Steen, A. T. M., de Kruijff, B., and de Gier, J., 1982, Glycophorin incorporation increases the bilayer permeability of large unilamellar vesicles in a lipid-dependent manner, *Biochim. Biophys. Acta* **691:**13–23.

Verkleij, A. V., 1984, Lipidic intramembranous particles, *Biochim. Biophys. Acta* **779:**43–63.

Weaver, J. C., Harrison, G. I., Bliss, J. G., Mourant, J. R., and Powell, K. T., 1988, Electroporation: High frequency of occurrence of a transient high-permeability state in erythrocytes and intact yeast, *FEBS Lett.* **229:**30–34.

Wilbrandt, W., 1941, Osmotische Natur sogenannter nicht osmotischer Hämolysen (Kolloidosmotische Hämolyse), *Pfluegers Arch. Gesamte Physiol. Menschen Tiere* **245:**23–52.

Wright, E. M., and Diamond, J. M., 1977, Anion selectivity in biological systems, *Physiol. Rev.* **57:**109–156.

Zimmermann, U., 1982, Electric field-mediated fusion and related electrical phenomena, *Biochim. Biophys. Acta* **694:**227–277.

Zimmermann, U., Pilwat, G., and Riemann, F., 1974, Dielectric breakdown of cell membranes, *Biophys. J.* **14:** 881–899.

Zimmermann, U., Pilwat, G., Beckers, F., and Riemann, F., 1976, Effects of external electrical fields on cell membranes, *Bioelectrochem. Bioenerg.* **3**:58–83.

Zimmermann, U., Vienken, J., and Pilwat, G., 1980, Development of drug carrier systems: Electrical field induced effects in cell membranes, *Bioelectrochem. Bioenerg.* **7**:553–574.

Chapter 9

Electroporation of Cell Membranes

Mechanisms and Applications

Tian Yow Tsong

1. INTRODUCTION

The recent success in applying the pulsed electric field (PEF) method to introduce gene material into living cells and to demonstrate the expression of these genes (Wong and Neumann, 1982; Neumann *et al.*, 1982; Potter *et al.*, 1984) has opened up new possibilities in molecular biology and genetic engineering research. Electric modification of cell membrane permeability, or "electroporation," is believed to be the basis of the DNA entry. No less an accomplishment is the use of PEF to induce fusion of cells when these cells are brought into contact before (Zimmermann and Pilwat, 1978; Zimmermann and Vienken, 1982; Teissie *et al.*, 1982; Berg *et al.*, 1983; Lo *et al.*, 1984) or after (Sowers, 1984, 1986) the PEF treatment. Electroporation, again, is believed to be one factor leading to membrane fusion. These findings have aroused the attention of cell biologists and biophysicists alike. One must realize, however, that the study of the effect of PEF on cell membranes has spanned more than three decades (Cole, 1972; see also Schwan, this volume). The present chapter reviews the physical characteristics of cell membranes that are essential for understanding the electroporation phenomenon and summarizes experiments done in my laboratory in this area.

Early studies of the effect of PEF on cell function led to the finding that PEF in the kV/cm range could cause the death and lysis of cells (Sale and Hamilton, 1967, 1968), and rupture of the cell membranes was recognized as the cause. Although recent studies have been conducted under ostensibly unrelated experimental circumstances, one is astonished with the thoroughness of Sale and Hamilton's work 20 years ago. For example, their work suggested that cell killing by PEF correlated with the field strength of the PEF but it did not correlate well with power input or joule heating. They showed that with microsecond PEF, the killing of yeast and hemolysis of erythroyctes occurred around 3 kV/cm. This could generate a transmembrane potential of approximately 1 V. Membrane rupture and the subsequent escape of cytoplasmic contents were monitored by hemoglobin and DNA leakage. All these results have been repeated and confirmed by work from several other laboratories (see references in Zimmermann *et al.*, 1981; Tsong, 1983).

TIAN YOW TSONG • Department of Biochemistry, University of Minnesota, St. Paul, Minnesota 55108.

2. STRUCTURE OF CELL MEMBRANE AND FIELD-INDUCED TRANSMEMBRANE POTENTIAL

The bilayer structure of a cell membrane is a dielectric. When it is overloaded with an electric field (i.e., a potential exceeding its dielectric strength) in the range of 1000 kV/cm, the barrier of the bilayer to permeation will break down (Cole, 1972; Tien, 1974). The breakdown voltage of a planar lipid membrane (BLM) is easily measured with electrodes. With a stationary transmembrane electric field, the breakdown potential of the lipid bilayer is in the range of 100–300 mV (Tien, 1974), depending on the lipid composition. Breakdown voltage increases with decreasing PEF pulse width (Benz *et al.*, 1979; Benz and Zimmermann, 1980). For example, with a pulse width of 500 nsec, the breakdown voltage measured by the charge-pulse technique was found to be around 1 V at 20°C (Benz *et al.*, 1979; Benz and Zimmermann, 1980). For cell membranes, the breakdown voltage was considerably higher. Under similar experimental conditions, the breakdown potential was 2.2 V for the giant cells of the alga *Valonia utricularis* (Coster and Zimmermann, 1975; Zimmermann and Benz, 1980).

When a spherical cell in suspension is exposed to an electric field, the Maxwell equation describing the current in the region near the cell is given by

$$\Delta \cdot J + \frac{\partial \rho}{\partial t} = 0$$

where J is the current density, ρ the resistivity, and t the time variable. If charges are not generated in a cell during the PEF treatment, this relation simplifies to

$$\Delta \cdot J = 0$$

Since the electric field is the gradient of the electric potential, the Maxwell equation becomes

$$\Delta^2 \psi = 0$$

where ψ denotes the electric potential. If the conductivity of a cell membrane is much smaller than that of the cytosol and external medium, and the thickness of the low-conductivity shell is small compared to the outer radius of the sphere, the solution to the Maxwell relation (the Laplacian) for a field-induced transmembrane potential, $\Delta\psi$, can be expressed as (e.g., see Pauly and Schwan, 1966; Cole, 1972; Tsong *et al.*, 1976; Lindner *et al.*, 1977)

$$\Delta\psi = 1.5\ a\ E \cos\theta \qquad (1)$$

where a, E, and θ are, respectively, the outer radius of the cell, the applied field strength, and the angle between the field line and the normal to the point of interest in the membrane. The maximum $\Delta\psi$ generated by E is simply $1.5\ a\ E$ where θ is either 0° or 180°. This relation assumes that the pulse width of PEF is much greater than the RC constant of the cell membrane, where C is the capacitance of the cell membrane (0.5–1.2 $\mu F/cm^2$) and R the resistance of the external medium. If an AC electric field is used, the Schwan equation should apply:

$$\Delta\psi = 1.5\ a\ E \cos\theta/[1 + (\omega\tau)^2]^{1/2} \qquad (2)$$

where ω and τ are, respectively, the angular frequency of the AC field and the dielectric relaxation constant (RC constant) of the cell membrane. For most of the experiments discussed here, $\omega\tau$ was much smaller than unity and the Schwan equation reduces to Eq. (1).

Equation (1) has been, in general, experimentally shown to be applicable to lipid vesicles. This was done by comparing the dielectric breakdown potential of BLM and uniformly sized lipid vesicles (diameter 95 ± 5 nm). The breakdown potential of vesicles was monitored by the entrapment or the leakage of Rb^+, Na^+, and sucrose tracers from these vesicles (Teissie and Tsong, 1981b; El-Mashak and Tsong, 1985). For dipalmitoylphosphatidylcholine vesicles, the field strength for the rupture of the lipid bilayer was 30.0 kV/cm, which according to Eq. (1) gives a maximum $\Delta\psi$ of 215 mV. This value agrees with the breakdown potential of BLM, 210 mV, directly measured by the electrode method (Tien, 1974).

If a membrane vesicle is not spherical, $\Delta\psi$ generated by an electric field may not follow Eq. (1). Kinosita and Tsong (1977c, 1979), as well as many others (e.g.,Gross *et al.*, 1986; Farkas *et al.*, 1984), have considered ellipsoidal structures and shown that the deviation from Eq. (1) is not severe. For all practical purposes, Eq. (1) can be used to estimate field-induced transmembrane potential. Gross *et al.* (1986) have used fluorescence imaging spectroscopy to measure the spatial distribution of the electric field-induced $\Delta\psi$ in several ellipsoidal cell types, using a potential-sensitive dye whose fluorescence intensity is calibrated against a known $\Delta\psi$. In the case of the human carcinoma A-431, the measured values agree well with values calculated from an extended form of Eq. (1) for ellipsoidal cell types. In other cell types, the agreement is not always good. It remains uncertain whether the $\Delta\psi$ estimated by Eq. (1), or its extended form, can be reliably applied to cells of heterogeneous surface structure. However, the optical imaging method should be able to directly measure the $\Delta\psi$ distribution on these cells, or on tissues and aggregated cell mass, or whenever such information needs to be directly determined (Gross *et al.*, 1986). The dielectric relaxation of thylakoid has been reported by Farkas *et al.* (1984).

3. ELECTROLYSIS OF CELLS AND ELECTROPORATION OF MEMBRANES

3.1. Electrolysis of Cells

The first sign of exposing cells in a suspension to kV/cm PEF is cell lysis, or for red blood cells hemolysis. The terms "dielectric breakdown" or "electric breakdown" had been used to describe the phenomenon up until the early 1980s. These two terms do not clearly distinguish global rupture and local modification of the cell membranes, although data available at the time suggested a limited rupture. For example, work of Zimmermann and co-workers on erythrocytes and other types of cells had shown that after the cells were lysed, the envelope or ghost thus produced could slowly regain the permeation barrier against large-size molecules or latex beads (Zimmermann *et al.*, 1981). Likewise, the earlier work of Sale and Hamilton (1967, 1968), and later of Neumann and Rosenheck (1972), also suggested that membrane ruptures by PEF were less than a global occurrence. Whether these PEF-treated cells, or PEF-produced cell envelopes and red cell ghosts, could restore the permeation barrier to Na^+, K^+, Ca^{2+}, and other ions was not investigated in these studies, however. Zimmermann and co-workers were among the first to suggest that cell *envelopes* or *ghosts* prepared by the PEF method could be used as drug carriers for clinical applications. Such ideas have been tested on many occasions (Zimmermann *et al.*, 1976, 1980b; Zimmermann and Pilwat, 1978). Charlmers (1985) used extreme conditions (20 kV/cm) to prepare erythrocyte membrane fragments and suggested that such types of cell debris are more superior drug carriers than the PEF-treated, drug-loaded erythrocytes which retained nearly 100% of the hemoglobin content (Kinosita and Tsong, 1978).

Neumann and co-workers used milder conditions in their studies of chromaffin granules, and suggested that reversible dielectric breakdown could be achieved in this system within 200 msec (Lindner *et al.*, 1977). They used optical signals (e.g., light scattering, turbidity) of the granule suspension to follow the breakdown and the resealing kinetics. However, in other types of cells or organelles, optical signals do not always measure pore opening and resealing events. (They perhaps reflect changes in cell size due to water influx or efflux.) For example, in human erythrocytes, optical signals are not in agreement with tracer measurements. Figure 1 shows light scattering signals of an erythrocyte suspension following PEF treatment (5 kV/cm, exponential decay time 3

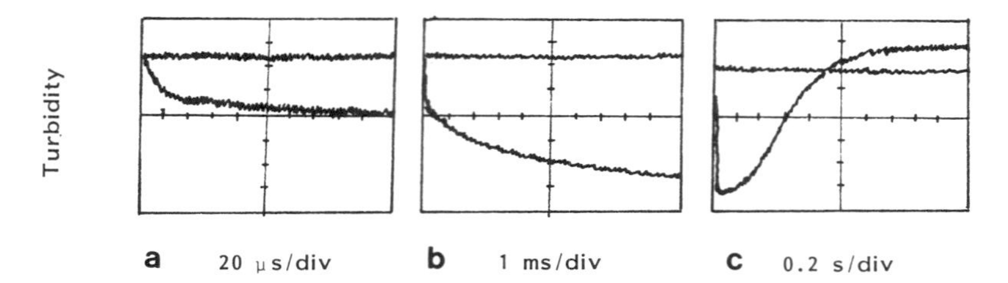

Figure 1. Turbidity changes of an isotonic saline suspension of human erythrocytes following an electric impulse of 5 kV/cm, with a decay constant of 3 μsec. The first kinetic phase occurred around 20 μsec (a), and the second phase occurred around 5 msec (b). Both phases were detected by the decrease in the turbidity, at 300 nm. Subsequently, there was a slow complex kinetic phase which nearly brought the turbidity of the suspension to its initial value (c). The first and second phases are interpreted as resulting from electric pore formation and the third phase from shrinking of the electric pores. Complete resealing of pores occurred in minute to hour time ranges (Fig. 4). See text for details.

μsec). This is to be compared with that of a similar sample in which recovery of the permeation barrier to Rb^+ was followed by the isotope tracer, $^{86}Rb^+$ (Fig. 4). Membrane conductance of PEF-treated red cells (see below) is also shown with time (Fig. 5). If light scattering signals were taken literally, the resealing, in this case, would have been completed within 2 sec. Yet, the tracer experiments indicated resealing in the 10 min range. The discrepancy between the light scattering and the tracer method was astounding, although the result of the conductance measurement was in general agreement with that of the tracer measurement (Kinosita and Tsong, 1979; Serpersu *et al.*, 1985). As will be shown later, use of tracers to measure the permeability of the PEF-treated cells is perhaps a more stringent criterion, and it is a reliable way of ascertaining the reversibility of electroperforated cell membranes. Other laboratories have studied different cell systems and several of these are represented in this volume.

3.2. Electric Pores and Colloid Osmotic Lysis of Cells

A surprise observation was made in 1977 (Kinosita and Tsong, 1977a) when we were investigating the kinetics of the leakage of different classes of ions or molecules in a PEF-treated human erythrocyte sample. We observed that after the sample was treated with a single 3.7 kV/cm electric pulse (squared wave, 20 μsec), a procedure known to cause 100% hemolysis of red cells in an isotonic saline, the internal K^+ of red cells leaked out within a few seconds. Likewise, the external Na^+ leaked in rapidly. However, the hemoglobin content of the red cells remained in the membrane-perforated cells for hours. These cells eventually lysed after a 6- to 20-hr incubation at 4°C. In fact, under these conditions, before the lysis of cells, cell membranes were still impermeable to a sucrose tracer. However, these cells continued to swell because of the colloid osmotic pressure of the cytosolic macromolecules. When the cell volume reached 155% of the initial volume, the cell membranes were torn and the swollen cells deflated. It was at this moment the cytoplasmic macromolecular contents escaped from the cells. Entrapment of large molecules, such as enzymes, into ghosts also occurred at this moment.

These experiments suggested to us that cell lysis was not due to the primary effect of the PEF, at least in the range of the field strength and the pulse width used in our experiments (field strength up to 10 kV/cm, pulse width up to 400 μsec). The main effect of the PEF was to electroperforate the cell membranes. Then as the osmotic pressure of the cytosol and the external medium became unbalanced and the cells swelled, the membrane was torn because of the overswelling. The perforation-to-swelling-to-hemolysis phenomenon is known as the colloid osmotic hemolysis. Figure 2 illustrates both the colloid osmotic lysis of a cell and its prevention.

Our subsequent experiments have shown that the size of molecular probes which can permeate into PEF-treated red cells depends on the following factors.

1. The field strength: The greater the applied field strength, the larger the probe molecules can permeate into the treated cells preceding cell lysis (Kinosita and Tsong, 1977a,b,c).
2. The pulse duration: The longer the PEF, the larger the probe molecules can permeate (Kinosita and Tsong, 1977b).
3. The ionic strength of the suspending medium: PEF treatment in a higher-ionic-strength medium, e.g., isotonic saline, leads to implantation of small pores, and in a lower-ionic-strength medium, e.g., isotonic sucrose, to bigger pores, when identical PEF treatment is used. In our experiments, this factor appears the most effective for controlling pore size with erythrocytes (Kinosita and Tsong, 1977b).
4. Temperature: As shown by Zimmermann and Benz (1980), PEF treatment of *Valonia utricularis* at higher temperatures leads to a lower critical voltage, implying that the induced pores could be larger. It is unclear whether or not one can take advantage of this property to load molecules of larger size into cells. The uncertainty is due to the fact that at a temperature close to the growing temperature of cells, PEF-induced pores tend to shrink or to reseal much faster than they would at a low temperature (Kinosita and Tsong, 1977b, 1978, 1979).

In 1977, during these experiments, we introduced the term "membrane pores" to distinguish the PEF modification of membrane permeability from the process of the hypotonic lysis of cells (Kinosita and Tsong, 1977a,b,c). Until now, I have used the term "pores' rather casually. How-

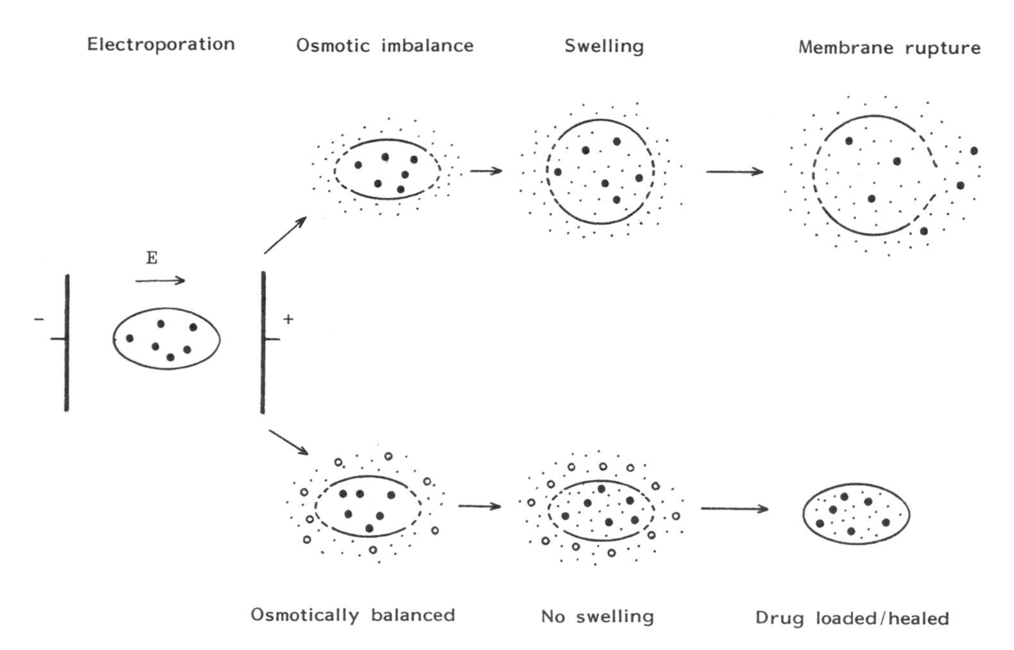

Figure 2. Colloid osmotic hemolysis and resealing of an electrically perforated red cell. (Upper path) When the cell membrane is electrically implanted with micropores, the membrane becomes permeable to ions and small probe molecules (shown as dots). These pores are small enough to block passage of hemoglobin (filled circles). As a result, the colloid osmotic pressure of hemoglobin will lead to influx of water. The cell swells, and the membrane is ruptured because of overswelling. (Lower path) If, on the other hand, molecules larger than the electric pores (open circles) are added to the external medium to counterbalance the osmotic pressure of hemoglobin, cell swelling can be prevented. Under such conditions, the electroperforated membrane will heal. The membrane-resealed cell now carries the drug in its cytoplasm. This cell behaves like a normal cell, except for the effect of the drug. See Fig. 3.

ever, the concept of pores, or "aqueous pores," as we define it, should gradually emerge as we continue. These pores are nonspecific as to the types of permeants they admit, except conceivably by size (Kinosita and Tsong, 1977a,b; Serpersu and Tsong, 1985).

4. LOADING AND RESEALING OF MOLECULES INTO FUNCTIONING CELLS

Loading of drugs or enzymes into PEF-prepared cell envelopes or erythrocyte ghosts had been reported before 1977 (Zimmermann *et al.*, 1976, 1978, 1980, 1981). However, at the time, no studies had been published which showed that exogenous molecules could be entrapped into "intact" or "functioning" cells by using the PEF method. If pores can be implanted in a cell membrane, and if small molecules can penetrate before cytosolic macromolecular contents leak out, there is a good likelihood that one can load molecules into PEF-treated cells and then reseal the perforated membranes before these cells reach the lysis stage. Despite the use of the term "reversible membrane breakdown," true reversibility had never been tested stringently before 1977 (e.g., by the criteria of: the complete repossession of ionic permeation barrier; cell survival, *in vivo*; and cell growth in cultures). Our other major concern was whether a complete resealing of PEF-perforated membranes can be easily achieved before cell lysis. Here, the benefit of working out the colloid osmotic mechanism of cell lysis became apparent. We designed the following protocols and succeeded in loading exogenous molecules into 100%-hemoglobin-retained erythrocytes (Kinosita and Tsong, 1977b, 1978; Serpersu *et al.*, 1985) (Fig. 2).

1. Pore size to be introduced in a cell membrane must be carefully controlled by adjusting the strength and the duration of PEF, and the ionic strength of the cell suspending medium. PEF-induced pores preferably should be smaller to avoid losing cytoplasmic contents. Loss of Na^+, K^+, and other, divalent ions can be minimized by using a suspending medium which mimics the ionic composition of the cytoplasmic fluid.
2. The molecular probe, or drug, to be loaded should be in the medium during the PEF treatment. We called this medium the *pulsing medium* (Kinosita and Tsong, 1977b, 1978; Serpersu *et al.*, 1985). Loading should be done at a low temperature to avoid the rapid shrinking of pores.
3. The most critical step is that after the PEF treatment, cell swelling due to colloidal pressure of the cytosolic macromolecules must be checked to prevent overswelling. To achieve this goal, molecules larger than the PEF-induced pores must be added to the medium a certain time after the PEF treatment. This osmotically balanced medium is termed the *resealing medium* (Kinosita and Tsong, 1977b, 1978; Serpersu *et al.*, 1985; Tsong and Kinosita, 1985). The resealing medium should also contain molecules to be loaded and other ingredients to aid cell recovery (Kinosita and Tsong, 1977b, 1978; Tsong and Kinosita, 1985). We have used oligosaccharides or small water-soluble proteins for osmotic balancing. For human erythrocytes, the osmotic balancing molecule is required at a concentration of 25–35 mM to halt the cell swelling, thus preventing cell lysis (Kinosita and Tsong, 1977b, 1978). In a well-balanced resealing medium, the electroperforated cells can be kept for 1–2 days without significant loss of cellular macromolecular contents, or of cells due to lysis.
4. Thereafter, the temperature of PEF-treated, probe-loaded cells in the resealing medium is raised to 37°C and incubated at this temperature until the resealing is complete. During this period, aliquots are drawn frequently to check the full recovery of the membrane permeation barrier using the Rb^+ permeation experiment.

An example using radiolabeled sucrose as a model drug is shown in Fig. 3. The resealing step is absolutely required for erythrocytes if the drug-loaded cells are to survive after returning to the circulation. However, other cell types may not require this step owing to the presence of a cell wall or a stronger membrane structure. Loading of long fibrous molecules such as DNA, or globular

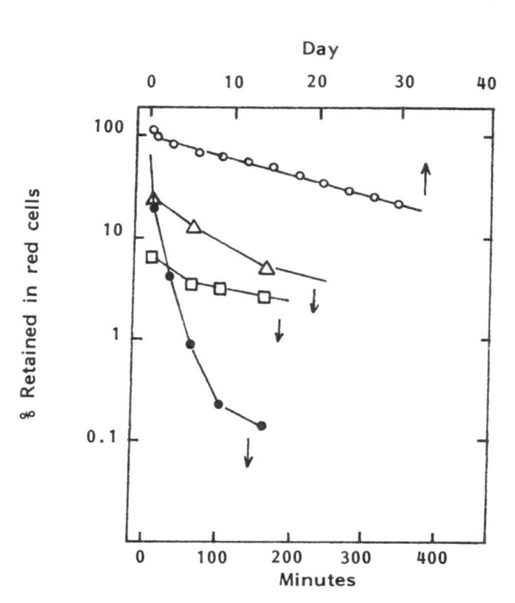

Figure 3. Survival of sucrose-loaded mouse red cells in the circulation. When radioactive sucrose was directly administered into the circulation (filled circles), it dissipated in 2 hr. However, when the sucrose was loaded in mouse erythrocytes and properly resealed before injection into the circulation, the sucrose was retained in the circulation with a half-life of 20 days (open circles). Erythrocyte membranes that were leaky to Rb^+, or erythrocytes that lost their hemoglobin (such as ghosts) did not survive in the circulation. Erythrocyte-loaded penicillin G (triangles) and actinomycin D (squares) were not retained in the circulation. This was not because of the improper resealing but rather because the drugs were found to be membrane permeants, and escaped in several hours. Arrows indicate the appropriate scale. Data from Tsong and Kinosita (1985).

proteins, or cell organelles may not follow the mechanisms described here. In these latter cases, binding of DNA on the membrane surface (i.e., by adding 2–5 mM Mg^{2+} before the PEF treatment) seems to facilitate DNA entry and promote expression of loaded genes (Neumann *et al.*, 1982). For the practical aspect of gene loading, see the chapter by Förster and Neumann.

5. PROPERTIES OF ELECTRIC PORES

5.1. Lipid Vesicles versus Cell Membranes

In lipid vesicles, *a priori,* there are no specific parts of the bilayer surface which are more susceptible to PEF puncture than any other part, except that due to the distribution of the transmembrane electric field as described by Eq. (1). However, most theories of electroporation of lipid bilayers (Hui *et al.*, 1981) assume there are thermally fluctuating, spontaneous aqueous pores on lipid bilayers, and that these pores, or dynamic lattice defects, on exposure to a transbilayer electric field, can expand, shrink, and reseal in the time scale of microseconds to milliseconds (El-Mashak and Tsong, 1985), according to their stability. Pore stability is determined by certain factors, e.g., edge free energy, or line tension (e.g.,Kanehisa and Tsong, 1978; Weaver and Powell, this volume) and might be adequately described by a stochastic process (see Sugar, this volume; Weaver and Powell, this volume). In cells, the structure of membranes is heterogeneous. There are channels, pores, and transport proteins. Each class of molecules may respond to PEF-induced transmembrane potential in different ways. Lipid domains, likewise, have their characteristics. These factors may not be adequately portrayed by a stochastic process. I will discuss "electroconformational change" mechanisms in Section 6.

5.2. Reversibility

Experimentally, PEF-induced electric pores in lipid vesicles are transient and their lifetime is short (Benz *et al.*, 1979; Teissie and Tsong, 1981b; El-Mashak and Tsong, 1985). Thus, ions or molecules entrapped within synthetic lipid vesicles will be retained in these vesicles unless the lipid

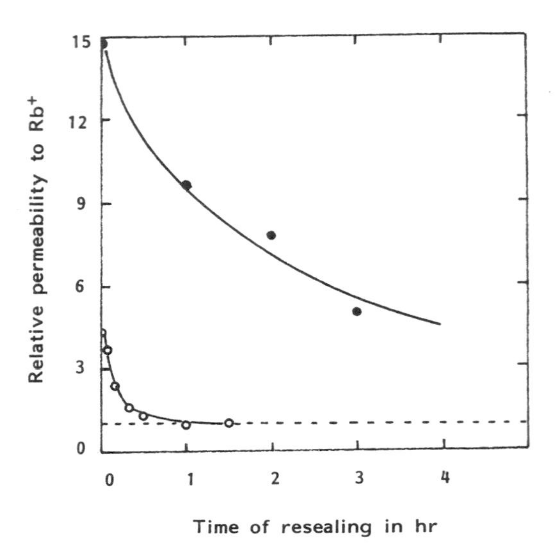

Figure 4. Restoration of the Rb^+ permeation barrier of electroperforated red cell membranes (adapted from Serpersu et al., 1985). (Lower curve) Red cells in isotonic saline were treated with a square wave of 4 kV/cm for 2 μsec. These cells became leaky to Rb^+ but not to sucrose. Restoration of the membrane permeation barrier against Rb^+ took more than 30 min. (Upper curve) If red cells were treated with 3.7 kV/cm PEF for 20 μsec (in a mixture of 10% isotonic sucrose and 90% isotonic saline), they became permeable to sucrose, but only transiently. Resealing of these red cells against Rb^+ leakage took more than 20 hr. See text for details.

bilayer is inherently permeable to these ions or molecules (Teissie and Tsong, 1981b; El-Mashak and Tsong, 1985). Electric pores in lipid vesicles are naturally reversible (not planar lipid bilayers, which may or may not be reversible depending on many factors; see Weaver and Powell, this volume). "Reversible dielectric breakdown" when applied to cell membranes has a different meaning. In fact, electric pores in permeabilized cell membranes turn out not to be so short-lived. With cell membranes, there are three possibilities. In the first case, after loading of moderate- to large-size molecules, e.g., DNA, small proteins, sucrose, or oligosaccharides, the PEF-perforated membranes rapidly regain the permeation barrier against these molecules. That is to say, as far as larger molecules are concerned, electroporation is a reversible process.

The second case is to consider reversibility with respect to ionic permeation. The ionic barrier of a cell membrane is especially important because cells having membranes leaky to K^+, Na^+, or Ca^{2+} cannot survive, grow, or proliferate under most circumstances. Nothing can be further from the truth than to assume that after PEF-treated cells have restored their permeation barrier to DNA, proteins, or other moderate-size molecules, they also have restored their normal barrier to ions or other small molecules. The lower curve in Fig. 4. indicates that even with a mild PEF treatment in which pores introduced were so small as to block sucrose penetration, it still took 30 min of a careful resealing step (discussed above) for these cells to regain their normal barrier to Rb^+, an ion usually used instead of K^+. The upper curve indicates another experiment with a slightly stronger PEF treatment for the red cells. These cells entrapped sucrose initially, and took less than 1 min to recover their barrier against sucrose from further entry or leak. However, it took nearly another 20 hr of "careful resealing" to salvage their normal barrier against Rb^+. Lipids of a cell membrane are mobile, the lateral diffusion coefficient being in the range of 1×10^{-7} to 1×10^{-9} cm^2/sec. Electric pores are less than or not much greater than about 1.0 nm in size after the rapid shrinking step (Kinosita and Tsong, 1977b, 1979). Pores that occur in the lipid domain should thus reseal within seconds. This estimate of pore resealing in the lipid bilayer was substantiated by our experiment with lipid vesicles (Teissie and Tsong, 1981b; El-Mashak and Tsong, 1985). However, the fact that it took a much longer time for the PEF-treated red cells to recover the normal permeation state of membranes suggested to us that electric "pores" in cell membranes did not occur only in the lipid domain. Proteins of cell membranes were likely involved.

In the third case, one must realize that cell membranes are permeable to most drugs and biological molecules. These molecules cannot be used to measure the extent of membrane resealing (Kaibara and Tsong, 1980; Tsong and Kinosita, 1985). After PEF loading, these molecules may still escape from a cell, but not because of failure to reseal the membrane properly. In fact, of the dozen drugs of various origins and clinical activities that we have tested, none were found to be membrane impermeant. For example, Kaibara and Tsong loaded several antigelation reagents into heterozygous sickle (SS) erythrocytes. The drug-loaded SS erythrocytes resisted sickling when oxygen was depleted from the suspensions. This indicated that these reagents were effective, *in vivo,* in their antigelation activity. However, they leaked out at 37°C within several hours (they were retained for days at 4°C), and the method was considered impractical (unpublished results). This being the case, drugs to be carried by cells must be modified, e.g., by attaching a carbohydrate or a highly charged moiety to its backbone as is suggested from the sucrose loading experiments of Kinosita and Tsong (1977b).

5.3. Kinetics of Electroporation

To monitor rapid kinetic events during electric pulse treatment of cells, we designed a sensitive means for detecting PEF-induced, or "pore-linked," membrane conductance. It was reasoned that once ion channels were burst open by a PEF greater than the threshold voltage, there would be an extra membrane conductance due to ion passage through these channels, as compared to the conductance of the same cell suspension treated with a subthreshold PEF. A differential circuit was constructed to measure this extra membrane conductance (Kinosita and Tsong, 1979; see Fig. 5). For human erythrocytes subjected to 3.7 kV/cm PEF, pore-linked membrane conductance (see Section 3) could be detected as fast as 0.5 μsec (Fig. 5). After this initial phase, a slower reaction in 50 μsec was detected (Kinosita and Tsong, 1979; Teissie and Tsong, 1980). If the same PEF were used for permeation of probe, Rb^+, K^+, Na^+ would permeate into the red cells with a pulse width less than 2 μsec, but sucrose would not permeate for a pulse width less than 10 μsec. Thus, pore initiation occurs in 1 μsec and pore expansion in 10 μsec time ranges. These results have been confirmed by the tracer permeation experiment of Serpersu *et al.* (1985; Fig. 5b). This latter experiment also verified the validity of using PEF-induced membrane conductance signals to monitor pore formation/resealing events. However, it is unclear whether the 0.5–1 μsec kinetic was limited by the *RC* constant of the red cell or by the pore initiation process.

6. ELECTROPORATION THROUGH MEMBRANE CHANNELS?

6.1. PEF-Induced Membrane Conductance

If the electroporation of lipid vesicles happens at thermally fluctuating dynamic pores or lattice defects, in cell membranes, there are other types of naturally occurring membrane pores which are perceptibly longer-lived. These protein channels or transport proteins may be subject to PEF penetration. The cytoskeletal network of a cell membrane may also be ruptured by a strong electric field. Once these protein components of a cell membrane are modified by an electric field, they may take a longer time to recover or to repair. To test this idea, we performed the following experiments.

Inhibitors to various transport proteins were added to determine whether any of these could block the PEF-induced membrane conductance. Both anion and cation transport inhibitors of red cells were tested (Teissie and Tsong, 1980, 1981a).

Na,K-ATPase of human erythrocytes. Most inhibitors had some effects, but none of the experiments were interpretable except the one using ouabain, a specific and potent inhibitor of Na,K-ATPase (Teissie and Tsong, 1980). Roughly 35% of the PEF-induced membrane conductance was blocked by the preincubation of red cell samples with ouabain for 30 min before the PEF treatment. The half point of inhibition effect was 0.15 μM, which agrees with the inhibition constant of the drug to the ATPase activity of the enzyme. This result suggests that, indeed, electroporation in cell membranes can occur at channels or pump proteins. Obviously, Na,K-ATPase was not the only site of electroporation since another 65% of membrane conductance could

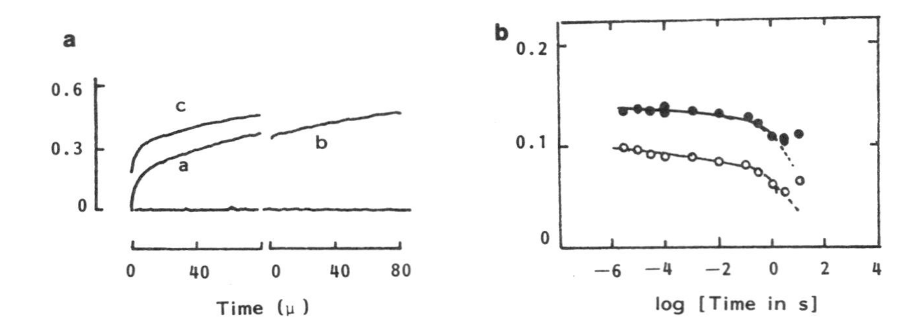

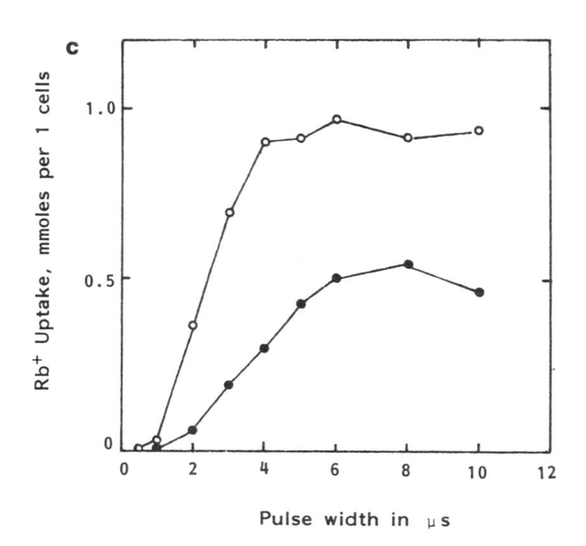

Figure 5. The kinetics of electroporation and resealing in human erythrocytes. (a) Electric pore-linked membrane conductance, with a single 80-μsec 3.1 kV/cm pulse, was used to follow pore formation kinetics in a mixture of 90% isotonic sucrose and 10% isotonic saline. Hematocrit of the suspension was 19%. Curve a indicates a rapid kinetic phase of pore initiation in 0.5 to 1.5 μsec, followed by a pore expansion kinetic phase around 20 μsec. Pore resealing was monitored by applying a second pulse at a certain time after the first pulse. Curve b indicates pore-linked conductance induced by a second pulse 5 μsec after the first pulse was terminated. The beginning of the trace is closed at the end of curve a, indicating that pores stayed open 5 μsec after. If a separation of 10 msec was allowed between the two pulses, curve c was obtained. Here the beginning of the trace is closer to the initial position, indicating that pores were halfway closed after 10 msec incubation. In this way, pore resealing kinetics can be measured. (b) The kinetics of pore resealing were measured for PEF-treated samples in an isotonic mixture of 90% saline and 10% sucrose. Filled circles: sample treated with one 20-μsec 3.7 kV/cm pulse; open circles: sample treated with one 2-μsec 3.7 kV/cm pulse. After 1 sec, pore-linked conductance began to increase again, indicating that electric pores never resealed reversibly. Dashed curves suggest that there is a slow pore resealing process in the 100 sec time range, although this process was never complete before the red cells began to swell. (c) The rapid pore initiation process was confirmed by the tracer experiments shown here. Erythrocytes in isotonic saline were PEF treated at 3.4 kV/cm (open circles) or at 2.8 kV/cm (filled circles) with different pulse widths. The relative permeability of membranes against Rb^+ was measured. For samples treated with a pulse width longer than 1 μsec, there was increased Rb^+ uptake, indicating that cell membranes were modified in these samples. See text for details.

not be traced to its origins. In this regard, the cytoskeletal network of 3T3 cells disintegrates when exposed to PEF and reorganizes after a long incubation period (Teissie, personal communication). This being the case, any theory dealing with the electroporation of cell membranes would have to take into account specific sites, in addition to random sites, for pore formation.

The next questions that came to our mind were: Was this channel opening an irreversible process? Would it be possible to find a condition where Na,K-ATPase could be reversibly activated by PEF? Na,K-ATPase is electrogenic, i.e., under physiological conditions it pumps three Na^+ out of the cytoplasm in exchange for two K^+. Since the transport of charge is asymmetric, the activation of this enzyme polarizes red cell membranes by about 10 mV. Therefore, a PEF which generates a $\Delta\psi$ in this range should activate the K^+ pump without denaturing the enzyme. In the electroporation experiment discussed above, the range of $\Delta\psi$ generated by the PEF was about 1000 mV, 100-fold greater than the normal range of the electric potential the enzyme is designed to handle. It is not surprising that when the enzyme is burst open by such an intense electric field, it would be irreversibly denatured. This line of thinking turned out to be correct.

Teissie and Tsong (1981a) used this differential conductance measurement with a low-amplitude AC field to detect a membrane conductance, 35% of which again was inhibited by ouabain. The strong evidence of Na,K-ATPase activation by a low-amplitude PEF came from our subsequent experiments using Na^+, K^+, and Rb^+ tracers (Serpersu and Tsong, 1983, 1984). Indeed, an AC field of 20 V/cm and 1 kHz induced K^+ or Rb^+ uptake by human erythrocytes. This uptake was an uphill transport, i.e., against the concentration gradient of the ion. There was no apparent consumption of ATP. The PEF-induced K^+ or Rb^+ uptake was completely inhibited by ouabain. Again, this indicated that the uptake was mediated by the Na,K-ATPase. These observations are especially important for understanding the mechanism of electroporation of cell membranes. The most significant observations of these experiments are that: (1) Certain membrane proteins or enzymes are capable of absorbing energy from a PEF and transducing it to other energy forms. (2) For this process to occur efficiently, there is an optimum field strength and an optimum frequency of the applied electric field. (3) These enzymes or proteins are structurally susceptible to an electric field, and are, thus, potential sites of electric modification or "electroporation."

6.2. Mechanisms of Electric Field Interaction

Electroporation experiments use kV/cm PEF which can induce a transmembrane potential of 500 to 1200 mV, or a trans-lipid bilayer electric field of 1000 to 2000 kV/cm (5 nm thickness). Endogenous transmembrane potential of cells is in the range of 10 to 250 mV, or a transbilayer electric field of 20 to 500kV/cm. In other experiments using low-amplitude electric fields, the range of transmembrane potentials generated spans a broad range, depending on cell size and tissue structure (e.g., see Adey, 1981). In all these cases, direct interaction of electric fields with cells or tissues most likely happens at the cell membrane, owing to the amplification of the electric field in and near the membrane surface [Eq. (1)]. How does an electric field interact with a membrane constituent, say a membrane protein? In other words, what is the primary effect(s) of the field interaction? Förster and Neumann's chapter deals specifically with this question. The discussion here is limited to the concept developed out of our study of PEF activation of Na,K-ATPase and mitochondrial F_0F_1 ATPase (Teissie *et al.*, 1981; Chauvin *et al.*, 1987).

Many studies of interactions of cells or tissues with low-amplitude electromagnetic fields fail to establish any relationship between the observed effect and the energy absorbed by the sample. Many theories, which completely disregard energy terms, are then proposed to explain these results. It is, however, odd to assume that any system or reaction which does not transfer energy from or to the applied electric field can derive any effect from it. Such would be in violation of basic thermodynamic principles. Consider the low-amplitude AC stimulation experiments discussed above. If we were to measure the power drawn by the red cell suspension and to correlate this quantity with the Rb^+ or K^+ uptake activity by the red cells, we would fail to see any correlation. This is not surprising. There are only 200–300 Na,K-ATPase molecules in a red cell, and the energy absorbed by these molecules would be miniscule compared to the total energy absorbed by a red cell or by the suspending medium. It is even more difficult to measure the fraction of energy absorbed by the enzyme in a sample in reference to the total energy applied to the sample. Difficult as it may sound, all indications point to the use of free energy contained in the AC field for the support of the active pumping of Rb^+ or K^+ by the Na,K-ATPase.

The basic relation governing the interaction of an electric field with a chemical reaction is given by a generalized van't Hoff equation,

$$\left[\partial(\ln K)/\partial E \right]_{p,V,T} = \Delta M/RT \tag{3}$$

in which the shift in the chemical equilibrium of a reaction is dependent on the difference in the macroscopic electric moment of the reactant and the product states, ΔM. (The symbols have the usual meaning.) Application of this relation to the understanding of the AC activation of Na,K-ATPase led us to propose the concept of "electroconformational change" and "electroconformational coupling" (Tsong and Astumian, 1986, 1987, 1988). Assume that the enzyme has two conformational states, E_1 and E_2, with an equilibrium constant at zero field, $K_0 = [E_1]_0/[E_2]_0$. E_1 has a permanent dipole of μ_1 and E_2 of μ_2. The polarizability of E_1 is α_1 and of E_2 is α_2. The ΔM of the two states under the electric field E is

$$\Delta M = (\mu_2 + \alpha_2 E) - (\mu_1 + \alpha_1 E) \tag{4}$$

If ΔM is positive, the field will shift the equilibrium toward E_2, and the concentration of $[E_1]_e < [E_1]_0$ and of $[E_2]_e > [E_2]_0$, where the subscript e denotes "under the influence of an electric field of E." If $[E_2]_e$ has a high affinity for the external K^+, and $[E_1]_e$ has a low affinity for the internal K^+, a four-state kinetic scheme will be able to absorb energy from an oscillating electric field for the pumping of K^+ from the external medium into the cytoplasmic medium (see Tsong and Astumian, 1986; Westerhoff *et al.*, 1986; Astumian *et al.*, 1987).

The mechanism proposed here goes beyond the proposal that charge accumulation in the two sides of a membrane is the primary effect of an electric field. Certainly, the increase in K^+ concentration near the outer surface of an erythrocyte is negligible when the low-amplitude AC field-induced transmembrane potential is only 12 mV. The K_m of outer $[K^+]$ for Na,K-ATPase is around 1.5 mM. The external medium contained 15 mM K^+, 10-fold higher than the K_m. Any slight increase in $[K^+]$ due to the AC field would have no effect on K^+ pumping by the enzyme. Thus, the uptake most likely was driven by the electroconformational coupling of the enzyme. Such an electroconformational change of a membrane protein is a plausible mechanism of electroporation.

Most membrane proteins have net charges at a neutral pH. They are also rich in α-helical structures. An α helix is a strong electrical dipole (Hol, 1985). These charges of functional groups and helices of the peptide backbone contribute to the electric moment of a protein molecule. The concept of electroporation occurring at both lipid domain and protein channels is consistent with our recent study of energy absorption by membrane-bound proteins (Tsong and Astumian, 1986, 1987, 1988), a process considered fundamental in the signal and energy transduction of cells.

7. PERSPECTIVE

The opening and closing of many membrane channels depend on the electric potential across the membrane. "Electroporation," or "electrochannel activity," or "electroactivation" of membrane enzymes or proteins are not only artificial processes, but also natural, functionally important reactions of cells and tissues. These processes are involved in signal and energy transductions, and the regulation of many cellular events. Nerve conduction, ion transport, and ATP synthesis are but a few of the most prominent examples. Electroporation by applied PEF can range from conditions which are very harsh and nonphysiological to conditions which mimic endogenous membrane potential. When high-intensity PEF is used, cell membranes can be punctured with artificial pores. These pores have a relatively long lifetime and exogenous molecules may penetrate into cells while these pores stay open. Resealing of these pores ensures the survival of PEF-treated cells. Most cells can reseal, although the time it takes is considerably longer than most investigators are willing to accept. Electrofusion of cells is by a similar process. One may try to distinguish the subtle difference between two scenarios in PEF-induced cell fusion. One portrays a membrane region implanted with many micropores, and the other portrays a membrane region modified to become "fusogenic" (Sowers, 1986). When such regions of two cells are brought into contact, the cells fuse. A complete fusion of two cells requires integration of their lipid bilayers, as well as their cytoskeletal networks. It would not be surprising if electroporation involves both lipids and proteins of the cell membrane. I see three directions for future investigations of electroporation phenomena. First, there will be continuous interest in the basic mechanisms. This interest may lead us to discovery of fundamental cellular processes yet unrealized. Second, there will be refinement and new design of experimental procedures, which will lead us to the study of single cell events, e.g., by using flow cytometry (see Weaver and Powell, this volume). Targeted electroporation or electrofusion to a specific population of cells in tissues or organs will also soon appear (see Lo and Tsong, this volume) and be practiced in clinical settings. And, third, it is obvious that applications of this method for molecular and genetic engineering and for agriculture and medical research will continue to provide the greatest impetus for moving the field ahead.

ACKNOWLEDGMENTS. This work has been supported by the U.S. Public Health Service and the Office of Naval Research. I thank my present and former colleagues, Drs. K. Kinosita, Jr., K. Makoto, J. Teissie, E. Serpersu, B. Knox, M. Tomita, and F. Chauvin, for their contributions to this project.

REFERENCES

Adey, W. R., 1981, Tissue interactions with nonionizing electromagnetic fields, *Physiol. Rev.* **61:**435–513.

Astumian, R. D., Chock, P. B., Tsong, T. Y., Chen, Y.-d., and Westerhoff, H. V., 1987, Can free energy be transduced from electric noise? *Proc. Natl. Acad. Sci. USA* **84:**434–438.

Benz, R., and Zimmermann, U., 1980, Relaxation studies on cell membranes and lipid bilayers in the high electric field range, Bioelectrochem. Bioenerg. **7:**723–739.

Benz, R., Beckers, F., and Zimmermann, U., 1979, Reversible electrical breakdown of lipid bilayer membranes: A charge-pulse relaxation study, *J. Membr. Biol.* **48:**181–204.

Berg, H., Schumann, I., and Steizner, A., 1983, Electrically stimulated fusion between myeloma cells and spleen cells, *Stud. Biophys.* **94:**101–102.

Charlmers, R. A., 1985, Comparison and potential of hypo-osmotic and iso-osmotic erythrocyte ghosts and carrier erythrocytes as drug and enzyme carriers, *Bibl. Haematol. (Basel)* **51:**15–24.

Chauvin, F., Astumian, R. D., and Tsong, T. Y., 1987, Voltage sensing of mitochondrial ATPase in pulsed electric field induced ATP synthesis, *Biophys. J.* **51:**243a.

Cole, K. S., 1972, *Membranes, Ions and Impulses,* University of California Press, Berkeley.

Coster, H. G. L., and Zimmermann, U., 1975, Dielectric breakdown in the membranes of *Valonia utricularis*: The role of energy dissipation, *Biochim. Biophys. Acta* **382:**410–418.

El-Mashak, M. E., and Tsong, T. Y., 1985, Ion selectivity of temperature-induced and electric field-induced pores in dipalmitoylphosphatidylcholine vesicles, *Biochemistry* **24:**2884–2888.

Farkas, D. L., Korenstein, R., and Malkin, S., 1984, Electroluminescence and the electrical properties of the photosynthetic membrane, *Biophys. J.* **45:**363–373.

Gross, D., Loew, L. M., and Webb, W. W., 1986, Optical imaging of cell membrane potential changes induced by applied electric fields, *Biophys. J.* **50:**339–348.

Hol, W. G. J., 1985, The role of the α-helix dipole in protein function and structure, *Prog. Biophys. Mol. Biol.* **45:**149–195.

Hui, S. W., Stewart, T. P., Boni, L. T., and Yeagle, P. L., 1981, Membrane fusion through point defects in bilayers, *Science* **212:**921–923.

Kaibara, M., and Tsong, T. Y., 1980, Voltage pulsation of sickle erythrocytes enhances membrane permeability to oxygen, *Biochim. Biophys. Acta* **595:**146–150.

Kanehisa, M. I., and Tsong, T. Y., 1978, Cluster model of lipid phase transitions with application to passive permeation of molecules and structure relaxations in lipid bilayers, *J. Am. Chem. Soc.* **100:**424–432.

Kinosita, K., Jr., and Tsong, T. Y., 1977a, Hemolysis of human erythrocytes by transient electric field, *Proc. Natl. Acad. Sci. USA* **74:**1923–1927.

Kinosita, K., Jr., and Tsong, T. Y., 1977b, Formation and resealing of pores of controlled sizes in human erythrocyte membrane, *Nature* **268:**438–441.

Kinosita, K., Jr., and Tsong, T. Y., 1977c, Voltage induced pore formation and hemolysis of human erythrocytes, *Biochim. Biophys. Acta* **471:**227–242.

Kinosita, K., Jr., and Tsong, T. Y., 1978. Survival of sucrose-loaded erythrocytes in circulation, *Nature* **272:** 258–260.

Kinosita, K., Jr., and Tsong, T. Y., 1979, Voltage-induced conductance in human erythrocyte membranes, *Biochim. Biophys. Acta* **554:**479–494.

Lindner, P., Neumann, E., and Rosenheck, K., 1977, Kinetics of permeability changes induced by electric impulses in chromaffin granules, *J. Membr. Biol.* **32:**231–254.

Lo, M. M. S., Tsong, T. Y., Conrad, M. K., Strittmatter, S. M., Hester, L. D., and Snyder, S., 1984, Monoclonal antibody production by receptor-mediated electrically induced cell fusion, *Nature* **310:**792–794.

Neumann, E., and Rosenheck, K., 1972, Permeability changes induced by electric impulses in vesicular membranes, *J. Membr. Biol.* **10:**279–290.

Neumann, E., Schaefer-Ridder, M., Wang, Y., and Hofschneider, P. H., 1982, Gene transfer into mouse lyoma cells by electroporation in high electric fields, *EMBO J.* **1:**841–845.

Pauly, H., and Schwan, W. P., 1966, Dielectric properties and ion mobility in erythrocytes, *Biophys. J.* **6:**621–639.

Potter, H., Weir, L., and Leder, P., 1984, Enhancer-dependent expression of the human kappa immunoglobulin genes introduced into mouse pre-B lymphocytes by electroporation, *Proc. Natl. Acad. Sci. USA* **81:**7161–7165.

Sale, A. J. H., and Hamilton, W. A., 1967, Effects of high electric fields on microorganisms. I. Killing of bacteria and yeasts, *Biochim. Biophys. Acta* **118:**781–788.

Sale, A. J. H., and Hamilton, W. A., 1968, Effects of high electric fields on microorganisms. III. Lysis of erythrocytes and protoplasts, *Biochim. Biophys. Acta* **163:**37–43.

Serpersu, E. H., and Tsong, T. Y., 1983, Stimulation of a ouabain-sensitive Rb^+ uptake in human erythrocytes with an external electric field, *J. Membr. Biol.* **74:**191–201.

Serpersu, E. H., and Tsong, T. Y., 1984, Activation of electrogenic Rb^+ transport of (Na,K)-ATPase by an electric field, *J. Biol. Chem.* **259:**7155–7162.

Serpersu, E. H., Kinosita, K., Jr., and Tsong, T. Y., 1985, Reversible and irreversible modification of erythrocyte membrane permeability by electric field, *Biochim. Biophys. Acta* **812:**779–785.

Sowers, A. E., 1984, Characterization of electric field induced fusion in erythrocyte ghost membranes, *J. Cell Biol* **99:**1989–1996.

Sowers, A. E., 1986, A long-lived fusogenic state is induced in erythrocyte ghosts by electric pulses, *J. Cell Biol.* **103:**1358–1362.

Teissie, J., and Tsong, T. Y., 1980, Evidence of voltage-induced channel opening in (Na,K)-ATPase of human erythrocyte membrane, *J. Membr. Biol.* **55:**133–140.

Teissie, J., and Tsong, T. Y., 1981a, Voltage modulation of Na^+/K^+ transport in human erythrocytes, *J. Physiol. Paris* **77:**1043–1053.

Teissie, J., and Tsong, T. Y., 1981b, Electric field induced transient pores in phospholipid bilayer vesicles, *Biochemistry* **20:**1548–1554.

Teissie, J., Knox, B. E., Tsong, T. Y., and Wehrle, J., 1981, Synthesis of ATP in respiration-inhibited submitochondrial particles induced by microsecond electric pulses, *Proc. Natl. Acad. Sci. USA* **78:**7473–7477.

Teissie, J., Knutson, V. P., Tsong, T. Y., and Lane, M. D., 1982, Electric pulse-induced fusion of 3T3 cells in monolayer culture, *Science* **216:**537–538.

Tien, H. T., 1974, *Bilayer Lipid Membranes (BLM)*, Dekker, New York.

Tsong, T. Y., 1983, Voltage modulation of membrane permeability and energy utilization in cells, *Biosci. Rep.* **3**:487–505.

Tsong, T. Y., and Astumian, R. D., 1986, Absorption and conversion of electric field energy by membrane bound ATPases, *Bioelectrochem. Bioenerg.* **211**:457–476.

Tsong, T. Y., and Astumian, R. D., 1987, Electroconformational coupling and membrane protein function, *Prog. Biophys. Mol. Biol.* **50**:1–45.

Tsong, T. Y., and Astumian, R. D., 1988, Electroconformational coupling: How membrane-bound ATPase transduces energy from dynamic electric fields, *Annu. Rev. Physiol.* **50**:273–290.

Tsong, T. Y., and Kinosita, K., Jr., 1985, Use of voltage pulses for the pore opening and drug loading, and the subsequent resealing of red blood cells, *Bibl. Haematol. (Basel)* **51**:108–114.

Tsong, T. Y., Tsong, T. T., Kingsley, E., and Siliciano, R., 1976, Relaxation phenomena in human erythrocyte suspensions, *Biophys J.* **16**:1091–1104.

Westerhoff, H. V., Tsong, T. Y., Chock, P. B., Chen, Y.-d., and Astumian, R. D., 1986, How enzymes can capture and transmit free energy from an oscillating electric field, *Proc. Natl. Acad. Sci. USA* **83**:4734–4738.

Wong, T.-K., and Neumann, E., 1982, Electric field mediated gene transfer, *Biochem. Biophys. Res. Commun.* **107**:584–587.

Zimmermann, U., and Benz, R., 1980, Dependence of the electrical breakdown voltage on the charging time in *Valonia utricularis, J. Membr. Biol.* **53**:33–43.

Zimmermann, U., and Pilwat, G., 1978, The relevance of electric field induced changes in the membrane structure to basic membrane research and clinical therapeutics and diagnosis, Sixth Int. Biophys. Congr., Kyoto, Abstr. IV-19 (H), p. 140.

Zimmermann, U., and Vienken, J., 1982, Electric field induced cell-to-cell fusion, *J. Membr. Biol.* **67**:165–182.

Zimmermann, U., Riemann, F., and Pilwat, G., 1976, Enzyme loading of electrically homogeneous human red blood cell ghosts prepared by dielectric breakdown, *Biochim. Biophys. Acta* **436**:460–474.

Zimmermann, U., Pilwat, G., and Esser, B., 1978, The effect of encapsulation in red blood cells in the distribution of methotrexate in mice, *J. Clin. Chem. Clin. Biochem.* **16**:135–144.

Zimmermann, U., Vienken, J., and Pilwat, G., 1980, Development of drug carrier system: Electric field induced effects in cell membranes, *Bioelectrochem. Bioenerg.* **7**:553–574.

Zimmermann, U., Scheurich, P., Pilwat, G., and Benz, R., 1981, Cells with manipulated functions: New perspectives for cell biology, medicine and technology, *Angew. Chem. Int. Ed. Engl.* **20**:325–345.

Part III

Electrofusion

Chapter 10

Electrofusion Kinetics

Studies Using Electron Microscopy and Fluorescence Contents Mixing

David A. Stenger and Sek Wen Hui

1. INTRODUCTION

Direct observations of membrane structural changes accompanying fusion events are invaluable to the understanding of membrane fusion mechanisms. In the ideal sense, sequences of morphological changes might be correlated with complementary data to assign kinetic and structure–function relationships during the fusion event. We have initiated this study by preserving the ultrastructural changes in the human erythrocyte membrane during electrofusion by rapid freeze quenching, followed by observation using freeze-fracture electron microscopy (EM). To provide an alternative means of monitoring the same events in a population of cells, we have used fluorescence contents-mixing fusion assays. In this chapter, we discuss how these techniques may be used to infer information regarding the mechanism(s) of electrofusion.

1.1. Predictions of Ultrastructural Changes

Until recently, no ultrastructural studies of electrofusion were reported. Several publications pertaining to electrofusion mechanisms imply, however, events which might be observable at the ultrastructural level. They may be categorized according to one of the necessary stages of fusion: the mutual approach and contact of apposed membranes, membrane destabilization, and the reorganization of membrane components.

The most common method of closely positioning membranes for electrofusion is dielectrophoresis (Pohl, 1978). In fact, high alternating current (AC) fields alone may cause erythrocyte fusion (Scheurich and Zimmermann, 1980). Fusion may also be caused by dielectrophoresis following pulse application (Sowers, 1984), and by continuous low-amplitude DC fields (Chang and Hunt, 1987). By definition, direct membrane contact must occur in all of these cases. However,

David A. Stenger and Sek Wen Hui • Electron Optics Laboratory, Biophysics Department, Roswell Park Memorial Institute, Buffalo, New York 14263.

direct contact can result only if the dielectrophoretic force (coupled with short-range forces) overcomes the electrostatic and hydration barriers (Rand, 1981) between the approaching membranes. It was suggested (Scheurich and Zimmermann, 1981) that the facilitation of erythrocyte electrofusion by Pronase or neuraminidase pretreatment is in part due to a reduction of the repulsive forces, allowing for closer apposition of the membranes at this stage. The role of high electric field pulses in obtaining direct membrane contact has not been analyzed.

The overall degree of membrane destabilization necessary for electrofusion is not well characterized. Electrically induced membrane "pores" (electropores) have most frequently been suggested as fusion intermediates, but have not been structurally observed in biological membranes. What is certainly known is that the human erythrocyte membrane is transiently permeabilized to molecules of different sizes, depending especially on electric pulse parameters. Molecules of approximately 17-nm diameter may penetrate the membrane following exponential decay pulse application (Sowers and Lieber, 1986). However, the diameters of permeable solutes are less than 2 nm when short (5–40 μsec) rectangular pulses of amplitude 2–5 kV/cm are used (Kinosita and Tsong, 1979; Schwister and Deuticke, 1985). Therefore, the membrane defects responsible for the electropermeation of some solutes might not be detectable by freeze-fracture EM.

Protein-depleted regions of membranes have been implicated in a number of artificially induced cell fusion mechanisms (Huang and Hui, 1986). Similarly, the electrofusion model proposed by Zimmermann and Vienken (1982) predicts that protein-depleted areas, which might facilitate lipid mixing, are also prerequisites for electrofusion. Such areas might be generated by an AC field-induced migration of membrane proteins, analogous to the rapid electrophoretic movement of membrane proteins observed by fluorescence microscopy (Poo, 1981) and by freeze-fracture EM (Sowers and Hackenbrock, 1981). The model also predicts that membrane fragmentation in the fusion regions may result in formation of inside-out vesicles, such as those observed at the ultrastructural level following electrofusion of inner mitochondrial membranes (Sowers, 1983).

1.2. Complementary Experiments: Fluorescence Fusion Assays

Attempts to relate ultrastructural changes to a fusogenic mechanism through the use of freeze-fracture EM alone will be complicated by three significant limitations of the technique. First, fusion intermediates may be very short-lived and thus might not be detected if the freezing step is too early or too late. Second, if ultrastructural changes are due to their unsynchronized development within the population of cells, then irregular specimen sampling will lead to statistical uncertainty regarding the actual sequence of the changes. Third, it is not possible to unequivocally associate certain ultrastructural features with the mixing of membrane components or of the contents of the fusion partners. Although the second problem can be countered by applying statistics to large numbers of samples, it becomes directly advantageous to monitor the behavior of the entire population of cells using complementary techniques.

Fluorescence fusion assays (Duzgunes and Bentz, 1986) provide a means to help resolve the above-mentioned problems. By monitoring the kinetics of fusion using fluorescence contents-mixing or lipid-mixing assays, the sampling times for freeze-fracture may be chosen to provide the maximum likelihood of capturing fusion intermediates or other structural changes. The fluorescence changes may also be used as complementary kinetic data to verify statements regarding the functional significance of observed structural changes.

2. ADAPTATION OF METHODS

2.1. Freeze-Fracture EM

The first requirement for the adaptation of this technique is the modification of the freeze-fracture specimen holder so that it also functions as a fusion chamber. The second requirement is that the applications of AC or pulsed electric fields need to be synchronized with the arrival of the sample with the rapid-freezing coolant, so that structural features are arrested at defined time points.

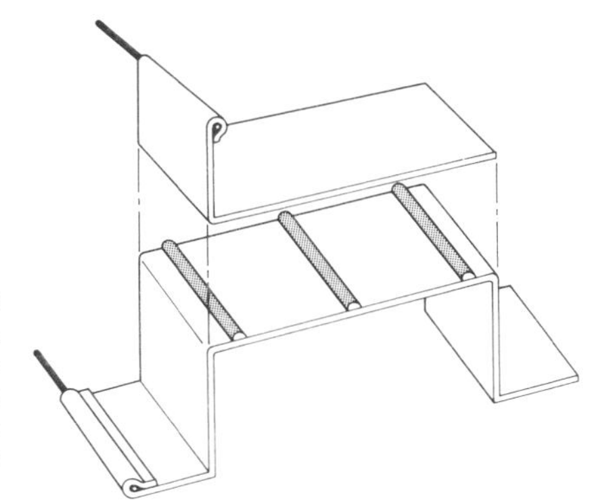

Figure 1. Rapid-freeze specimen holder. Electrical contact is provided by very thin (#38 AWG) insulated wires which are crimped into each half of the copper sandwich. Electrode separation (approx. 25 μm) is provided by three insulation strips which are painted on the bottom surface. The surface tension of the aqueous sample tightly draws the two surfaces together until the top surface contacts the insulation strips.

To study electric pulse-induced intramembranous particle (IMP) movement, Sowers and Hackenbrock (1981) applied electric fields across Balzers specimen holders. They synchronized the rapid freezing with a triggered electromechanical propane carrier. This technique should also prove to be quite suitable for electropermeation/electrofusion studies.

The specimen holder design used in this laboratory (Fig. 1) is a conventional copper sandwich holder. Thin wires provide electrical contact to the top and bottom surfaces, which function as electrodes and define the dimensions of the fusion chamber. Painted epoxy strips set the electrode spacing and provide electrical insulation between the surfaces. The samples are frozen without cryoprotectant by the plunger method (Costello and Corless, 1978). Pulse applications are synchronized in relation to the time of contact with the coolant.

2.2. Fluorescence Fusion Assays

We are currently using the terbium (Tb)/dipicolinic acid (DPA) assay as developed for human erythrocyte ghost fusion (Hoekstra *et al.*, 1985). Human erythrocyte ghosts, prepared by hypotonic lysis, are encapsulated with 10 mM Hepes buffer, pH 7.2, containing either Tb (7.5 mM) and citrate (75 mM), or DPA (75 mM) and NaCl (50 mM). Equal amounts of Tb- and DPA-containing ghosts are resuspended in fusion medium (0.28 M sucrose, 0.2 mM EDTA, pH 7.2), and then subjected to various combinations of AC field and pulse conditions. Contents mixing results in the formation of the fluorescent Tb–DPA complex. Fluorescence from leaked contents is quenched by the EDTA in the fusion medium. The total amount of Tb–DPA leakage during fusion is measured in separate experiments as the decrease of the maximal fluorescence produced by ghosts which are coencapsulated with equal volumes of the Tb and DPA stock solutions.

The fluorometric fusion chamber design used is shown in Fig. 2. Because of the high optical

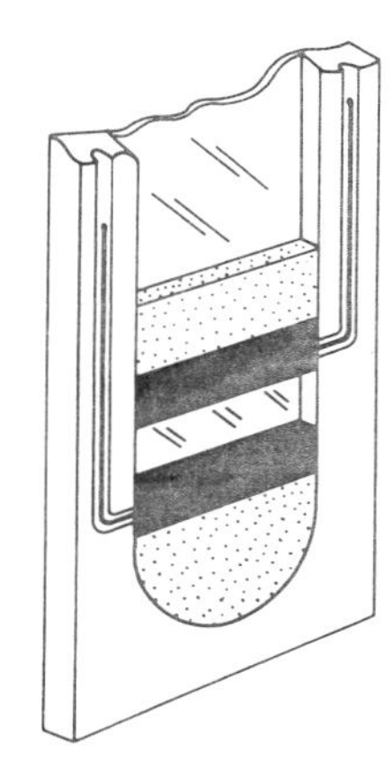

Figure 2. Fluorometric fusion chamber. Two rectangular stainless steel electrodes (dark bars) are horizontally mounted to the rear surface of a demountable spectrophotometer cuvette at a distance of 0.6 mm. The electrodes and the surrounding epoxy (dot pattern) become flush with the back surface of the front window when the cuvette is closed. Application of vacuum grease to the surrounding contact sites ensures tight sealing of the sample material inside the chamber. Courtesy of *Biophysical Journal.*

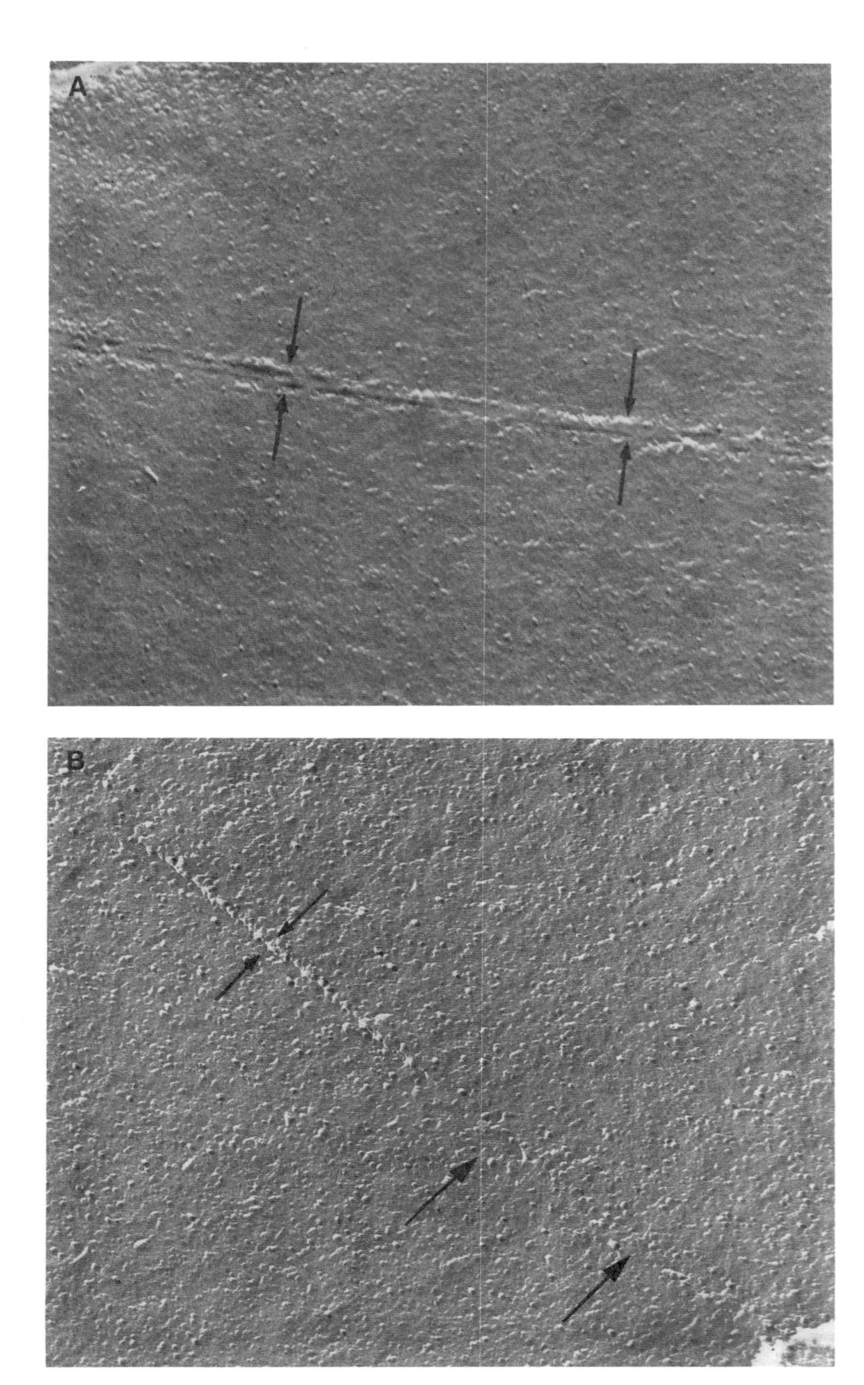

Figure 3. Freeze-fracture electron micrographs of (A) cross fracture of Pronase-treated human erythrocytes aligned in a 1.5 MHz, 0.5 kV/cm field. Arrows denote the well-defined aqueous boundary. (B) Cross fracture of aligned cells within 100 msec following application of a single 15 μsec, 5.0 kV/cm electric pulse. The aqueous boundary (small arrows) is not readily distinguished in some areas (larger arrows). (C) Cross-fractured cells at 10 sec following application of the same type of pulse. Arrow indicates cytoplasmic bridging. Bar = 250 nm.

Figure 3. *(Continued)*

density of ghost suspensions at the concentrations we used, a front-surface fluorescence measurement was necessary. The additional components of the system filter reflected light, and ensured that only the entire chamber contents were in the light path.

3. EXPERIMENTAL RESULTS

The detailed methods involved with the ultrastructural and fluorescence techniques are described elsewhere (Stenger and Hui, 1986, 1988, respectively). The results of these studies are summarized in this section.

3.1. Observation/Quantitation of Ultrastructural Changes

The conditions for the electrofusion of pearl-chained erythrocytes by a single pulse were optimized using phase microscopy. We found that a 15 μsec, 5.0 kV/cm rectangular pulse caused nearly 100% fusion of Pronase-treated cells which were resuspended in 0.28 M sucrose and aligned in a 1.5 MHz, 0.5 kV/cm sine wave AC field. Unless specifically stated, the ultrastructural observations were made using these conditions.

Cross fractures of adjacent cells provided information about intercellular separation and cytoplasmic bridging. Several hundred pairs of adjacent cross-fractured cells were examined from specimens which were frozen during dielectrophoresis. All pairs were observed to be separated by a clearly defined aqueous boundary. The minimum-observed separation between intact (non-enzymatically treated), dielectrophoretically aligned cells was about 25 nm. Although Pronase-treated cells exposed to the same AC field were also separated by an aqueous boundary, they had a minimum-observed separation of about 15 nm (Fig. 3A).

Samples frozen at 0.1 and 2 sec following pulse application contained pairs or groups of cross-

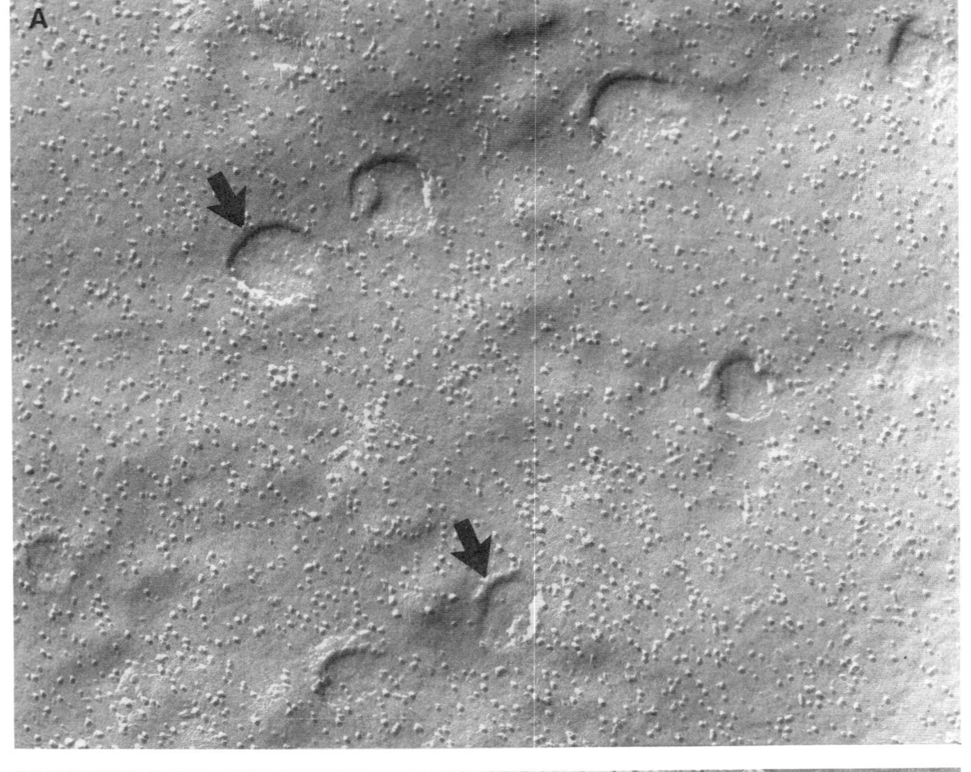
A

B

fractured cells having disrupted, and at some points, indistinguishable aqueous boundaries (Fig. 3B). Because this small separation approached the resolution limits of freeze-fracture EM, it could not be determined if coalescence of internal contents had occurred within 2 sec following pulse application. Samples frozen at 10 and 60 sec following pulse application contained pairs of cells having definitive cytoplasmic bridges (Fig. 3c): except at the sites of bridge formation, the minimum intercellular separation observed remained at about 15 nm.

For quantitation purposes, only those cells which were separated by 25 nm or less (presumably cell pairs which were cross-fractured at the point of closest approach) were included in the cross-fracture category. Fifteen to twenty percent of such cells had disrupted aqueous boundaries and minimum-observed separations of 5 nm or less at 0.1 and 2 sec following pulse application. This percentage was sharply decreased at later time points. In contrast, definitive cytoplasmic bridges were not observed prior to 2 sec following pulse application, but were observed in about 15% of the adjacent cross-fractured cells at 10 and 60 sec.

The face fractures of Pronase-treated cells which were dielectrophoretically aligned for 30 sec showed no apparent differences from control cells. No large-scale IMP segregation was observed. However, more than 20% of the membranes frozen at 0.1 sec following pulse application contained localized groups of discontinuous areas (DA) projecting from the fracture planes (Fig. 4A). Many DA were IMP-free, and were always asymmetric in the sense that they were continuous with the membrane at one point. Their formation did not require AC field alignment or Pronase treatment, but depended only on sufficiently high pulse amplitude. Curiously, DA were not observed in the membranes of cells which were suspended in sucrose for longer than 20 min, when the specific conductivity of the suspension rose from the $10^{-4}\ \Omega^{-1}\ cm^{-1}$ to the $10^{-3}\ \Omega^{-1}\ cm^{-1}$ range.

At 2 sec following pulse application, the percentage of cells exhibiting DA decreased sharply, but approximately 20% of the face-fractured membranes contained localized groups of point defects (PD) (Fig. 4B) which always projected from the P-face and into the E-face fracture planes. Unlike DA, PD formation required dielectrophoretic alignment prior to pulse application. This, together with the outward orientation of defects, and their coincidental occurrence with very close membrane positioning observed by cross fractures, suggested that they were related to a transient form of membrane contact or adhesion.

At 10 sec following pulse application, virtually no DA or PD were observed. Only large disruptions (Fig. 5A) occurred, being present in about 5% of all face fractures. Although their dimensions (250–500 nm across) approximated those of the cytoplasmic bridges seen at this time point, they could not be further characterized. At 60 sec following pulse application, the only abnormal face fracture features observed were broad zones of fusion between P- and E-face fractures of different membranes (Fig. 5B). The areas of membrane fusion were typically linear zones which were noticeably IMP-free.

3.2. Kinetics of Contents Mixing and Leakage

The Tb- and DPA-containing ghosts formed rigid pearl chains in the fluorometer fusion chamber (Fig. 6A). A single 15 μsec, 5.0 kV/cm pulse caused a variable fraction of the pearl chains to fuse (Fig. 6B). The extent of fusion could not be determined by phase microscopy. If the ghosts were made from neuraminidase-treated erythrocytes, the same fusion protocol produced large fusion products (not shown). However, neuraminidase treatment was not required for large fluorescence changes. In all cases presented here, we refer only to ghosts prepared from intact erythrocytes (non-enzymatically treated).

Figure 4. Freeze-fracture electron micrographs of (A) face fracture of erythrocyte membrane frozen within 100 msec of a 15 μsec, 5.0 kV/cm pulse application. Arrows denote predominantly IMP-free DA. (B) Face fracture at 2 sec following pulse application. Solid arrows point to PD which were the most frequently observed membrane perturbation at this time point. Hollow arrow indicates a loss of hemoglobin into the surrounding aqueous medium. Bar = 250 nm.

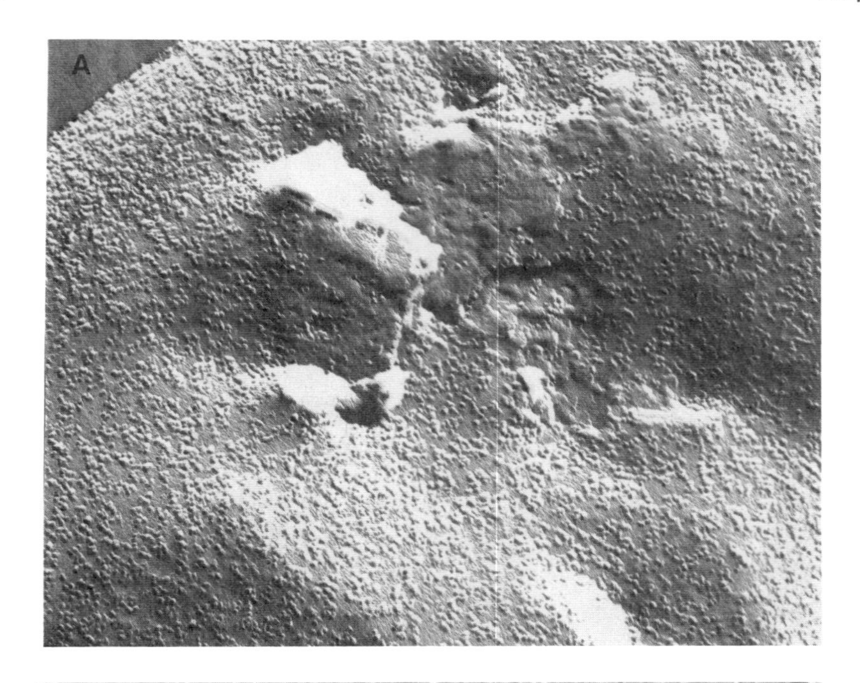
A

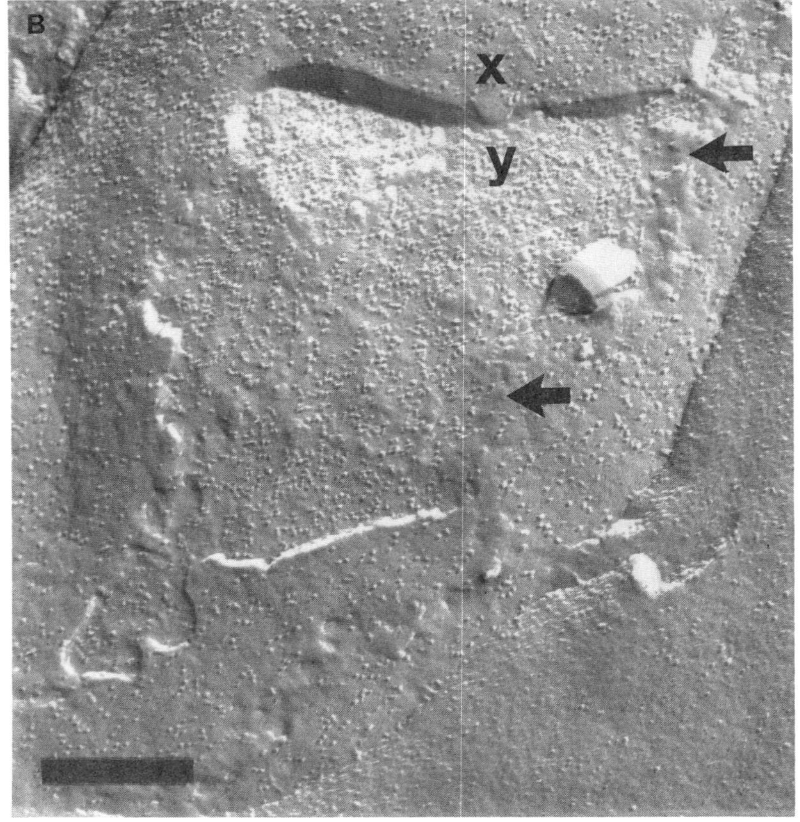
B
x
y

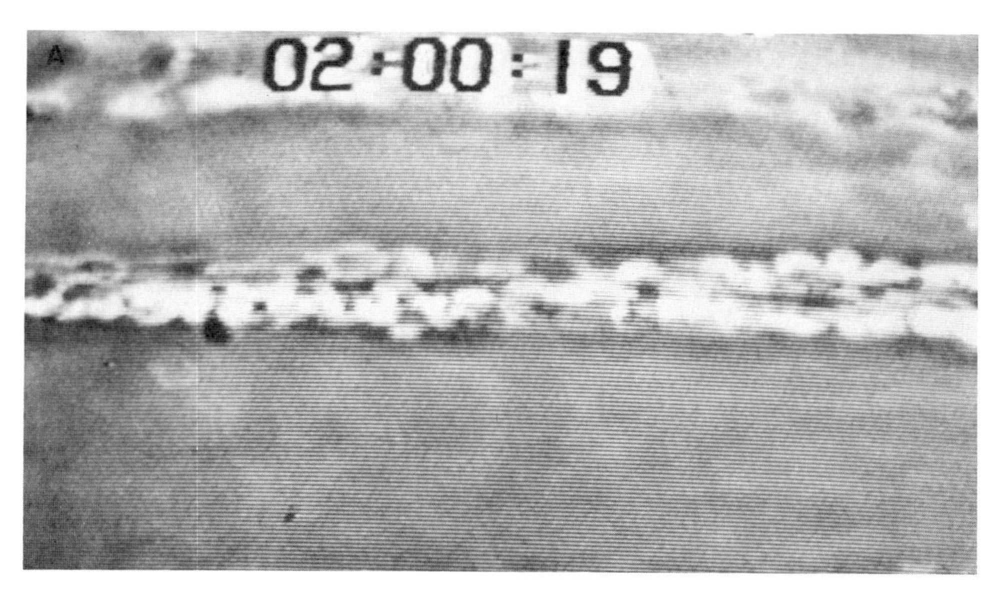

Figure 6. Phase contrast micrographs of (A) a 1 : 1 mixture of Tb- and DPA-containing ghosts aligned in the fluorometric fusion chamber by a 3 MHz, 350 V/cm alternating field; (B) fusion products formed from the same pearl chains observed at 40 sec following application of a single 15 μsec, 5.0 kV/cm pulse. Bar = 25 μm. Courtesy of *Biophysical Journal*.

Figure 5. Freeze-fracture electron micrographs of (A) face fracture at 10 sec following pulse application. The only abnormal morphological features observed at this time point were uncharacterized disruptions in a small percentage of cells. (B) The contact area between a P-face membrane (y) and the E-face of another cell (x), as viewed along the pearl chain axis at 60 sec following application of a 15 μsec, 5.0 kV/cm pulse. Arrows point to the regions of membrane continuity. Bar = 250 nm. Courtesy of *Journal of Membrane Biology*.

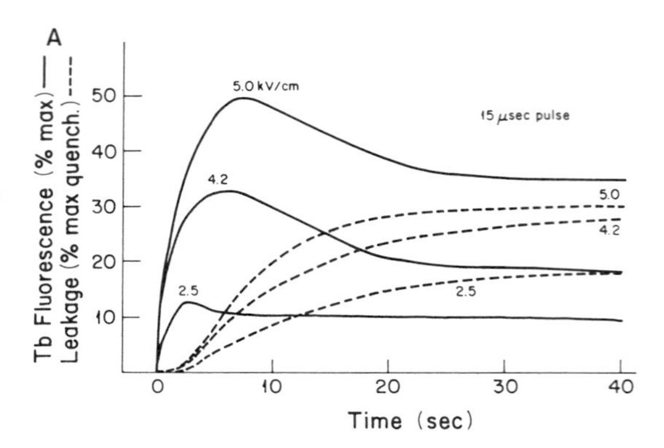

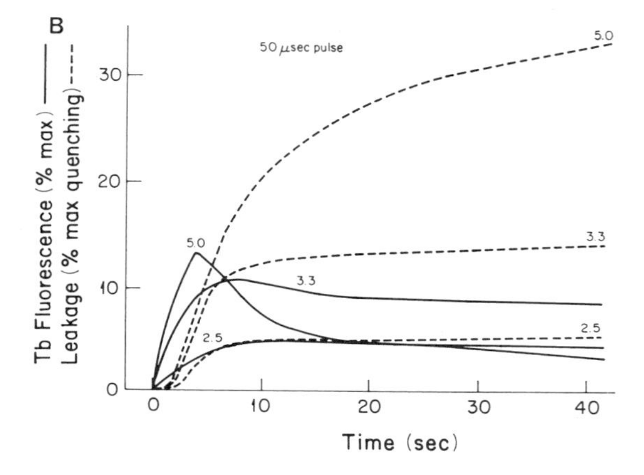

Figure 7. Contents-mixing and leakage kinetics of a 1 : 1 mixture of Tb- and DPA-containing ghosts. The ghosts were aligned in a 3 MHz, 350 V/cm alternating field for 15 sec, then subjected to (A) a single 15 μsec pulse or (B) a 50 μsec pulse of varying amplitude at $t = 0$. Fusion was measured as the Tb fluorescence (solid lines) produced by contents-mixing of separate Tb- and DPA-containing ghosts. Total leakage (dashed lines) was measured in separate experiments as the quenching of fluorescence of coencapsulated Tb and DPA by external EDTA in the fusion medium. Courtesy of *Biophysical Journal*.

Figure 7A shows the changes in Tb fluorescence following application of single 15 μsec pulses of increasing amplitude to ghosts aligned in a 1.5 MHz, 0.35 kV/cm AC field. An immediate increase due to fusion (solid lines) was observed for each case if the pulse amplitude exceeded 2.0 kV/cm. The peak fusion fluorescence increased by more than fourfold as the pulse amplitude was doubled from 2.5 to 5.0 kV/cm. The total leakage of the Tb–DPA complex (dashed lines) became apparent only after a 2–3 sec lag period following pulse application, but increased by only about 50% over the same range of pulse amplitudes. Neither fusion nor leakage was observed if the ghosts were not first aligned in an AC field.

Phase microscopy results (not shown) revealed that the ghost pearl chains were disrupted within 1 sec following application of a 50 μsec pulse. The severity of the disruption increased with increasing pulse amplitude. The percentage of maximum Tb fluorescence due to fusion (Fig. 7B, solid lines) was sharply reduced. The corresponding leakage (dashed lines) was also reduced, except for the case of the 50 μsec, 5.0 kV/cm pulse, which caused the maximum observed leakage of all experiments. Again, neither fusion nor leakage was observed when using a pulse amplitude of less than 2.0 kV/cm, or when the AC field alignment step was omitted.

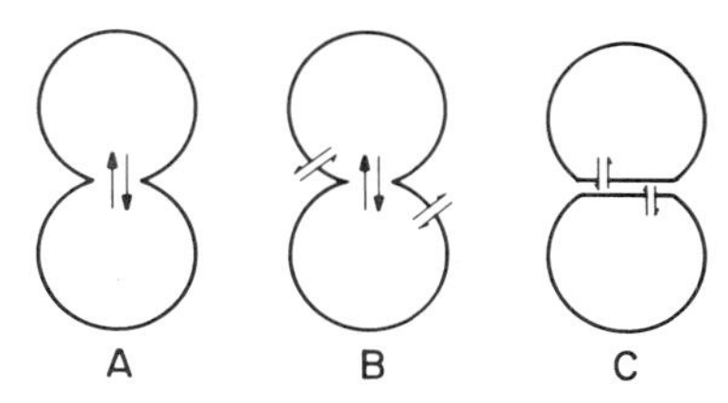

Figure 8. Simplified model used to account for the relative fusion and leakage kinetics. "A" represents fusion products which have lost only small amounts (< 10%) of the encapsulated contents. "B" represents fusion products which undergo a relatively large loss of contents. "C" represents a contact-related leakage from unfused ghosts. Courtesy of *Biophysical Journal*.

4. DISCUSSION AND CONCLUSIONS

4.1. Ultrastructural Model

Our data do not provide conclusive evidence regarding the formation or the involvement of "electropores" in electrofusion. Only localized groups of transient DA, most of which had lifetimes of less than 2 sec, were observed at 0.1 sec following the pulse. However, the formation of defects having a diameter of less than 2 nm, as predicted for these experimental conditions by previous studies (Kinosita and Tsong, 1979; Schwister and Deuticke, 1985), would not be detected by freeze-fracture EM. "Electroporation" defects might also have ultrastructurally observable dimensions for << 100 msec following pulse application. In contrast to various forms of chemically induced fusion, large-scale IMP segregation was not observed, although the DA, as mentioned before, were predominantly depleted of IMP. We also did not observe the vesiculation of membrane fragments in the fusion regions.

A model (Fig. 8) was constructed from the sequences of observed ultrastructural changes. It should be emphasized that the model applies only to changes observed under the described conditions. Under these conditions, the apposed membranes remain separated by an aqueous boundary that is approximately 15 nm wide. Direct membrane contact is caused by pulse application, which might occur because attractive polarization forces exceed the electrostatic and hydration repulsions. The reduction of the intercellular separation during dielectrophoresis is consistent with previous predictions concerning Pronase treatment (Scheurich and Zimmermann, 1981). Our observations may be useful in calculations concerning the rate of the aqueous boundary disruption during electrofusion (Dimitrov and Jain, 1984; Sugar *et al.*, 1987).

4.2. Fluorescence Kinetics Model

The relative amounts of fusion and leakage of erythrocyte ghosts, as measured by the fluorescence contents mixing data, may be interpreted in terms of the simplified model depicted in Fig. 9. "A" represents binary and higher-order fusion events which are accompanied by only a minor loss (< 10%) of encapsulated contents, producing the sharp rise in fluorescence prior to the onset of observable leakage. "B" represents a possible later stage of "A," i.e., fusion events which are followed by a relatively large loss of contents. "B" events become apparent as the decrease from the peak fluorescence intensity due to fusion. "C" represents contact-induced leakage from unfused ghosts, appearing as the amount of total leakage which exceeds that required to reduce the peak Tb fluorescence, even in the limiting case of exclusively binary fusion events (see Stenger and Hui, 1988, for complete discussion).

Accordingly, a 2.5 kV/cm pulse of 15 μsec duration (Fig. 7A) produces: (1) a relatively small amount of fusion products with only minor leakage (type A) and (2) a distinct subpopulation undergoing type C leakage. When the pulse amplitude is increased to 4.2 and 5.0 kV/cm, an increasing amount of type A fusion products are again formed for the first several seconds following

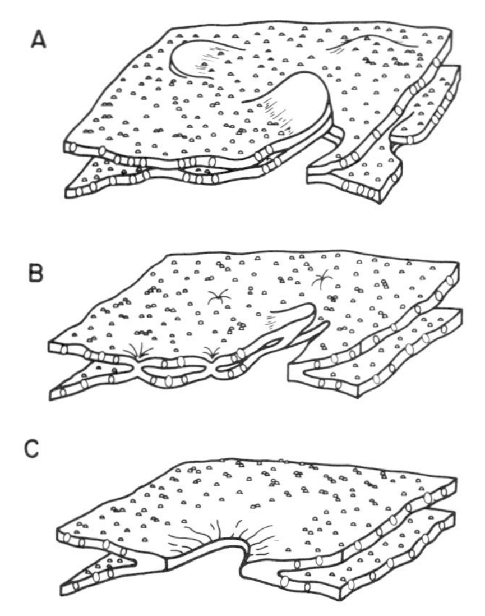

Figure 9. Proposed ultrastructural model of the observed electrofusion events. (A) Within 100 msec of pulse application, the aqueous boundary is perturbed and direct membrane contact results. Cells simultaneously exhibit membrane perturbations (DA). When a DA contacts a destabilized area of the opposite membrane, fusion is initiated. (B) By 2 sec, membrane continuity is established. (C) Within 10 sec following pulse application, permanent lumina are formed. Courtesy of *Journal of Membrane Biology*.

pulse application, but much larger percentage undergo type B rather than type C leakage. The A/C and B/C ratios increase with increasing pulse amplitude.

Pulses of 50 μsec caused increasing pearl chain disruption with increasing pulse amplitudes causing a corresponding decrease in fusion fluorescence. The relative changes (Fig. 7B) reflect: (1) that small percentages of stable type A products are again formed with lower amplitude (2.5 and 3.3 kV/cm) pulses, and (2) in contrast to that observed when using 15 μsec pulses, the relative amount of leakage from unfused ghosts (type C) was maximal for the 50 μsec, 5.0 kV/cm pulse, when pearl chain disruption was most pronounced.

An important observation is that no pulse-induced Tb–DPA leakage occurred from the ghosts unless the pulse was applied when the ghosts were aligned by dielectrophoresis. Our results are not in disagreement with previous observations (Sowers and Lieber, 1986), because their pulse parameters (7 kV/cm, 1.2 msec exponential half-decay time), probe molecules, and fusion medium (20–30 mM phosphate buffer) were quite different.

4.3. Correlation of Results

The significance of the DA is not clear, and more repetitions of the experiments are necessary to confirm their validity. Sugar *et al.* (1987) have suggested that the observed DA may arise from the coalescence of randomly generated electropores. An alternative explanation may be that the DA represent sites where the rate of electrically imposed deformation exceeds that permitted by the viscoelastic properties of the membrane. Even if DA are necessary for fusion, it cannot be said that they interact directly, indirectly, or whether they are merely indicators of a broader scale of membrane destabilization. DA were observed in cells which were not dielectrophoretically aligned, but alignment was a prerequisite for Tb–DPA leakage. Therefore, if DA formation is required for ghost electrofusion by the protocol used, it is unlikely that they represent or contain perforations large enough to permit passage of the Tb–DPA complex.

Because contents mixing between AC field-aligned ghosts starts immediately following pulse application, electrofusion appears to occur more rapidly than the emergence of definitive cytoplasmic bridges as observed by EM at 10 sec (Fig. 3C), suggesting that contents mixing begins at the instant of aqueous boundary disruptions (Fig. 3A).

The observation that leakage from the ghost membranes required AC field alignment prior to pulse application implies that large-scale destabilization of the membrane is contact and/or fusion dependent. The delay of type B leakage for 2–3 sec after the initiation of contents mixing (Fig. 7A,B, dashed lines) corresponds to the temporary adhesion of two membranes in this time period shown by the ultrastructural model.

Most of the probe leakage due to type C events begins at 10–15 sec after fusion is initiated, indicating that this type of event occurs later in time than A and B type events. Sowers (1984) showed that a significant fraction of the ghosts exposed to fusion-inducing pulses may adhere and exchange lipid-soluble probes without the exchange of aqueous contents. It is possible that the later-occurring type of membrane destabilization may be related to a physical separation of adhered membranes.

In summary, we have demonstrated that information regarding the kinetics of erythrocyte membrane electrofusion may be obtained using two complementary methods: freeze-fracture EM, and fluorescence fusion assays. A direct quantitative comparison between the respective results should be possible if the experiments are performed using identical experimental conditions. It is hoped that a more thorough understanding of erythrocyte electrofusion will be of fundamental relevance, and contribute to the better application of this important technology.

ACKNOWLEDGMENT. This work was supported by NIH Grant GM 30969 awarded to S.W.H.

REFERENCES

Chang, D. C., and Hunt, J. R., 1987, Effects of pH on electrofusion of human red blood cells and HL-60 cells, in: *Proceedings of International Symposium: Molecular Mechanisms of Membrane Fusion* (S. Ohki, ed.), Plenum Press, New York.

Costello, M. J., and Corless, J. M., 1978, The direct measurement of temperature changes within freeze-fracture specimens during rapid quenching in liquid coolants, *J. Microsc. (Oxford)* **112:**17–37.

Dimitrov, D. S., and Jain, R. K., 1984, Membrane stability, *Biochim. Biophys. Acta* **779:**437–468.

Duzgunes, N., and Benz, J., 1986, Fluorescence assays for membrane fusion, in: *Spectroscopic Membrane Probes* (L. M. Loew, ed.), CRC Press, Boca Raton, Fla.

Hoekstra, D., Wischut, J., and Scherpof, G., 1985, Fusion of erythrocyte ghosts induced by calcium phosphate, *Eur. J. Biochem.* **146:**131–140.

Huang, S. K., and Hui, S. W., 1986, Chemical co-treatments and intramembrane particle patching in the poly(ethylene glycol)-induced fusion of turkey and human erythrocytes, *Biochim. Biophys. Acta* **860:**539–548.

Kinosita, K., and Tsong, T. Y., 1979, Voltage-induced conductance in human erythrocyte membranes, *Biochim. Biophys. Acta* **554:**479–497.

Pohl, H. A., 1978, *Dielectrophoresis,* Cambridge University Press, London.

Poo, M., 1981, In situ electrophoresis of membrane components, *Annu. Rev. Biophys. Bioeng.* **10:**245–276.

Rand, R. P., 1981, Interacting phospholipid bilayers: Measured forces and induced instructural changes, *Annu. Rev. Biophys. Bioeng.* **10:**277–314.

Scheurich, P., and Zimmermann, U., 1980, Membrane fusion and deformation of red blood cells by electric fields, *Z. Naturforsch.* **35c:**1081–1085.

Scheurich, P., and Zimmermann, U., 1981, Giant human erythrocytes by electric-field-induced cell-to-cell fusion, *Naturwissenschaften* **68:**45–47.

Schwister, K., and Deuticke, B., 1985, Formation and properties of aqueous leaks induced in human erythrocytes by electrical breakdown, *Biochim. Biophys. Acta* **816:**332–348.

Sowers, A. E., 1983, Fusion of mitochondrial inner membranes by electric field produces inside-out vesicles: Visualization by freeze-fracture electron microscopy. *Biochim. Biophys. Acta* **735:**426–428.

Sowers, A. E., 1984, Characterization of electric-field mediated fusion in erythrocyte ghost membranes, *J. Cell Biol.* **99:**1989–1996.

Sowers, A. E., and Hackenbrock, C. R., 1981, Rate of lateral diffusion of intramembrane particles: Measurement of electrophoretic displacement and rerandomization, *Proc. Natl. Acad. Sci. USA* **78:**6246–6250.

Sowers, A. E., and Lieber, M., 1986, Electropore diameters, lifetimes, numbers, and locations in individual erythrocyte ghosts, *FEBS Lett.* **205:**179–184.

Stenger, D. A., and Hui, S. W., 1986, Kinetics of ultrastructural changes during electrically-induced fusion of human erythrocytes, *J. Membr. Biol.* **93:**43–53.
Stenger, D. A., and Hui, S. W., 1988, Human erythrocyte electrofusion kinetics monitored by aqueous contents mixing, *Biophys. J.* **53:**833–838.
Sugar, I. P., Forster, W., and Neumann, E., 1987, Model of cell electrofusion: Membrane electroporation, pore coalescence and percolation, *Biophys. Chem.* **26:**321–335.
Zimmermann, U., and Vienken, J., 1982, Electric field-induced cell-to-cell fusion, *J. Membr. Biol.* **67:**165–182.

Chapter 11

Electrofusion of Lipid Bilayers

Grigory B. Melikyan and Leonid V. Chernomordik

1. INTRODUCTION

Electrostimulated fusion (electrofusion) is one of the promising methods in the study of the somatic hybridization of cells. In spite of a great number of studies in which this phenomenon is used, its mechanism has not been elucidated. Electrofusion, and generally fusion, presupposes the joining of cell membranes and the volumes limited by them. Therefore, for analyzing the mechanism of electrofusion, it is reasonable to turn not only to the universal phenomenon of membrane fusion which occurs under the action of polyethylene glycol (Pontecorvo, 1975), Sendai virus (Poste and Pasternak, 1978), and a host of other fusogens, but also to the *in vivo* course of biogenesis of muscular fibers (Bischoff, 1978) and of fertilization (Bedford and Cooper, 1978). At the subcellular level, membrane fusion is the key event in the processes of endocytosis and exocytosis (Belitser and Zaalishvili, 1983).

We note that the elucidation of the mechanism(s) of membrane fusion appears to be important not only from a fundamental standpoint but also for enhancing the efficiency of somatic hybridization realized, in particular, by applying an electric field to cells.

It is evident that for the fusion of cell membranes, the joining of their lipid matrices is necessary. Moreover, according to numerous data (Lucy and Ahkong, 1986; Ahkong *et al.*, 1975; Pinto da Silva and Nogueira, 1977; Belitser and Zaalishvili, 1983), fusion is preceded by the displacement of protein molecules from the contact region of membranes and the formation of purely lipid domains of small area in this region. This circumstance speaks in favor of the significant role of lipid matrices in the structural reorganizations of biomembranes leading to fusion. Therefore, the fusion mechanism is being actively investigated in lipid bilayer model systems. The most popular models are: fusion of liposomes with each other (Nir *et al.*, 1983), fusion of liposomes with planar bilayers (Cohen *et al.*, 1984), as well as fusion of two contacting planar bilayers (Neher, 1974; Melikyan *et al.*, 1983a,b).

In the present review we consider in more detail the results obtained using two planar bilayers as the model since this experimental system makes it possible to investigate single interaction events

Grigory B. Melikyan and Leonid V. Chernomordik • The A. N. Frumkin Institute of Electrochemistry, Academy of Sciences of the USSR, 117071 Moscow, USSR.

in membranes and the mechanisms of structural reorganizations in the course of intermediate stages of their fusion. We summarize the main results obtained from the investigation of the mechanism of successive stages of fusion of lipid bilayers and dwell upon the mechanism of their electrofusion.

2. THE INTERACTION OF PLANAR LIPID BILAYERS AS A MODEL FOR STUDYING THE STAGES OF FUSION

We briefly discuss the experimental system of two lipid bilayers and illustrate the main stages of fusion. Figure 1a shows one of the most universal variants of a sample cell for investigating the interaction of two bilayers (Melikyan *et al.*, 1983a,b). The cell compartments are filled with electrolyte solutions. Membranes are formed by the Mueller–Rudin method (Mueller *et al.*, 1962) on openings (1 and 2). Teflon rods (3 and 4) enable one to control the levels of solutions in compartments I and III and, consequently, to push the membranes toward each other. Visual observation of the interacting bilayers is carried out by means of microscopes through glass windows.

The procedures for measuring the electrical parameters of separate membranes and for monitoring their interaction are described in Chernomordik *et al.* (1987a). A fairly simple setup, consisting of two current amplifiers, a generator, and an oscillograph (Fig. 1a), makes it possible to monitor easily quite a number of parameters of bilayers, in particular, the conductance, capacitance, charge, and electromechanical stability of each of the membranes and of the region of their contact.

Figure 1b–d shows the successive stages of fusion of two lipid bilayers. As a result of monolayer fusion of the contacting membranes in the contact region, a bilayer is formed which will be called the contact bilayer (Fig. 1c). The complete fusion of membranes involves joining of both the bilayers and the aqueous volumes limited by them. In the model system, fusion corresponds to the formation of a membrane tube (Fig. 1d). This tube fissions stepwise into two bilayers when, for instance, the hydrostatic pressure outside the membrane tube is increased. Thus, the system returns to its initial state (Fig. 1b).

The method of capacitance monitoring (Neher, 1974) makes it possible to distinguish clearly the main stages of interaction of the two membranes. In the upper part of Fig. 1b are oscillograms of currents, recorded at the output of an operational current amplifier CA_2, which are characteristic of the successive stages of fusion. Specifically, monolayer fusion of bilayers is indicated by a qualitative change in the oscillogram: an alternating square capacitive signal appears. Complete fusion leads to a large conduction current through the electrolyte solution in the membrane tube.

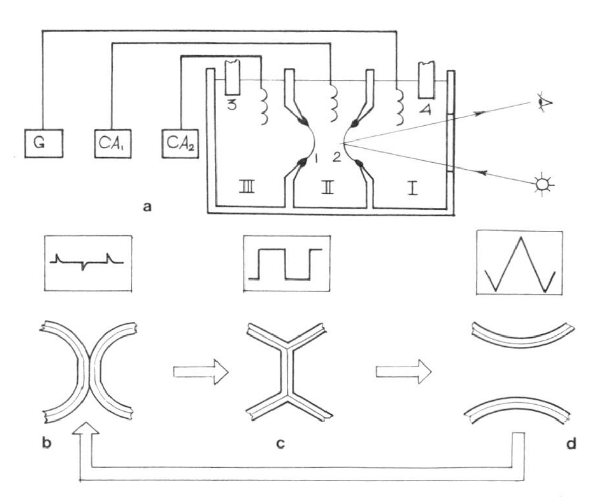

Figure 1. Construction of the measuring cell and its setup (a), and the principle of capacitance monitoring in investigating the interaction between lipid bilayers (b–d). (a) I, II, III, cell compartments; 1, 2, openings in the Teflon walls; 3, 4, Teflon rods used to create gradients of the hydrostatic pressure when bringing membranes into contact. G is a linear voltage sweep generator; CA_1 and CA_2 are current amplifiers connected to the Ag–AgCl electrodes in the cell compartments. (b, c, d) The cycle of interaction of bilayers (characteristic oscillograms of currents at the output of CA_2 are displayed in the upper part of these panels). (b) Membranes in contact; (c) the formation of a single bilayer in the contact region; (d) the membrane tube.

3. APPROACH AND CONTACT OF BILAYERS

It is evident that the necessary condition for membrane fusion is the approach of the membranes and the establishment of close contact between them. The contact of bilayers and its stabilizing forces on multilayer systems were studied comprehensively (Rand, 1981). Here, a change in the interbilayer distance occurs under the action of forces (e.g., osmotic) which press the adjacent bilayers against each other. In the case of two planar bilayers, the approach is achieved by a low hydrostatic pressure. The interaction of liposomes with each other or with a planar bilayer is a result of random collisions in the process of the Brownian motion. The shortest distance between the bilayers is determined by the equilibrium of the molecular attraction forces and the external forces (e.g., hydrostatic pressure) on the one hand and by electrostatic repulsion arising from the interaction of like charged membranes (Rand, 1981) on the other. When the thickness of the aqueous interlayer becomes less than 30 Å, the further approach of membranes is hindered by powerful hydration forces (Gruen *et al.*, 1984). These forces exponentially increase with decreasing distance between the bilayers.

The decrease in the repulsion between membranes leads to their approach. Therefore, the decrease or screening of the charge of lipids causes aggregation of liposomes in suspension or a decrease in the equilibrium distance between the planar bilayers. This effect is achieved, for example, by an increase in the ionic strength of solution (Nir *et al.*, 1983; Abidor *et al.*, 1987). In this case the process is reversible. A decrease in the ionic strength of the solution, conversely, increases the distances between membranes.

4. MONOLAYER FUSION OF LIPID BILAYERS

4.1. The Formation of a Contact Bilayer

The step in fusion which follows the approach and contact of bilayers is monolayer fusion. In other words, two apposed bilayers (four leaflets) (Fig. 1b) condense to a single bilayer (two leaflets) (Fig. 1c). The formation of a contact bilayer during the interaction of liposomes with a planar bilayer was demonstrated by Babunashvili *et al.* (1981). The possibility of the formation of a contact bilayer between the liposome and planar bilayer has also been suggested (Cohen *et al.*, 1984). The prefusion state described in that work may be treated as monolayer fusion of the liposome bilayer with a planar membrane. This accounts, in particular, for the simultaneous rupture of both bilayers in the contact region under the action of the osmotic pressure gradient (Chernomordik *et al.*, 1987a,b). The investigation of unilamellar and multilamellar liposomes indicated that there is monolayer fusion in the contact region of bilayers (Bondeson and Sundler, 1985; Bentz *et al.*, 1985; Duzgunes *et al.*, 1985; Markin *et al.*, 1984).

The formation of a contact bilayer was proven most convincingly in the experimental system of two interacting bilayers. Liberman and Nenashev (1972) and Melikyan *et al.* (1983a,b) studied the structure of contact of membranes using the channel former nystatin. This drug causes a considerable increase in the conductance only in the case of its symmetrical addition to a membrane. Introducing nystatin into the side compartments of the cell (I and II in Fig. 1a) does not lead to a significant change in conductance through two membranes. However, after bringing them into contact, one observes a drastic five to six order of magnitude increase in the conductance in the contact region. This reflects the formation of a single bilayer in which both monolayers are modified by nystatin.

The mean lifetimes of gramicidin channels in a conventional membrane are similar to those in a contact bilayer of the same composition. Also, these lifetimes have a well-known dependence on the membrane thickness, suggesting the formation of a single bilayer in the zone of interaction of membranes. Optical measurements of the bilayer thickness in the contact region have also proved that this bilayer and the conventional membrane have the same thickness (Berestovsky and Gyulkhandanyan, 1976).

The most direct proof of the formation of a single bilayer in the contact region of membranes is, in our opinion, the coincidence of the values of the specific capacitance of the contact region with the specific capacitance of the conventional membrane of the same composition (Melikyan *et al.*, 1983a,b; Fisher and Parker, 1984). Thus, the existence of the stage of monolayer fusion of lipid bilayers has been proved reliably by a wide variety of independent methods.

The structure arising from monolayer fusion of two membranes (Fig. 1c) is sufficiently stable under the usual conditions where the presence of a solvent in the membranes is not the necessary condition for its stabilization (Schindler and Feher, 1976; Chernomordik *et al.*, 1985). Note that in some cases, monolayer fusion was also discovered in the process of biomembrane fusion (Pinto da Silva and Nogueira, 1977; Kalderon and Gilula, 1979; Belitser and Zaalishvili, 1983; Lucy and Ahkong, 1986). Thus, the stage of monolayer fusion, revealed and thoroughly explored during the interaction of planar bilayers, is not a specific feature of just this experimental system but is probably observable in the process of fusion of any membrane. We therefore assume that in the interaction of planar bilayers, the formation of a contact bilayer is the mandatory intermediate stage of membrane fusion (Gingell and Ginsberg, 1978; Lucy and Ahkong, 1986). Therefore, the investigation of the mechanism of monolayer fusion appears to be an important task.

4.2. The Mechanism of Monolayer Fusion of Bilayers

In studying the interaction of solvent-free planar bilayers, there is an opportunity to record sufficiently accurately the waiting time of monolayer fusion t_{mf}, i.e., the time interval from the instant of formation of a close plane-parallel contact to the initiation of a contact bilayer (Chernomordik *et al.*, 1985). At the same value of the contact area, the waiting time of monolayer fusion is dependent on the lipid composition of the membrane. In order to understand the cause of such a difference, it is necessary to consider the notion of the effective shape of lipid molecules, introduced by Helfrich (1973) and by Petrov and Derzhanski (1976). Formally, this parameter reflects the relation between the size of the polar head group and the width of the hydrophobic part of a molecule (Fig. 2). By the shape of lipid molecules, we imply not merely the conformation of an individual molecule but also the effective shape of a lipid in the monolayer built from it. Different reorganizations are characteristic for the membranes built from lipids of different shape (Fig. 2). If the lipids forming a bilayer have the effective shape of a cone, e.g., lysolecithin (Fig. 2a, I), then the formation of a hydrophilic pore is energetically favorable (Fig. 2b, I). In contrast, the appearance of local bulgings in a membrane (Fig. 2b, III) is characteristic of lipids having the shape of an inverted cone (unsaturated phosphatidylethanolamine) (Fig. 2a, III).

Markin *et al.* (1984) proposed a stalk mechanism of membrane fusion (see Fig. 3a–d). According to this mechanism, there appear bulging defects in the contacting monolayer representing local ruptures of monolayers. Once close to each other, these bulgings form a stalk (Fig. 3d); the

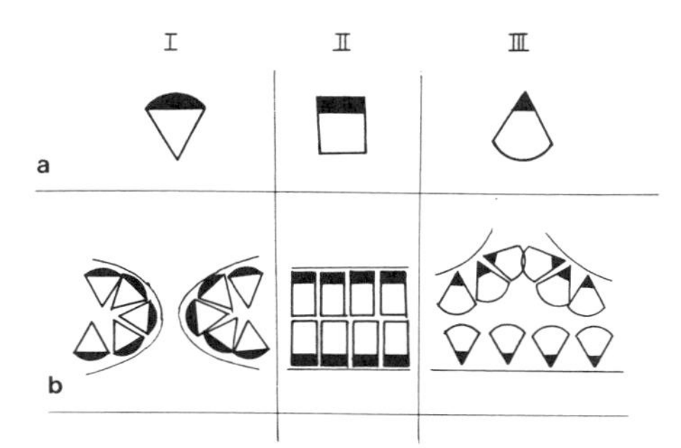

Figure 2. Effective shape of lipid molecules (a: I, II, III) and probable structural defects in the lipid bilayer [b: a hydrophilic pore (I) and local bulging (III)].

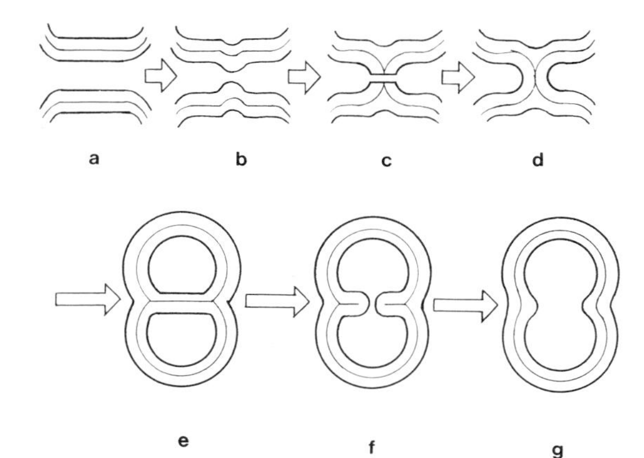

Figure 3. Successive stages of membrane fusion according to the stalk mechanism (Markin *et al.*, 1984; Chernomordik *et al.*, 1987b): a, membranes at the equilibrium distance; b, local approaching; c, rupture of the contacting monolayers; d, the formation of a stalk; e, increase in the area of the contact bilayer; f, rupture of the contact bilayer, which completes the fusion process (g).

increase in the diameter of the latter gives rise to a contact bilayer. It was assumed that the formation of a stalk is the limiting stage of monolayer fusion.

Chernomordik *et al.* (1985) further developed the stalk theory and gave experimental proofs. The monolayers forming a stalk are considerably bent and consequently they possess a marked bending energy. The more asymmetry and the smaller the polar head group in the lipid molecule compared with the width of the hydrocarbon tail, the lower is this energy (Fig. 2a, III). Therefore, the stalk energy and, correspondingly, the waiting time of monolayer fusion are determined by the effective shape of lipid molecules in a bilayer.

For the experimental verification of the stalk hypothesis, it is convenient to have a two-component membrane where the components have different spontaneous curvature and the relative concentration is continuously variable. Chernomordik *et al.* (1985) chose solvent-free phosphatidylethanolamine membranes in which lysolecithin could be added to the aqueous medium.

The logarithm of the waiting time of the stalk was found to be dependent on several measurable parameters according to:

$$\log t_{\mathrm{mf}} = \mathrm{const} + 6.15 \frac{B\delta}{kT} (\zeta_{S1} - \zeta_{S2}) A C_0 \tag{1}$$

where C_0 is the concentration of lipid molecules with the spontaneous curvature ζ_{S1} in the membrane bathing solution; A is the coefficient of distribution of this lipid among the membrane and the solution (AC_0 is the fraction of molecules of the given type in the monolayer); ζ_{S2} is the spontaneous curvature of molecules forming the initial monolayer; δ is the monolayer thickness; and B is the monolayer bending modulus. Thus, the dependence $\log t_{\mathrm{mf}}(C_0)$ should be of a linear character: the mean waiting time should increase with increasing C_0. Chernomordik *et al.* (1985) found a method to determine the value of the combination parameter occurring in Eq. (1) as a factor multiplied by C_0. The point is that the spontaneous curvature of the membrane monolayer should influence not only the formation of a stalk but also another reorganization of the lipid bilayer: the formation of a hydrophilic pore (Fig. 2b, I). The monolayer at the pore edge is bent and has an elastic energy. The sign of curvature of the hydrophilic pore is opposite to the sign of curvature of the stalk surface. For a membrane consisting of lipids of two types, the dependence of the energy of the pore edge per unit length, i.e., the linear tension of the pore, on the bulk concentration of the first component is represented by:

$$\gamma = \frac{\pi B}{\delta} - 2\pi B\zeta_{S2} - \frac{2\pi B}{\delta} (\zeta_{S1} - \zeta_{S2})\, \delta A C_0 \tag{2}$$

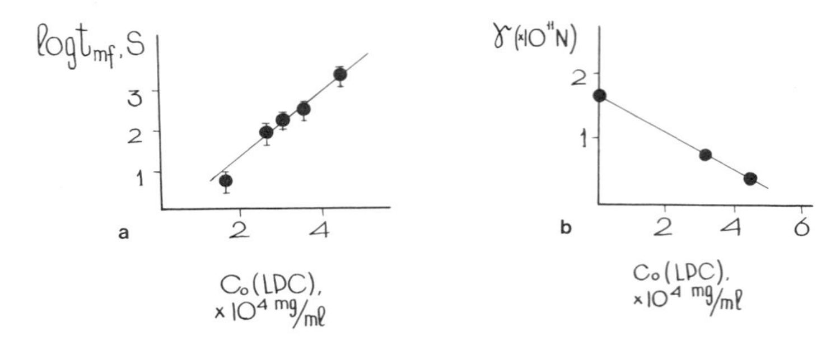

Figure 4. Influence of the lysolecithin concentration C_0 in the membrane bathing solution on the linear tension of the edge of a hydrophilic pore γ (b) and the waiting time of monolayer function t_{mf} (a). The experimental points were obtained for bilayers of phosphatidylethanolamine in squalene; 10^{-1} M KCl. The theoretical curves were obtained from Eqs. (1) and (2).

It is evident from Eq. (2) that the energy of the pore edge, like the energy of the monolayer stalk, decreases linearly with increasing C_0. Figure 4a presents the γ values determined from the experimental study of the electromechanical stability of membranes at different concentrations of lysolecithin in the electrolyte solution (Chernomordik *et al.*, 1985). The dependence of γ on C_0 is linear in accordance with Eq. (2). Figure 4b gives the experimental points which reflect the dependence of log t_{mf} on C_0. The linear character of this dependence is described by Eq. (1). The solid curve in the same figure shows the theoretical dependence log t_{mf} (C_o) obtained if, instead of the combination parameter $B(\zeta_{S1} - \zeta_{S2})\delta A$, its value is determined experimentally from the slope of the dependence $\gamma(C_0)$, and Eq. (2) is substituted into Eq. (1). It is seen that the experimental points fit the theoretical dependence of log t_{mf} (C_0). It is important to emphasize that in this comparison of theory and experiment, not a single free fit parameter was used. Consequently, the fact that all the experimental points were on the same straight line with the same slope as that obtained theoretically argues strongly in favor of the stalk mechanism for the monolayer fusion.

It is to be noted that structures similar to stalks have been discovered on multilamellar liposomes of a certain composition (Borovjagin *et al.*, 1982). The appearance of stalks between the adjacent bilayers of such liposomes leads to the formation of local regions of the contact bilayer. Interestingly, such structures are observed in liposomes containing molecules which are shaped like an "inverted cone" (Fig. 2a, III). According to our ideas, this promotes the formation of stalks.

5. MEMBRANE FUSION

5.1. The Complete Fusion of Lipid Bilayers

The structure arising from the monolayer fusion of membranes (Fig. 1c) is sufficiently stable and the spontaneous transfer of the monolayer fusion to the complete one is a rather rare event. Therefore, special conditions are necessary for the formation of a membrane tube in the system of two planar bilayers (Melikyan *et al.*, 1983a,b). It is important to emphasize that in the complete fusion of planar bilayers, the formation of a contact bilayer necessarily precedes the formation of a membrane tube.

Fission of the membrane tube (Fig. 1d–b), which occurs spontaneously during the formation of a definite bilayer configuration (Melikyan *et al.*, 1984), should be considered as a special case of bilayer fusion. This process does not require any additional conditions. It seems to be determined by a special geometry of the system: by the formation of a narrow cylindrical channel at the initial stages of fission of the tube.

Among the numerous mechanisms presently proposed for the phenomenology of fusion observed on bilayers (Ohki, 1984; Cullis *et al.*, 1986; Siegel, 1984; Raudino, 1986), we will discuss briefly the approach developed by Leikin *et al.* (1986), Kozlov *et al.* (1987) and Chernomordik *et al.* (1987b) (Fig. 3).

After the distance between the bilayers is within about 25 Å, the further approach of the membranes is drastically hindered by the presence of powerful hydration forces. The further approaching of bilayers in lipid multilayers or of bilayers deposited on the surface of mica cylinders necessitates huge pressures of hundreds and thousands of atmospheres (Rand, 1981; Horn, 1984). On the other hand, the planar membranes (or liposomes) are able to overcome the hydration barrier practically without an external pressure apparently due to thermal fluctuations (Leikin *et al.*, 1986; Chernomordik *et al.*, 1987b). This is accompanied by the local approaching of membranes to a small distance of about 5 Å (Fig. 3b). As a result of the competition of the hydration and hydrophobic interactions, the structure of the contacting monolayers is disturbed and a stalk is formed (Fig. 3c,d). For the extension of the stalk diameter, the formation of a contact bilayer and its rupture, i.e., the complete fusion, it is necessary that the molecules of the external monolayers of membranes have the effective shape of an "inverted cone" (Fig. 2a, III) and the molecules of the internal monolayers, of a cone (Fig. 2a, I) (Kozlov *et al.*, 1987; Chernomordik *et al.*, 1987b). Such opposite shapes of lipid molecules in different monolayers are likely to be actually realized in experimental systems in the presence of various fusogens, e.g., Ca^{2+} ions (Duzgunes, 1985) and phospholipases C and D (Schragin *et al.*, 1985) which modify the external monolayers of membranes.

In accordance with the predictions of the hypothesis considered, complete fusion was observed when the internal monolayers of the planar membranes were modified with lysophosphatidylcholine (Chernomordik *et al.*, 1987b). Formally, the action of fusogens should, within the framework of the mechanism proposed, reduce to an increase in the probability of monolayer fusion and to rupture of the contact bilayer thus formed through the formation of a hydrophilic pore in it (Fig. 3f). We note that the mechanical stresses in the contact bilayer (arising, in particular, with the presence of the osmotic pressure gradient on a liposome membrane or from tension of planar lipid membranes) are expected to promote an additional destabilization of the contact bilayer and an increase in the diameter of the hydrophilic pore formed in it.

5.2. The Mechanism of Bilayer Electrofusion

When an electric field is applied to a system of planar bilayers after monolayer fusion, in every instance the formation of a membrane tube (i.e., membrane fusion) is observed (Melikyan *et al.*, 1983a; Fisher and Parker, 1984). Electrofusion is also realized in the case of giant liposomes (Zimmermann, 1982). In contrast to the other ways of getting complete fusion, the electrofusion of liposomes and planar bilayers is observed for all the lipid compositions investigated irrespective of the presence of a solvent in planar membranes.

The system of two planar membranes turned out to be very convenient for examining the mechanism of bilayer electrofusion. First, in contrast to the electrofusion of liposomes and cells, the values of voltages which drop across the contact bilayer and noncontacting regions of planar membranes are known exactly. Second, fusion can be realized not only by applying an external electric field but also due to the diffusion potential or the intramembranous field in the case of monolayer fusion of bilayers with different surface charge (Melikyan *et al.*, 1983a). Third, the instant of fusion is measured extremely accurately and sensitively by current methods. Fourth, it is possible to check reliably the stage of interaction between membranes at the instant of application of a pulse (before monolayer fusion or after it). There is no such possibility in all the other model systems.

It is important that electrofusion, as well as the other methods of obtaining a membrane tube, necessitates the intermediate formation of a contact bilayer. The application of an electric field to membranes at the stage of a plane-parallel contact, even at very small distances between the membranes, causes irreversible breakdown of both membranes but does not induce fusion (Melikyan *et al.*, 1983a; Chernomordik *et al.*, 1987a).

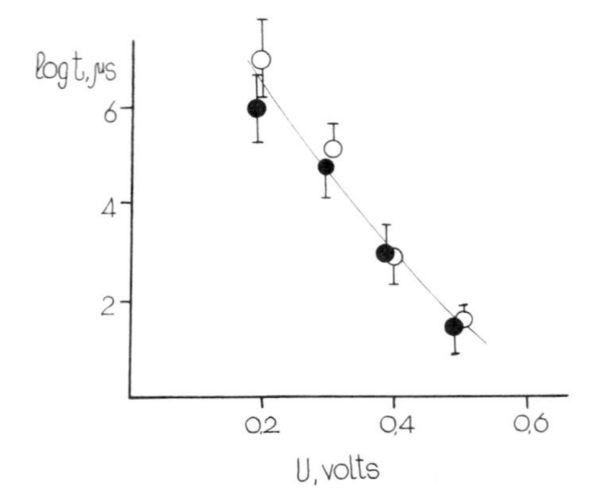

Figure 5. Voltage dependences of the average duration of a pulse required for the induction of fusion (t_f) (●), and of the mean lifetime of the conventional membrane (t_l) (○). Bilayers of a mixture of dioleoylphosphatidylcholine/cholesterol in decane; 0.1 M KCl.

What is the mechanism of electrofusion of bilayers in the given system? In order to answer this question, Melikyan *et al.* (1983a) investigated the dependence of the duration of a square pulse, required for the induction of membrane fusion, on the amplitude of this pulse U (closed circles in Fig. 5). It is evident from Fig. 5 that there is a strong dependence of the fusion time on the pulse amplitude. The shape of the curve is similar to the exponential one and resembles the dependence of the mean lifetime of single membranes on the amplitude of a voltage pulse applied to them, which is observed in the studies on electrical breakdown of bilayers (Abidor *et al.*, 1979; Chernomordik *et al.*, 1983).

For verifying the assumption of the similarity of the mechanisms of electrofusion and breakdown of membranes, we have investigated the voltage dependence of the logarithm of the mean lifetime of bilayers of the same lipid composition and area as those of the contact bilayer (open circles in Fig. 5). The coincidence of both curves in the whole voltage range covered proves that the same mechanism underlies electrofusion and irreversible breakdown of membranes in an electric field.

It has been established (Abidor *et al.*, 1979) that irreversible breakdown of bilayers in an electric field results from the development of hydrophilic pores of the overcritical radius, which tend to spontaneous expansion. Consequently, the development of such pores in the contact bilayer causes the irreversible breakdown and the formation of a membrane tube during electrofusion. The exponential character of the voltage dependence of the electrofusion time (t_f) of bilayers and of the lifetime of the conventional membrane (t_l) in the range covered accounts for nearly 100% of the formation of a membrane tube in experiments on electrofusion. In fact, if the voltage across the contact bilayer is U, then the voltage drop across the noncontacting regions of membranes is only $U/2$. The relation between the time ratio t_l/t_f is estimated to be approximately $\exp(\alpha U/2)$, where α is the slope of the curves in Fig. 5. The higher the U values, the higher is the probability that the contact bilayer will disintegrate earlier than the noncontacting regions.

6. FUSION OF LIPID BILAYERS AND ELECTROFUSION OF CELLS

We have considered the successive stages of fusion of lipid bilayers: approaching and contact, monolayer fusion and complete fusion. There are data which suggest that in the case of biomembrane fusion, the same sequence of events takes place (Pinto da Silva and Nogueira, 1977; Belitser and Zaalishvili, 1983). Let us determine at what stage of cell fusion the electric field effect is realized.

Approaching. Application of an electric field to the contacting cells causes their transient additional approach (Weber *et al.*, 1981; Stenger and Hui, 1986). This effect, as suggested by Weber *et al.* (1981), is due to attraction of unlike charges induced on the contacting membranes with the application of a pulse. In addition, an electric field of sufficiently high intensity may give rise in membranes to lipid domains free of protein molecules (Zimmermann, 1982). As a result, electrostatic repulsion and the steric factors which hinder the approaching of cells may be eliminated.

Monolayer fusion. The stage of monolayer fusion is observed in some cases of biomembrane fusion. It is reasonable to assume that electrofusion also involves the stage of formation of a contact bilayer. In fact, Stenger and Hui (1986) discovered the formation of local regions of the contact bilayer in the regions of contact of lipid domains during electrofusion of cells in dielectrophoresis chains. The displacement of proteins from the contact region may uncover the lipid domains and, possibly, change the effective shape of lipid molecules in the external monolayers and may thus promote monolayer fusion of membranes.

Complete fusion. The electrofusion mechanism proposed by Zimmermann (1982) presupposes local disordering of the structure of the contacting membranes and their further rejoining. In our opinion there are a few unexplained problems. First, the exposure of the hydrophobic tails of lipids to the aqueous solution is extremely unfavorable energetically. Second, the probability for the contact of such defect sites upon membrane approach appears to be very low (Sowers, 1986). Therefore, the mechanism for joining of the contacting membranes in the zone of these defects remains unclear.

The mechanism of the field effect at the stage of complete fusion can be explained by assuming that electrofusion is preceded by monolayer membrane fusion. In this case the voltage drop across the regions of the contact bilayer will be twice as high as that across the adjacent contacting regions of cell membranes. Consequently, electrical breakdown will occur with a higher probability in the regions of the contact bilayer similarly to electrofusion described above using the model of two planar membranes. The electrofusion model proposed avoids the energetically unfavorable stage of disordering of the structure. Note that irreversible breakdown of the contact bilayer results from the development of a hydrophilic pore in it. Destabilization of the contact bilayer may be determined not only by the direct breakdown effect of an electric pulse on the contact bilayer. Osmotic swelling of cells as a result of electroporation of membranes, which was observed during electrofusion (Sowers, 1984), is able to create in the cell membranes a mechanical tension which additionally destabilizes the contact bilayer and promotes expansion of the hydrophilic pore formed in it. It is noted that in the system of planar bilayers, due to the existence of tension, the stage of growth of an overcritical pore in the contact bilayer occurs very rapidly. For the fusion of cells the osmotic swelling may become a mandatory stage required for extending the diameter of a cytoplasmic bridge between the cells (Lucy and Ahkong, 1986).

Thus, electrical treatment may influence all the stages of membrane fusion. One might also assume that the key effect of the field in electrofusion is the direct action only on one of the stages. Specifically, Zimmermann (1982) advocated the opinion that the field effect is realized at the stage of complete fusion. At present, such an interpretation appears to be far too simplified. Sowers (1984, 1986), Teissie and Rols (1986), and Zimmermann *et al.* (1985) showed the existence of a relatively long-lived prefusable state of cells arising from electrical treatment. When the cells are brought into contact some time after the field has been applied, the induction of fusion occurs. Thus, the field effect, which is the key event of electrofusion, does not require the presence of contacting membranes during the field action. Zimmermann (1982), Sowers (1986), and Teissie and Rols (1986) found a similarity in the conditions of electrical treatment which cause electrofusion and electroporation. Teissie and Rols (1986) assumed that it is just the contact of pores during the subsequent approaching of membranes that leads to fusion. It seems to us, however, that this interpretation also needs significant refinement. The point is that the area of pores and their number are insufficiently large to ensure a sufficient probability for meeting of two pores when the membranes are brought into contact. The mechanism of closing of the edges of such pores to each other also remains unclear. In our opinion, the following hypothesis appears to be most probable at present. Under the effect of an electric field in the region of the poles of cells, pores appear in the membranes. In the course of resealing of pores or as a result of temporary loss of integrity of the membrane skeleton, protein-free domains may arise in these regions of membranes. The area of these domains may significantly exceed the area of the pores themselves by the time the cells are brought into contact. Therefore, the probability for the approach of the lipid domains of the interacting membranes may be sufficiently high. The contact of such domains just after the electrical treatment may initiate monolayer fusion and then the complete fusion of cell membranes after the destabilization of the contact bilayer. Possibly at this stage an additional tension develops from electroporation due to processes of a colloid-osmotic nature.

ACKNOWLEDGMENTS. We express our deep gratitude to our friend and colleague Dr. M. M. Kozlov for many valuable comments and critical reading of the manuscript. We are also grateful to Drs. Y. A. Chizmadzhev, V. S. Markin, and S. L. Leikin for fruitful discussions, and to Mrs. G. F. Voronina for assistance in preparing the manuscript.

REFERENCES

Abidor, I. G., Arakelyan, V. B., Chernomordik, L. V., Chizmadzhev, Y. A., Pastushenko, V. F., and Tarasevich, M. R., 1979, Electric breakdown of bilayer lipid membrane. I. The main experimental facts and their qualitative discussion, *Bioelectrochem. Bioenerg.* **6**:37–52.

Abidor, I. G., Pastushenko, V. F., Osipova, E. M., Melikyan, G. B., Kuzmin, P. I., and Fedotov, S. V., 1987, Relaxation studies of the plane contact of bilayer membranes, *Biol. Membr.* **4**:67–76.

Ahkong, Q. F., Fisher, D., Tampion, W., and Lucy, J. A., 1975, Mechanisms of cell fusion, *Nature* **253**:194–195.

Babunashvili, I. N., Silberstein, A. Y., and Nenashev, V. A., 1981, The investigation of the interaction between the liposomes and planar lipid membranes by the method of ionophore blocking, *Stud. Biophys.* **83**:131–137.

Bedford, M., and Cooper, G. W., 1978, Membrane fusion events in the fertilization of vertebrate eggs, in: *Membrane Fusion* (G. Poste and G. Nicolson, eds.), Elsevier/North-Holland, Amsterdam, pp. 66–126.

Belitser, N. V., and Zaalishvili, I. V., 1983, Electron microscopic visualization of membrane Ca^{2+}-binding sites in barley root tip cells: Induction in vivo by seed germination in the presence of $CaCl_2$, *Protoplasma* **115**:222–227.

Bentz, J., Ellens, H., Lai, M. Z., and Szoka, F. C., 1985, On the correlation between H_{II} phase and the contact-induced destabilization of phosphatidylethanolamine-containing membranes, *Proc. Natl. Acad. Sci. USA* **82**:5742–5745.

Berestovsky, G. N., and Gyulkhandanyan, M. Z., 1976, Contact interaction of bilayer lipid membranes, *Stud. Biophys.* **56**:19–20.

Bischoff, R., 1978, Fusion during muscle embryogenesis, in: *Membrane Fusion* (G. Poste and G. Nicolson, eds.), Elsevier/North-Holland, Amsterdam, pp. 127–179.

Bondeson, J., and Sundler, R., 1985, Lysine peptides induce lipid intermixing but not fusion between phosphatidic acid-containing vesicles, *FEBS Lett.* **190**:283–287.

Borovjagin, V. L., Vergara, J. A., and McIntosh, T. J., 1982, Morphology of the intermediate stages in the lamellar to hexagonal lipid phase transition, *J. Membr. Biol.* **69**:199–212.

Chernomordik, L. V., Sukharev, S. I., Abidor, I. G., and Chizmadzhev, Y. A., 1983, Breakdown of lipid bilayer membranes in an electric field, *Biochim. Biophys. Acta* **736**:203–213.

Chernomordik, L. V., Kozlov, M. M., Melikyan, G. B., Abidor, I. G., Markin, V. S., and Chizmadzhev, Y. A. 1985, The shape of lipid molecules and monolayer membrane fusion, *Biochim. Biophys. Acta* **812**: 643–655.

Chernomordik, L. V., Kozlov, M. M., Leikin, S. L., Melikyan, G. B., Markin, V. S., and Chizmadzhev, Y. A., 1987a, Lipid membrane fusion through stalk formation, *Biophys. J.* (submitted for publication).

Chernomordik, L. V., Melikyan, G. B., and Chizmadzhev, Y. A., 1987b, Biomembrane fusion. A new concept derived from model studies using two interacting planar lipid bilayers, *Biochim. Biophys. Acta* **906**:309–352.

Cohen, F. S., Akabas, M. H., Zimmerberg,J., and Finkelstein, A., 1984, Parameters affecting the fusion of unilamellar phospholipid vesicles with planar bilayer membranes, *J. Cell. Biol.* **98**:1054–1062.

Cullis, P. R., Hophe, M. G., and Tilcock, C. P. S., 1986, Lipid polymorphism and the role of lipids in membranes, *Chem. Phys. Lipids* **40:**127–144.

Duzgunes, N., 1985, Membrane fusion, *Subcell. Biochem.* **11:**195–285.

Duzgunes, N., Straubinger, R. M., Baldwin, P. A., Friend, D. S., and Papahadjopoulos, D., 1985, Proton-induced fusion of oleic acid–phosphatidylethanolamine liposomes, *Biochemistry* **24:**3091–3098.

Fisher, L. R., and Parker, N. S., 1984, Osmotic control of bilayer fusion, *Biophys. J.* **46:**253–258.

Gingell, D., and Ginsberg, L., 1978, Problems in the physical interpretation of membrane interaction and fusion, in: *Membrane Fusion* (G. Poste and G. Nicolson, eds.), Elsevier/North-Holland, Amsterdam, pp. 791–833.

Gruen, D. W. R., Marcelja, S., and Parsegian, V. A., 1984, Water structure near the membrane surface, in: *Cell Surface Dynamics* (A. S. Perelson, C. De Lisi, and F. W. Wiegel, eds.), Dekker, New York, pp. 59–91.

Helfrich, W., 1973, Elastic properties of lipid bilayers: Theory and possible experiments, *Z. Naturforsch.* **28c:** 693–703.

Horn, R. G., 1984, Direct measurement of the force between two lipid bilayers and observation of their fusion, *Biochim. Biophys. Acta* **778:**224–228.

Kalderon, N., and Gilula, N. B., 1979, Membrane events involved in myoblast fusion, *J. Cell Biol.* **81:**411–425.

Kozlov, M. M., Leikin, S. L., Chernomordik, L. V., Markin, V. S., and Chizmadzhev, Y. A., 1987, Stalk mechanism of membrane fusion: Joining of aqueous volumes, *Biol. Membr.* **4:**57–69.

Leikin, S. L., Kozlov, M. M., Chernomordik, L. V., Markin, V. S., and Chizmadzhev, Y. A., 1986, Membrane fusion: Overcoming of hydration barrier and local reorganizations of the structure, *Biol. Membr.* **3:**1159–1171.

Liberman, E. A., and Nenashev, V. A., 1972, Bilayer phospholipid membranes as a model for investigation of the changes of cells contact permeability, *Biophysika* **17:**231–238.

Lucy, J. A., and Ahkong, Q. F., 1986, An osmotic model for the fusion of biological membranes, *FEBS Lett.* **199:**1–11.

Markin, V. S., Kozlov, M. M., and Borovjagin, V. L., 1984, On the theory of membrane fusion: The stalk mechanism, *J. Physiol. Biophys.* **5:**361–377.

Melikyan, G. B., Abidor, I. G., Chernomordik, L. V., and Chailakhyan, L. M., 1983a, Electrofusion and fission of lipid bilayers, *Biochim. Biophys. Acta* **730:**395–398.

Melikyan, G. B., Chernomordik, L. V., Abidor, I. G., Chailakhyan, L. M., and Chizmadzhev, Y. A., 1983b, Ca^{2+}-induced fusion of solvent-free bilayer lipid membranes, *Dokl. Akad. Nauk, SSSR* **269:**1221–1225.

Melikyan, G. B., Kozlov, M. M., Chernomordik, L. V., and Markin, V. S., 1984, Fission of membrane tube, *Biochim. Biophys. Acta* **776:**169–175.

Mueller, P., Rudin, D. O., Tien, H. T., and Wescott, W. C., 1962, Reconstitution of cell membrane structure in vitro and its transformation into an excitable system, *Nature* **194:**979–980.

Neher, E., 1974, Asymmetric membranes resulting from the fusion of two black lipid bilayers, *Biochim. Biophys. Acta* **373:**327–336.

Nir, S., Bentz, J., Wilschut, J., and Düzgünes, N., 1983, Aggregation and fusion of phospholipid vesicles, *Prog. Surf. Sci.* **13:**1–124.

Ohki, S., 1984, Effect of divalent cations, temperature, osmotic pressure gradients and vesicle curvature on phosphatidylserine vesicle fusion, *J. Membr. Biol.* **77:**265–275.

Petrov, A. G., and Derzhanski, A., 1976, On some problems in the theory of elastic and flexo-electric effects in bilayer lipid membranes and biomembranes, *J. Phys. C* **37**(suppl. 3):155–160.

Pinto da Silva, P., and Nogueira, M. L., 1977, Membrane fusion during secretion, *J. Cell Biol.* **73:**161–181.

Pontecorvo, G., 1975, Production of mammalian somatic cell hybrids by means of polyethylene glycol treatment, *Somatic Cell Genet.* **1:**397–400.

Poste, G., and Pasternak, C. A., 1978, Virus-induced cell fusion, in: *Membrane Fusion* (G. Poste and G. Nicholson, eds.), Elsevier/North-Holland, Amsterdam, pp. 305–367.

Rand, R. P., 1981, Interacting phospholipid bilayers: Measured forces and induced structural changes, *Annu. Rev. Biophys. Bioeng.* **10:**277–314.

Raudino, A., 1986, A theoretical study of the rate of fusion between lipid vesicles, *J. Phys. Chem.* **90:**1916–1921.

Schindler, H., and Feher, G., 1976, Branched bimolecular lipid membranes, *Biophys. J.* **19:**1109–1113.

Schragin, A. S., Vasilenko, I. A., Selisheva, A. A., and Shvets, V. I., 1905, The influence of phospholipid metabolites on the fusion of model membranes of different compositions, *Biol. Membr.* **2:**789–794.

Siegel, D. P., 1984, Inverted micellar structures in bilayer lipid membranes: Formation, rates and half-lives, *Biophys. J.* **45:**399–420.
Sowers, A. E., 1984, Characterization of electric field-induced fusion in erythrocyte ghost membranes, *J. Cell Biol.* **99:**1989–1996.
Sowers, A. E., 1986, A long-lived fusogenic state is induced in erythrocyte ghosts by electric pulses, *J. Cell Biol.* **102:**1358–1362.
Stenger, D. A., and Hui, S. W., 1986, Kinetics of ultrastructural changes during electrically induced fusion of human erythrocytes, *J. Membr. Biol.* **93:**43–53.
Teissie, I., and Rols, M. P., 1986, Fusion of mammalian cells in culture is obtained by creating the contact between cells after their electropermeabilization, *Biochem. Biophys. Res. Commun.* **140:**258–266.
Weber, H., Förster, W., Berg, H., and Jacob, H.-E., 1981, Parasexual hybridization of yeasts by electric field stimulated fusion of protoplasts, *Curr. Genet.* **4:**165–166.
Zimmermann, U., 1982, Electric field-mediated fusion and related electrical phenomena, *Biochim. Biophys. Acta* **694:**227–277.
Zimmermann, U., Vienken, J., Halfmann, J., and Emeis, C. C., 1985, Electrofusion: A novel hybridization technique. *Adv. Biotechnol. Processes* **4:**79–150.

Chapter 12

Role of Proteases in Electrofusion of Mammalian Cells

Takako Ohno-Shosaku and Yasunobu Okada

1. INTRODUCTION

It has been reported that electrical pulsation produces a high frequency of fusion (> 50%) of mammalian cells in suspension if the cells are pretreated with proteases (Pilwat *et al.*, 1981; Zimmermann *et al.*, 1981). This result contrasts with the electrofusion of protoplasts (Scheurich *et al.*, 1981; Zimmermann and Scheurich, 1981a,b) and with that of mammalian cells in confluent monolayers (Blangero and Teissie, 1983). Zimmermann and his colleagues have suggested that the enzyme effect is not directly linked to the proteolytic activities (Zimmermann, 1982; Vienken *et al.*, 1983).

Facilitating effects of exogenous proteases are also known in the fusion of erythrocytes induced by some chemical agents (Hartmann *et al.*, 1976; Ahkong *et al.*, 1978). In addition, it has been reported that proteolytic degradation of membrane proteins by endogenous proteases, especially Ca-activated thiol-protease (calpain: Murachi *et al.*, 1981), is involved in the chemically induced fusion of erythrocytes (Quirk *et al.*, 1978; Ahkong *et al.*, 1980; Kosower *et al.*, 1983). Endogenous proteases (metalloendoprotease and Ca-activated protease) may play an essential role in spontaneous fusion of myoblasts (Couch and Strittmatter, 1983, 1984; Schollmeyer, 1986).

The question arose as to whether facilitation of electrofusion of mammalian cells by exogenous proteases is due to the proteolytic function, or whether some endogenous proteases (especially calpain) are involved in electrofusion. The purpose of the present chapter is to answer this question.

2. METHODS

Details of the methods employed in this study have been described in a previous paper (Ohno-Shosaku and Okada, 1985). In short, mouse T lymphoma cells (L5178Y) were suspended at a

TAKAKO OHNO-SHOSAKU AND YASUNOBU OKADA • Department of Physiology, Kyoto University Faculty of Medicine, Kyoto 606, Japan. *Present address of T.O.-S.:* Department of Physiology, Osaka Medical College, Osaka 569, Japan.

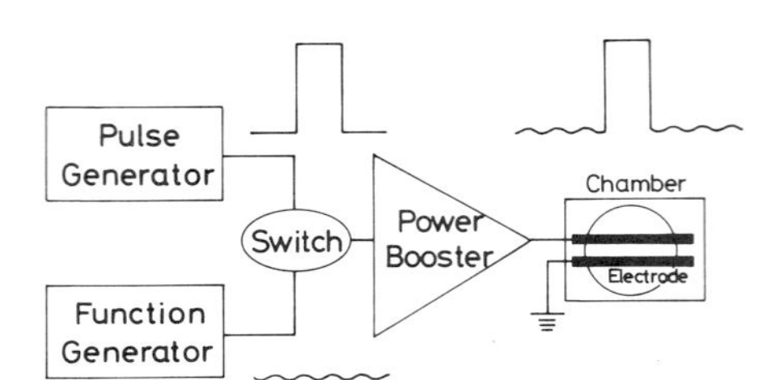

Figure 1. Arrangement of apparatus for electrofusion of cells in suspension under dielectrophoresis. Alternating fields were disconnected during the pulse application by an automatic switch.

density of about 1.5×10^6/ml in the fusion medium, and transferred into a fusion chamber equipped with two parallel platinum wires (0.2 mm in diameter) at a distance of 0.2 mm (Fig. 1). After attaining cell-to-cell contact by dielectrophoresis (0.8 kV/cm, 0.1 MHz), cell fusion was induced by applying four successive rectangular pulses (5, 6, 7, and 8 kV/cm, 20 μsec) at an interval of about 2 sec. Alternating electrical fields for dielectrophoresis and rectangular electric pulses for cell fusion were generated from a function generator and a pulse generator, respectively (Fig. 1). A booster amplifier circuit was employed to increase the output voltage (up to 300 V) and to reduce the output impedance (down to 10 ohms).

In some experiments, another cell line (mouse T lymphoma MCN151, mouse myeloma NS1, human B lymphoblast Raji, or human lymphoblastoma HSC93) or human peripheral lymphocytes isolated from freshly drawn blood of healthy donors (kindly prepared by Dr. Hiroko Hama-Inaba) were used. For the electrofusion of these cells, two square wave pulses of 3.3 and 5 kV/cm (10 μsec) were employed.

To examine the effect of exogenous proteases, a variety of proteases were added directly to the cell suspension before and during the application of electric fields. The effect of prior treatment with pronase E, dispase, or neuraminidase was examined after removing the enzyme from the cell suspension before administration of pulsations.

The fusion yield (%) was determined by counting the cells participating in cell fusion among 100–400 viable cells 5–15 min after pulsations. The percent of dead cells exposed to electrical fields was estimated by counting irreversibly damaged cells with characteristic phase-contrast images, as described previously (Ohno-Shosaku and Okada, 1984).

The chemicals employed were as follows: trypsin (E. Merck AG), α-chymotrypsin (Sigma Chemical Co.), papain (Type III, Sigma), collagenase (Type I, Sigma), protease Type I (Sigma), Type IV (Sigma), Type XI (proteinase K, Sigma), Type XIV (pronase E, Kaken Chem.), dispase (Godo Shusei), neuraminidase (Type V, Sigma), phenylmethylsulfonylfluoride (PMSF, Sigma), *p*-tosyl-L-lysine chloromethylketone hydrochloride (TLCK, Nakarai Chem.), aprotinin (Sigma), *N*-ethylmaleimide (NEM, Nakarai), iodoacetamide (IAA, Nakarai), *p*-chloromercuribenzoic acid (PCMB, Nakarai), [L-3-*trans*-carboxyoxiran-2-carbonyl]-L-leu-agmatin (E-64, a gift from Professor Takashi Murachi), leupeptin (a gift from Professor Takashi Murachi), 2-mercaptoethanol (Nakarai), and EGTA (Nakarai).

Free Ca^{2+} concentrations were measured with Ca^{2+}-selective electrodes filled with the neutral Ca^{2+} sensor ETH 1001 (kindly fabricated by Dr. Shigetoshi Oiki). Calibration of the electrodes was carried out, as described previously (Ueda *et al.*, 1986), using Ca-EGTA buffer solutions (Oiki and Okada, 1987).

The control fusion medium was composed of 300 mM mannitol, 1 mM $MgCl_2$, 0.1 mM $CaCl_2$, and 5 mM Tris-HCl (pH 7.0–7.4). In some experiments, $CaCl_2$ was removed, or 1 mM EGTA and an appropriate amount of $CaCl_2$ were added.

All experiments were performed at room temperature (20–25°C). Temperature increases due to joule heating during dielectrophoresis were less than 4°C.

All data are expressed as the mean ± S.E.M.

3. ELECTROFUSION WITHOUT PROTEASE TREATMENT: REQUIREMENT FOR Ca^{2+}

When L5178Y cells were suspended in fusion medium lacking $CaCl_2$ (nominally Ca^{2+}-free), more than 60% of the cells were irreversibly damaged and only a small number (< 10%) of viable cells participated in cell fusion after the administration of a train of square pulses under dielectrophoresis with alternating fields (Ohno-Shosaku and Okada, 1984). In the presence of 30 μM Ca^{2+}, however, relatively high fusion yields (up to 30%) and cell viability (up to 70%) were obtained (Ohno-Shosaku and Okada, 1984). These results are at variance with previous reports of Zimmermann and Teissie and their respective co-workers, wherein a high efficiency of electrofusion of mammalian cells was obtained even in nominally Ca^{2+}-free media (Pilwat *et al.*, 1981; Scheurich and Zimmermann, 1981; Zimmermann *et al.*, 1982; Teissie *et al.*, 1982; Blangero and Teissie, 1983, 1985). Zimmermann's group exposed suspension cells to pronase that might have been contaminated with minute amounts of Ca^{2+} (see Section 4). Teissie's group employed confluent monolayer cells. It was possible that some Ca^{2+} might have been trapped in the space between the cell layer and the substrate to which the cells were attaching, unless a chelating agent was used. In addition, Ca^{2+} was usually present in micromolar concentrations in distilled water. Therefore, the effects of Ca^{2+} on electrofusion should quantitatively be examined, using a Ca-buffering system. Such was done in L5178Y cells in suspensions by monitoring free Ca^{2+} concentrations via Ca^{2+}-selective electrodes (Okada *et al.*, 1984; Ohno-Shosaku and Okada, 1985). Increases in the extracellular free Ca^{2+} concentration from submicromolar to submillimolar ranges enhanced in a dose-dependent manner not only cell viability but also fusion efficiency. Cell viability reached a plateau of about 90% at more than 0.1 mM Ca^{2+}. The fusion yield reached a maximum value (about 40%) at 0.1–0.5 mM and leveled off thereafter. Inhibitory effects of millimolar Ca^{2+} were also found in the electrofusion of mammalian cells in monolayer cultures (Blangero and Teissie, 1985). In the presence of exogenous dispase, such facilitating and stabilizing effects of submillimolar Ca^{2+} were more prominent. Very high fusion yields (about 80%) and cell viability (about 90%) were consistently achieved in the presence of 0.1 mM Ca^{2+} and 10 μg/ml dispase. However, it should be noted that almost all the cells could not fuse at extracellular free Ca^{2+} concentrations below 10 μM and were irreversibly damaged below 1 μM Ca^{2+} after exposure to the electrical fields. The requirement for Ca^{2+} in the electrofusion of mammalian cells is consistent with previous reports that Ca^{2+} plays a role in virus-induced (Okada, 1969; Poste and Pasternak, 1978) and chemical agent-induced cell fusion (Lucy, 1978), as well as in spontaneously occurring myoblast fusion (David and Higginbotham, 1981).

4. ROLE OF EXOGENOUS PROTEASES

Protease treatment facilitates the fusion of erythrocytes or their ghosts induced by several chemical agents (Hartmann *et al.*, 1976; Ahkong *et al.*, 1978; Laster *et al.*, 1979). It has been suggested that virus-associated protease activities are involved in Sendai virus-induced fusion of erythrocytes (Israel *et al.*, 1983). When relatively high concentrations (1–3 mg/ml) of pronase P were used, Zimmermann and co-workers found that exogenous proteases facilitated electrofusion and stabilized mammalian cells exposed to electrical fields (Scheurich and Zimmermann, 1981; Pilwat *et al.*, 1981; Zimmermann *et al.*, 1981, 1982; Zimmermann, 1982). Using mouse lymphoma L5178Y cells, we tested eight commercially available proteases of different species. As shown in Fig. 2, all but collagenase markedly enhanced the yield of electrofusion without affecting pearl chain formation in the presence of 0.1 mM Ca^{2+}, irrespective of the species (serine, thiol, or metal protease). The viability of L5178Y cells exposed to electrical fields was also improved by the addition of most proteases (except for chymotrypsin and proteinase K) (Fig. 2). Similar facilitating

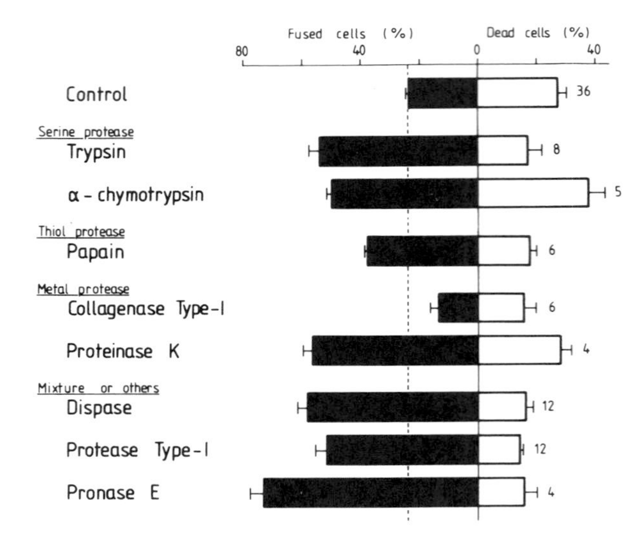

Figure 2. Effects of proteases on the fusion yield (solid bars) and the percentage of dead cells (open bars) after exposure to electrical fields. L5178Y cells were suspended in the control fusion medium. Conditions of protease treatment were as follows: trypsin 0.3–3 mg/ml for 4 min; chymotrypsin 0.3–1 mg/ml for 4 min; papain 10 μg/ml for 5–29 min; collagenase 1–3 mg/ml for 10 min; proteinase K (protease Type XI) 50–500 μg/ml for 4 min; dispase 10 μg/ml for 26–110 min; protease Type I 50 μg/ml for 5 min; and pronase E (protease Type XIV) 50 μg/ml for 4 min. Thin horizontal bars indicate S.E.M. Also shown are the number of experiments.

and stabilizing effects of pronase E and dispase were observed in the electrofusion of mouse myeloma NS1 cells and human lymphoblast Raji cells.

Zimmermann and his collaborators suggested that the effects of exogenous proteases are due to their physicochemical properties (e.g., charges and hydrophobic structures), which are common to other proteins, rather than their proteolytic activity (Zimmermann, 1982; Vienken *et al.*, 1983). In their electrofusion system, a protease inhibitor, PMSF, failed to prevent the effects of pronase, and prior treatment of cells with pronase was ineffective. They carried out experiments with relatively high doses of pronase added to a nominally Ca^{2+}-free medium. Thus, a small amount of Ca^{2+} possibly present as a contaminant in the pronase preparation may affect the results. We directly checked this possibility by measuring free Ca^{2+} concentrations in the enzyme solutions with Ca^{2+}-selective electrodes. As shown in Table 1, most protease preparations were found to contain a sizable amount of Ca^{2+}. Since a small amount of free Ca^{2+} should affect the fusion yield and cell viability, the effects of pronase observed under nominally Ca^{2+}-free conditions may be due, at least in part, to the contaminating Ca^{2+}. In fact, it was found that high concentrations (> 0.3 mg/ml) of pronase E were required to facilitate electrofusion of L5178Y cells in a nominally Ca^{2+}-free fusion medium, while the enzyme at a much lower dose (5 μg/ml) was sufficient to stimulate electrofusion in the presence of 0.1 mM Ca^{2+} (Ohno-Shosaku and Okada, 1985). Thus, we reexamined the

Table 1. Ca^{2+} Contamination in Protease Preparations[a]

Protease	Free Ca^{2+} concentration
Trypsin	1.11×10^{-4} M
α-Chymotrypsin	1.55×10^{-5} M
Protease Type I	1.44×10^{-4} M
Protease Type IV	1.18×10^{-3} M
Protease Type XI (proteinase K)	6.14×10^{-4} M
Protease Type XIV (pronase E)	1.00×10^{-3} M[b]
Dispase	2.90×10^{-3} M[b]

[a]All the proteases were dissolved at a concentration of 10 mg/ml in a solution of 0.3 M mannitol and 5 mM Tris-HCl (pH 7.2).
[b]In agreement with these data, pronase E and dispase are noted to contain about 15 and 50% Ca-acetate, respectively, by the commercial source.

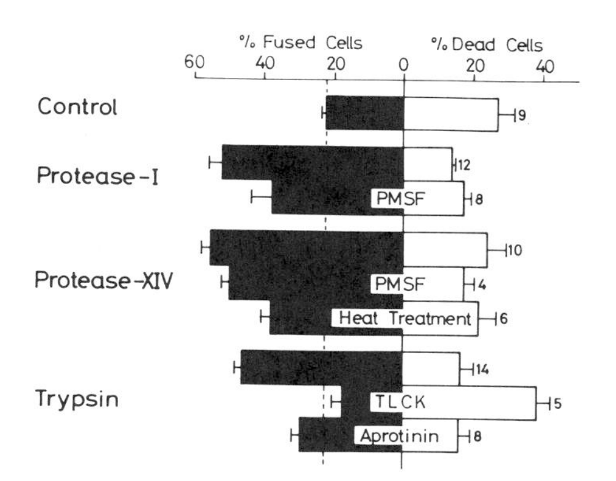

Figure 3. Effects of protease inhibitors in the presence of proteases, and effects of heat-inactivated pronase E on the fusion yield (solid bars) and the percentage of dead cells (open bars) after exposure to electrical fields. L5178Y cells suspended in the control fusion medium were exposed for several minutes to proteases and drugs at the following concentrations: protease Type I 50 μg/ml; protease Type XIV (protease E) 50–500 μg/ml; trypsin 0.5–1 mg/ml; PMSF 2 mM; TLCK 5 mM; and aprotinin 0.125 mg/ml. Thin horizontal bars and numbers as in Fig. 2.

mechanism by which exogenous proteases facilitate the electrofusion of L5178Y cells in the presence of an ample amount of Ca^{2+} to cancel the effect of contaminating Ca^{2+} (Ohno-Shosaku and Okada, 1984). The serine protease inhibitor PMSF remarkably inhibited the facilitating effect of protease Type I, as shown in Fig. 3. This inhibitor, however, only slightly suppressed the facilitating effect of pronase E (protease Type XIV) (Fig. 3), which is an unusually nonspecific protease preparation composed of several proteolytic enzymes. It seems likely that PMSF could not inhibit all the constituent enzyme activities. In fact, when pronase E was exposed to 90°C for 30 min, it was no longer able to produce marked facilitation (Fig. 3). Another serine protease inhibitor, aprotinin, and an alkylating proteinase inhibitor, TLCK, suppressed the facilitating effect of trypsin (Fig. 3). Taken together, the stimulatory effect of proteases on electrofusion can be attributed to their proteolytic activity.

Supporting this conclusion, prior treatment of L5178Y cells with pronase E or dispase produced essentially the same facilitating effect on electrofusion even following removal of the proteases before administration of pulsations (Fig. 4). This fact rules out the possibility that proteases exert their effects by passing into cells via the membrane pores induced by electrical breakdown (electropores). The close apposition of the two membranes is known to be essential for membrane fusion (Lucy and Ahkong, 1986). Thus, it is likely that exogenous proteases may act outside the cell membrane and degrade certain membrane-integrated or membrane-attached proteins, thereby enabling the two cell membranes adjoining (presumably the lipid domains) to come into very close contact. This concept is consistent with two facts. Without proteases, high yields of electrofusion were obtained in the case of protoplasts (Scheurich *et al.*, 1981; Zimmermann and Scheurich, 1981a,b), which had been treated with some cellulase (also contaminated with proteases) during the procedures to prepare them from plants. Exogenous proteases were not required for high fusion

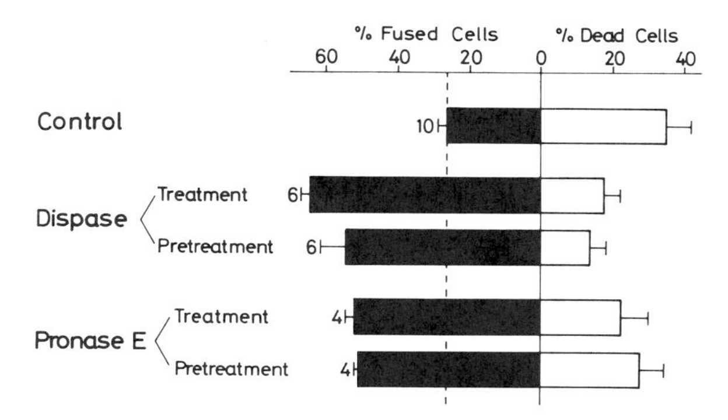

Figure 4. Effects of exposure to proteases and of prior treatment with proteases on the fusion yield (solid bars) and the percentage of dead cells (open bars). L5178Y cells were suspended in the control fusion medium. Treatment: dispase (10 μg/ml, 26–110 min) or pronase E (50 μg/ml, 3–5 min) was added before and during the pulse application. Pretreatment: dispase (10 μg/ml, 20 min) or pronase E (50 μg/ml, 5 min) was added and then removed prior to the application of electrical fields. Thin horizontal bars and numbers as in Fig. 2.

efficiency in the case of electrofusion of mammalian cells in confluent monolayer cultures (Blangero and Teissie, 1983), in which close cell-to-cell contacts were readily formed.

Negatively charged sialic acid residues on plasma membrane surfaces may prevent cells from coming into close contact, thereby inhibiting electrofusion. In L5178Y, Raji, and NS1 cells, however, neuraminidase treatment (0.25 mg/ml, 30 min) did not stimulate electrofusion in the presence or absence of exogenous proteases. In contrast, prior treatment with neuraminidase, in addition to the protease treatment, was actually necessary to bring about high yields of electrofusion of human peripheral lymphocytes (Hama-Inaba *et al.*, 1985), human lymphoblastoma HSC93 cells (Ohno-Shosaku *et al.*, 1984), and mouse mammary carcinoma FM3A cells (Okada and Hama-Inaba, 1986). Even in the presence of pronase E, these cells could only be in loose contact under dielectrophoresis and fused with low efficiency (less than 20%). Prior treatment with neuraminidase was found to improve the cell-to-cell contact under dielectrophoresis and to increase the fusion efficiency up to 40%. In the absence of exogenous proteases, the neuraminidase treatment improved the dielectrophoretic cell adhesion, but not the fusion yield so much. Thus, it can be deduced from these observations that depending on the properties of the cell surface coats, sialic acid residues of sialoglycoproteins and/or sialoglycolipids in certain cell species provide additional hindrance to the close cell contact. The removal by neuraminidase leads to higher fusion yields in protease-treated cells. However, neuraminidase cannot, by itself, facilitate electrofusion, in contrast to protease treatment.

5. POSSIBLE ROLE OF ENDOGENOUS PROTEASES

Endogenous protease activity has been found in many cell species, including lymphoid cells (Grayzel *et al.*, 1975). A growing body of evidence suggests that endogenous protease activity is involved in a variety of natural (Lui and Meizel, 1979; Couch and Strittmatter, 1983, 1984; Mundy and Strittmatter, 1985; Schollmeyer, 1986) and experimental membrane fusion events (Quirk *et al.*, 1978; Ahkong *et al.*, 1980; Kosower *et al.*, 1983; Nakornchai *et al.*, 1983; Lang *et al.*, 1984).

As shown in Fig. 3, the alkylating protease inhibitor TLCK, to which both serine and thiol proteases are sensitive, inhibited electrofusion of L5178Y cells exposed to a serine protease (trypsin) to a level of fusion yield less than the control level (no trypsin treatment). In contrast, serine protease inhibitors (PMSF and aprotinin) suppressed the protease-induced increase in fusion efficiency but the fusion level was still greater than for the control (Fig. 3). This fact suggests that endogenous thiol (but not serine) protease activities are associated with the efficiency of electrofusion of the cells. Supporting this inference, electrofusion of L5178Y cells was markedly inhibited by TLCK, but not by aprotinin, in the absence of exogneous proteases (Table 2). A thiol reagent (NEM) also suppressed electrofusion, although others (IAA and PCMB) did not, probably because of their low membrane permeability (Table 2).

Since a calcium ionophore (A23187) stimulated electrofusion (Okada *et al.*, 1984), it is likely that Ca^{2+} plays a primary role in electrofusion within the cell. Therefore, there is a possibility that a Ca-activated thiol proteinase (calpain: Murachi *et al.*, 1981) is involved in electrofusion. It was actually suggested that benzyl alcohol-induced erythrocyte fusion is mediated by the membrane-bound, Ca-activated thiol proteinase activity, which may degrade spectrin-binding proteins, thereby giving lateral movement to intramembranous particles (Ahkong *et al.*, 1980). However, the specific calpain inhibitor E-64 (Sugita *et al.*, 1980) could only slightly suppress electrofusion of L5178Y cells not treated with exogenous proteases (Table 2) and of those pretreated with pronase E (Table 3). Another putative calpain inhibitor, leupeptin (Toyo-Oka *et al.*, 1978), also failed to produce prominent inhibition of electrofusion of cells pretreated with pronase E (Table 3). Since a sufficient amount of E-64 might not permeate into the cells, the cells were exposed to four electrical pulses (5 to 8 kV/cm, 20 μsec) in the presence of E-64 (100 μg/ml) prior to applying dielectrophoretic and fusion-triggering fields to introduce the drug into the cells. However, again, marked inhibition was not observed (data not shown). Furthermore, the SH-reducing agent 2-mercaptoethanol, which

Table 2. Effects of Protease Inhibitors on the Fusion Yield of L5178Y Cells and the Percentage of Dead L5178Y Cells Exposed to Electrical Pulses in the Control Fusion Medium without Added Proteases[a]

	Concentration	Fused cells (%)	Dead cells (%)	Number of experiments
Control	—	27.9 ± 1.5	17.1 ± 2.6	25
TLCK	5 mM	5.9 ± 1.6	28.6 ± 4.8	6
Aprotinin	0.125 mg/ml	28.5 ± 2.1	16.1 ± 1.5	8
NEM	1–3 mM	13.5 ± 1.6	18.3 ± 4.1	8
IAA	10 mM	26.8 ± 2.5	16.7 ± 3.3	4
PCMB	0.3 mM	31.4 ± 2.6	24.7 ± 3.9	3
E-64	1 μg/ml	20.8 ± 3.2	2.9 ± 0.5	3

[a]The cells were treated with each drug for 5 to 30 min prior to pulse application.

activates calpain (Murachi, 1983), did not increase the electrofusion yield of L5178Y cells pretreated with pronase E (Table 3). In light of these observations, it is likely that thiol, but not serine, proteases may be at least partly involved in electrofusion. However, no compelling evidence for the involvement of calpain was obtained.

6. CONCLUSION

Ca^{2+} ions at greater than micromolar concentrations were required for electrofusion of L5178Y cells adhered to each other under dielectrophoresis. Treatment of the cells with exogenous proteases markedly raised fusion efficiency without affecting dielectrophoretic cell-to-cell contact. This effect of proteases should be discriminated from the effect of Ca^{2+} contamination of the protease preparations. The protease effect *per se* is actually due to its proteolytic activity, because a number of protease inhibitors blocked the effect and because heat-inactivated proteases were no longer effective. Prior treatment of the cells with proteases was also effective in facilitating electrofusion even after removing the enzyme before pulsations. Thus, it is likely that exogenous proteases may act from outside the cell membrane and degrade certain membrane-associated proteins, which sterically prevent bilayer–bilayer interactions, thereby bringing the lipid domains of

Table 3. Effects of Inhibitors and an Activator of Calpain on the Fusion Yield of L5178Y Cells and the Percentage of Dead L5178Y Cells Exposed to Electrical Pulses in the Control Fusion Medium after Prior Treatment with Pronase E (100 μg/ml, 10 min)[a]

	Concentration	Fused cells (%)	Dead cells (%)	Number of experiments
Control	—	64.8 ± 2.4	13.5 ± 2.7	14
Leupeptin	1 μg/ml	53.9 ± 2.7	12.8 ± 8.3	6
E-64	1 μg/ml	64.7 ± 3.5	17.3 ± 5.8	4
	100 μg/ml	58.6 ± 3.7	14.2 ± 1.1	4
2-Mercaptoethanol	1.3 mM	45.9 ± 3.2	20.1 ± 3.8	5
2-Mercaptoethanol and E-64	1.3 mM and 100 μg/ml	47.0 ± 2.7	23.0 ± 6.6	5

[a]The cells were treated with each drug for 5 to 30 min prior to pulse application.

two adjoining cell membranes very close together. In several cell species, including human lymphocytes, prior treatment with neuraminidase was necessary, in addition to treatment with the protease, to facilitate electrofusion. In these cells, the close cell-to-cell contact could not be achieved without neuraminidase treatment. Therefore, it is likely that there are two barriers to electrofusion: first, a neuraminidase-sensitive glycocalyx, which prevents the close adherence of cells; and second, protease-sensitive membrane-associated proteins which hinder the lipid membranes from coming into close proximity.

Thiol protease inhibitors (TLCK and NEM), but not serine protease inhibitors (aprotinin and PMSF), suppressed electrofusion of lymphoma cells. On the other hand, putative inhibitors for Ca^{2+}-activated thiol proteinase (calpain), E-64 and leupeptin, were much less effective in inhibiting electrofusion. The SH-reducing agent 2-mercaptoethanol never enhanced the fusion yields. Although a definite conclusion cannot be drawn solely from pharmacological studies, the involvement of endogenous serine proteases and calpain seems unlikely. However, it is possible that some endogenous thiol proteases play a role in electrofusion.

ACKNOWLEDGMENTS. The authors are grateful to Professor Motoy Kuno for reading the manuscript, and to Professor Takashi Murachi for a generous gift of calpain inhibitors and for helpful discussions. Thanks are also due to Drs. Shigetoshi Oiki and Hiroko Hama-Inaba for their collaboration in some of the experiments. This work was supported by Grants-in-Aid for Scientific Research from the Japanese Ministry of Education, Science and Culture.

REFERENCES

Ahkong, Q. F., Blow, A. M. J., Botham, G. M., Launder, J. M., Quirk, S. J., and Lucy, J. A., 1978, Proteinases and cell fusion, *FEBS Lett.* **95:**147–152.

Ahkong, Q. F., Botham, G. M., Woodward, A. W., and Lucy, J. A., 1980, Calcium-activated thiol-proteinase activity in the fusion of rat erythrocytes induced by benzyl alcohol, *Biochem. J.* **192:**829–836.

Blangero, C., and Teissie, J., 1983, Homokaryon production by electrofusion: A convenient way to produce a large number of viable mammalian fused cells, *Biochem. Biophys. Res. Commun.* **114:**663–669.

Blangero, C., and Teissie, J., 1985, Ionic modulation of electrically induced fusion of mammalian cells, *J. Membr. Biol.* **86:**247–253.

Couch, C. B., and Strittmatter, W. J., 1983, Rat myoblast fusion requires metalloendoprotease activity, *Cell* **32:** 257–265.

Couch, C. B., and Strittmatter, W. J., 1984, Specific blockers of myoblast fusion inhibit a soluble but not the membrane-associated metalloendoprotease in myoblasts, *J. Biol. Chem.* **259:**5396–5399.

David, J. D., and Higginbotham, C.-A., 1981, Fusion of chick embryo skeletal myoblasts: Interactions of prostaglandin E_1, adenosine 3′:5′ monophosphate, and calcium influx, *Dev. Biol.* **82:**308–316.

Grayzel, A. I., Hatcher, V. B., and Lazarus, G. S., 1975, Protease activity of normal and PHA stimulated human lymphocytes, *Cell. Immunol.* **18:**210–219.

Hama-Inaba, H., Ohno-Shosaku, T., and Okada, Y., 1985, Electric pulse-induced heterokaryon hybridization using human lymphocytes, in: *The China–Japanese Bilateral Symposium on Biophysics* (Biophys. Soc. China, ed.), China Academic Press, pp. 194–197.

Hartmann, J. X., Galla, J. D., and Emma, D. A., 1976, The fusion of erythrocytes by treatment with proteolytic enzymes and polyethylene glycol, *Can. J. Genet. Cytol.* **18:**503–512.

Israel, S., Ginsberg, D., Laster, Y., Zakai, N., Milner, Y., and Loyter, A., 1983, A possible involvement of virus-associated protease in the fusion of Sendai virus envelopes with human erythrocytes, *Biochim. Biophys. Acta* **732:**337–346.

Kosower, N. S., Glaser, T., and Kosower, E. M., 1983, Membrane-mobility agent-promoted fusion of erythrocytes: Fusibility is correlated with attack by calcium-activated cytoplasmic proteases on membrane proteins, *Proc. Natl. Acad. Sci. USA* **80:**7542–7546.

Lang, R. D. A., Wickenden, C., Wynne, J., and Lucy, J. A., 1984, Proteolysis of ankyrin and of band 3 protein in chemically induced cell fusion: Ca^{2+} is not mandatory for fusion, *Biochem. J.* **218:**295–305.

Laster, Y., Lalazar, A., and Loyter, A., 1979, Viral and non-viral induced fusion of pronase-digested human erythrocyte ghosts, *Biochim. Biophys. Acta* **551**:282–294.

Lucy, J. A., 1978, Mechanisms of chemically induced cell fusion, in: *Membrane Fusion* (G. Poste and G. L. Nicolson, eds.), North-Holland, Amsterdam, pp. 267–304.

Lucy, J. A., and Ahkong, Q. F., 1986, An osmotic model for the fusion of biological membranes, *FEBS Lett.* **199**:1–11.

Lui, C. W., and Meizel, S., 1979, Further evidence in support of a role for hamster sperm hydrolytic enzymes in the acrosome reaction, *J. Exp. Zool.* **207**:173–186.

Mundy, D. I., and Strittmatter, W. J., 1985, Requirement for metalloendoprotease in exocytosis: Evidence in mast cells and adrenal chromaffin cells, *Cell* **40**:645–656.

Murachi, T., 1983, Intracellular Ca^{2+} protease and its inhibitor protein: Calpain and calpastatin, in: *Calcium and Cell Function,* Volume IV (W. Y. Cheung, ed.), Academic Press, New York, pp. 377–410.

Murachi, T., Tanaka, K., Hatanaka, M., and Murakami, T. 1981, Intracellular Ca^{2+}-dependent protease (calpain) and its high-molecular-weight endogenous inhibitor (calpastatin), *Adv. Enzyme Regul.* **19**:407–424.

Nakornchai, S., Sathitudsahakorn, C., Chongchirasiri, S., and Yuthavong, Y., 1983, Mechanism of enhanced fusion capacity of mouse red cells infected with *Plasmodium berghei, J. Cell Sci.* **63**:147–154.

Ohno-Shosaku, T., and Okada, Y., 1984, Facilitation of electrofusion of mouse lymphoma cells by the proteolytic action of proteases, *Biochem. Biophys. Res. Commun.* **120**:138–143.

Ohno-Shosaku, T., and Okada, Y., 1985, Electric pulse-induced fusion of mouse lymphoma cells: Roles of divalent cations and membrane lipid domains, *J. Membr. Biol.* **85**:269–280.

Ohno-Shosaku, T., Hama-Inaba, H., and Okada, Y., 1984, Somatic hybridization between human and mouse lymphoblast cells produced by an electric pulse-induced fusion technique, *Cell Struct. Funct.* **9**:193–196.

Oiki, S., and Okada, Y., 1987, Ca-EGTA buffer in physiological solutions, *Seitai No Kagaku* **38**:79–83 (in Japanese).

Okada, Y., 1969, Factors in fusion of cells by HVJ, *Curr. Top. Microbiol. Immunol.* **48**:102–128.

Okada, Y., and Hama-Inaba, H., 1986, Electric pulse-induced gene transfer and cell fusion, *Metabolism and Disease* **23**:211–219 (in Japanese).

Okada, Y., Ohno-Shosaku, T., and Oiki, S., 1984, Ca^{2+} is prerequisite for cell fusion induced by electric pulses, *Biomed. Res.* **5**:511–516.

Pilwat, G., Richter, H.-P., and Zimmermann, U., 1981, Giant culture cells by electric field-induced fusion, *FEBS Lett.* **133**:169–174.

Poste, G., and Pasternak, C. A., 1978, Virus-induced cell fusion, in: *Membrane Fusion* (G. Poste and G. L. Nicolson, eds.), North-Holland, Amsterdam, pp. 305–367.

Quirk, S. J., Ahkong, Q. F., Botham, G. M., Vos, J., and Lucy, J. A., 1978, Membrane proteins in human erythrocytes during cell fusion induced by oleoylglycerol, *Biochem. J.* **176**:159–167.

Scheurich, P., and Zimmermann, U., 1981, Giant human erythrocytes by electric field-induced cell-to-cell fusion, *Naturwissenschaften* **68**:45–46.

Scheurich, P., Zimmermann, U., and Schnable, H., 1981, Electrically stimulated fusion of different plant cell protoplasts, *Plant Physiol.* **67**:849–853.

Schollmeyer, J. E., 1986, Possible role of calpain I and calpain II in differentiating muscle, *Exp. Cell Res.* **163**:413–422.

Sugita, H., Ishiura, S., Suzuki, K., and Imahori, K., 1980, Inhibition of epoxide derivatives on chicken calcium-activated neutral protease (CANP) in vitro and in vivo, *J. Biochem.* **87**:339–341.

Teissie, J., Knutson, V. P., Tsong, T. Y., and Lane, M. D., 1982, Electric pulse-induced fusion of 3T3 cells in monolayer culture, *Science* **216**:537–538.

Toyo-Oka, T., Shimizu, T., and Masaki, T., 1978, Inhibition of proteolytic activity of calcium activated neutral protease by leupeptin and antipain, *Biochem. Biophys. Res. Commun.* **82**:484–491.

Ueda, S., Oiki, S., and Okada, Y., 1986, Oscillations of cytoplasmic concentrations of Ca^{2+} and K^{+} in fused L cells, *J. Membr. Biol.* **91**:65–72.

Vienken, J., Zimmermann, U., Fouchard, M., and Zagury, D., 1983, Electrofusion of myeloma cells on the single cell level: Fusion under sterile conditions without proteolytic enzyme treatment, *FEBS Lett.* **163**:54–56.

Zimmermann, U., 1982, Electric field-mediated fusion and related electrical phenomena, *Biochim. Biophys. Acta* **694**:227–277.

Zimmermann, U., and Scheurich, P., 1981a, Fusion of *Avena sativa* mesophyll cell protoplasts by electrical breakdown, *Biochim. Biophys. Acta* **641**:160–165.

Zimmermann, U., and Scheurich, P., 1981b, High frequency fusion of plant protoplasts by electric fields, *Planta* **151:**26–32.

Zimmermann, U., Pilwat, G., and Richter, H. P., 1981, Electric-field-stimulated fusion: Increased field stability of cells induced by protease, *Naturwissenschaften* **68:**577–579.

Zimmermann, U., Pilwat, G., and Pohl, H. A., 1982, Electric field-mediated fusion, *J. Biol. Phys.* **10:**43–50.

Chapter 13

Electrofusion of Mammalian Cells and Giant Unilamellar Vesicles

J. Teissié, M. P. Rols, and C. Blangero

1. INTRODUCTION

Cell electrofusion is one of the most striking observations linked to the effect of electric fields on viable cells or protoplasts. Since its first applications in the early 1980s, this methodology has been developed in many fields of cell biology and biotechnology. Different applications are reviewed in this book.

Electrofusion appears to be linked to electropermeabilization because cells to be fused are always pulsed by fields which are strong enough to induce the permeabilization. A direct link between the two processes has never been described. Further the early descriptions of electrofusion presumed that fusion was induced during the field pulse (Zimmermann, 1982) and that intracellular contact during the pulse was needed for the induction of electrofusion. This concept has been shown not to be true (Sowers, 1986; Teissié and Rols, 1986). Permeabilized cells are fusogenic and in order for fusion to occur, contacts can be created *after* the pulses.

Different approaches have been proposed for the induction of the cell–cell contacts:

- Mechanical when cells are large enough
- Natural with slime molds (Neumann *et al.*, 1980)
- By dielectrophoresis after proteolytic enzyme treatment
- By immunological "targeting" (see Lo and Tsong, this volume)
- Chemical by use of aggregative agents (Weber *et al.*, 1981; Chapel *et al.*, 1986)
- By use of a high density of cells [this can be obtained by pelleting the suspension (Teissié and Rols, 1986)
- By use of the contact inhibition of plated cells (Teissié *et al.*, 1982)

The last approach is of considerable power when one wants to study the mechanisms involved in electrofusion. Any chemical, enzymatic, or physical treatment of the cells is avoided. Artifacts

J. Teissié, M. P. Rols, and C. Blangero • Center for Biochemistry and Cellular Genetics, CNRS, 31062 Toulouse, France.

associated with their effects are then not present and the behavior is going to be affected only by electrical pulsations. One of the limits is that, of course, this kind of approach is restricted to cells able to grow on dishes. It should be emphasized that this way of fusing cells was shown to be highly efficient in obtaining viable stable hybrids (Finaz *et al.*, 1984).

In this chapter, our experimental procedure for electrofusing plated cells is described in complete detail. We observed that electrofusion was under the control of parameters which could be easily changed. Their observations provide the experimental background needed for a description of the involved molecular processes.

2. PHYSICAL METHODS

Electric fields are generated across aqueous ionic suspensions by applying a voltage between two electrodes in contact with the solution. From work by the groups of Tsong (Kinosita and Tsong, 1979) and Chidmadzhev (Chernomordik *et al.*, 1983), it was clear that the induction of permeant structures in the plasma membrane was under the control of:

- The intensity of the field
- The duration of the pulse
- The number of pulses
- The nature of the external buffer
- The cell strain

In order to have a better understanding of the processes underlying the electric field-induced fusion, we designed an electropulsator where each of the five parameters listed above was under our control (Fig. 1).

All cells were pulsed with the same field intensity. For that purpose, we chose pulses of a constant and uniform field intensity. This was obtained by applying a constant voltage to two parallel flat stainless steel electrodes. The pulse duration was electronically selected between 5 and 300 μsec. The number of pulses was either automatically or manually selected.

By means of a voltage divider, a part of the applied pulse V was observed on line with an oscilloscope. The shape (square wave) and intensity (V/d, d being the width between the electrodes) were thus monitored during the experiment. This observation showed that the rise-time (on and off) of the pulse was very short (< 1 μsec) and that the intensity of the field was kept constant (within 10%) during the pulsation. Furthermore, this showed that the pulsing conditions were highly reproducible between experiments.

Changing the nature and the salt content of the pulsing buffer would change its conductance and thus the resistance of the solution between the electrodes. Owing to the high power of the generator (10 kW), we were nevertheless able to obtain high field pulses even when working with 0.15 M monovalent salt solutions. By assuming that all the electrical energy is converted in joule heating, we computed that the temperature increase of the sample was very small (about 1°C) under all experimental conditions.

3. BIOLOGICAL METHODS

Chinese hamster ovary (CHO) cells have been adopted for use in a large number of somatic cell genetics laboratories (Gottesman, 1985). We selected the WTT clone which was kindly supplied by Professor Zalta (this institute).

This strain, which grows in monolayer (generation time of 16–18 hr), has been adapted for suspension culture (generation time of 22–24 hr). Cells grown in suspension can be replaced very easily (in less than 2 hr).

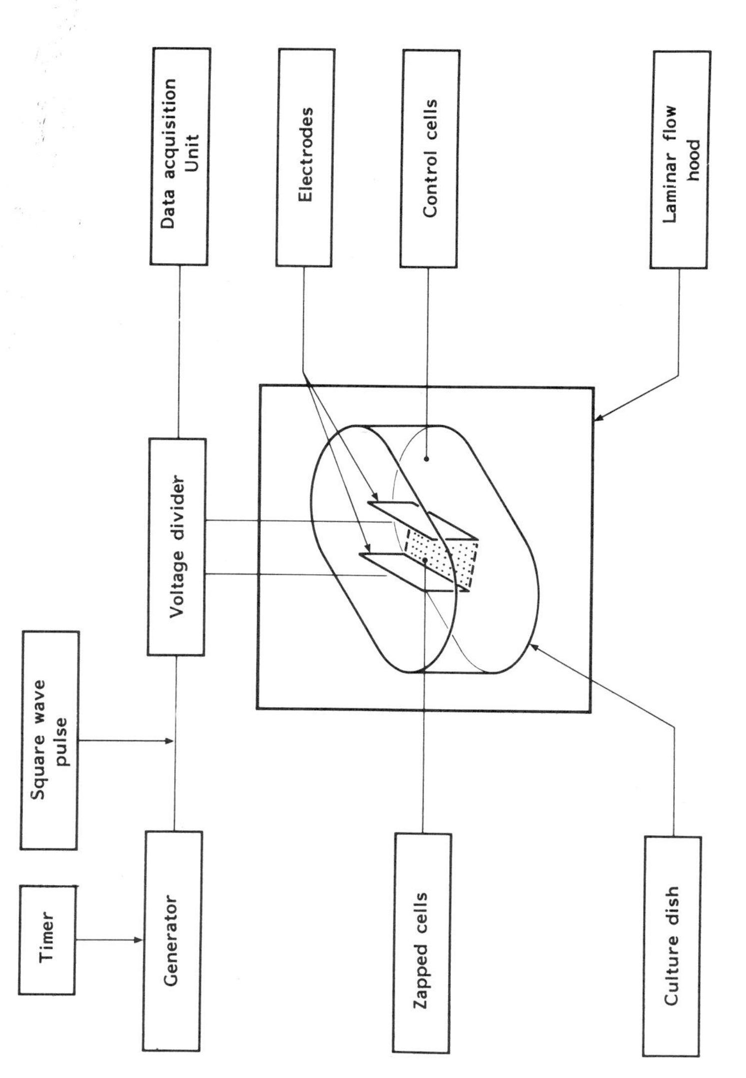

Figure 1. Basic design of the electropulsator.

3.1. Culture Conditions

Cells were grown in suspension of 37°C in a 5% CO_2/95% air atmosphere (Heraeus incubator, Germany) in Eagle's minimum essential medium (MEM OIII, Eurobio, France). This solution was supplemented with 6–8% newborn calf serum, penicillin (100 U/ml), streptomycin (100 μg/ml), and L-glutamine (1.16 mg/ml). The cell density was kept between 4 and 10 × 10^5 per ml by daily dilution.

The cells were then plated on culture dishes (Nunc, Denmark) and kept for 24 hr in the CO_2 atmosphere. The cell density was adjusted to be about 1000 cells/mm² in order to obtain numerous cell–cell contacts. CHO cell growth is known to be inhibited when such contacts occur.

CHO cells are able to grow on glass. Cells were grown on clean and sterile microscope slides (18 × 18 mm, Polylabo, France) when their viability was controlled after electropulsation. The entire surface of the slide was pulsed and the slide was taken out of the dish. It was washed with culture medium and placed in a new clean culture dish. The growth of the cells in the new dish was monitored for several weeks. When the cell density was high, the culture was diluted by trypsinization and replating. The behavior of the pulsed cells was similar to that of the control (unpulsed cells).

3.2. Procedure for Plated Cell Electrofusion

Cells are grown to a confluent density (more than 1000 cells/mm²). The culture medium is removed by pipeting and substituted with a pulsing buffer. This solution should be: (1) isosmolar, (2) low in salt content, (3) low in Mg^{2+} (mM range), and (4) of physiological pH. In most of our studies, we selected 10 mM phosphate buffer (pH 7.2), 1 mM $MgCl_2$, 250 mM sucrose. This buffer was controlled to not affect the growth of the unpulsed CHO cells when removed after 2 hr contact. The electrodes were brought into contact with the bottom of the dish by lifting the plate. Cells which were between the electrodes were pulsed; other cells on the dish were control samples (see Fig. 2). We must emphasize that these controls were from the same culture. They were treated exactly as the pulsed cells except for the electric field intensity, pulse duration, pulse number, and delay between the pulses (as short as 0.1 sec, if needed). The oscilloscope allowed the shape (i.e., the strength and duration) of the electric field pulses to be monitored on line and it showed whether the experiment was correctly performed. A clean square wave shape should be observed.

The electrodes were then removed and the dish was placed in a 37°C incubator for 1 hr after substituting the pulsing buffer with the culture medium. Thereafter, the culture was observed under an inverted microscope.

When sterile conditions were needed (long-term postpulsation culture), the solutions were sterilized by use of Millex filters (Millipore, France) and all experiments were performed under a laminar flow hood (ESI, France). The electrodes were treated with ethanol and sterile water before pulsation. No contamination was observed after the pulsation even during several months of long-term culture.

The occurrence of fusion was quantitated by different means. The simplest way was to observe the cells after pulsation and incubation. This was easily done by an inverted microscope with a video system. With our electrofusion technique on plated cells, unpulsed control cells were also present on the same dish. The mark drawn by the electrodes against the bottom of the dish determined the limit of the pulsed area. We used as a fusion index the extent of polynucleation:

$$I = \sum_{i=2} i\, C_i \Big/ \sum_{i=1} i\, C_i$$

where C_i is the number of cells with i nuclei in their cytoplasm. We compared the value of the unpulsed samples (a basic polynucleation was always present) with that of the pulsed samples.

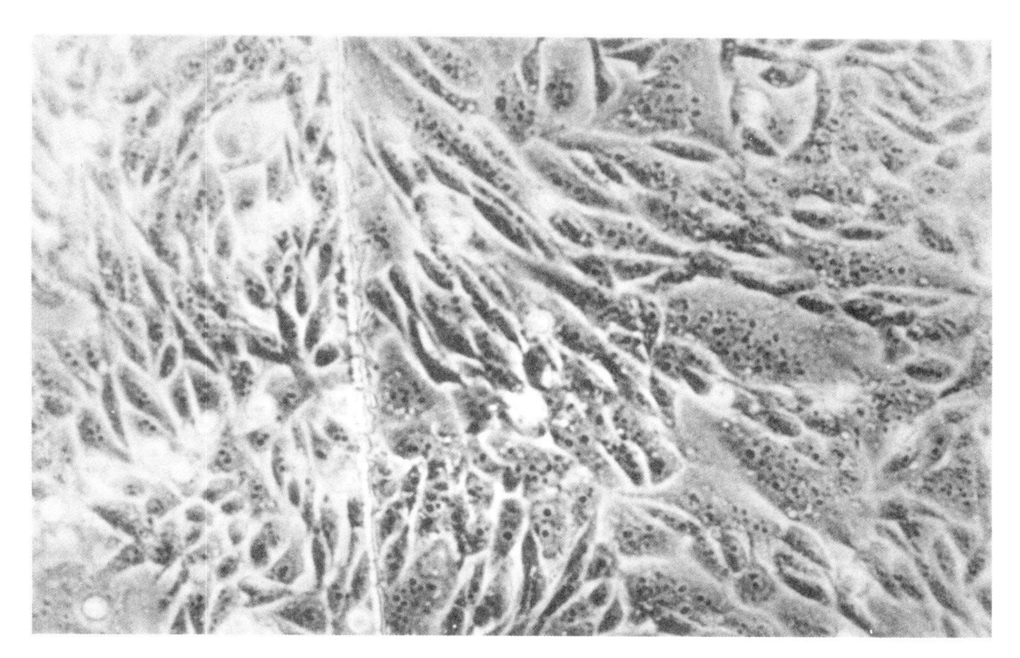

Figure 2. Electrofused and control plated CHO cells. The line is the trace of the contact of the electrode with the bottom of the dish. Control cells (unpulsed because they lie outside the electrodes), where the level of polynucleation is low, are on the left. Pulsed cells display a high polynucleation index. Courtesy of Academic Press.

Another way to document the occurrence of fusion was to plot the polynucleation histogram. A large number of highly polynucleated cells would be present.

A third way was to pulse fluorescent-labeled cells. This was conveniently obtained by treating CHO cells in suspension separately with fluorescein isothiocyanate (green fluorescence) and rhodamine isothiocyanate (red fluorescence). A borate buffer with a pH of 8.5 was not toxic to the cells; a 30 min incubation at room temperature was sufficient to obtain strongly fluorescent cells after washing repeatedly by centrifugation. The fluorescent cells were then mixed and plated. The two color population was then pulsed and incubated. Cells with the two colors mixed together were then observed.

A more definitive way was to work on cells with genetic markers. TK^- and $HPGRT^-$ cells were cocultured on the dish. The population was pulsed and then incubated. After 72 hr, the culture medium was replaced by a selective HAT one (Finaz *et al.*, 1984). The obtainment of resistant cells was the evidence that the fusion and the production of viable hybrids were a consequence of the electrical treatment.

4. GIANT UNILAMELLAR VESICLES (GUV)

4.1. Chemicals

Phospholipids and cholesterol were obtained from Sigma (USA). Trichloroacetic acid (TCA) and 3-morpholinopropane sulfonic acid (MOPS) were purchased from Merck (West Germany). Salts and sucrose were analytical grade. Ultrapure water obtained from a MilliQ system (Millipore, France) was used.

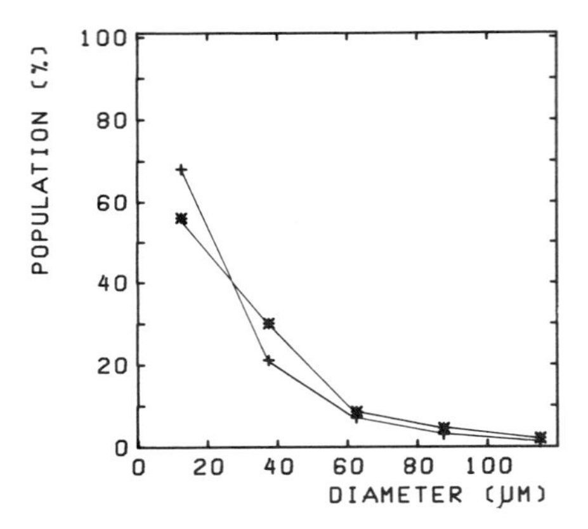

Figure 3. Change in the size distribution of GUV after pulsation. The unpulsed sample is (+); the pulsed one (1 kV/cm, 100 μs, 5 times) is (*).

4.2. Preparation

The procedure described by Oku and MacDonald (1983) was followed with some modifications. The lipid mixture (soybean lecithin, cholesterol, phosphatidic acid, 68:23:9) was chosen in order to mimic the composition in a biological membrane. TCA was used in acid form.

The vesicles were observed under an inverted microscope using phase-contrast optics. Their size distribution is shown in Fig. 3. Clearly, many large vesicles (up to 100 μm in diameter) are present.

4.3. Electropulsation of GUV

A culture dish was put on the movable stage of an inverted microscope (Leitz Diavert, West Germany). Two parallel stainless steel electrodes were brought into contact with the dish. The distance between the electrodes was 5 mm. The spacer between the electrodes was hollow to allow illumination of the volume in between. Two hundred and fifty microliters of the GUV suspension [diluted to half in the pulsing buffer containing PEG 2% (w/v) or spermine (200 μM)] was pipeted between the electrodes and allowed to settle for 5 min. The vesicles were easily observed by use of the 20× or 32× objective.

The electrodes were connected to the same generator as described above.

The preparation was observed with a video system connected to a magnetoscope. This technology permitted replay of the video film at reduced speed or picture by picture without any loss of quality.

Different places on the GUV preparation were first observed and video recorded. The size distribution was obtained later by playback and observation picture by picture. The pulsations were applied and controlled on line with the oscilloscope. The fusion was either directly observed and recorded on line, or the record was operated later in order to obtain the size distribution after the pulsations.

5. ELECTROFUSION OF MAMMALIAN CELLS

The electropulsator we were using permitted us to independently change every parameter of the pulsation, i.e., field intensity, pulse duration, ionic content of the pulsing buffer, ionic composition of the postpulse incubation buffer, temperature of the sample during the pulsation, and temperature of the sample after the pulsation. It should be noted that washing the cell culture and changing the external medium required only a few seconds because we were working on the plated cells.

5.1. Viability of the Cells

Under our pulsing conditions, CHO cells were observed to withstand repetitive applications of strong square wave pulses without any loss of viability. A treatment involving 20 pulses of 2 kV/cm with a duration of 100 μsec (delay 1 sec) did not affect the growth of the culture. This is a great advantage over other systems (proteolytic enzyme-treated cells in suspension, ion-free buffer, need of dielectrophoresis, exponentially decaying field).

5.2. Effect of Electrical Parameters

All these experiments were performed using the standard pulsing buffer which was described in the pulsation procedures. Electrofusion of plated CHO cells could be obtained only under well-defined electrical conditions (Fig. 2).

No fusion was observed if the field intensity was below a reproducible threshold (0.6 kV/cm) (Fig. 4). The extent of fusion was then observed to increase with an increase in the magnitude of the field up to the plateau value of 1.2 kV/cm. Using stronger field intensities did not improve the yield of fusion. The pulse duration was set at 100 μsec and five successive pulses were applied with a delay of 1 sec.

The pulse duration was observed to be a very critical parameter. Using a high field intensity that we had first shown to be highly efficient for cell fusion (1.5 kV/cm, five times), we observed that fusion was obtained only when the pulse duration was long enough. The use of pulses less than 10 μsec was ineffective. For pulse duration between 10 and 80 μsec, the fusion yield was increased with an increase in field intensity. If the pulse duration was longer than 100 μsec (up to 300 μsec), no further improvement in the fusion yield was observed.

Accumulation of successive pulsations with optimized intensity and duration (1.5 kV/cm, 100 μsec) increased the yield of fusion to some extent. But no improvement was observed when more than four pulses were applied.

This observation of a limited advantage of the accumulation of pulses must be corrected by the observation that a change in the direction of the field between successive pulses did improve the fusion yield. If eight pulses were applied, four in one direction, four in the perpendicular one, then the fusion yield increased 20% in relative value (Teissié and Blangero, 1984). Changing the direction of the field is easily obtained with plated cells, by simply rotating the dish and keeping the electrodes in the same direction.

5.3. Effect of Physical Parameters

When the cell density was high enough (> 400 cells/mm^2), it had no effect on fusion yield. This indicated that cell–cell contacts must be present in order to induce fusion.

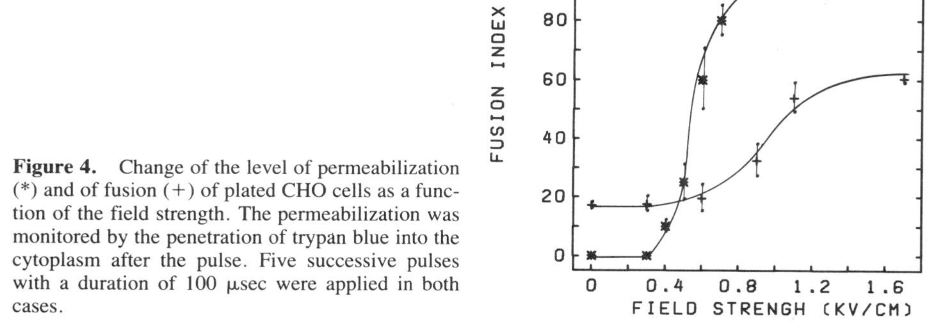

Figure 4. Change of the level of permeabilization (*) and of fusion (+) of plated CHO cells as a function of the field strength. The permeabilization was monitored by the penetration of trypan blue into the cytoplasm after the pulse. Five successive pulses with a duration of 100 μsec were applied in both cases.

The temperature of the sample during the pulsation (4, 21, 37°C) was not observed to affect the level of electrofusion. As a consequence, all further experiments were performed at 21°C.

The pH of the pulsing buffer was not a decisive factor in the electrofusion process when kept in a physiological range ($5.5 < pH < 8.5$).

The osmolarity of the pulsing buffer was changed by using different sucrose concentrations. The same fusion yield was observed under the same electrical conditions regardless of whether the buffer was hypoosmotic, isosmotic, or hyperosmotic.

5.4. Effect of the Composition of the Pulsing Buffer

The osmolarity (change in sucrose content) and the pH of the pulsing buffer did not affect the electrofusion process when all other parameters were kept constant.

The ionic composition of the pulsing buffer was observed to be a decisive factor in electrofusion (Blangero and Teissié, 1985). Increasing the concentration of monovalent cations (Na^+, K^+, choline, Li^+) induced a dramatic decrease in the fusion yield without affecting the cell viability. When the pulsing buffer contained NaCl (or KCl) at 50 mM, no fusion was induced by electrical pulsation of the culture. This inhibition was observed to be a function of the nature of the cation; the decrease in the fusion yield was observed in the order $Li^+ <$ choline $< Na^+ = K^+$. As a consequence, when working LiCl 100 mM, some fusions were still observed. In contrast, the nature of the counterion (Cl^- or SO_4^{2-}) did not affect the fusion yield.

Mg^{2+} was observed to have a beneficial effect on the fusion process when used at a low concentration (< 4 mM). Ca^{2+} had a lytic effect even when used at a free concentration of 0.2 mM (value corrected of the complexation by phosphate).

5.5. Effect of the Postpulse Incubation

In the procedure described in Methods, pulsation of the cell culture is followed by a 1 hr incubation at 37°C. This step is critical for obtaining a high number of fused cells.

The fusion yield was in fact observed to increase during the incubation period at 37°C. A 10 min lag was present at the beginning and was followed by a steep increase, a final plateau value being observed after 40 min.

The incubation temperature drastically affected the extent of fusion. In fact, we observed that the kinetics of the increase in fusion was strongly dependent on the incubation temperature. When working at 21°C, a final level of polynucleation was not observed 4 hr after the pulsation. If the cells were kept on ice after the electrical treatment, no fusion was observed 1 hr after the pulses. But if they were then brought to 37°C, the fusion process occurred normally during the incubation. In other words, the pulsed cells kept their ability to fuse during their incubation at 4°C but were not able to express it.

5.6. Effect of Drugs and Ethanol

The organization of the membrane and its dynamical properties could be affected by different treatments.

Fluidization could be obtained by incubating the cells with a buffer containing a large percentage of ethanol. Such a treatment was observed to dramatically decrease the fusion yield. After a 30 min incubation with 400 mM ethanol at 37°C, no fusion was obtained by pulsing the cells under conditions known to be highly effective otherwise (1.5 kV/cm, 100 μsec, five times) (Orgambide *et al.*, 1985).

Membrane component mobility was shown to be controlled by the organization of the cytoskeleton. Polymerization of microtubules could be altered by treating the cells by colchicine (depolymerization) or by taxol (polymerization). Microfilaments of actin could be modified by the action of cytochalasin B. Electrofusion of CHO cells was affected in two ways by drug treatment.

Postpulse treatment by cytochalasin B induced lysis of the cells. A most interesting observation was found for cells treated with colchicine. A prepulse incubation gave a 20% relative improvement in fusion yield. A postpulse incubation was associated with a decrease of a similar magnitude.

The colchicine treatment was too short to induce a synchronization of the cell culture but did affect the organization of the cytoskeleton by depolymerizing the microtubule network. Our observation that the fusion yield was increased by a prepulse treatment is thus more direct evidence that the cytoskeletal organization is implicated in the cell electrofusion process.

The direct observation of this organization by means of immunofluorescent staining of the microtubules showed that 10 min after the pulsations (at 37°C), the microtubule network was reduced to a ring around the nuclei. After clustering of the fused cell nuclei in the middle of the huge cytoplasm, the network was re-formed to a shape similar to what was observed in control cells.

6. GUV

The methodology described by Oku and MacDonald (1983) gives cell-sized GUV which are able to withstand physiological buffer conditions. They are thus suitable tools for the comparison of the electrofusion processes in cells and in pure lipid systems.

6.1. Size Distribution of the Native GUV

The vesicles proved to be unilamellar by the fluorescence method described by Tsong (1975). They were observed under phase contrast with an inverted microscope. Due to the problems of observation, we took into account only GUV with diameters larger than 10 μm. The histogram of size distribution is shown in Fig. 3. But it has to be reemphasized that a large number of smaller vesicles, which should be considered as giant when compared to those obtained by classical procedures, were present but not taken into account in this distribution due to the distribution threshold we selected.

6.2. Creation of Contact between GUV

No specific binding systems were present on the surface of GUV. As a consequence, no spontaneous contact was observed. As in our previous work on the electrofusion of plant protoplasts (Chapel *et al.*, 1986), we induced the contact between the species by adding agglutinating agents (spermine 0.2 mM or PEG 1500 2% w/v) in the pulsing buffer. The GUV morphology was not altered by the polyamine but the vesicles were slightly crenated when PEG was present. Contacts were easily created by such treatments as observed under the microscope.

6.3. GUV Electrofusion

The application of successive pulses on the GUV suspension (1.2 kV/cm, 100 μsec duration) induced their fusion. The process was observed either directly under the microscope or by plotting the histogram of size distribution after electric pulsation. Comparison of this histogram with the one of the same GUV preparation before pulsation indicated a shift toward the larger sizes (Fig. 3). The proportion of vesicles with diameters between 10 and 25 μm was clearly decreased. Another index is representative of vesicle fusion. The mean radius $\langle R \rangle$ was obtained from the distribution by:

$$\langle R \rangle = \Sigma n_i R_i \ / \ \Sigma \ n_i$$

where R_i is the average radius of the size class (i.e., 17.5/2 for the first class, 10–25 μm) and n_i is the population of the size class.

$\langle R \rangle$ was observed to increase from 14 μm for unpulsed GUV to 16 μm after pulsations.

More information was provided by the direct observation of the fusion events. As described earlier, this was conveniently obtained by means of the video system. The information was stored on a videotape which could be replayed picture by picture (time resolution: 40 msec in continental Europe). The most striking observation was that the fusion process was very fast. In order to quantitate these events, we measured the time needed for the rounding process. The rounding process was the reaction which gave one larger vesicle from two vesicles in contact. The time resolution for this determination was very good because of the low-speed display mode of our magnetoscope. The first step was a flattening of the contact area followed by the disappearance of this contact. The shapes of the interacting vesicles then changed very quickly to one of a new larger spherical vesicle. These events occurred within 2 to 10 sec following the composition of the buffer.

Another conclusion from direct observation of the fusion was that the process was vectorial. Fusion occurred only when the contact between the vesicles was facing the electrodes. If three vesicles were clustered together in line with the direction of the field, then the simultaneous fusion of the three vesicles into one final vesicle was observed.

6.4. Conclusions

A suspension of chemically aggregated vesicles was pulsed with fields whose intensity was greater than 1 kV/cm. As was the case for biological membranes, electropulsed lipid membranes acquired a fusogenic character. At the present state of investigation, it was not possible to directly correlate this character to the permeabilization induced by the field. Similarities between cell electrofusion and vesicle electrofusion were clearly present. Both processes displayed a vectorial character which led us to conclude that the field triggered the fusion (Teissié and Blangero, 1984). Field intensities were stronger in the case of GUV. As noticed in their preparation, they appeared to be leaky (fast transmembranous ANS diffusion). Consequently, the permeability of their membrane should affect the magnitude of their external field induced membrane potential change (see Neumann, this volume). The duration of the pulses was of the same order for the two systems. The major difference in the electrofusion of cells and lipid vesicles was the duration of the reaction. It was on the second time range for GUV and on the hour time range for plated CHO cells (at 21°C).

7. DISCUSSION

Cell fusion is the set of events leading to the formation of polykaryons. We can describe it as follows:

1. Cell contact
2. Membrane fusion
3. Cytoplasmic reorganization

In our approach, step 1 is spontaneous and due to the contact inhibition induced by the growth conditions.

Membrane fusion, step 2, is clearly induced by the electric pulses. Threshold values of field intensity and pulse duration are present (Fig. 4). The dependence on the relative orientation between field and cells is another feature of the process. It remains to be determined if fusion is a consequence of permeabilization. The pulse duration dependences of electrofusion and electropermeabilization are similar; both processes display a vectorial character in the relationship between field and orientation of the cells (Teissié and Blangero, 1984; Sowers, 1986; Sowers and Lieber, 1986). It was shown that electropermeabilization of cells occurred in the area of the cell facing the electrodes, i.e., where the induced potential was large ($\cos \theta = \pm 1$). The major difference between fusion and permeation is the field intensity threshold needed to induce the process (Fig. 4). Different explanations were proposed (i.e., requirement of larger "pores" for fusion, need of larger per-

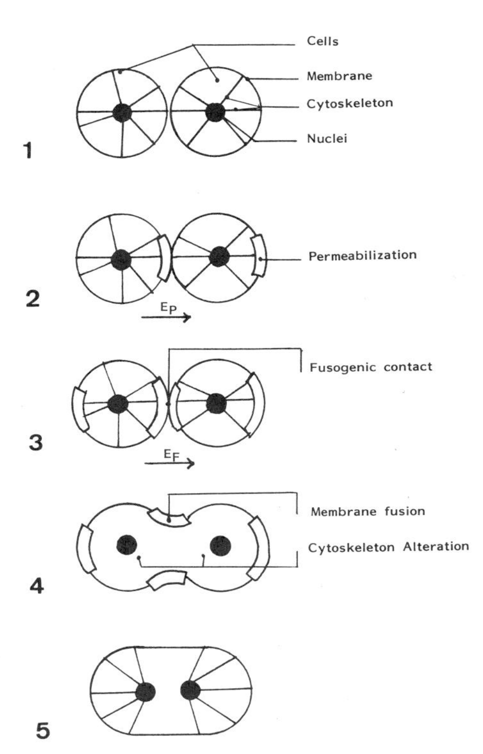

Figure 5. Tentative description of the events occurring during cell electrofusion. (1) Cells are in contact. (2) By use of a moderated field strength (0.4–0.5 kV/cm), the cell is permeabilized. The process is very fast (microsecond time range) and leads to a new fusogenic organization of the membrane. No fusion occurs because one membrane in the contact remains unpermeabilized. (3) With a stronger field (> 0.6 kV/cm), the two membranes in contact are permeabilized (microsecond time range) and can fuse. (4) Membrane fusion occurs (this may be a fast process) and a cytoskeleton disorganization is triggered. (5) A new organization of the membrane (impermeable again) and of the cytoskeleton occurs (hour time range) leading to a clustering of the nuclei in the middle of the fused cell.

meabilized areas) but without any experimental support. We propose another alternative. The field induces a modification of the membrane potential from its resting value which is not zero in mammalian cells. Due to the vectorial character of the field, this gives an increase in membrane potential on one side of the cell but a decrease on the opposite side. When the magnitude of the external field increases, the "permeabilizing" membrane potential (see Neumann, this volume; Chernomordik and Chizmadzhev, this volume) is then not reached simultaneously on the two sides (Fig. 5). But if the contact areas on the two cells must be permeabilized in order to induce fusion, then when the threshold is reached only one membrane in the contact area is permeabilized and fusion cannot occur. A further increase in field is needed to permeabilize the opposite side and to induce fusion. This explanation is in agreement with the correlation between electropermeabilization and electrofusion. This can be associated with the induction of a fusogenic state of the permeabilized membrane as we observe with cells grown in suspension (Teissié and Rols, 1986).

In our experimental procedure where cells are not treated with proteolytic enzymes, electropermeabilization is not sufficient to induce fusion. The organization of the membrane as shown by the inhibiting effect of ethanol, a fluidizing agent, remains a decisive factor. Another interesting observation is the modulation of the extent of fusion by the ionic content of the medium. Electrofusion is inhibited by a high salt content (but keeping the osmolarity unchanged). This effect cannot be explained simply by electrostatic interactions. The nature of the ion is modulating the decrease in fusion yield. We propose that two conflicting phenomena are involved in this observation. Cell–cell contact is mainly due to the interaction between surface proteins (key–lock concept). This is due to electrostatic interactions and as such decreases with an increase in the ionic content. But the organization of interfacial water is playing a role in the cell–cell repulsion. "Water structure breakers" (K^+, Na^+) are more potent in the inhibition than "water structure makers" (Li^+, Mg^{2+}). The postpulse temperature can play a role either in membrane fusion or in cytoplasmic reorganization. But whatever the step involved, we can now conclude from our studies that a target other than the membrane is implicated in electrofusion and, maybe, in electropermeabilization: the cytoskeleton. Its reorganization is the kinetic limiting step in electrofusion. When the vesicle has no cytoskeleton (GUV), the process is very fast (a few seconds) as compared to mammalian cells (several hours at 21°C).

Precise description of the molecular events involved remains to be done.

ACKNOWLEDGMENTS. We thank the scientists and technical assistants of this institute for their help in this study. Financial support was received from the CNRS, the ANVAR, the French minister for research, and the association pour la recherche sur le cancer (ARC).

REFERENCES

Blangero, C., and Teissié, J., 1985, Ionic modulation of electrically induced fusion of mammalian cells, *J. Membr. Biol.* **86:**247.

Chapel, M., Montané, M. H., Ranty, B., Teissié, J., and Alibert, G., 1986, Viable somatic hybrids are obtained by direct current electrofusion of chemically aggregated plant protoplasts, *FEBS Lett.* **196:**79.

Chernomordik, L. V., Sukharev, S. I., Abidor, I. G., and Chizmadzhev, Y. A., 1983, Breakdown of lipid bilayer membranes in an electric field, *Biochim. Biophys. Acta* **736:**203.

Finaz, C., Lefèvre, A., and Teissié, J., 1984, Electrofusion: A new, highly efficient technique for generating somatic cell hybrids, *Exp. Cell. Res.* **150:**477.

Gottesman, M. M. (ed.), 1985, *Molecular Cell Genetics,* Wiley, New York.

Kinosita, K., and Tsong, T. Y., 1979, Voltage induced conductance in human erythrocyte membranes, *Biochim. Biophys. Acta* **554:**479.

Neumann, E., Gerisch, G., and Opatz, K., 1980, Cell fusion induced by high electric impulses applied to Dictyostelium, *Naturwissenschaften* **67:**414.

Oku, N., and MacDonald, R. C., 1983, Formation of giant liposomes from lipids in chaotropic ion solutions, *Biochim. Biophys. Acta* **734:**54.

Orgambide, G., Blangero, C., and Teissié, J., 1985, Electrofusion of Chinese ovary cells after ethanol incubation, *Biochim. Biophys. Acta* **820:**58.

Sowers, A. E., 1986, A long lived fusogenic state is induced in erythrocyte ghosts by electric pulses, *J. Cell Biol.* **102:**1358.

Sowers, A. E., and Lieber, M. R., 1986, Electropore diameters, lifetimes, numbers and locations in individual erythrocyte ghosts, *FEBS Lett.* **205:**179.

Teissié, J., and Blangero, C., 1984, Direct experimental evidence of the vectorial character of the interaction between electric pulses and cells in cell electrofusion, *Biochim. Biophys. Acta* **775:**446.

Teissié, J., and Rols, M. P., 1986, Fusion of mammalian cells in culture is obtained by creating the contact between the cells after their electropermeabilization, *Biochem. Biophys. Res. Commun.* **140:**258.

Teissié, J., Knutson, V. P., Tsong, T. Y., and Lane, M. D., 1982, Electric pulse-induced fusion of 3T3 cells in monolayer culture, *Science* **216:**537.

Tsong, T. Y., 1975, Effect of phase transition on the kinetics of dye transport in phospholipid bilayer structures, *Biochemistry* **14:**5409.

Weber, H., Forster, W., Jacob, H. E., and Berg, H., 1981, Microbiological applications of electric field effects, *Z. Allg. Mikrobiol.* **21:**555.

Zimmermann, U., 1982, Electric field-mediated fusion and related electrical phenomena, *Biochim. Biophys. Acta* **694:**227.

Chapter 14

Cell Fusion and Cell Poration by Pulsed Radio-Frequency Electric Fields

Donald C. Chang

1. INTRODUCTION

Cell fusion plays a very important role in modern biotechnology. For example, one key procedure in genetic engineering is the introduction of exogenous genetic material into a host cell. Such insertion of genes is accomplished by either permeabilizing the cell membrane to allow entry of genetic material, or fusing the host cell with a cell containing the desired genetic material. Furthermore, cell fusion is important in the production of monoclonal antibodies, which requires fusion of antibody-producing cells with continuously dividing cancer cells such as myeloma cells (Galfre *et al.*, 1977; Lo *et al.*, 1984). Also, cell fusion can be used as a microinjection technique to deliver drugs which normally cannot enter a cell. One can simply fuse the cell with liposomes or red blood cell ghosts that have been preloaded with specific drugs (Schlegel and Lieber, 1987).

The conventional techniques of cell fusion today utilize three basic approaches: (1) virus-mediated fusion (White *et al.*, 1981), (2) chemical-mediated fusion (e.g., using PEG) (Davidson *et al.*, 1976), and (3) electrofusion (Neumann *et al.*, 1980; Zimmermann and Vienken, 1982; Bates *et al.*, 1987). The virus and chemical methods of fusion were developed earlier than electrofusion, and have thus been used more widely. However, they are known to have certain shortcomings; typically their fusion yields are very low and they sometimes produce undesirable side effects on the cells. The electrofusion method can avoid some of these problems. This technique uses a pulsed high-intensity DC electric field to induce electrical breakdown in the cell membrane and thus avoids many of the side effects of earlier methods. Also, the yield of fused cells is typically higher using electrofusion.

However, there are still many limitations to the use of electrofusion. First, not all cell types can be electrofused with the same ease. In fact, many cell types are extremely difficult to fuse with DC pulses. Second, there are many unknown factors which influence fusion yield. Fusion of one cell type may be successful in one laboratory but not in others. The DC fusion method is still more of an

DONALD C. CHANG • Department of Physiology and Molecular Biophysics, Baylor College of Medicine, Houston, Texas 77030, and Marine Biological Laboratory, Woods Hole, Massachusetts 02543.

art than a well-understood scientific procedure. Third, it is very difficult to use the DC pulse method to fuse cells of different sizes. This difficulty is probably due to the fact that the membrane potential induced by an external DC field is proportional to the diameter of the cell (see Section 2). Although at this point it is not clear what is the molecular nature of the fusogenic site (Sowers, 1986), it is generally assumed that fusion is caused by a localized membrane breakdown (Zimmermann and Vienken, 1982; Stenger and Hui, 1986). When one tries to fuse cells of two different sizes, it is difficult to choose a proper field strength of the pulse for both cell types. When the field is just large enough to cause membrane breakdown in the larger cell, it is not sufficient to induce a critical membrane potential to break down the membrane of the smaller cell. On the other hand, if the field strength is increased so that it can cause membrane breakdown in the smaller cell, the potential induced in the larger cell is so high that it will cause irreversible damage and destroy the cell.

This chapter discusses a new method of electrofusion that is designed to overcome some of the above problems. Unlike the conventional electrofusion methods which employ DC pulses to induce membrane breakdown, we use radio-frequency (RF) pulses to permeabilize the cell membrane and to induce cell fusion. The RF pulse acts not only by inducing a large membrane potential to cause electrical breakdown of the membrane, it also produces ultrasonic oscillation in the membrane. As discussed in the following section, this induced ultrasonic energy is focused at the poles of the cell (i.e., the points of the cell facing the electrodes). Thus, the sonication of the membrane induced by the pulsed RF field is localized. Such localized sonication can be used to fuse or porate cells.

2. PRINCIPLES OF TECHNIQUE

2.1. Electromechanical Breakdown of the Cell Membrane by Localized Sonication

It is well known that the cell membrane, which is composed mainly of a phospholipid bilayer, can be disrupted by forced oscillation at high frequency (> 10 kHz). Thus, it is possible to break down a cell membrane by high-power sonication. However, ordinary means of sonication which apply ultrasonic energy to the cell by mechanical oscillation are not useful for controlled cell poration or cell fusion. In such a method, the ultrasonic energy generates a violent mechanical vibration which destroys the whole structure of the cell. In contrast, our RF method induces a localized oscillation only at a very small, discrete area of the cell membrane. The oscillation is only applied for a very brief time, typically a fraction of a millisecond. When a small membrane pore is created, it can reseal quickly, thus maintaining the integrity of the cell structure.

When a cell is placed in an electric field, an electrical potential is induced across the plasma membrane. The magnitude of this induced potential follows roughly the relation (for a spherical cell)

$$V = (3/2)\, r \cdot E \cdot \cos\theta \tag{1}$$

where V is the induced membrane potential, r is the radius of the cell, E is the external electric field, and θ is the angle between the electric field vector and the normal vector of the cell membrane (Cole, 1968). [Note: Eq. (1) is valid when the external field is changed slowly. The situation is more complicated when the external field is changed rapidly. See Section 2.3.] From Eq. (1), it is easily seen that the induced membrane potential is not uniform over the cell surface; V is maximum at the two cell poles (i.e., at $\theta = 0°$ and $180°$). Since the cell membrane is extremely thin ($\sim 6 \times 10^{-7}$ cm), the induced electric field within the membrane, E_m, is much larger than the externally applied field, E, since:

$$E_m = V/d = (3/2)\, r/d \cdot E \cdot \cos\theta \tag{2}$$

where d is the thickness of the membrane. For a typical cell ($r \approx 20\ \mu$m), E_m/E is approximately $3/2\ r/d = 5000$ at the poles of the cell.

The large electric field within the membrane produces two effects: (1) it exerts a strong force on the phosphate head groups of the lipid molecules of the bilayer (which are charged), and tends to move them along the field lines; (2) it compresses the membrane. When the external field oscillates, the lipid molecules of the membrane also undergo an oscillating motion. This oscillation produces a "sonicating" effect on the lipid bilayer. In such an arrangement, the cell itself functions as an antenna which absorbs the RF energy from the external field, and the membrane acts as a transducer to convert RF electrical energy into mechanical energy. Since the sonication can be localized at the points of the cell at which the induced field E_m is maximum, one can control the area of membrane breakdown by adjusting the external field strength so that sufficient sonication occurs only at the poles of the cell but nowhere else.

Another advantage of RF sonication over mechanical ultrasonic devices is that the ultrasonic energy can be applied in very short bursts (which typically last only a few tens of microseconds). In mechanical ultrasonic devices, the power applied to the cell cannot be turned on or off quickly. It takes time to build up and dissipate the energy stored in the medium. Such a problem is avoided in our method because energy is transferred to the cell membrane in the form of electrical impulses which can be switched on or off in microseconds.

2.2. The RF Field Is More Efficient in Transmitting Energy to the Cell Membrane Than the DC Field

The efficiency of energy transfer from one system to another, in general, is frequency-dependent. The frequency spectrum of a DC pulse is limited to a narrow bandwidth. A RF pulse, on the other hand, can be tuned to a wide range of bandwidths. The cell membrane is composed of macromolecules, which have characteristic frequencies of thermal motion. When the frequency of the applied external RF field matches one of these inherent frequencies, a condition of resonance is reached, and the efficiency of energy transfer from the external field to the membrane is greatly enhanced. In real biological cells, where the structure of the membrane is very complicated, the spectrum of molecular motion can be very broad. The RF frequency can be carefully varied to optimize the efficiency of energy transfer to the particular cells of interest. The consequence of this frequency tuning is that less power is required to induce membrane breakdown using a pulsed RF field as compared to a pulsed DC field. This means that there is less risk of irreversibly damaging the cell.

2.3. The Pulsed RF Field Can Be Used to Porate or Fuse Cells of Different Sizes

In order to produce an electrical breakdown of the cell membrane using a DC pulse, the field-induced membrane potential must exceed a certain critical value, V_c (typically 1 volt). Such a breakdown is usually reversible. The membrane will reseal after the external field is turned off, and cells can normally remain viable. On the other hand, if the induced potential is much higher than V_c, the breakdown is not reversible, the membrane does not reseal, and the cell will not remain viable.

From Eq. (1) it can be seen that when cells of different sizes are placed in an external field, the induced membrane potential is higher for the larger cell than for the smaller cell. This size-dependence of membrane potential causes a problem when trying to fuse cells of differing sizes using a DC field. Assume that two cells, A and B, are to be fused and that the radius of cell A, r_A, is about twice the radius of cell B, r_B. In order to cause a reversible membrane breakdown in cell B, the applied external field must be sufficient so that $(3/2)\ r_B \cdot E$ is greater than V_c. However, the same applied electric field will induce a membrane potential much larger than V_c in cell A, and will thus damage that cell.

This problem can be overcome using a RF pulse instead of a DC pulse. When the applied field oscillates, the amplitude of the induced membrane potential is a function of the frequency. The membrane potential predicted in Eq. (1) is derived from a steady-state condition. The induced potential does not arise instantly upon the application of the external field. If the applied field is constant (i.e., a DC field), the membrane potential will reach V given a sufficient amount of time. The time required to establish this steady-state membrane potential is called "relaxation time," or τ, which is given by:

$$1/\tau = 1/(R_m \cdot C_m) + 1/[r \cdot C_m \cdot (\rho_i + 0.5\ \rho_e)] \tag{3}$$

where R_m and C_m are the specific resistance and specific capacitance of the membrane, and ρ_i and ρ_e are the specific resistances of the intracellular and extracellular media, respectively (Holzapfel *et al.*, 1982). Since R_m in most cells is very large, for practical purposes, Eq. (3) can be simplified to:

$$\tau = r \cdot C_m \cdot (\rho_i + 0.5\ \rho_e) \tag{4}$$

Thus, the relaxation time is approximately proportional to the radius of the cell. For a cell of several micrometers in diameter, τ is typically less than 1 μsec.

Because of the finite value of τ, the membrane potential induced by a RF field is frequency dependent. When the frequency of the applied RF field is smaller than $1/\tau$, the membrane potential has no problem in following the external field. The induced membrane potential reaches 100% of the value predicted in Eq. (1). On the other hand, if the frequency of the applied RF field is greater than $1/\tau$, the membrane potential cannot "catch up" with the changes in the applied field, and the response of the membrane potential will be less than 100%. In general, the maximum membrane potential induced by a RF field is (Holzapfel *et al.*, 1982):

$$V(\omega) = (3/2)\ r \cdot E \cdot \cos\theta\ X(\omega) \tag{5}$$

where r, E and θ have the same meanings as in Eq. (1), ω is the angular frequency, and $X(\omega)$ is a function of the frequency such that:

$$X(\omega) = 1/[1 + (\omega\tau)^2]^{1/2} \tag{6}$$

when $\omega < 1/\tau$, $X(\omega)$ is near unity. When $\omega > 1/\tau$, $X(\omega)$ decreases very rapidly with increasing frequency.

This frequency-dependent property of the induced potential can be used to fuse cells of different sizes. Suppose one wants to fuse a large cell (cell A) with a small cell (cell B). Since τ is roughly proportional to r, from Eq. (4), the relaxation time of cell A (τ_A) will be longer than that of the smaller cell B (τ_B). To fuse cells A and B, one can apply a pulsed RF field with a frequency ω such that:

$$1/\tau_A < \omega < 1/\tau_B \tag{7}$$

Because the applied frequency is less than $1/\tau_B$, $X(\omega)$ approaches unity for the smaller cell B. On the other hand, since the applied frequency is greater than $1/\tau_A$, $X(\omega)$ in the larger cell A is less than unity. According to Eq. (5), one can choose a proper frequency to counterbalance the effect of r on the induced membrane potential with the factor $X(\omega)$. Consequently, a reversible breakdown of the membrane can be induced in both cells without irreversibly damaging the larger cell A.

3. EXPERIMENTAL METHODS

A typical design of our fusion chambers is shown in Fig. 1. Two glass coverslips were separated by spacers of approximately 0.3 mm thickness. The cell suspension was sandwiched between the coverslips. Two parallel Pt wires (about 0.4 mm apart) served as electrodes, which in turn were connected to a high-power function generator specially built for the cell fusion studies.

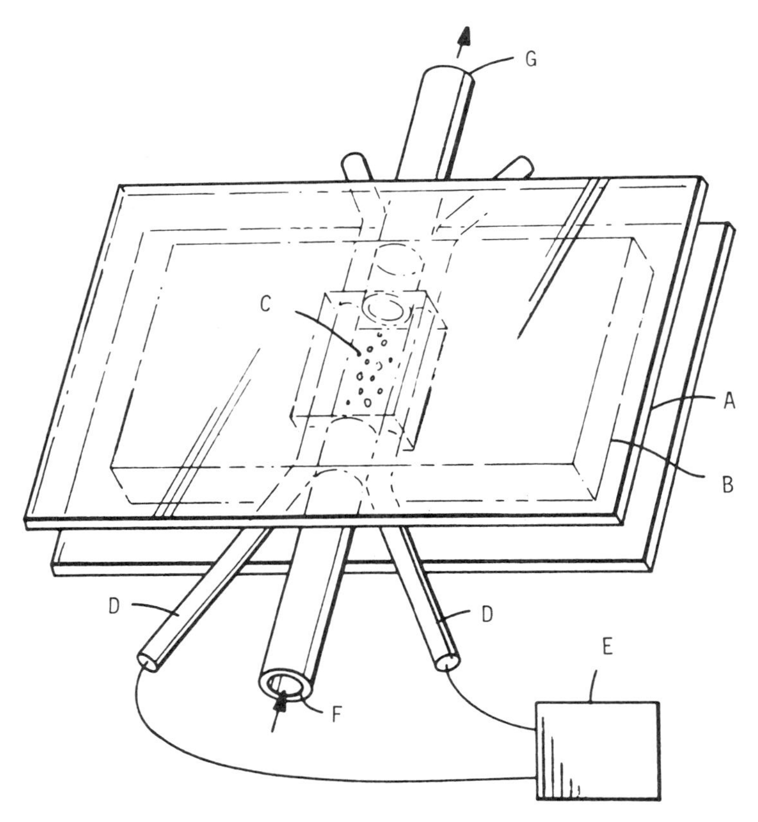

Figure 1. Schematic diagram of a chamber for cell poration and/or cell fusion for optical microscopic observation. Letters indicate: (A) thin glass plate, (B) spacer, (C) cell suspension, (D) Pt electrodes, (E) high-power function generator, (F) inlet for cell suspension, (G) outlet for cell suspension. Note that the cell suspension is in direct contact with the electrode.

In order to observe the events of fusion, the chamber was mounted in an optical microscope (Zeiss Axiophot). A small volume of cell suspension (roughly 1 μl) was used. The cell medium was a low-ionic-strength solution which was composed of 27 mM Na-phosphate buffer (pH 7.5) and 150 mM sucrose.

Before cells could be fused, they were brought into close contact by a dielectrophoretic process (Schwan and Sher, 1969; Pohl *et al.*, 1981). Dielectrophoresis was accomplished by applying a continuous AC electric field of comparatively low strength (typically of 60 Hz to several kHz in frequency and 100–400 V/cm in field strength). Under this low-intensity AC field, the cells behaved as dipoles and lined up parallel to the field, giving the appearance of "pearl chains."

Once the cells were closely aligned, the high-power RF pulse was applied through the Pt electrodes. The pulse configuration was one of those shown in Fig. 2. In Fig. 2A the pulse is a pure RF oscillation with a single frequency. In Fig. 2B the RF pulse consists of a single frequency mixed with a DC component. In Fig. 2C the RF pulse contains a mixture of multiple frequencies (in this example, two frequencies). In Fig. 2D alternating pulses of different frequencies are used. The pulse in Fig. 2B was most commonly used in our experiments.

The parameters of the RF pulse can be chosen to accommodate the properties of different biological samples. The frequency may be varied over the entire RF spectrum. Typically we have used RF of 0.01–1 MHz. The width of the pulse may vary from about 1 μsec to 10 msec. In most

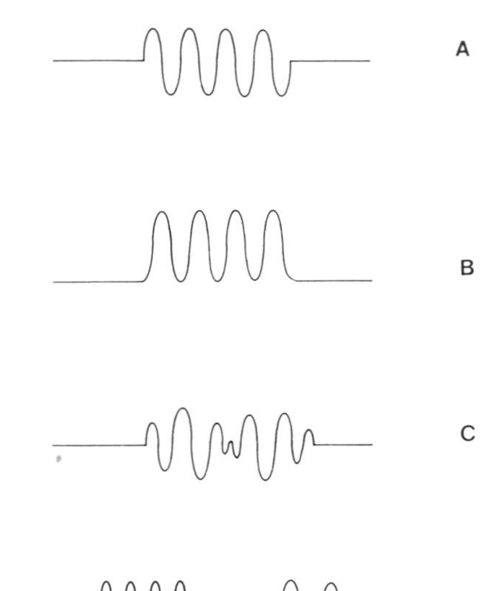

Figure 2. Four sample types of RF pulses: (A) RF pulse of single frequency, (B) pulse consisting of a mixture of an RF component and a DC shift, (C) pulse consisting of multiple RF components, (D) consecutive RF pulses of different frequencies.

experiments we have used pulse lengths of 20 to 100 μsec. The typical amplitude of the RF field which we have used is 1 to 4 kV/cm.

The high-power function generator which provided both the AC field for dielectrophoresis and the RF pulses for fusion was designed by the author and built in our laboratory. The block diagram in Fig. 3 shows the configuration of our electronic apparatus. Before a fusion pulse is applied, the mercury-wetted relay connects the sinusoidal wave generator with the fusion chamber to produce an AC field for dielectrophoresis. When the triggering circuit is energized, the relay switches to connect the power amplifier to the chamber so that the RF pulse can be applied. The RF pulse is generated by using a rectangular pulse (produced by a DC pulse generator) to gate the output of an RF oscillator. This RF pulse is then amplified by the power amplifier before passing to the chamber. After the RF pulse has passed, the relay is switched again to the unenergized mode, which

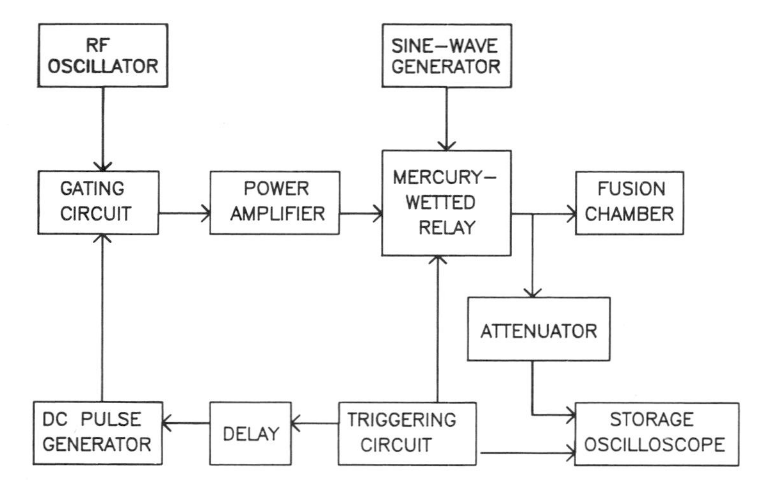

Figure 3. Block diagram of apparatus in our cell fusion experiments using pulsed RF fields.

disconnects the power amplifier from the chamber, and reconnects the AC sinusoidal field. The output is monitored with a storage oscilloscope (Tektronix model 5111A). The pulse width is varied by the settings of the DC pulse generator.

4. RESULTS

As examples for demonstrating the usefulness of this new method of cell fusion, we would like to briefly summarize the preliminary results of two ongoing studies in our laboratory.

4.1. Fusion of Human Erythrocytes

Red cells were obtained from whole human blood treated with heparin. They were washed twice in isotonic Na-phosphate buffer (Dodge *et al.*, 1963), and then washed and resuspended in fusion medium. This medium contained 27 mM Na-phosphate buffer and 150 mM sucrose (pH 7.5). The cell suspension was injected into the fusion chamber, which was placed under a Zeiss Axiophot microscope (equipped with an epifluorescence attachment) for optical observation.

Before the RF pulses were applied, the cells were first aligned into "pearl chains" by dielectrophoresis using a low-intensity AC electric field (Fig. 4A). Once the pearl chains were stabilized (in roughly half a minute), high-power RF pulses were applied. Immediately following the RF pulses, neighboring cells within the pearl chains began to fuse. Within a few minutes, one could observe not only many pairs of fusing cells, but also chains of multiple cells fusing together to become large, elongated cells (see Fig. 4B).

Figure 4A and B are light micrographs obtained by modified DIC (differential interference contrast) optics, which gives an impression of three-dimensional images. Fusion events can be clearly detected when the cytoplasms of the fusing cells start to merge. However, sometimes neighboring cells may fuse without merging their cytoplasms. A better method of studying RF-induced cell fusion is to use a fluorescent probe to assay fusion yield. Following the procedure of Sowers (1984), we labeled a small fraction (about 10%) of the red cells with a lipophilic fluorescent dye, 1,1′-dihexadecyl-3,3,3′,3′-tetramethylindocarbacyanine perchlorate (DiI). This compound provides brilliant fluorescent staining of the red cell membrane when observed under the fluorescence microscope. Because of the low proportion of labeled cells to unlabeled cells, each labeled cell was usually separated from other labeled cells within the pearl chains. Thus, one seldom saw two labeled cells as nearest neighbors (see Fig. 5A). Once fusion was induced by application of the RF pulse, many cells started to fuse. Under fluorescence observation, one could see dye passing from the labeled cells to the unlabeled cells. Eventually, neighboring fused cells became equally strongly labeled with fluorescent dye (see Fig. 5B). In many cases, the newly labeled cell also fused with its next neighboring cell. The dye could spread as far along the pearl chain as the contiguous fusion events allowed, and resulted in long chains of labeled cells (see Fig. 5B).

One might expect that, given enough time, all chains of fused cells may coalesce into rounded cells. This was not always observed. Many times, individual cells within a long, fluorescently labeled chain would maintain their shape, without merging. There seem to be two types of cell fusion in this system. The first type is a complete fusion where the cytoplasms of the fused cells totally merge into a large, single cell. The second type is a membrane fusion (as indicated by the dye movement) in which the cytoskeletons of the individual cells remain intact, and thus prevent the cells from changing shape. Apparently little or no exchange of cytoplasm occurs in the second type of fusion. Evidence of these two types of fusion has also been noted by Sowers (1984) in a study of human red cell ghosts, using a DC pulse fusion method.

By examining the pattern of DiI spreading, one can study the effectiveness of the RF pulse in causing fusion. Using human red blood cells as a model system, we have compared the fusion yield obtained by our RF pulse method with that obtained with a conventional DC pulse electrofusion

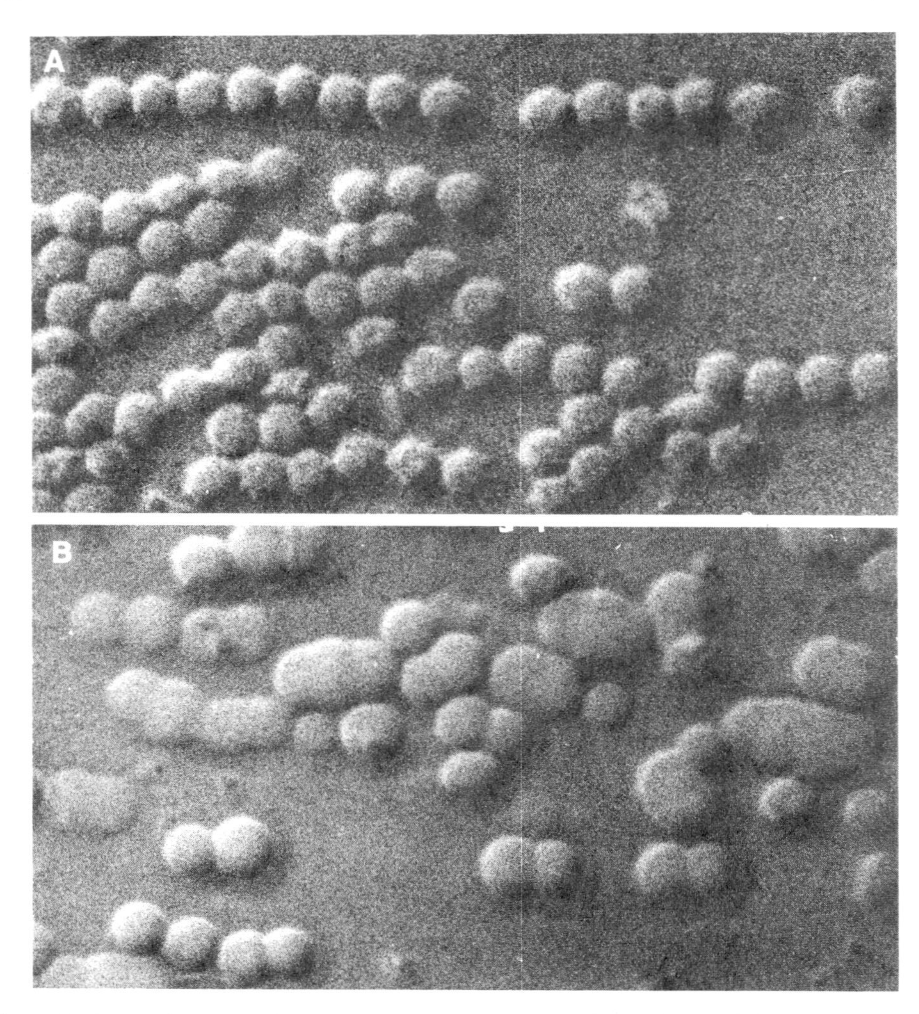

Figure 4. Light micrographs of human erythrocytes obtained using DIC optics. (A) Before application of RF pulse. The cells were aligned by dielectrophoresis. (B) Several minutes after application of three RF pulses (40 μsec wide, 400 kHz, 5 kV/cm). Many neighboring cells were seen to fuse together.

method. Using a single DC pulse of 20 μsec and field strengths up to 3.7 kV/cm, we observed essentially no fusion events in one batch of our cell preparation. On the other hand, a single RF pulse of 40 μsec duration and 400 kHz frequency, at a field strength of 1.5 kV/cm was able to induce a significant number of fusion events. (Note that although the width of the RF pulse was twice that of the DC pulse, the power generated by the RF field was actually less than that of the DC pulse.) With the RF pulse, the fusion yield reached a maximum as the field strength approached 2.6 kV/cm. A sample result of fusion yield versus field strength is shown in Fig. 6. It can be readily seen that with a single-pulse protocol, the RF pulse method is much more effective than the DC pulse method at inducing fusion in erythrocytes. We also found in later studies that the difference in yield between the RF and DC pulse methods becomes less dramatic when three or more consecutive pulses are used to fuse the cells, since applying multiple pulses can enhance the fusion yield in both methods.

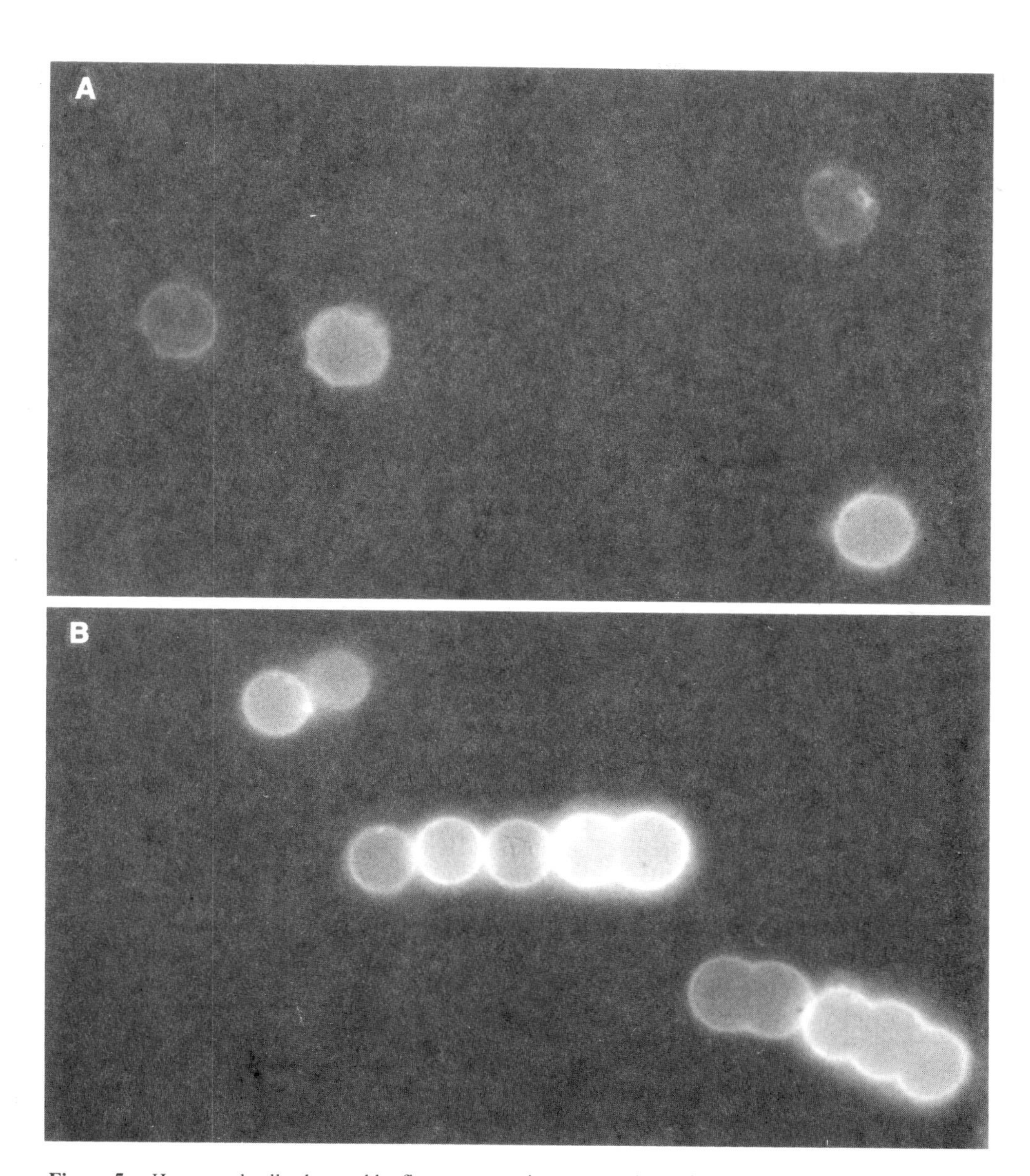

Figure 5. Human red cells observed by fluorescence microscopy. About 10% of the cells were labeled with a lipophilic fluorescent dye, DiI, which stains the cell membrane to produce a bright fluorescent image. The unlabeled cells do not fluoresce and cannot be seen. (A) Before application of RF pulses, the labeled cells appeared only as single isolated cells. (B) After application of RF pulses, many neighboring cells were fused together. Fluorescent dye diffused from labeled cells to newly fused unlabeled cells, making these cells also fluorescent.

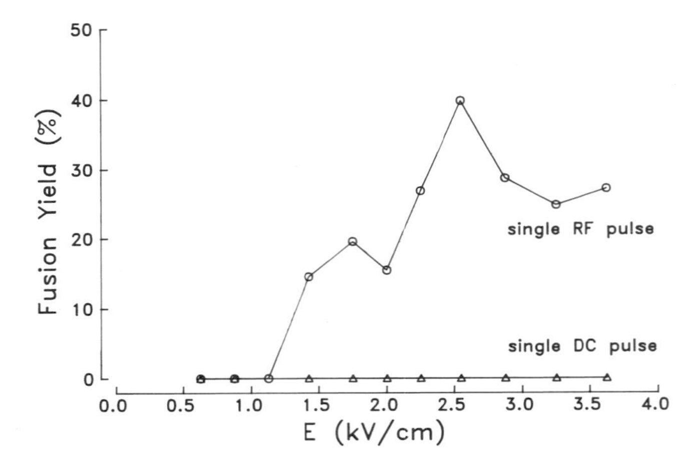

Figure 6. Comparison of the fusion yield using the RF pulse method with the DC pulse method. In a typical batch of human red cells, we found no fusion after a single DC pulse (20 μsec wide) was applied. Using cells from the same batch and identically treated, a single RF pulse (40 μsec wide, 400 kHz) was able to induce a significant number of cells to fuse. The events of fusion were assayed using the DiI labeling method.

4.2. Cell Hybridization

Another potential application of the RF fusion method is the hybridization of different types of cells. For example, we have used the RF pulse method to fuse goldfish chromatophore cells with a goldfish tumor cell line. The bright orange color of goldfish skin is produced by carotenoid pigment droplets within xanthophore cells of the dermis. Translocation of pigment droplets can disperse or aggregate pigment within the cell to alter the coloration of the skin. Pigment translocation is microtubule-dependent and controlled by hormones (Tchen *et al.*, 1986; Lynch *et al.*, 1986). Because of the easy visualization of pigment droplets and the ability to control their movements with appropriate hormones, chromatophore cells are a favorable model for studying organelle movement. Tumor cell lines which share a common origin with chromatophore cells have been derived (Matsumoto *et al.*, 1984). These tumor cell lines can be continuously cultured but do not express pigment. In collaboration with Dr. T. T. Tchen of Wayne State University, we have used the RF pulse method to fuse goldfish xanthophore cells with pigment-cell-derived tumor cells. The objective of this study is to produce a hybrid cell line that can retain the pigment expression and translocation properties of xanthophore cells. Because xanthophore cells have a built-in histochemical marker (the carotenoid droplets), it is comparatively easy to assay their fusion with nonpigmented tumor cells. Figure 7 shows an example of fusion between a xanthophore cell and a xanthophore-derived tumor cell. In Fig. 7A the cells were brought into close contact by dielectrophoresis. Three pulses of RF field (width 40 μsec, frequency 400 kHz, field strength 3.3 kV/cm) were then applied. Within 2 min the cytoplasms of the two cells were seen to merge (Fig. 7B). After 8 min the cells completely coalesced into a single giant pigment cell (Fig. 7C).

5. DISCUSSION

The electrofusion technique is very useful in many cell biology studies. In an earlier series of experiments we have used a conventional DC pulse method to study the effects of pH on fusion of erythrocytes (Chang *et al.*, 1989). By comparison, this new technique which uses pulsed RF electric fields to induce cell fusion seems to be more effective. In principle, the RF pulse method is more

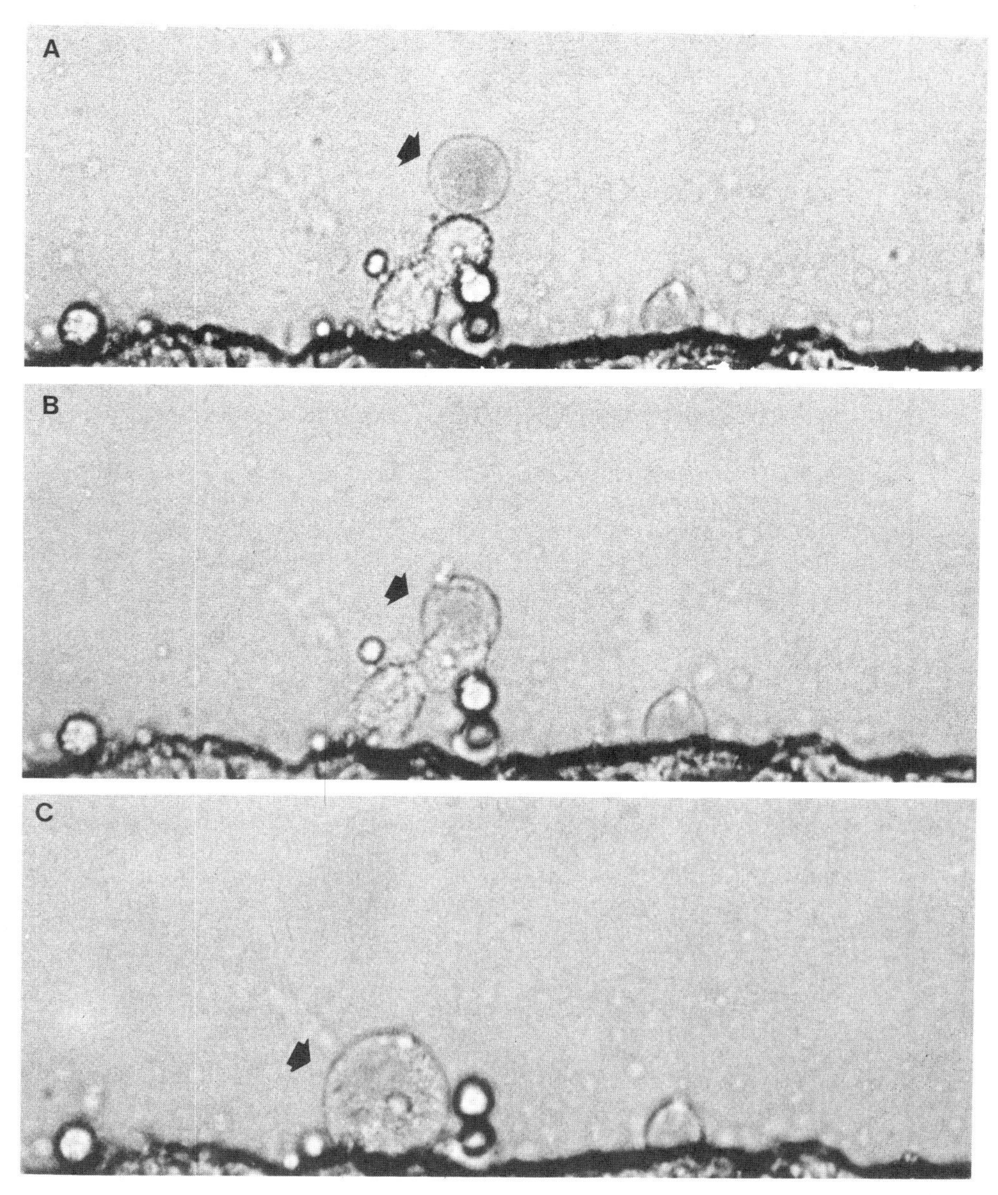

Figure 7. Fusion of a xanthophore cell with fish tumor cells induced by pulsed RF fields. (A) Before fusion. The xanthophore (arrow) was brought into close contact with two tumor cells by dielectrophoresis. (B) Two minutes after application of the RF pulses. The xanthophore (arrow) is already in the process of fusing with the tumor cells. (C) Eight minutes after application of the RF pulses. The xanthophore and tumor cells have completely merged into a single round cell (arrow).

efficient in inducing a reversible breakdown of the cell membrane, and is less damaging to the cell. As demonstrated in Section 4, the RF pulse method has been successfully used to induce membrane fusion and cytoplasmic fusion in human erythrocytes and to fuse fish chromatophore cells with tumor cells.

In a certain sense, the RF pulse method can be considered the most generalized method of electrofusion. The DC pulse can be considered a special case in which the frequency of the oscillating field is less than the inverse of the pulse width. As pointed out in Section 2, the RF pulse method utilizes both a localized sonication and an electrical breakdown to porate and fuse cells. At this point the detailed mechanism of cell poration and cell fusion is not well understood. Further experiments are needed to determine whether reversible permeabilization of the membrane is caused primarily by a pure electrical breakdown or by a mechanical shock resulting from the powerful electrical pulses. Our opinion is that both electrical and mechanical disturbances to the membrane are likely involved in the poration (and fusion) of cells.

We envision that in future applications of electrofusion, the RF and DC pulse methods will become complementary techniques, since each may prove advantageous to particular applications. The DC pulse method has the advantage of simplicity; it requires much less equipment to do the work. In some applications, one can simply discharge a capacitor to obtain the pulsed field. On the other hand, in more complicated experimental systems, such as those involving the fusion of cells of significantly different sizes, the RF field method is expected to provide more satisfactory results.

ACKNOWLEDGMENTS. I am grateful to John Hunt and P. Q. Gao for their technical assistance, Dr. T. T. Tchen for providing the xanthophore and tumor cells, and the staff of the Marine Biological Laboratory, Woods Hole, for their generous research support. The author is supported by a grant from the National Science Foundation (BNS-84-06932) and a grant from the Texas Advanced Technology Program.

REFERENCES

Bates, G. W., Saunders, J. A., and Sowers, A. E., 1987, Electrofusion: Principles and applications, in: *Cell Fusion* (A. E. Sowers, ed.), Plenum Press, New York, pp. 367–395.

Chang, D. C., Hunt, J. R., and Gao, P. Q., 1989, Effects of pH on cell fusion induced by electric fields, *Cell Biophysics* (in press).

Cole, K. S., 1968, *Membranes, Ions and Impulses: A Chapter of Classical Biophysics*, University of California Press, Berkeley, pp. 12–18.

Davidson, R. L., O'Malley, K. A., and Wheeler, T. B., 1976, Polyethylene glycol-induced mammalian cell hybridization: Effect of polyethylene glycol molecular weight and concentration, *Somatic Cell Genet.* **2:** 271–280.

Dodge, J. T., Mitchell, C., and Hanahan, D. J., 1963, The preparation and chemical characterization of hemoglobin-free ghosts of human erythrocytes, *Arch. Biochem. Biophys.* **100:**119–130.

Galfre, G., Howe, S. C., Milstein, C., Butcher, G. W., and Howard, J. C., 1977, Antibodies to major histocompatability antigens produced by hybrid cell lines, *Nature* **266:**550–552.

Holzapfel, C., Vienken, J., and Zimmermann, U., 1982, Rotation of cells in an alternating electric field: Theory and experimental proof, *J. Membr. Biol.* **67:**13–26.

Lo, M. M. S., Tsong, T. Y., Conrad, M. K., Strittmatter, S. M., Hester, L. D., and Snyder, S. H., 1984, Monoclonal antibody production by receptor-mediated electrically induced cell fusion, *Nature* **310:**794–796.

Lynch, T. J., Taylor, J. D., and Tchen, T. T., 1986, Regulation of pigment organelle translocation. I. Phosphorylation of the organelle associated protein p57, *J. Biol. Chem.* **261:**4204–4211.

Matsumoto, J., Ishikawa, T., Masahito, P., Takayama, S., Taylor, J. D., and Tchen, T. T., 1984, Clonal heterogeneity in physiological properties of melanized cells induced from goldfish erythrophorma cell lines, *Differentiation* **27:**36–45.

Neumann, E., Gerisch, G., and Opatz, K., 1980, Cell fusion induced by high electric impulses applied to *Dictyostelium*, *Naturwissenschaften* **67:**414–415.

Pohl, H. A., Kaler, K., and Pollock, K. 1981, The continuous positive and negative dielectrophoresis of microorganisms, *J. Biol. Phys.* **9:**67–86.

Schwan, H. P., and Sher, L. D., 1969, Alternating-current field-induced forces and their biological implications, *J. Electrochem. Soc.* **116:**22C–26C.

Schlegel, R. A., and Lieber, M. R., 1987, Microinjection of culture cells via fusion with loaded erythrocytes, in: *Cell Fusion* (A. E. Sowers, ed.), Plenum Press, New York, pp. 457–478.

Sowers, A. E., 1984, Characterization of electric field-induced fusion in erythrocyte ghost membranes. *J. Cell Biol.* **99:**1989–1996.

Sowers, A. E., 1986, A long-lived fusogenic state is induced in erythrocyte ghosts by electric pulses, *J. Cell Biol.* **102:**1358.

Stenger, D. A., and Hui, S. W., 1986, Kinetics of ultrastructural changes during electrically induced fusion of human erythrocytes, *J. Membr. Biol.* **93:**43–53.

Tchen, T. T., Allen, R. D., Lo, S. J., Lynch, T. J., Palazzo, R. E., Hayden, J., Walker, G. R., and Taylor, J. D., 1986, Role of microtubules in the formation of carotenoid droplet aggregate in goldfish xanthophores, *Ann. N.Y. Acad. Sci.* **466:**887–894.

White, J., Matlin, K., and Helenius, A., 1981, Cell fusion by Semliki Forest, influenza, and vesicular stomatitis viruses, *J. Cell Biol.* **89:**674–679.

Zimmermann, U., and Vienken, J., 1982, Electric field-induced cell-to-cell fusion, *J. Membr. Biol.* **67:**165–182.

Chapter 15

The Mechanism of Electroporation and Electrofusion in Erythrocyte Membranes

Arthur E. Sowers

1. INTRODUCTION

It has been known for nearly a decade that electric field pulses can be used to induce membrane fusion. Fusion induced in this way is now almost universally called "electrofusion." Most uses for electrofusion involve applications and both the methodology and examples of applications using electrofusion have been reviewed (Zimmermann, 1982; Berg, 1982, 1987; Berg *et al.*, 1983; Pohl *et al.*, 1984; Hofmann and Evans, 1986; Bates *et al.*, 1987b). Dr. Zimmermann and co-workers contributed much to the early literature and authored numerous review papers which emphasize applications. The most recent of his reviews (Zimmermann, 1986) also cites the earlier reviews and covers much new relevant literature. As is relevant to this book, the same electrical pulses which will induce fusion in membranes will also induce a transient increase in membrane permeability. The permeability increase is most likely due to the induction of pores (called electropores) since their size can be probed with soluble molecules with known shapes and sizes. Early hypotheses which explain electrofusion (see below) considered electropores as part of a membrane intermediate structure which mediates the conversion of two close-spaced but unfused membranes into one fused membrane. This chapter assumes that the reader is familiar with at least one of the earlier reviews (e.g., Zimmermann, 1982). Discussion of much peripheral material will not be repeated here.

The idea that electrofusion can be used to study the mechanism of membrane fusion is relatively new and is the primary purpose of our studies. The human erythrocyte ghost membrane is used in our laboratory to study the electrofusion mechanism because it is easy to prepare (Dodge *et al.*, 1963), much is understood about its structure and properties, and many procedures have been worked out for experimental manipulation. Since ghost membranes are free of their normal cytoplasmic contents and cannot undergo further growth or development, there is no need to consider problems of lysis or viability in connection with fusion. Also, electropore induction is better characterized in erythrocyte membranes than in any other membrane system. Our earlier

ARTHUR E. SOWERS • Jerome H. Holland Laboratory for the Biomedical Sciences, Rockville, Maryland 20855.

reviews cover the electrofusion mechanism in erythrocyte ghosts (Sowers and Kapoor, 1987a), emphasize the role of electropore collisions in the fusion mechanism (Sowers and Kapoor, 1987b), and emphasize the molecular mechanism of electrofusion (Sowers and Kapoor, 1988). Parts of the review paper on membrane stability by Dimitrov and Jain (1984) are also relevant to the electrofusion mechanism. This chapter integrates relevant information in the broader literature, covers recent developments from our laboratory, and describes methodology and procedures.

Regardless of the way fusion is induced, the mechanism of membrane fusion is not well understood in any system. Consequently, it is essential that any investigation of the mechanism of electrofusion proceed with at least an appreciation of the problem of fusion in other *in vitro* systems (Wilschut and Hoekstra, 1988; Ohki *et al.*, 1988; Ohki, 1987; Sowers, 1987a; Beers and Bassett, 1984; Evered and Whelan, 1984; Poste and Nicolson, 1978; Ringertz and Savage, 1976).

2. METHODOLOGY

2.1. General Requirements

The usual electrofusion procedure involves the application of one or more electric field pulses to membranes in close contact. The direction of the field must be perpendicular to the planes of the two membranes at the site where they are in contact. Membranes can be brought into close contact by ligands such as Con A (Lo *et al.*, 1984; Conrad *et al.*, 1987), chemicals such as PEG (Berg *et al.*, 1983), gravity to sediment cells into a monolayer (Sowers, 1985a), growth to confluence in a monolayer (Teissié *et al.*, 1982), or can be held together by micromanipulation (Senda *et al.*, 1979). The most convenient method, however, utilizes a physical effect known as dielectrophoresis (Pohl, 1978). Dielectrophoresis occurs when a weak AC passes through the membrane suspension to generate a corresponding electric field in the medium. This causes all membranes to become aligned into numerous pearl chains. Dielectrophoresis is a significant way to induce membrane–membrane contact because it is a mild, reversible, and nonchemical effect with undetectable or nearly undetectable side effects on cells. The points of close membrane–membrane contact during pearl chain alignment are such that the same electrodes used for dielectrophoresis can also be used to carry the fusion-inducing DC pulse. In short, the electrofusion protocol is essentially an exposure to a weak alternating electric current until pearl chains are formed which is then followed by one or more strong electric field pulses.

The dependence of fusion on the pulse is complicated by several factors. First, a fusion-inducing electric field pulse must have a strength (E = V/mm) that is midway between two poorly characterized thresholds. Little or no fusion occurs if the pulse strength is below the lower threshold. However, if pulses are above the upper threshold, then membrane fragmentation can take place instead of fusion. Second, over a range of field strengths, it is possible to get the same fusion yield by using a lower-strength square wave pulse with a longer pulse width or a lower-strength exponentially decaying pulse with a longer decay halftime. Hence, there is some reciprocity between the "duration" that current is flowing and the field strength of the pulse. Third, any qualitative effects due to the pulse waveform are, however, unknown. Unfortunately, not all investigators make a distinction between pulses with square or exponentially decaying waveforms. This contributes to the confusion about the role of a poorly understood factor. While the most commonly used waveform types are square and exponentially decaying, literally any arbitrary waveform can, in theory, be synthesized. It happens that the above two types of commonly used waveforms are, respectively, either available in commercial electronic signal generators or easily produced as a consequence of physical laws that govern how a capacitor discharges through a resistive load. The two waveforms cannot be easily or simply compared except in terms of total power (area under the curve) transferred to the chamber.

A fourth factor is the actual exposure of the specimen membranes to a given electric field treatment. Reproducibility in the data appears to depend on reproducibility in pulse voltage, decay halftime, and pulse waveform. We use a storage screen oscilloscope to continuously monitor the pulse characteristics. It has been found that control of pulse voltage and, in our case, decay halftime to within 3–5% is necessary for experimentally reproducible fusion yields. Thus, fusion yields measured in a second sample exposed to an identical pulse treatment will produce a fusion yield which will be within an average of 2–3% of that measured in the first sample. For unknown reasons, some membrane systems are resistant to electrofusion and other membrane systems display highly variable fusion yields (Bates *et al.*, 1987b).

2.2. Procedures and Apparatus Needed

Electrofusion of membranes requires a chamber and a pulse generator. Most early workers either constructed their own apparatus or used combinations of home-built and commercial devices. Commercially available instruments designed for electrofusion exist but are still relatively expensive.

The disadvantage of commercial devices is that they may have pulse or other device characteristics that cannot be adjusted and design features that may not match the goals of the investigator. Since electrofusion is not yet well understood, at least some commercial devices may not be optimally designed for electrofusion applications. Some devices involve such a high level of complexity and sophistication that field repair or modification is impractical or impossible. This is especially true if the circuit diagram is withheld for proprietary reasons by the manufacturer from the user's documentation. Lastly, if a manufacturer withdraws from the market, it will be difficult to replace specialized parts which become unavailable or obsolete.

The construction of pulse generators requires a level of practical electronics knowledge comparable to that needed by a graduate student of electrophysiology, instrumentation biophysics, or an undergraduate in electrical engineering or physics. Design and construction of pulse generators is convenient and practical for experimental applications involving small chamber volumes but less so for large preparative chamber volumes. Indeed, a number of papers have been published describing in considerable detail the necessary instrumentation (Sowers, 1984; Mischke *et al.*, 1986; Miles and Hochmuth, 1987). It should also be appreciated, however, that no matter how crude some improvised devices may be, valuable results, particularly in genetics experiments involving electroporation, have been achieved (Potter *et al.*, 1984; Fromm *et al.*, 1985). Additional relevant background may be found in the chapter by Hofmann. Perusal of that chapter, some of the published literature, and this chapter, together with serious reflection, should be helpful in deciding whether efforts should be directed toward homemade construction, obtaining a pulse generator from commercial sources as a complete unit, or assembling a mixture of commercial and homemade devices. The hazard of accidental electric shock must also be considered and investigators who are not comfortable with homemade instrumentation or do not have the required mechanical dexterity or are accident-prone would be well advised to not become involved in instrument construction.

Homemade apparatus: Homemade chambers and pulse generators were used in many early investigations. Recent reviews (Neumann, 1984; Neumann *et al.*, 1982; Tsong, 1983; Berg, 1982; Baker and Knight, 1983; Knight and Scrutton, 1986) describe or otherwise show simplified drawings of some of the chambers previously built and used by investigators. In all cases, the chamber permits an electric current to pass from the power source to the membrane suspension. Our original chamber (Sowers, 1984) was designed for continuous observation by light microscopy and continues to be the most used and most practical of all we have tried. Based on our experience, the only new recommendation we make is that the wires used to carry the electrical current to the suspension be changed about once per day to eliminate the effects of electrochemical corrosion that occurs at the electrode/suspension interface. Such corrosion leads to bubble formation which significantly increases field strength in the immediate vicinity of the electrodes and thus reduces the pulse field strength in the part of the chamber where observations take place. The electric field induced in a given chamber is calculated by dividing the voltage applied to the electrodes by the distance between electrodes. Our chamber and pulse generator (Sowers, 1984) permit up to 1800 volts to be

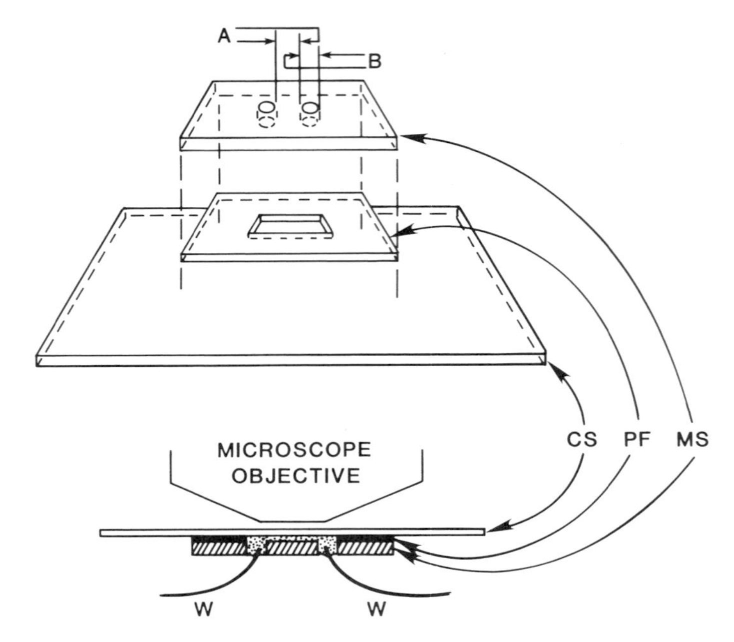

Figure 1. Chamber for long-term observation of electrofusion and related phenomena. Plastic (Plexiglas) microslide (MS) is at least 2.4 mm thick with two holes (diameter, B, of 1.6 mm) and edge-to-edge spacing (A) of 2.0 mm is anchored to a standard glass coverslip (CS) by a sheet of parafilm (PF) using heat from a dry mounting press or a hot plate. Wire electrodes (W) are inserted into the holes as shown after the chamber is loaded with the membrane suspension.

generated. When placed across a chamber length of about 2 mm, this generates an electric field strength of up to 900 volts/mm. The pool of buffer on each end of the chamber is subject to uneven drying and room air currents which result in a drift in position of the membranes being observed. This prevents long-term observation. In a modified design (Fig. 1), the pools are surrounded by the chamber. This permits drift-free observation of membranes over long time intervals. This chamber is more difficult to construct and is designed to be reused, whereas the previous chamber (Sowers, 1984) was designed to be easy to construct and is disposed after use.

The pulse generators used by most investigators generate a pulse with either a square waveform or an exponentially decaying waveform. A pulse width (square wave) must not be confused or compared with a decay halftime (exponentially decaying). The former usually requires complex circuitry and for that reason is usually available only in commercial instruments. The latter is easier to obtain since it only requires that a charged capacitor be discharged through the chamber. Our experiments with human erythrocyte ghosts have utilized pulses obtained from such an instrument.

While most of our work has involved pulse field strengths of 500–700 V/mm and pulse decay halftimes of 0.6–1.0 msec, we have made modifications in our pulse generator and found that lower field strength pulses with longer decay halftimes will also lead to fusion (Fig. 2). This general feature of electrofusion must be appreciated in planning not only pulse generator and chamber design but also experiments. It should be pointed out that at lower pulse strengths, the maximum possible fusion yield is lower, however, and control of the fusion yield is much more critically dependent on pulse decay halftime for a given pulse field strength. Also, the fusion yield has a strong dependence (but see Blangero and Teissié, 1985) on buffer strength (which relates to both ionic strength and conductivity). Since higher ionic strengths generally show higher electrical conductivity, it is necessary to monitor the decay halftime of the pulses on an oscilloscope and

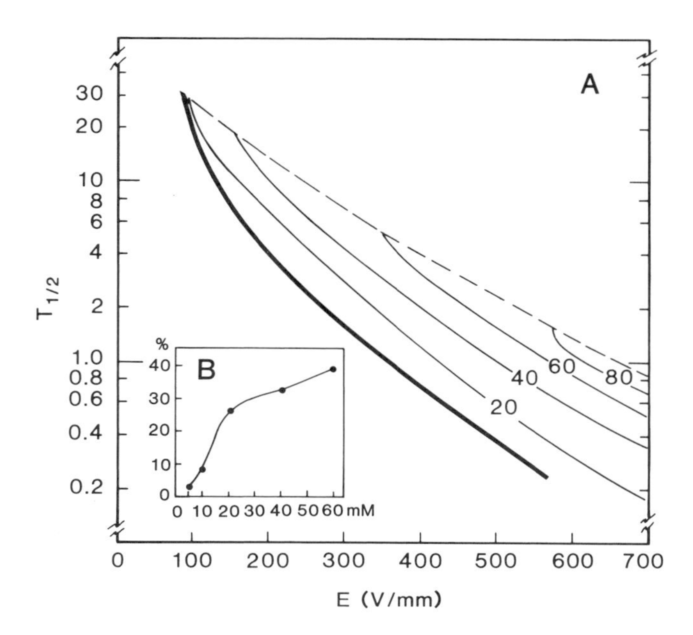

Figure 2. Fusion yield as induced with a single field pulse in erythrocytes: A, as a function of electric field strength (E) and decay halftime $T_{1/2}$ (in msec) in 60 mM sodium phosphate buffer (pH 8.5); B, as a function of strength of the buffer. Fusion yield measured using DiI as the fluorescent label and the CF protocol (see text). Results in A for given pulse parameters: *above dashed line*, fragmented membranes; *below lower (thick) line*, no effects; *between lower line and dashed line*, both fusion (isofusion contours in percent fusion yield) and electroporation (as revealed by loss of mw = 10 kD FITC-dextran from cytoplasmic compartments with application of a single pulse). Ghosts were isolated and labeled as previously described (Sowers, 1984). Electroporation-caused loss of FITC-dextran is the same at all locations on lower (thick) line. Results in B are for a single pulse of 600 V/mm and 0.6 msec decay halftime.

compensate for the shorter decay halftime caused by the higher conductivity by adding capacitance or decreasing the auxiliary load resistance (see below).

Our original pulse generator (Sowers, 1984) was very simple in design and operation. It required that the "switch" used to connect the charged capacitor to the chamber electrodes be ideal. In contrast, when real switches "close" they allow (1) a spark to form just before contact is made if the voltage is high enough and (2) a phenomenon called "contact bounce" to take place. Both will cause an uncontrollable variability in pulse waveform fidelity and will lead to irreproducibility in the data. To get around this problem, our original pulse generator (Sowers, 1984) used simple and inexpensive mercury-wetted relays. Contact bounce is highly suppressed or nonexistent in mercury switches and mercury-wetted relays. These devices are inexpensive and permit exponentially decaying pulses with excellent waveforms to be produced. Unfortunately, our use of these relays at voltages higher than they were rated caused an accelerated rate of wear and eventually led to spurious distortions in the waveform as monitored on a storage screen oscilloscope and eventual mechanical failure. This necessitated continual replacement. Minor nonuniformities in production sometimes resulted in a "double pulse" each having half the voltage for which the circuit was designed. Once analyzed, this could be corrected by exchanging one relay for another. Unfortunately, the manufacturer superseded the original series of relays with a new series which did not perform as well at overvoltages. This forced us to abandon mercury-wetted relays as a switch element in this application.

Our present pulse generator is based on a type 5557 mercury vapor thyratron as a switch element (Fig. 3). Thyratrons become conductive when the grid voltage—normally held at a relatively high negative voltage—is made to fall below a lower limit. This permits the tube to become conductive and

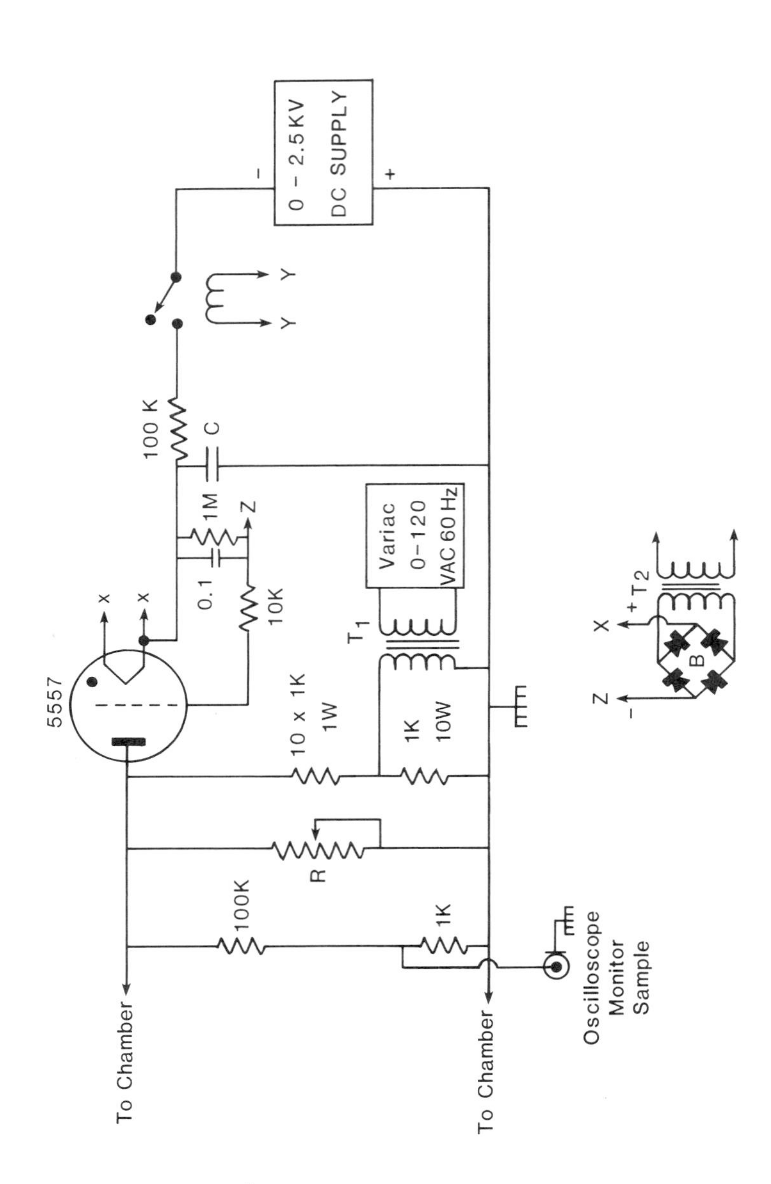
0 – 2.5 KV
DC SUPPLY
–
+
Y
Y
100 K
C
1M
Z
0.1
10K
x
x
5557
Variac
0–120
VAC 60 Hz
T1
10 x 1K
1W
1K
10W
R
100K
1K
To Chamber
To Chamber
Oscilloscope
Monitor
Sample
X
Z
+
–
T2
B

can permit a large current to flow. Manufacturer ratings for the 5557 include a peak plate-cathode voltage of 5000 V and can repetitively switch on a current of 2.0 A. Nonrepetitive current can be up to 40 A. Once the tube becomes conductive, it cannot be shut off. Shutting off the current flow must be accomplished from some other point in the circuit which is in series with the tube. This automatically occurs, however, when the capacitor becomes completely discharged. The 5557 not only operates well within all electrical ratings but is naturally resistant to wearout and failure. Additional thyratron background information may be found in Malmstadt *et al.* (1963) and Dawes (1942). We have built an inexpensive and simple but effective pulse generator around this device using new parts for less than $1000 (including an expensive cabinet). This circuit was designed with simplicity and low cost in mind so that it could be easily duplicated and understood with a minimum knowledge of physics and electronics. Many variations are possible and only one version (Fig. 3) is shown here. We make minor changes in accordance with the particular experiment and as we accumulate experience with the unit. Cautions to be observed include: (1) operation of the tube in an upright position, (2) a preuse conditioning of the tube through a long warmup and application of gradually increasing high voltage to charge the capacitor, and (3) a choice of component parts with a large margin of power and voltage handling capability over apparent power needs.

We also designed a square wave pulse generator utilizing mercury-wetted relays (Fig. 4) and used it to obtain preliminary data (see below) by which a comparison could be made with exponentially decaying pulses. The circuit is simple in that no tubes or transistors are used, but its operation requires constant monitoring by an oscilloscope and its adjustment is complex. We do not have long-term experience with its use but include it for completeness and to show the versatility of these relays. For pulses in the low voltage range, we expect that it should be quite reliable (these relays are generally rated for about 1–10 million switching operations before failure).

References are available which may also be useful in construction of instruments (e.g., Malmstadt *et al.*, 1963, 1981; Moore *et al.*, 1983). A storage screen oscilloscope is essential to continuously monitor the pulse waveform characteristics (voltage and decay halftime) and the effect of the load on the pulse characteristics. The equivalent resistance of the chamber depends on its size and geometry and the conductivity of the membrane suspension. Lastly, complex electrochemical reactions may take place, even with platinum electrodes (Rosenberg *et al.*, 1965). However, this is of minimal concern in our chamber design since the fusion assay takes place before the electrochemical products generated at the electrodes can diffuse to the observation site in the chamber (Sowers, 1983a). It is not often appreciated that both the chamber and the pulse generator must be matched for efficient transfer of energy from the generator to the chamber (see Hofmann, this volume). Sometimes unexpected problems develop. For example, we have observed that 0.6–0.8 msec decay halftime pulses used in a chamber with close-spaced (100–200 μm) parallel wires

Figure 3. Essential details of pulse generator circuit which uses a type 5557 mercury vapor thyratron. Resistances are in ohms, capacitances are in microfarads, and heat dissipations are in watts. A 0–2500 volt DC power supply charges capacitor, C, through a 100 k resistor when high-voltage relay coil is activated with current at connections Y–Y. Thyratron is nonconducting when the grid is held at negative voltage (200–300 V) with respect to filament (X–X) with DC obtained using bridge rectifier, B, and AC from secondary of transformer T_2. When current in primary of this transformer is turned off, the grid voltage falls (at a rate determined by the time constant of the 0.1 μF capacitor and the 1 M resistor in parallel with it) until the tube fires. This discharges the capacitor, C, through the chamber, the variable resistor, R, the resistor network which allows the dielectrophoresis-inducing AC to be carried to the chamber without an additional switch element, and the resistor network (100 k + 1 k) which samples a small part of the pulse for oscilloscope monitoring. After capacitor, C, is discharged, the thyratron returns to the nonconducting state. Then the primary of transformer T_2 can be connected to power to hold the grid negative. The relay coil may then be reactivated to recharge the capacitor. Isolation transformer, T_1, has a 1 : 1 turns ratio and is designed for 110 V 60 Hz input. The Variac is connected to the utility power lines and feeds 0–110 V to the primary of T_1. The 5557 requires a filament voltage of 2.5 V at 5 A which can be supplied by a filament transformer. However, it and transformer T_2, and the high-voltage relay coil must be insulated for at least 2500 V to prevent short circuits. Resistor, R, and capacitor, C, must be adjusted to control the decay halftime of the pulse (details are omitted for clarity).

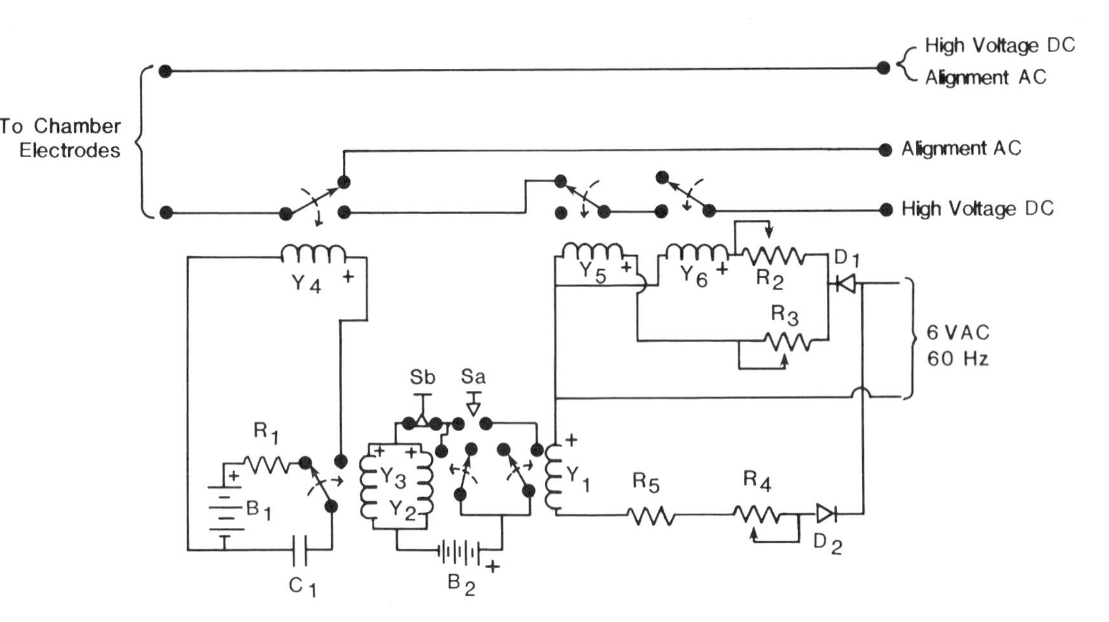

Figure 4. Circuit for pulse generator which provides single square wave pulses with a magnitude of up to several hundred volts. Relays Y_5 and Y_6 open and close 60 times per sec but with a phase shift controlled by variable resistances R_2 and R_3. Thus, 60 pulses of a controllable pulse width are generated each second. Depressing switch Sa causes a single pulse to be delivered to the chamber only when relay Y_1 permits the switched current to pass to relay Y_3 and Y_2. Y_2 latches the current "on" while relay Y_3 allows capacitor C_1 to discharge through relay Y_4. Proper choice of component values permits the contacts to close transiently at a precise time with respect to the generated pulses. Thus, only one pulse is permitted to pass to the chamber while the dielectrophoresis-inducing AC is removed from the chamber electrodes. The circuit must then be "reset" for another pulse by pressing switch Sb.

(similar to Fig. 4 in Bates *et al.*, 1987a, or Fig. 25 in Zimmermann, 1986) result in disruptive bubble formation.

2.3. Misconceptions

Even though the basic electrofusion protocol is simple in concept and will lead to fusion in many applications, the persistence of certain dogmas and misconceptions has constrained our understanding of the electrofusion mechanism and may have limited further applications. Below are examples of experimental observations which contradict previously implied or stated principles.

1. *Dielectrophoresis does not necessarily require a chamber and electrode geometry which generates a bulk nonhomogeneous electric field.* Dielectrophoresis in a bulk homogeneous electric field has been experimentally observed by numerous investigators (Zachrisson and Bornman, 1984; Sowers, 1983a, 1984) and theoretically analyzed (Sauer, 1983). It is possible that the presence of the cells in a bulk homogeneous electric field produces a local perturbation of the electric field to cause a local *in*homogeneous field within a bulk homogeneous field. Further experimental exploration of this problem may permit electrofusion to be studied or utilized under wider sets of conditions.

2. *Frequencies in the range of 10^4 to 10^7 Hz are not necessarily needed for dielectrophoresis.* The commonly available utility power line frequency (60 Hz) is sufficient to align erythrocyte ghosts (Sowers, 1984), and mitochondrial inner membranes (Sowers, 1983a) into pearl chains for electrofusion experiments.

Although electrochemical reactions occur even with platinum electrodes at low AC frequencies (Rosenberg *et al.*, 1965), the limitations in pulse generator voltage output may have forced investigators to use a chamber design involving close-spaced electrodes which therefore places cells near if not in contact with the electrodes. Such close-spaced electrodes expose the membranes to electrochemical reaction products induced by the high-strength DC pulse (e.g., Zimmermann, 1982). Part of the reason for the use of high-frequency AC to induce dielectrophoresis may be to minimize the electrochemical reactions during dielectrophoresis (Rosenberg *et al.*, 1965). However, their formation will not be prevented during the pulse. Our approach to avoiding this problem uses a chamber design which places the electrodes as far as practical from the site at which membrane observation takes place and uses procedures which minimize the time interval between pulse treatment and the experimental observations (Fig. 1, and Fig. 1 of Sowers, 1984).

3. *Dielectrophoresis does not necessarily require a medium of low conductivity (Zimmermann, 1982, 1986).* Usually a low-conductivity medium such as distilled water is used with an impermeant molecule such as sucrose to maintain osmotic balance across the plasma membranes of cells. We have, on the other hand, observed dielectrophoresis in phosphate buffers up to 60 mM in strength (Sowers, 1986a) and ammonium acetate buffers up to 120 mM (Sowers and Kapoor, 1988). Indeed, it was also reported in a recent review (Zimmermann *et al.*, 1985) that dielectrophoresis will occur in RPMI 1640, a growth medium with a considerable ionic concentration. This was not acknowledged as contradicting the previously stated need for low-conductivity media.

Part of the need for low conductivity is that during dielectrophoretic alignment, the medium is exposed to a continuous AC in which the product of current and voltage can be high enough to cause significant heating. Calculations can be easily performed which will show significant deposition of energy into the medium either during the alignment or during the pulse. These simple calculations do not account for the cooling effect provided by the chamber wall material. In very small chambers (Fig. 1, and Fig. 1 of Sowers, 1984), the distance from the center of the chamber to the wall is small enough to buffer the medium thermally. Experiments with small shavings of low-melting-temperature waxes (Sowers, 1986a) suggest that this thermal buffering may be quite high.

4. *The general protocol in which membrane–membrane contact is induced first and then, and only then, are the fusion-inducing pulses applied (Zimmermann, 1982, 1986) has become a central dogma which has played a crucial role in our understanding of electrofusion.* In contrast to this original membrane-contact-before-pulse [referred to hereafter as the "contact-first" (CF)] protocol, it has been found that fusion will occur even if pulses are first applied to membranes while they are

in random positions and the membranes are then brought into contact (Sowers, 1983b, 1984, 1985b, 1986a,b). We refer to this procedure as the ''pulse-first'' (PF) protocol to both name and contrast this procedure with the original CF protocol. Although conditions which permit PF electrofusion have not yet been found for mitochondrial inner membranes or chicken erythrocyte ghosts, it has, on the other hand, been reported that PF electrofusion can occur in Chinese hamster ovary cells (Teissié and Rols, 1986). The achievement of fusion by this protocol appears to have been observed by others in some membrane systems (Bates *et al.*, 1987b; Zimmermann *et al.*, 1984) but has neither been fully recognized as possibly revealing a fundamentally different fusion mechanism nor acknowledged as an important variation in the normal protocol.

5. *Fusion of membranes is often recognized as a process in which two equal-diameter spherical cells in contact undergo a change which is observed as an increase in the diameter of the lumen or hourglass constriction until the two spheres become one large sphere.* Fusion in cells which have irregular shapes is sometimes measured, for example, by the number of nuclei that are surrounded by a plasma membrane-delimited mass of cytoplasm. Although cells can be observed by brightfield microscopy to fuse in these ways, it is possible for fusion to occur without this change in morphology (Wojcieszyn *et al.*, 1983). Thus, fusion needs to be scored on the basis of more rigorous criteria.

As human erythrocyte ghosts have only a plasma membrane bound by a glycocalyx on the outside and a rudimentary cytoskeletal system on the inside, observation of fusion by light microscopy requires: (1) at least phase optics to resolve the plasma membrane and (2) distinct changes in morphology. The membrane area of the ghost membranes cannot change but the enclosed volume can increase after fusion if, and only if, the lumen diameter increases. This takes place frequently in sodium phosphate buffer generally at about or above 20–40 mM (pH 8.5). But at or below 20 mM, lumina generally do not increase and cannot, therefore, be detected by phase optics (Sowers, 1984). However, fusion of membranes permits intermixing of all laterally mobile membrane components and can be rigorously shown if the fusion assay uses mixtures of unlabeled ghosts and ghosts labeled with, for example, the fluorescent lipid analogue, 1,1′-dihexadecyl-3,3,3′,3′-tetramethylindo carbocyanine perchlorate (DiI). Fusion thus results in a movement of fluorescence from the labeled membrane to one or more unlabeled membranes at rates consistent with the lateral diffusion coefficient for lipids (Sowers, 1985a). This method of detecting fusion requires fluorescence optics for FITC wavelengths.

If a soluble molecule is inside the enclosed space of one of several membranes in a pearl chain, the formation of a lumen during fusion will permit the label to move by aqueous diffusion to the interior space of the adjacent but originally unlabeled membrane. Likewise, fusion of two membranes into one membrane will allow a lipid-soluble molecule in one to laterally diffuse without restrictions throughout all of the area of both membranes. These possibilities are commonly recognized as operational criteria for fusion and are referred to as ''contents mixing'' and ''membrane mixing.'' Most fusion studies up to now have utilized large populations of vesicular membranes in solution and fluorescent-tagged water and lipid-soluble molecules as labels to trace the two kinds of mixing. Sometimes leakage of molecules from the surrounding to the interior (or vice versa) of one of the membrane-enclosed spaces is observed during the fusion process. These experimental approaches and the interpretation of the data obtained with them are straightforward in principle but not in practice. Additional discussion is beyond the scope of this chapter and the reader is referred to several reviews for further information (Blumenthal, 1987; Duzgunes *et al.*, 1987; Szoka, 1987; Morris *et al.*, 1988).

The use of fluorescent markers, either in lipid-soluble molecules to label the membrane or in water-soluble molecules to label the cytoplasmic compartment, is necessary to positively demonstrate fusion. Fusion is indicated as the label moves to an unlabeled membrane in a pearl chain adjacent to a labeled membrane following a pulse. Fusion yield can be calculated in many ways. We prefer to count the number of prefusion labeled membranes, N_u, and then count the number of occurrences (= fusion events) in which the label moved to at least one adjacent but originally unlabeled membrane, N_f, and calculate $N_f/(N_f + N_u)$. This number will always be between 0 and 1 which can be multiplied by 100 to give an answer in percent.

Our use of fluorescent-tagged labels, the above mixing criteria, and studies of single fusion events between two membranes induced by electric pulses showed that phase optics can underestimate fusion yields because a fraction of all fusion products do not produce detectable lumina (Sowers, 1984). Also, pulses will induce attachment between membranes in close contact. This attachment represents fused membranes. However, a discrepancy was observed when comparing fusion yields using a fluorescent-tagged contents mixing indicator (FITC-dextran) with fusion yields using a membrane mixing indicator (DiI). The use of fluorescent soluble labels overestimated fusion yield (Sowers, 1988). This discrepancy was interpreted by recognizing that the fluorescent indicators trace different parts of the fusion process. Indeed it is likely that FITC-dextran movements indicate the possible existence of an intermediate state (see below).

3. THE ELECTROFUSION MECHANISM

3.1. Membrane Fusion in General

The mechanism of electrofusion must be considered within the context of membrane fusion as it occurs in nature and in other *in vitro* systems. Most membrane fusion studies involve model membranes (vesicles) composed of only one or two phospholipid species and no membrane proteins. While fusion in these artificial membranes is followed by indirect methods in large populations of membranes, fusion in natural systems is very difficult to study because they are much more complex. Indeed, Rand and Parsegian (1986) have suggested that fusion in artificial membranes cannot be useful for learning about fusion in natural membranes. Nevertheless, the problem of how fusion occurs in model systems served as a starting point for their studies. They postulated that, at least for lipid bilayers, six sequential steps or events should be involved in membrane fusion. They are:

(i)	(ii)	(iii)	(iv)	(v)	(vi)
Stable membrane apposition →	Triggering →	Contact →	Focused destabilization →	Membrane coalescence →	Restabilization

Attention is called to this scheme since numerous hypothetical mechanisms of membrane fusion involve many fewer steps. For example, the possible involvement of the hexagonal II lipid structure in membrane fusion is a three-step process (Siegel, 1987):

(i)	(ii)	(iii)
Unfused membranes →	H_{II} intermediate →	Fused membrane

The possibility that fusion may involve more than three steps offers additional mechanistic elements for which experiments can be designed.

Other investigators have invoked a variety of specific factors, processes, conditions, or requirements to explain or understand fusion mechanism (for discussion and additional references, see, e.g., Plattner, 1987; Ohki, 1987). All of these fusion proposals involve *specific* microscopic processes and are unrelated to membrane fusion as can be induced by a gross membrane disturbance such as microsurgery (Diacumakos and Tatum, 1972).

A permeability disturbance is often associated with membrane fusions but its significance is unknown. The observation of a pulse-induced permeability increase reflects a different cause, but is otherwise analogous to this permeability increase (for discussions, see Schlegel, 1987; Siegel, 1987; Szoka, 1987; Duzgunes *et al.*, 1987). The role of leakiness is contentious but may be an additional clue to the molecular rearrangements possibly occurring in the bilayer as part of the fusion process. Indeed, the experimental detection of leaky fusion may be a reflection of a reversible early step which does not go to completion (see below).

It should also be noted that laser light pulses can induce membrane fusion (Schierenberg, 1987). Considering that laser light pulses can have extraordinary power, it is natural to imagine that the fusion mechanism may be a trivial consequence of a gross membrane disturbance comparable to the re-formation of molecules into stable membranes after microsurgery (Diacumakos and Tatum, 1972). However, the electric vector of light is perpendicular to the direction it propagates and thus the membrane may experience a voltage pulse from the light pulse. The fact that membranes fail to fuse if light pulses are too high is reminiscent of the fact that electrofusion does not take place if electric pulses are too strong. The mechanism of laser pulse fusion deserves more serious attention and this may shed light on the electrofusion mechanism.

3.2. Electrofusion in Specific

At present, all electrofusion mechanisms fall into two categories: electropore hypotheses and point defect hypotheses.

Fusion mechanisms involving electropores: Electropores have been invoked as part of fusion mechanisms at least partly because much is known about their properties and they are known to be induced by the same pulses which induce fusion. Permeabilization (= electroporation) of membranes by electric field pulses has recently been reviewed by workers in several laboratories (Tsong, 1983; Baker and Knight, 1983; Neumann *et al.*, 1982; Knight and Scrutton, 1986; and this volume). The simplest explanation for this permeability increase is that the pulse induces pores or holes in the membrane. However, relatively little is known about the relationship between the characteristics of the external field pulse and the shape, numbers, effective diameter, lifetime, and location of these pores. Their effective diameter can be probed by the use of labeled molecules with known diameter, however.

Membrane electropore induction is also considered to be related to membrane breakdown which occurs above a threshold in induced transmembrane potentials. The mathematical relationship between the membrane potential induced in a spherical membrane by a bulk uniform electric field induced in the medium by an electric current (Fig. 5) is often discussed in connection with electroporation (see Fricke, 1953, and reviews) and is given in Eq. (1):

$$V = 1.5\, E\, r \tag{1}$$

where V is the peak induced transmembrane potential (volts) at the points that are perpendicular to the field direction, E is the induced electric field in the medium (volts/mm), and r is the radius of the spherical membrane (mm). Although widely quoted, Eq. (1) makes many assumptions that may be unwarranted under some circumstances.

This equation has been verified at least up to induced transmembrane voltages close to membrane potentials obtained in the course of normal metabolism (Gross *et al.*, 1986; Ehrenberg *et al.*, 1987; Farkas, this volume). Further understanding will require that this equation include terms that

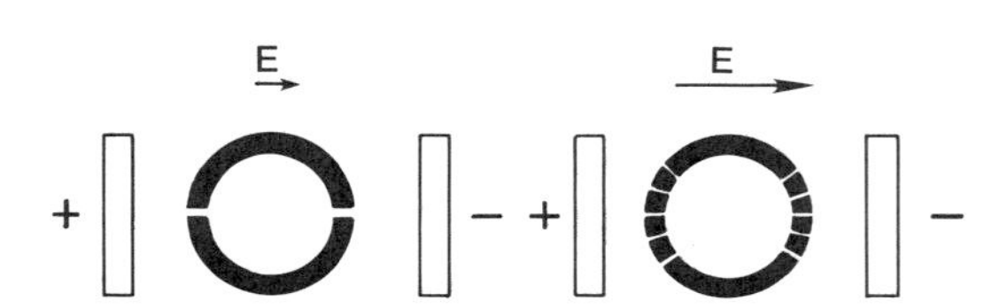

Figure 5. Electropore induction in the spherical membrane in a homogeneous electric field. (Left) At induced transmembrane voltage threshold for electroporation, only one pore is induced at each pole. (Right) At induced transmembrane voltages higher than the threshold, additional pores are expected to be induced at locations farther from the poles. Adapted from Zimmermann (1983).

account for the time- and capacitance-dependent rate of voltage increase across the membrane in the early time after exposure to a pulse.

The pulse-induced permeabilization which occurs when the induced membrane potential exceeds certain thresholds of $V = 0.3–0.6$ (Benz and Zimmermann, 1981) may or may not be reversible depending on experimental conditions. This permeabilization is rapidly and completely reversible in phospholipid vesicles (Teissié and Tsong, 1981) but slowly and possibly incompletely reversible in natural membranes (Tsong, 1983; Sowers and Lieber, 1986; Schwister and Deuticke, 1985).

Equation (1) and the knowledge of the threshold implies that, for cells of at least roughly spherical shape, the bulk field strength needs to be higher for bacterial protoplasts than for typical mammalian cells. Although the pulse field strength needed to induce electropores may be estimated, there is little information available to predict the effective diameter, number, lifetime, shape, or location of these electropores. From numerous experimental studies, however, it has been generally observed that at a minimum transmembrane electric field strength, electropores appear to open independently of temperature and have a maximum radius which in inversely related to ionic strength. Also, electropores appear to reseal faster at higher temperature, or slower near 0–4°C. Lastly, they reseal faster or sooner at high ionic strengths. Resealing is complete in artificial phospholipid bilayers (Teissié and Tsong, 1981) but incomplete and/or slow in erythrocyte membranes (Serpersu *et al.*, 1985; Schwister and Deuticke, 1985). Studies indicate that electropores have effective diameters of 0.3–8 nm and number 1–4000 per membrane (Sowers, 1986a; Sowers and Lieber, 1986; Schwister and Deuticke, 1985; Neumann and Rosenheck, 1972; Tsong, 1983; Teissié and Tsong, 1981; Knight and Baker, 1982; Benz and Zimmermann, 1981). Most of these papers utilized the erythrocyte membrane which may be different in important ways from other mammalian cells. It should be pointed out that low-ionic-strength-induced hemolytic pores have been rigorously and accurately measured and have been found to be reversible (resealable) down to a radius of about 0.7 nm, but cannot be further resealed (Lieber and Steck, 1982a,b). This may have factors in common with our observation of residual electropores with an effective radius of about 0.5 nm (Sowers and Lieber, 1986).

Electropores are part of two similar hypothetical electrofusion mechanisms (Fig. 6). Pilwat *et al.* (1981) pictured the electrofusion mechanism as a process in which the electric field pulses induce short-lived pairs of electropores as an intermediate membrane structure. The individual lipid molecules continually change in topology and position to form a fused membrane. In the original proposal and in subsequent reviews (Zimmermann *et al.*, 1985), the space surrounded by the pore pair is proposed to be transiently occupied by a cloud composed of individual lipid molecules. The dispersed form that these molecules take is evidently necessary to help bridge the gap between each of the close-spaced but originally separate membranes. However, the basis for expecting lipids to be dispersed is unclear; calculations indicate that heating should not be excessive (Pastushenko, 1983). Pohl *et al.* (1984) apparently favored a similar sequence but included the possibility that one of the two pores opened first, while Pilwat *et al.* (1981) pictured the mechanism as involving concentric pairs of pores which open simultaneously as a single entity. The mechanism presented by Pohl *et al.* (1984) was unfortunately not accompanied by illustrative diagrams.

Dimitrov and Jain (1984) proposed a four-step mechanism for electrofusion. To bridge the space between the membranes, they proposed that the pores developed a "flare" so that the hydrophobic free edges of the pores can touch each other and then fuse. The two pores are not accompanied by the cloud of lipid molecules as presented by Pilwat *et al.* (1981). Sugar *et al.* (1987) proposed a model for fusion which includes the propagation of a cracklike defect as part of the fusion process. We consider the model proposed by Sugar *et al.* (1987) as a variation on the Dimitrov and Jain (1984) model rather than the Pilwat *et al.* (1981) model since concentric pore pairs are involved but, in contrast, do not include a cloud of dispersed lipid molecules.

Fusion involving point defects in membranes: Stenger and Hui (1986) conducted an interesting freeze-fracture electron microscopy study of electrofusion in human erythrocytes (see also their chapter in this volume). Since their results are also published in detail in this volume, we will refer

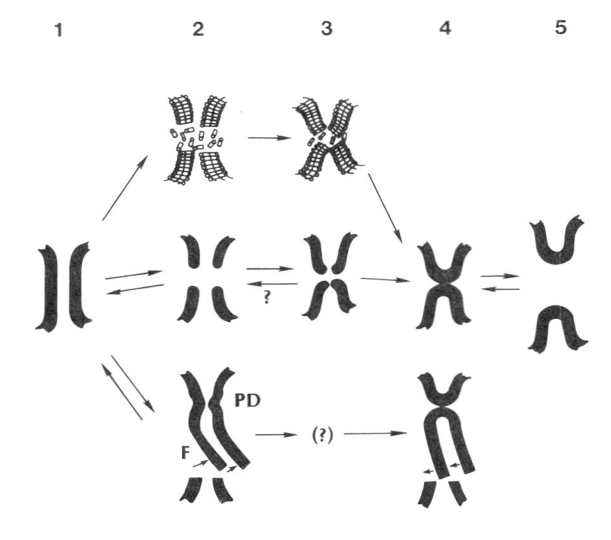

Figure 6. Multistep (1–4) models for the mechanism of electrofusion. Unfused but close-spaced membranes (1) become fused (4) and may be followed by expansion of the lumen (5). Fusion product at step (4) could have a microscopic lumen which may subsequently contract and not permit contents mixing or may permit contents mixing only transiently. Upper path, adapted from Pilwat *et al.* (1981). Pulse-induced concentric pore formation (2) (pores filled with dispersed lipid molecules) proceeds to fusion product (4) by passing through an intermediate stage (3) in which the lipid molecules change position in an orderly but apparently continuous formward process which reforms the two membranes into one. Middle path, adapted from Dimitrov and Jain (1984). Induction of concentric pairs of pores (2) proceeds to a stage (3) in which the edges of the pores become flared before ending with a fusion product (4). Lower path, adapted from Stenger and Hui (1986). Fusion involves pulse-induced point defects (PD) and pulse-induced "flaps" (F) which follow separate paths. The point defects lead to fusion while the flaps reseal after providing a large cytoplasmic–cytoplasmic conductance pathway for soluble molecules to move from one compartment to another [the intermediate structure in the model represented by the lower path was placed at stage (2) since the authors' description of the process did not have additional steps]. Note that all pathways can permit contents mixing events. However, contents mixing events without membrane connections would make the path to at least stage (2) and possibly stage (3) reversible (note arrows added in reverse direction). The process in the upper pathway is implied in original paper (Pilwat *et al.*, 1981) to be self-completing (i.e., irreversible) since fused membranes are thermodynamically stable low-energy structures. Lumen expansion ($4 \leftrightarrow 5$) is reversible in erythrocyte ghosts as previously reported (Sowers, 1984).

to the salient aspects which are relevant to our results. Indeed, their electron micrographs show both point defect-like structures and flaplike structures but their interpretation of the micrographs is that membrane fusion occurs as a consequence of the formation of the point defects in the membrane (Fig. 6). The flaplike structures appear to be reversibly inducible structures that should have electropore properties even though they do not have the cylindrical geometry implied by either the Pilwat *et al.* (1981) model or the Dimitrov and Jain (1984) models. It is important to consider this model as one which does not utilize electropores for part of the intermediate structure (see below).

4. COMPARISONS BETWEEN THEORY AND EXPERIMENT

General observations. Within the appropriate range of effective values, fusion yield is proportional to pulse field strength, pulse decay halftime, and ionic strength of the medium (Fig. 2). This applies to the CF protocol (as shown in Fig. 2) or the PF protocol for at least the range of values as previously explored (Sowers, 1986a, 1987b). It should be noted from Fig. 2 that use of lower pulse voltage and longer decay halftime will lead to a decrease in the maximum possible fusion yield and that experimental control of the appropriate variables will consequently be more difficult. Although square wave pulses and exponentially decaying pulses are commonly used, little information is available on the effects of the two waveforms. During the preliminary study conducted with the mercury-wetted relay pulse generator (Fig. 4), we obtained data (Table 1) suggesting no major qualitative differences between the two waveforms. Square wave pulses appear to need a pulse width shorter than the decay halftime scale of exponentially decaying pulses, but the two time scales

Table 1. Effect of Pulse[a] Waveform on Fusion Yields (Percent)

		Waveform							
		Exponentially decaying		Square					
				40 μsec		120 μsec		400 μsec	
Buffer strength	*N*	CF	PF	CF	PF	CF	PF	CF	PF
		Lumen-producing fusion products only[b]							
20 mM	3	24	1	0	0	0	0	7	0
	10	35	2	21	10	23	11	4	3
60 mM	3	35	22	0	0	0	0	0	0
	10	76	46	0	0	11	7	4	0
		All fusion products (nonlumen plus lumen)							
20 mM	3	nd	nd	10	2	36	1	71	8
	10	nd	nd	51	20	65	17	94	35
60 mM	3	nd	nd	10	1	47	0	45	11
	10	nd	nd	40	4	61	13	89	50

[a]Buffer was sodium phosphate and the pulse field strength was 700 V/mm. *N* is the number of pulses, applied at the rate of two per sec. The rectangular chamber, the preparation of the human erythrocyte ghosts used in the assay, and the exponentially decaying waveform pulse generator were as previously described (Sowers, 1984). Exponentially decaying pulses had a decay halftime of 0.6 msec. When the PF protocol was used, the time interval between pulses and induction of membrane–membrane contact was 15 sec. The square waveform pulse generator is shown in Fig. 4.
[b]Lumen and non-lumen-producing fusion products are described in Sowers (1984).

should not be compared with each other. Our observations indicate that fusion nearly always coincides with the application of a pulse when membranes are held in contact by dielectrophoresis at the time the pulse is applied. Conversely, fusion rarely, if ever, occurs any finite time after a pulse. Consequently, when the CF protocol is used, we think of the pulse-induced membrane change leading to fusion as involving a transient or short-lived membrane change. Fusion never occurs a significant time after a pulse when this protocol is used. On the other hand, when the PF protocol is used, fusion occurs only at the moment the membranes are brought into contact. This clearly reflects the presence of a long-lived fusogenic membrane modification. Multiple pulses may be used in electrofusion applications to increase fusion yields, but we now recognize that application of a train of pulses may cause some fusion events to occur as a result of the long-lived fusogenic state in addition to the short-lived fusogenic state. When membranes were held in contact at the time of pulse application (CF protocol), the rate of pulse application appeared to have little effect on fusion yield. However, the long-lived fusogenic state appears to have a strong dependence on pulse rate and is presently being intensively studied in our laboratory. We have also found that human erythrocyte ghosts gradually lose their ability to be electrofused as they age during storage. This requires that fusion experiments be conducted no later than 1 day after membrane preparation.

Implications of the long-lived fusogenic state. Our experimental observation of fusion in membranes using the PF protocol was not predictable from any published hypothesis. The widely accepted protocol (CF) calls for membrane–membrane contact to be established before and maintained during the pulse treatment. A full paper from another laboratory confirms the observation of fusion with the PF protocol (Teissié and Rols, 1986). This experimental fact showed that the pulses induced a long-lived fusogenic state in the membranes. Thus, it is impossible for concentric pore pairs to be formed upon pulse treatment as single entities. However, as two pulse-treated membranes are aligned into close contact, it is possible for a head-on collision to take place between a pore on one membrane and a pore on a second membrane. This could form a concentric pore pair which would then lead to a fusion event. Our preliminary calculations, however, indicate that this would be too unlikely to account for the observed fusion yields (Sowers, 1986a,b; Sowers and

Kapoor, 1987a,b, 1988; Sowers and Lieber, 1986). On the other hand, the calculations are sensitive to the assumptions and work is in progress to make refinements in both the measurements and the theory behind the calculations. Nevertheless, the idea that a concentric pore pair fusion intermediate might be assembled from individual pores implies that electropores could be fusogenic sites.

Induction of the long-lived fusogenic state requires more stringent conditions than for the case where pulses are applied to membranes in close membrane–membrane contact. We have found no evidence that the long-lived fusogenic state can be induced with only one pulse. Supplementing original observations (Sowers, 1984, 1986a), we have found two variables which play a significant role in the induction of this state (Sowers, 1987b). Normally, the induction of the long-lived fusogenic state requires the application of pulses and then the application of the AC to induce contact at some time afterwards. Proper measurement of the long-lived fusogenic state actually requires that a "background" fusion yield be determined by treating a membrane suspension with a given pulse treatment (without the alignment-inducing AC present). Then a "gross" fusion yield is determined by repeating the measurement with only the AC turned on to bring all membranes into contact via pearl chain formation (Sowers, 1987b). The actual fusion yield is then found from the difference between the gross and background fusion yields. The background fusion yield is very sensitive to membrane concentration and has its origin in the fact that a pearl-chaining effect takes place if a series of pulses are applied to a membrane suspension. Whether the pearl chaining which occurs during the pulse treatment is due to true dielectrophoresis or charge-induced attraction seems unclear. In either case, a fraction of the membranes will come into contact as additional pulses are applied. We begin to see the long-lived fusogenic state when the pulse treatment includes a total of at least 5–7 pulses, but then the fusogenicity begins to fall off if the pulse treatment includes more than 10–12 pulses.

The second important factor involving induction of the long-lived fusogenic state is that it is sensitive to pulse rate. Pulses applied at less than one per second do not induce sufficient fusogenicity to be measurable. Pulses need to be applied at about two or three per second. Studies are in progress to further characterize and elucidate this fusion pathway.

We have observed that the fusogenic areas induced by the multiple pulse treatment using the PF protocol are relatively small and that the fusogenic property does not laterally diffuse into areas where it is not induced (Sowers, 1986a, 1987b). A similar observation was made by Teissié and Blangero (1984) for cells already in contact at the time of the pulse application. In this sense, the fusogenic property does not behave as if it were composed of membrane pores (1) composed exclusively of lipid molecules and (2) totally surrounded by lipid molecules. Pores made of lipids and totally surrounded by other lipids would laterally diffuse in the plane of the membrane at rates comparable to laterally mobile membrane components (Sowers, 1985a; Sowers and Hackenbrock, 1981).

The pulse-induced effects which lead to the long-lived fusogenic state are very obscure but of potential significance in terms of factors that control the stability of membranes. Long-lived electric field pulse-induced conformational changes in biopolymers (Neumann and Katchalsky, 1972; Porschke, 1985) and spectroscopically detectable structural parameters in lipid bilayers (Stulen, 1981; Chernomordik *et al.*, 1985) have been observed. Electric fields have been reported to elongate and finally open lecithin vesicles (Harbich and Helfrich, 1979) and induce lipid domain growth in lipid monolayers (Heckl *et al.*, 1986). It is also possible that ferroelectric effects (Lovinger, 1983; Goodby, 1986) are involved.

Electropore induction does not appear to correlate with electrofusion. Pulse treatment of membranes in suspension led to a transient permeabilization in which some combination of the pore lifetime, diameter, and numbers *decreased* as the ionic strength of the medium was increased (Sowers, 1986a). This is in agreement with what has been generally found as the effect on electropore induction with changes in ionic strength (see reviews). However, when the same membranes were aligned into close contact, there was an increase in fusion yield which followed the increase in ionic strength (Fig. 2). This is opposite to what would be expected if there were a simple relationship between pore formation and fusion. The situation as we now understand it, is more complex than this because pulse-induced pores open to large sizes and then immediately reclose within 100–200 msec (Sowers and Lieber, 1986). However, the reclosing is incomplete and a residual pore of about 1 nm diameter is formed. Consequently, when the PF protocol is used, the electropores have residual rather than peak diameters. It is difficult to see how such a small pore could undergo the large topological changes necessary to span the relatively large interbilayer distances in a fusion event. However, in comparison to the PF protocol, fusion yields are generally higher using the CF protocol. This is consistent with a large peak pore diameter being present when membranes are in close contact compared to the residual pores that would be present at the time membrane–membrane close contact is induced when the PF protocol is used.

Contents mixing events may reflect a reversibly inducible intermediate structure. The conventional notion is that fusion is rigorously shown if and only if evidence of both membrane continuity and communication between compartments can be obtained. In the case of electrofusion of erythrocyte ghosts, this is apparent through the use of FITC-dextran and DiI as indicators of contents mixing and membrane mixing, respectively. It follows that fusion should not be leaky; otherwise, it might interfere with the measurement of fusion events when FITC-dextran is the label. The conventional expectation is that each fusion event requires both one contents mixing event and one membrane mixing event to occur. However, measurements of CF-protocol fusion events under a variety of conditions showed significantly more occurrences of fluorescence movement into unlabeled membranes when FITC-dextran was used than when DiI was used as a label (Sowers, 1988).

The interpretation of this apparent discrepancy was aided by three additional observations. First, a large difference in pulse-induced permeabilization between the cathode-facing and the anode-facing hemispheres was reported from three laboratories (Mehrle *et al.*, 1985; Rossignol *et al.*, 1983; Sowers and Lieber, 1986). Second, we observed that the fusion-inducing pulse caused the movement of FITC-dextran from labeled membranes to adjacent unlabeled membranes held in pearl chains to occur predominantly (70–95% of all events) toward the negative electrode (Fig. 7). This movement rarely occurred toward the positive direction or in both directions. These movements were usually complete within 20–100 msec. Lastly, we have observed that the application of electric pulses induced a shape change in the erythrocyte ghosts regardless of whether they were in pearl chains or in random positions in the suspension (Fig. 8).

Nearly all erythrocyte ghosts in sodium phosphate buffer (pH 8.5) will be perfect spheres at or below a buffer concentration of about 15 mM. From 15 to about 25–30 mM, the geometry of the population will convert to an irregular collapsed sphere or discoidlike shape. Discoids and cup shapes predominate above about 30 mM. When pulses are applied one at a time to suspensions of ghosts which show the irregular shape, a "wave" propagates from the positive-facing hemisphere toward the negative-facing hemisphere in a fraction of a second (Sowers, 1984). Additional pulses caused a stepwise transformation of the shape until the perfect sphere geometry was attained (Fig. 8). If ghosts start with the perfect sphere geometry, the pulses have no effect on shape (except see below).

The amount of shape change per pulse was found to be greatest when pulse field strength and pulse decay halftime coincided with that needed for fusion. Longer decay halftime pulses above a certain threshold caused fragmentation of the membrane but pulses with much shorter decay halftime or much lower voltage caused neither fusion nor shape change (Fig. 2). We also noticed that at low ionic strength (20 mM), application of pulses with field strengths and decay halftimes just under the threshold for fragmentation caused spherical ghosts to convert to stomatocytes with the concave surface facing the negative electrode. If ghosts were not spherical, then pulse application caused them to end in stomatocytes instead of spheres. Conversely, once the ghosts were in the stomatocyte shape, application of a pulse with a shorter decay halftime caused the shape to return to that of a spherocyte (Fig. 9). By switching back and forth between long and short decay halftime on alternate pulses, it was possible to change shape alternately and reversibly at least several times before the reversibility was lost.

The above three observations were puzzling until we asked two fundamental questions about fusion events reported by FITC-dextran movement from labeled to adjacent unlabeled ghosts. First,

Figure 7. Example of contents mixing without attachment (i.e., no membrane fusion). Events are indicated by MW = 10 kD FITC-dextran moving from labeled membrane (arrow) to an adjacent unlabeled (not visible until labeled) membrane when pulse is applied to membranes held in pearl chains by dielectrophoresis and then AC is turned off immediately after the pulse. Buffer is sodium phosphate (20 mM, pH 8.5) and pulse strength is 600 V/mm and 0.6 msec decay halftime. Alphanumerics are: upper—month, day, year; lower—min, sec, hundredths of a sec. Single pulse is applied at 17 : 17 : 61. Note that distinct fluorescence equilibrium is attained at 17 : 17 : 70 or about 90 msec after the pulse. Fluorescence movement is detectable at 17 : 17 : 63, or about 30 msec after the pulse. Separation, initially detected at 17 : 17 : 78, is distinct at 17 : 19 : 33 or 1.7 sec after the pulse. Electric field is parallel to the plane of the micrograph and oriented left–right with positive electrode at right and negative electrode at left. Photo width: 50 μm.

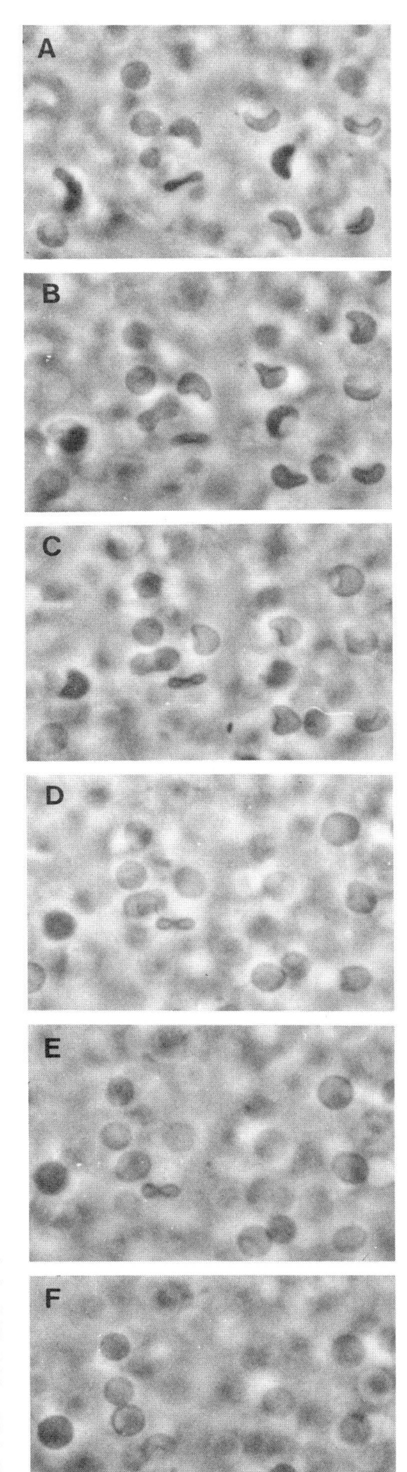

Figure 8. Pulse-induced shape change of discoid and stomatocytic human erythrocyte ghosts to the spherical shape after application of pulses. Ghosts are in 20 mM sodium phosphate buffer (pH 8.5) and were prepared as previously described (Sowers, 1984). (A) Start of sequence; no pulses applied; (B–F) two pulses (600 V/mm, 0.6 msec decay halftime) are applied after each frame. Field is parallel to page and oriented left–right. Note that concave surface of pulse-induced stomatocytes in C and D face toward the left (the negative electrode). Phase optics micrographs made with a xenon microflash system. Photo width: 82 μm.

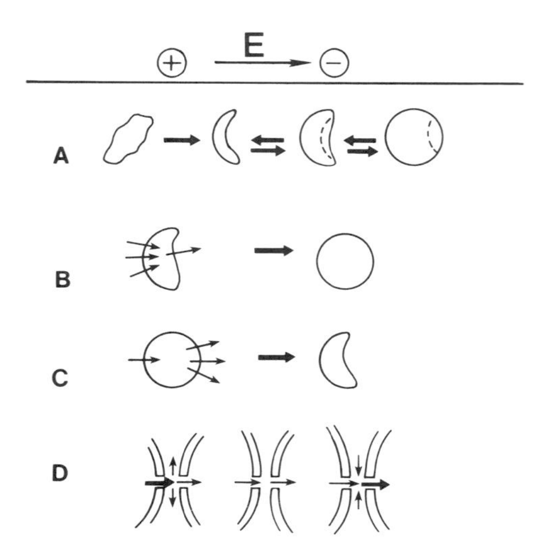

Figure 9. Interpretation of pulse-induced erythrocyte ghost shape change with respect to participation of electroosmosis-induced flow (small arrows) (see Fig. 10). Field direction for all is as indicated. (A) Shape change as shown in Fig. 8. Pulse application in direction shown causes irregular shapes (left) to become spherical (right) by passing through a cup-shaped intermediate (middle two steps). Process is reversible for a finite number of cycles (see text). (B) The pulse-induced volume increase at low ionic strengths with short decay halftime pulses can take place if a greater influx occurs in the positive-facing hemisphere than outflux at the negative-facing hemisphere. (C) The pulse-induced decrease in volume (as induced at low ionic strength with longer decay halftime pulses) as would occur with a greater outflux at the negative-facing hemisphere. (D) Effect of uneven electroosmosis flow through concentric pore pairs: Left—net inflow to intermembrane space may push membranes apart. Middle—no effect. Right—net removal of medium from intermembrane space may bring membranes closer together thus favoring fusion.

we wondered if the occurrences of pulse-induced FITC-dextran label movement into unlabeled cytoplasmic compartments in a pearl chain actually represented fusion events. This was tested by simply applying pulses to membranes in pearl chains and then turning off the AC immediately after the pulse. The result was that linear and continuous groups of two or three (occasionally more) labeled ghosts sometimes separated into individual labeled ghosts and diffused apart while other linear groups remained connected to one another in linear groups. The only change was that the axes of the linear groups became disoriented with time by Brownian motion.

In phase optics, ghosts in pearl chains which were treated with pulses and then released from pearl chains by turning off the AC afterwards showed a fraction of the ghosts returning to random positions and another fraction which stayed in linear groups. The axes of these groups also eventually became disoriented by Brownian motion (Sowers, 1984). However, in fluorescence optics, at least 99% of the groups that contained one DiI-labeled ghost also showed DiI laterally diffusing into unlabeled membranes. Taking the laterally diffusing DiI as evidence that fusion had taken place leads to the conclusion that every point of pulse-induced attachment between membranes actually represents fusion of those membranes.

When combined with the observation of contents mixing events that are followed by either separation (Fig. 7) or nonseparation after the AC was turned off, then all of the separations must represent unfused membranes! Fused membranes are, by definition, attached by membrane continuity which will permit the lateral diffusion of the lipid-soluble label, DiI. Similarly, such attachments would be detected when the AC was turned off after a pulse. Thus, the contents mixing event reported by FITC-dextran movement must be interpreted to represent the induction by the pulse of a transient high-conductance (permeability) channel which is a fusion precursor or intermediate stage. Since the induction of this channel is transient and is not accompanied by significant loss of fluorescence to the external medium, it must have a short lifetime (on the order of 20–50 msec). The induction of this channel must also be reversible since it does not always lead to fusion but neither does it lead to subsequent loss of FITC-dextran to the medium.

The second question was prompted by the observation of predominantly movement of FITC-dextran toward the negative electrode (Fig. 7). The clue to the answer to this question was found when ghost membranes were aligned into pearl chains with AC and the AC was turned off for a few seconds just before applying a pulse until the membranes separated from each other by Brownian motion. After many tries, alternative viewing by phase and fluorescence optics eventually allowed both an FITC-dextran labeled and an unlabeled ghost to drift apart (1–3 μm) on the axis of the electric field pulse. Focus and immediate application of the pulse allowed the movement of FITC-dextran fluorescence to be followed out of the negative-facing hemisphere of the labeled ghost and form a transient cloud outside but between both ghosts. Immediately after the cloud diffused into the background, an enhanced level of fluorescence was seen in the ghost which was not labeled with FITC-dextran. Low-light-level video microscopy and faint DiI labeling of ghosts not labeled with FITC-dextran were used to capture these events on magnetic tape for subsequent analysis (Fig. 10).

This observation indicated that the soluble marker must enter the cytoplasmic compartment of the originally unlabeled ghost through a transient permeabilization in the positive-facing hemisphere. Entry through the permeabilization which we know occurs in the negative hemisphere (Sowers and Lieber, 1986) is unlikely since the concentration of FITC-dextran is high in the transient cloud of fluorescence between the two ghosts but very low in the vicinity of the negative-facing hemisphere. This observation then suggests that significant permeabilization does occur in both hemispheres. Whether both locations are identically permeabilized cannot, however, be answered at this time. All reported observations (Mehrle *et al.*, 1985; Rossignol *et al.*, 1983; Sowers and Lieber, 1986) are consistent with each other if electroosmosis (Balasubraminian and McLaughlin, 1982) is assumed to occur in each electropore (Fig. 11). For the reports from the other laboratories, the molecule or ion which was used to trace the permeabilization actually started from a location *outside* the cytoplasmic compartment and moved *inward* with the pulse treatment. In contrast, in our study (Sowers and Lieber, 1986) the FITC-dextran label started from *inside* the cytoplasmic compartment and moved *outward* with the pulse treatment.

The reversible pulse-induced shape change therefore must be a manifestation of an electroosmosis-based aqueous flow *into* the cytoplasmic compartment at the positive-facing hemisphere and *out* of the cytoplasmic compartment at the negative-facing hemisphere. Change in shape from a collapsed sphere to a perfect sphere must occur by virtue of a volume increase. This can only occur with a net influx of aqueous medium which would mean a greater influx than outflux. When outflux is greater, stomatocytes are formed and this is consistent with the fact that fusion does not occur.

Relationship between experimental observations and models. Our interpretation (see above) that the pulse-induced FITC-dextran movements represent the reversible induction of an intermediate structure favors the idea that fusion involves a multistep and reversible process. This is not as compatible with the Pilwat *et al.* (1981) mechanism, which appears to be a continuous and irreversible process, as with the Dimitrov and Jain (1984) model. The Dimitrov and Jain model has two distinct substeps in the interval between unfused and fused membranes: (1) concentric pore pairs and (2) concentric pore pairs with flares. Our FITC-dextran movements cannot distinguish between these substeps but do appear to support the notion that there is some relationship between electroporation and fusion since the number of fusion events is always lower than the number of contents mixing events. However, the Stenger and Hui (1986) model physically separates the fusion-initiating structure (point defects) from the membrane structure (the "flaps") which would be responsible for contents mixing. Their interpretation does allow for concentric pore pairs, but they have a flaplike structure. Their study did not include experiments to determine the effect of a change in ionic strength although our experimental results would predict that an increase in ionic strength may lead to a decrease in the number, size, or lifetime in their population of flaps and a possible increase in the number of point defects (assuming a one-to-one relationship between point defects and fusion events). Additional resolution in time between pulse and quick freezing might allow this problem to be resolved.

Ultrastructural evidence for different forces in two hemispheres. Occasional tethers and electropores were directly observed in mitochondrial inner membranes in experiments that employed a relatively low-strength (70–100 V/mm) but long-duration (1–3 sec) DC square wave pulse to induce a lateral reorganization of membrane components (Sowers and Hackenbrock, 1981). They were always located in the part of the membrane which was facing the negative electrode (Fig. 12). This may be an important additional clue to further elucidating the forces on the membrane caused by the pulse. It also suggests, along with the observation of the induction of different surface geometries (convex and concave) in the two hemispheres (Fig. 9) during pulse-induced shape

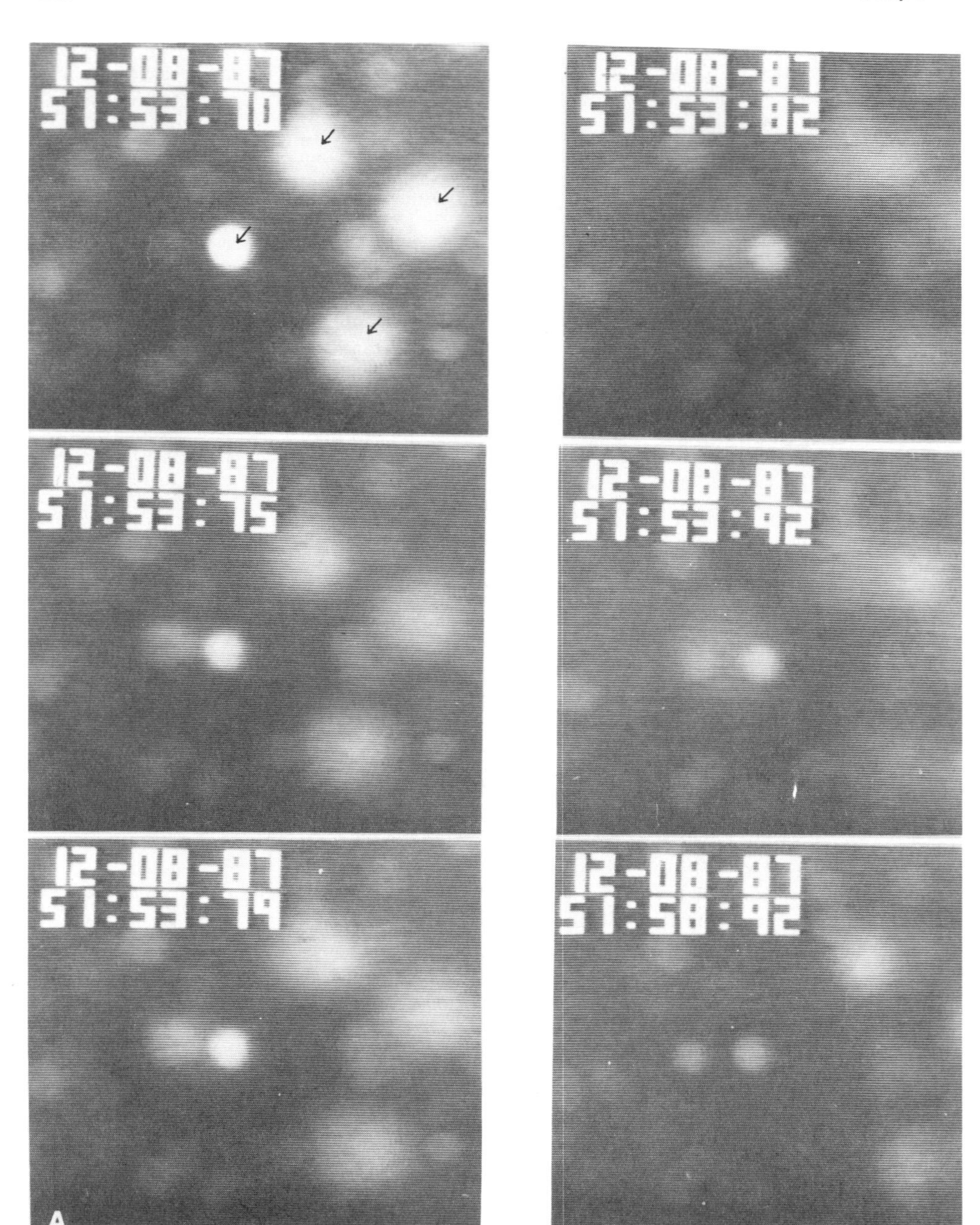
12-08-87
51:53:70
12-08-87
51:53:75
12-08-87
51:53:79
12-08-87
51:53:82
12-08-87
51:53:92
12-08-87
51:58:92
A

B

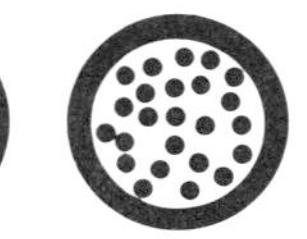

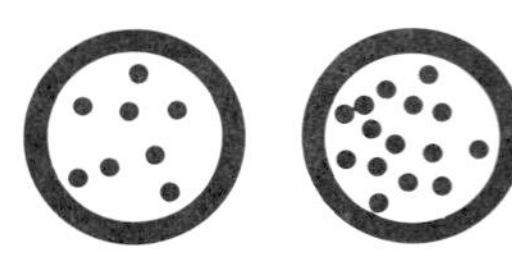

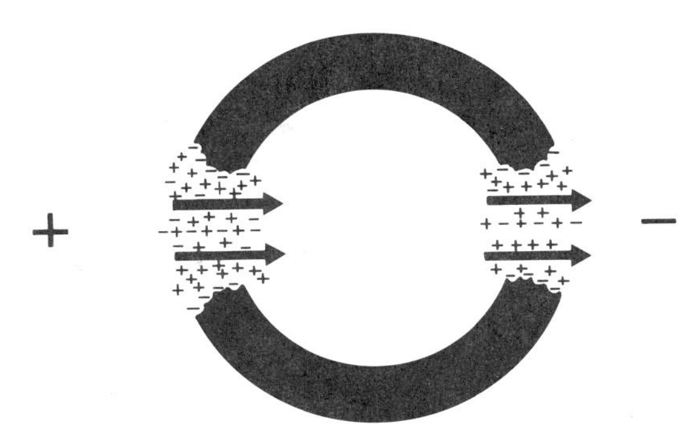

Figure 11. Effect of electroosmosis in vicinity of conventional pores at center of both hemispheres of a membrane in an electric field. Field direction is left to right and in plane of page. Excess of protons and cation counterions over anions is induced near the negative-charged surface of the membrane. An electric field parallel to this surface would induce an electrophoretic flow toward the negative electrode because of the excess number of mobile positive charges. Hence, net flow would be *into* sphere at anode-facing hemisphere and *outward* at cathode-facing hemisphere.

change, that membrane processes and events in the two hemispheres may be qualitatively different in subtle ways.

This may also be relevant to speculation that low-voltage long pulses, compared to high-voltage short pulses, will also permeabilize membranes (Sukharev *et al.*, 1985; Chu *et al.*, 1987) and also fuse membranes (Fig. 2). Asymmetric effects have also been observed to be related to current polarity in gel-collapse phenomena (Tanaka *et al.*, 1982) and may have elements in common with electrofusion.

5. CONCLUSIONS

Our characterization of the electrofusion mechanism has led to a greater understanding of the interplay of forces and phenomena which accompany electrofusion. The long-lived fusogenic state induced by the PF protocol is different in many ways from that induced by the CF protocol. This suggests that very different mechanisms may be involved. The long-lived fusogenic state is a laterally immobile membrane property, and reflects one or more locally induced domains which may be linked to membrane proteins and the cytoskeletal system.

Fusion induced by the CF protocol can now be interpreted to involve an experimentally detectable intermediate state which has a short lifetime. Electroosmosis may accompany the electrofusion process but whether it directly or indirectly participates in fusion is not clear. While its induction is reversible, the movement of intermediate states forward to fusion products rather than backward to unfused membranes is promoted by higher ionic strength. On the other hand, it is possible that the movement of FITC-dextran through electropores by electroosmosis is contributing to the excess number of contents mixing events. This possibility tends to both refute the hypothesis of a highly populated intermediate state and reveal that measurement of contents mixing events will

Figure 10. Experiment to demonstrate the induction of electropores in the positive-facing hemispheres of erythrocyte ghosts. (A) All conditions including micrograph alphanumerics are as in Fig. 7, except that the ghost membranes are either brightly labeled with FITC-dextran (arrows pointing to four bright spheres) or faintly labeled with DiI (all others). The DiI-labeled ghosts are present to permit their simultaneous observation with the FITC-dextran-labeled ghosts. Their labeling intensity is near the lower sensitivity limit of the low light level camera in order to maximize the detectability of any fluorescence increase caused by trapping of FITC-dextran in the interior of the ghost following pulse-induced release from the nearby sharply focused ghost image. Pulse applied at 51 : 53 : 75. Note: (1) no change in fluorescence of any DiI-labeled ghost except the one near the center, (2) large loss of fluorescence from all ghosts originally labeled with FITC-dextran, (3) pulse-induced transient clouds of fluorescence just outside and immediately to left of all ghosts originally labeled with FITC-dextran, and (4) slight increase in background fluorescence between all ghosts compared to initial micrograph. (B) Illustration of events which resulted in A. Left—Prepulse position of ghost not containing FITC-dextran and ghost containing FITC-dextran. Middle—Pulse causes transient induction of electropores in both hemispheres but aqueous flow caused motion of label only to left (see Figs. 7 and 11). Right—After electropores reseal, some fluorescence is trapped inside ghost that was not labeled with the FITC-dextran. From Sowers (1988).

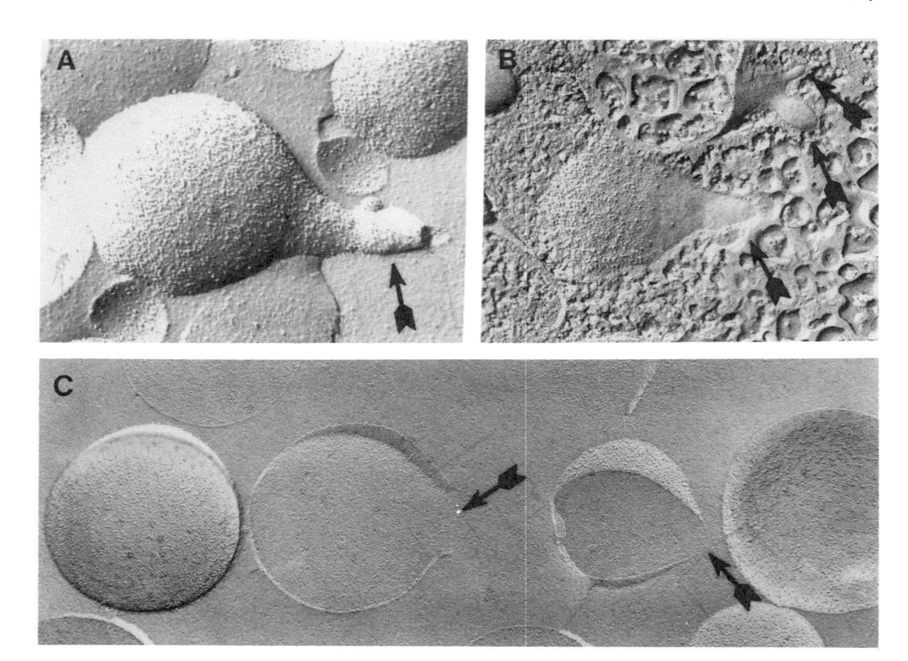

Figure 12. Asymmetrical pulse electric field effects on spherical mitochondrial inner membranes (Sowers and Hackenbrock, 1981). (A, B) Numerous membranes showed tethers (arrows) pointing toward the negative electrode. Photo widths: A, 2.2 μm; B, 2.8 μm. (C) Cross fractures through two membranes show large electropores in only one hemisphere facing the negative electrode. Photo width: 4.1 μm.

require additional fundamental study and methodological care. Regardless of whether the PF or CF protocol is used, the data do not appear to support a simple relationship between electropore induction and membrane fusion. The data suggest a more complex process and that fusion involves a multistep process, including a distinct intermediate stage, which may have some features in common with fusion in natural systems. This is especially relevant since naturally occurring fusion in at least some membrane systems also appears to be reversible and to have a time scale (Breckenridge and Almers, 1987; Fernandez *et al.*, 1984; Zimmerberg *et al.*, 1987) which is compatible with what we found for electrofusion in erythrocyte ghosts.

NOTE ADDED IN PROOF

Good evidence has now been published (Chernomordik *et al.*, 1987) that electropore resealing is much slower in both cell membranes and lipid bilayers. This is additional evidence in favor of electroosmosis as the basis of movement of media through electropores because electroosmosis can move media faster than diffusion. Another report shows a correlation between electroporation and electrofusion (Zhelev *et al.*, in press).

ACKNOWLEDGMENTS. The expert technical assistance of Ms. Veena Kapoor is greatly appreciated. Supported by ONR Contract N00014-87-K-0199.

REFERENCES

Baker, P. F., and Knight, D. E., 1983, High voltage techniques for gaining access to the interior of cells: Application to the study of exocytosis and membrane turnover, *Methods Enzymol.* **98:**28–37.

Balasubraminian, A., and McLaughlin, S., 1982, Electroosmosis at the surface of phospholipid bilayer membranes, *Biochim. Biophys. Acta* **685:**1–5.

Bates, G., Nea, L. J., and Hasenkampf, C. A., 1987a, Electrofusion and plant somatic hybridization, in: *Cell Fusion* (A. E. Sowers, ed.), Plenum Press, New York, pp. 479–496.

Bates, G., Saunders, J., and Sowers, A. E., 1987b, Electrofusion: Principles and applications, in: *Cell Fusion* (A. E. Sowers, ed.), Plenum Press, New York, pp. 367–395.

Beers, R. F., and Bassett, E. G., 1984, *Cell Fusion: Gene Transfer and Transformation,* Raven Press, New York.

Benz, R., and Zimmermann, U., 1981, The resealing process of lipid bilayers after reversible electrical breakdown, *Biochim. Biophys. Acta* **640:**169–178.

Berg, H., 1982, Molecular biological implications of electric-field effects, *Stud. Biophys.* **90:**169–176.

Berg, H., 1987, Electrotransfection and electrofusion of cells and electrostimulation of their metabolism, *Stud. Biophys.* **119:**17–29.

Berg, H., Bauer, E., Berg, D., Forster, W., Hamann, M., Jacob, H.-E., Kurischko, A., Muhlig, P., and Weber, H., 1983, Cell fusion by electric fields, *Stud. Biophys.* **94:**93–96.

Blangero, C., and Teissié, J., 1985, Ionic modulation of electrically-induced fusion of mammalian cells, *J. Membr. Biol.* **86:**247–253.

Blumenthal, R., 1987, Membrane fusion, *Curr. Top. Membr. Transp.* **29:**203–254.

Breckenridge, L. J., and Almers, W., 1987. Final steps in exocytosis observed in a cell with giant secretory granules, *Proc. Natl. Acad. Sci. USA* **84:**1945–1949.

Chernomordik, L. V., Sukharev, I. G., and Abidor, I. G., 1985, Long-living defects in BLM after reversible electrical breakdown, *Biol. Membr.* **2:**87–94.

Chu, G., Hayakawa, H., and Berg, P., 1987, Electroporation for the efficient transfection of mammalian cells with DNA, *Nucleic Acids Res.* **15:**1311–1326.

Conrad, M. K., Lo, M. M. S., Tsong, T. Y., and Snyder, S. H., 1987, Bioselective cell–cell fusion for antibody production, in: *Cell Fusion* (A. E. Sowers, ed.), Plenum Press, New York, pp. 427–439.

Dawes, C. L., 1942, *Industrial Electricity,* Part II, 2nd ed., McGraw–Hill, New York.

Diacumakos, E. G., and Tatum, E. L., 1972, Fusion of mammalian somatic cells by microsurgery, *Proc. Natl. Acad. Sci. USA* **69:**2959–2962.

Dimitrov, D. S., and Jain, R. K., 1984, Membrane stability, *Biochim. Biophys. Acta* **779:**437–468.

Dodge, J. T., Mitchell, C., and Hanahan, D. J., 1963, The preparation and chemical characterization of hemoglobin-free ghosts of human erythrocytes, *Arch. Biochem. Biophys.* **100:**119–130.

Duzgunes, N., Hong, K., Baldwin, P. A., Bentz, J., Nir, S., and Papahadjopoulos, D., 1987, Fusion of phospholipid vesicles induced by divalent cations and protons: Modulation by phase transitions, free fatty acids, monovalent cations, and polyamines, in: *Cell Fusion* (A. E. Sowers, ed.), Plenum Press, New York, pp. 241–267.

Ehrenberg, B., Farkas, D. L., Fluhler, E. N., Lojewska, Z., and Loew, L. M., 1987, Membrane potential induced by external electric field pulses can be followed with a potentiometric dye, *Biophys. J.* **51:**833–837.

Evered, D., and Whelan, J., 1984, *Cell Fusion,* Pitman, London.

Fernandez, J. M., Neher, E., and Gomperts, B. D., 1984, Capacitance measurements reveal stepwise fusion events in degranulating mast cells, *Nature* **312:**453–455.

Fricke, H., 1953, The electric permittivity of a dilute suspension of membrane covered ellipsoids, *J. Appl. Phys.* **24:**644–646.

Fromm, M., Taylor, L. P., and Walbot, V., 1985, Expression of genes transferred into monocot and dicot plant cells by electroporation, *Proc. Natl. Acad. Sci. USA* **82:**5824–5828.

Goodby, J. W., 1986, Optical activity and ferroelectricity in liquid crystal, *Science* **231:**350–355.

Gross, D., Loew, L. M., and Webb, W., 1986, Optical imaging of cell membrane potential changes by applied electric fields, *Biophys. J.* **50:**339–348.

Harbich, W., and Helfrich, W., 1979, Alignment and opening of giant lecithin vesicles by electric fields, *Z. Naturforsch.* **34a:**1063–1065.

Heckl, W., Losche, M., Cadenhead, D. A., and Mohwald, H., 1986, Electrostatically induced growth of spiral lipid domains in the presence of cholesterol, *Eur. Biophys. J.* **14:**11–17.

Hofmann, G. A., and Evans, G. A., 1986, Electronic genetics—Physical and biological aspects of cellular electromanipulation, *IEEE Eng. Med. Biol.* **5:**6–25.

Knight, D. E., and Baker, P. F., 1982, Calcium-dependence of catecholamine release from bovine adrenal medullary cells after exposure to intense electric fields, *J. Membr. Biol.* **68:**107–140.

Knight, D. E., and Scrutton, M. C., 1986, Gaining access to the cytosol: The technique and some applications of electropermeabilization, *Biochem. J.* **234**:497–506.

Lieber, M. R., and Steck, T. L., 1982a, A description of the holes in human erythrocyte membrane ghosts, *J. Biol. Chem.* **257**:11651–11659.

Lieber, M. R., and Steck, T. L., 1982b, Dynamics of the holes in human erythrocyte membrane ghosts, *J. Biol. Chem.* **257**:11660–11666.

Lo, M. M. S., Tsong, T. Y., Conrad, M. K., Strittmatter, S. M., Hester, L. D., and Snyder, S. H., 1984, Monoclonal antibody production by receptor-mediated electrically-induced cell fusion, *Nature* **310**:792–794.

Lovinger, A. J., 1983, Ferroelectric polymers, *Science* **220**:1115–1121.

Malmstadt, H. V., Enke, C. G., and Toren, E. C., 1963, *Electronics for Scientists,* Benjamin, New York.

Malmstadt, H. V., Enke, C. G., and Crouch, S. R., 1981, *Electronics and Instrumentation for Scientists,* Benjamin/Cummings, Menlo Park.

Mehrle, W., Zimmermann, U., and Hampp, R., 1985, Evidence for asymmetrical uptake of fluorescent dyes through electropermeabilized membranes of *Avena* mesophyll protoplasts, *FEBS Lett.* **185**:89–94.

Miles, D. M., and Hochmuth, R. M., 1987, Micromanipulation and elastic response of electrically fused red cells, in: *Cell Fusion* (A. Sowers, ed.), Plenum Press, New York, pp. 441–456.

Mischke, S., Saunders, J. A., and Owens, L., 1986, A versatile low-cost apparatus for cell electrofusion and other physiological treatments, *J. Biochem. Biophys. Methods* **13**:65–75.

Moore, J. H., Davis, C. C., and Coplan, M. A., 1983, *Building Scientific Apparatus,* Addison–Wesley, London.

Morris, S. J., Bradley, D., Gibson, C. C., Smith, P. D., and Blumenthal, R., 1988, Use of membrane-associated fluorescence probes to monitor fusion of bilayer vesicles: Application to rapid kinetics using pyrene excimer/monomer fluorescence, in: *Spectroscopic Membrane Probes* (L. M. Loew, ed.), CRC Press, Boca Ratan, Fla., pp. 161–191.

Neumann, E., 1984, Electric gene transfer into culture cells, *Bioelectrochem. Bioenerg.* **13**:219–223.

Neumann, E., and Katchalsky, A., 1972, Long-lived conformational changes induced by electric impulses in biopolymers, *Proc. Natl. Acad. Sci. USA* **69**:993–997.

Neumann, E., and Rosenheck, K., 1972, Permeability changes induced by electric impulses in vesicular membranes, *J. Membr. Biol.* **10**:279–290.

Neumann, E., Schaefer-Ridder, M., Wang, Y., and Hofschneider, P. H., 1982, Gene transfer into mouse lyoma cells by electroporation in high electric fields, *EMBO J.* **1**:841–845.

Ohki, S., 1987, Physicochemical factors underlying lipid membrane fusion, in: *Cell Fusion* (A. E. Sowers, ed.), Plenum Press, New York, pp. 331–352.

Ohki, S., Doyle, D., Flanagan, T. D., Hui, S.-W., and Mayhew, E., 1988, *Membrane Fusion,* Plenum Press, New York.

Pastushenko, V. F., 1983, Pore heating during electrical breakdown of bilayer lipid membranes, *Biol. Membr.* **1**:176–181.

Pilwat, G., Richter, H.-P., and Zimmermann, U., 1981, Giant culture cells by electric field-induced fusion, *FEBS Lett.* **133**:169–174.

Plattner, H., 1987, Synchronous exocytosis in paramecium cells, in: *Cell Fusion* (A. E. Sowers, ed.), Plenum Press, New York, pp. 69–98.

Pohl, H. A., 1978, *Dielectrophoresis,* Cambridge University Press, London.

Pohl, H. A., Pollock, K., and Rivera, H., 1984, The electrofusion of cells, *Int. J. Quant. Chem. Quant. Biol. Symp.* **11**:327–345.

Poste, G., and Nicolson, G. L., 1978, *Membrane Fusion,* Elsevier/North-Holland, Amsterdam.

Porschke, D., 1985, Effects of electric fields on biopolymers, *Annu. Rev. Phys. Chem.* **36**:159–178.

Potter, H., Weir, L., and Leder, P., 1984, Enhancer-dependent expression of human κ immunoglobulin genes introduced into mouse pre-B lymphocytes by electroporation, *Proc. Natl. Acad. Sci. USA* **81**:7161–7165.

Rand, R. P., and Parsegian, V. A., 1986, Mimicry and mechanism in phospholipid models of membrane fusion, *Annu. Rev. Physiol.* **48**:201–212.

Ringertz, N. R., and Savage, R. E., 1976, *Cell Hybrids,* Academic Press, New York.

Rosenberg, B., Van Camp, L., and Krigas, T., 1965, Inhibition of cell division in *Escherichia coli* by electrolysis products from a platinum electrode, *Nature* **205**:698–699.

Rossignol, D. P., Decker, G. L., Lennarz, W. J., Tsong, T. Y., and Teissié, J., 1983, Induction of calcium-dependent, localized cortical granule breakdown in sea-urchin eggs by voltage pulsation, *Biochim. Biophys. Acta* **763**:346–355.

Sauer, F. A., 1983, Forces on suspended particles in the electromagnetic field, in: *Coherent Excitations in Biological Systems* (H. Frohlich and F. Kremer, eds.), Springer-Verlag, Berlin, pp. 134–144.

Schierenberg, E., 1987, Laser-induced cell fusion, in: *Cell Fusion* (A. E. Sowers, ed.), Plenum Press, New York, pp. 409–418.

Schlegel, R., 1987, Probing the function of viral proteins with synthetic peptides, in: *Cell Fusion* (A. E. Sowers, ed.), Plenum Press, New York, pp. 33–43.

Schwister, K., and Deuticke, B., 1985, Formation and properties of aqueous leaks induced in human erythrocytes by electrical breakdown, *Biochim. Biophys. Acta* **816:**332–348.

Senda, M., Takeda, J., Abe, S., and Nakamura, T., 1979, Induction of cell fusion of plant protoplasts by electrical stimulation, *Plant Cell Physiol.* **20:**1441–1443.

Serpersu, E. H., Kinosita, K., Jr., and Tsong, T. Y., 1985, Reversible and irreversible modification of erythrocyte membrane permeability by electric field, *Biochim. Biophys. Acta* **812:**779–785.

Siegel, D. P., 1987, Membrane–membrane interactions via intermediates in lamellar-to-inverted hexagonal phase transitions, in: *Cell Fusion* (A. E. Sowers, ed.), Plenum Press, New York, pp. 181–203.

Sowers, A. E., 1983a, Fusion of mitochondrial inner membranes by electric fields produces inside out vesicles: Visualization by freeze-fracture electron microscopy, *Biochim. Biophys. Acta* **735:**426–428.

Sowers, A. E., 1983b, Red cell and red cell ghost membrane shape changes accompanying the application of electric fields for induction fusion, *J. Cell Biol.* **97:**179a.

Sowers, A. E., 1984, Characterization of electric field-induced fusion in erythrocyte ghost membranes, *J. Cell Biol.* **99:**1989–1996.

Sowers, A. E., 1985a, Movement of a fluorescent lipid label from a labeled erythrocyte membrane to an unlabeled erythrocyte membrane following electric field-induced fusion, *Biophys. J.* **47:**519–525.

Sowers, A. E., 1985b, Electric field-induced membrane fusion in erythrocyte ghosts: Evidence that pulses induce a long-lived fusogenic state and that fusion may not involve pore formation, *Biophys. J.* **47:**171a.

Sowers, A. E., 1986a, A long-lived fusogenic state is induced in erythrocyte ghosts by electric pulses, *J. Cell Biol.* **102:**1358–1362.

Sowers, A. E., 1986b, Long-lived fusogenic membrane sites induced by electric field pulses are not free to diffuse laterally in the plane of the membrane, *Biophys. J.* **49:**132a.

Sowers, A. E., (ed.), 1987a, *Cell Fusion,* Plenum Press, New York.

Sowers, A. E., 1987b, The long-lived fusogenic state induced in erythrocyte ghosts by electric pulses is not laterally mobile, *Biophys. J.* **52:**1015–1020.

Sowers, A. E., 1988, Fusion events and nonfusion contents mixing events induced in erythrocyte ghosts by an electrical pulse, *Biophys. J.* **54:**619–626.

Sowers, A. E., and Hackenbrock, C. R., 1981, Rates of lateral diffusion of intramembrane particles: Measurements by electrophoretic displacement and rerandomization, *Proc. Natl. Acad. Sci. USA* **78:**6246–6250.

Sowers, A. E., and Kapoor, V., 1987a, The electrofusion mechanism in erythrocyte ghosts, in: *Cell Fusion* (A. E. Sowers, ed.), Plenum Press, New York, pp. 397–408.

Sowers, A. E., and Kapoor, V., 1987b, Fusogenic membrane alterations induced by electric field pulses, in: *Mechanistic Approaches to Interactions of Electromagnetic Fields with Living Systems* (M. Blank and E. Findl, eds.), Plenum Press, New York, pp. 325–337.

Sowers, A. E., and Kapoor, V., 1988, The mechanism of erythrocyte ghost fusion by electric field pulses, in: *Proceedings of the International Symposium on Molecular Mechanisms of Membrane Fusion* (S. Ohki, ed.), Plenum Press, New York, pp. 237–254.

Sowers, A. E., and Lieber, M. L., 1986, Electropores in individual erythrocyte ghosts: Diameters, lifetimes, numbers, and locations, *FEBS Lett.* **205:**179–184.

Stenger, D. A., and Hui, S. W., 1986, Kinetics of ultrastructural changes during electrically induced fusion of human erythrocytes, *J. Membr. Biol.* **93:**43–53.

Stulen, G., 1981, Electric field effects on lipid membrane structure, *Biochim. Biophys. Acta* **640:**621–627.

Sugar, I. P., Förster, W., and Neumann, E., 1987, Model of cell electrofusion: Membrane electroporation, pore coalescence and percolation, *Biophys. Chem.* **26:**321–335.

Sukharev, S. I., Popov, S. V., Chernomordik, L. V., and Abidor, I. G., 1985, A patch-clamp study of electrical breakdown of cell membranes, *Biol. Membr.* **2:**77–86.

Szoka, F. C., 1987, Lipid vesicles: Model systems to study membrane–membrane destabilization and fusion, in: *Cell Fusion* (A. E. Sowers, ed.), Plenum Press, New York, pp. 209–240.

Tanaka, T., Nishio, I., Sun, S.-T., and Ueno-Nishio, S., 1982, Collapse of gels in an electric field, *Science* **218:**467–469.

Teissié, J., and Blangero, C. 1984, Direct experimental evidence of the vectorial character of the interaction between electric pulses and cells in cell electrofusion, *Biochim. Biophys. Acta* **775:**446–448.
Teissié, J., and Rols, M. P., 1986, Fusion of mammalian cells in culture is obtained by creating the contact between cells after their electropermeabilization, *Biochem. Biophys. Res. Commun.* **140:**258–264.
Teissié, J., and Tsong, T. Y., 1981, Electric field induced transient pores in phospholipid bilayer vesicles, *Biochemistry* **20:**1548–1554.
Teissié, J., Knutson, V. P., Tsong, T. Y., and Lane, M. D., 1982, Electric pulse-induced fusion in 3T3 cells in monolayer culture, *Science* **216:**537–538.
Tsong, T. Y., 1983, Voltage modulation of membrane permeability and energy utilization in cells, *Biosci. Rep.* **3:**487–505.
Wilschut, J., and Hoekstra, D., 1988, *Cellular Membrane Fusion: Fundamental Mechanisms and Applications of Membrane Fusion Techniques,* Decker, New York.
Wojcieszyn, J. W., Schlegel, R. A., Lumley-Sapanski, K., and Jacobson, K. A., 1983, Studies on the mechanism of polyethylene glycol-mediated cell fusion using fluorescent membrane and cytoplasmic probes, *J. Cell Biol.* **96:**151–159.
Zachrisson, A., and Bornman, C. H., 1984, Application of electric field fusion in plant tissue culture, *Physiol. Plant.* **61:**314–320.
Zimmerberg, J., Curran, M., Cohen, F. S., and Broderick, M., 1987, Simultaneous electrical and optical measurements show that membrane fusion preceeds secretory granule swelling during exocytosis of beige mouse mast cells, *Proc. Natl. Acad. Sci. USA* **84:**1585–1589.
Zimmermann, U., 1982, Electric field-mediated fusion and related electrical phenomena, *Biochim. Biophys. Acta* **694:**227–277.
Zimmermann, U., 1983, Electrofusion of cells: Principles and industrial potential, *Trends Biotech.* **1:**149–155.
Zimmermann, U., 1986, Electrical breakdown, electropermeabilization and electrofusion, *Rev. Physiol. Biochem. Pharmacol.* **105:**175–256.
Zimmermann, U., Buchner, K.-H., and Arnold, W. M., 1984, Electrofusion of cells: Recent developments and relevance for evolution, in: *Charge and Field Effects in Biosystems* (M. J. Allen and P. N. R. Usherwood, eds.), Abacus Press, Normal, Ill., pp. 293–318.
Zimmermann, U., Vienken, J., Halfmann, J., and Emeis, C. C., 1985, Electrofusion: A novel hybridization technique, in: *Advances in Biotechnological Processes,* Volume 4 (A. Mizrahi and A. L. van Wezel, eds.), Liss, New York, pp. 79–150.

Part IV

Applications

Chapter 16

Producing Monoclonal Antibodies by Electrofusion

Mathew M. S. Lo and Tian Yow Tsong

1. INTRODUCTION

The technology of antibody production from hybridomas pioneered by Milstein and co-workers (Kohler and Milstein, 1975; Galfre *et al.*, 1977) has provided an important research tool for the biomedical sciences. Naturally, any technique that improves the production of hybridomas would arouse widespread interest. In recent years, much attention has been given to the technique of electrofusion (Tsong, 1983; Neumann *et al.*, 1980; Zimmermann *et al.*, 1985). Electrofusion can be stringently controlled, and is typically more efficient than conventional methods of cell fusion induced by chemicals. However, increasing cell fusion efficiency alone does not appear to improve the overall efficiency of producing hybridomas. It is important to understand other mechanisms involved in the process of hybridoma production.

1.1. Humoral Immune Response and B-Cell Differentiation

The immune system typically consists of about 10^8 lymphocytes which are capable of producing up to 10^{10} different antibodies of different specificities (Golub, 1987). B-cell populations derived from bone marrow stem cells differentiate and express cell surface antigen receptors (Fig. 1). Antigen binding to its receptor stimulates B cells to proliferate. Plasma cells, formed after further cell stimulation and maturation, secrete a unique monoclonal antibody. Some of the cells derived from this subpopulation are memory cells, which are long-lived and are not destined to become terminally differentiated. Some of these memory cells undergo somatic mutations, resulting in permanent genetic changes that alter the specificity and affinity of the antigen receptor (Rudikoff *et al.*, 1984; Griffiths *et al.*, 1984; Berek *et al.*, 1985). Repeated immunizations induce further somatic mutation. Subsequent exposure of memory cells to the same antigen results in a selective stimulation of cells that produce antigen receptors which have the best fit to that antigen. This clonal selection of memory cells (Burnet, 1959) results in the production of specific antibodies with very

MATHEW M. S. LO • Molecular Biology and Genetics Unit, Neuroscience Branch, Addiction Research Center, National Institute on Drug Abuse, Baltimore, Maryland 21224. TIAN YOW TSONG • Department of Biochemistry, University of Minnesota, St. Paul, Minnesota 55108.

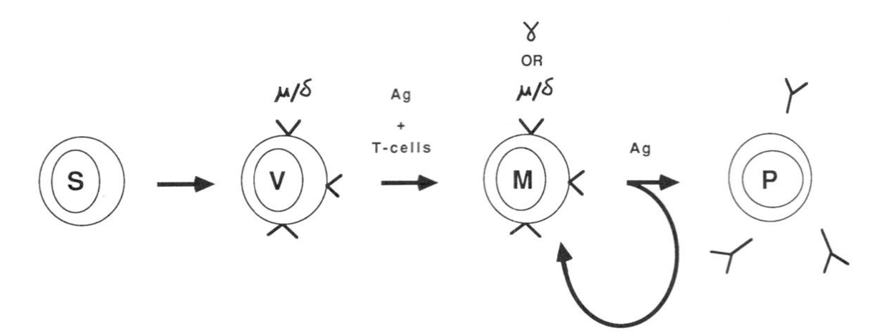

Figure 1. B-cell differentiation and maturation. Stem cells (S) originating from the bone marrow differentiate into pre-B-cell precursors (V) expressing cell surface monomeric receptor immunoglobulin of μ and/or δ heavy-chain subclass. Antigenic stimulation and T-cell factors trigger cells to divide and differentiate into antibody-producing plasma cells (P) and memory cells (M). Further antigenic stimulation of memory cells induces cell division and somatic mutations leading to formation of high-affinity receptors which are predominantly of the γ heavy-chain subclass.

high affinities. These memory cells comprise only a minute percentage of the potential immune repertoire. Fusion of these few selected B cells with myeloma cells produces desirable hybridomas secreting specific high-affinity monoclonal antibodies.

1.2. Surface Receptors and Targeting

The presence of cell surface antigen receptors on B cells provides the basis of antigen targeted or bioselective fusions in which these selected B cells are fused to myeloma cells (Bankert *et al.*, 1980; Lo *et al.*, 1984). The advantage of these targeting techniques is the preselection of B cells and the formation of only a few hybrids, almost all of which are producing the desired antibody. Previously, targeted cell adhesion alone was reported to promote the fusion of B cells without the use of any fusogen (Bankert *et al.*, 1980). In this method, myeloma cells were coated with the antigen of interest. The absence of current widespread use of this method suggests that there are insurmountable difficulties in its practical application. Other methods of targeting have utilized covalent conjugation between, especially, biotin and avidin (Green, 1975). The practically irreversible binding between avidin and biotin has been applied in numerous other systems (Godfrey *et al.*, 1981; Wormmeester *et al.*, 1984) although it had never been applied to cell fusion. We have utilized these techniques in developing a specific linkage between proliferating myeloma cells and pertinent B cells (Lo *et al.*, 1984). This is achieved by the attachment (1) of biotin to the surface of the myeloma cells, and (2) by the covalent attachment of avidin to antigen, which binds to the surface receptors on previously stimulated B cells. The high-affinity interaction between biotin and avidin creates specific adherence between avidin-antigen-bound B cells and biotinylated myeloma cells when they are mixed together, thus achieving the critical close cell–cell contact required for cell fusion (Fig. 2). In this preselected cell fusion method, antigen have to be chemically conjugated with avidin using an appropriate cross-linking reagent. In some instance, when biotinylation of the antigen is more appropriate, myeloma cells may be avidinylated to complete the cell–cell cross-link.

1.3. Electrofusion

Hybridoma production by electrofusion was first reported by Zimmermann and co-workers (Vienken and Zimmermann, 1982; Bischoff *et al.*, 1982). Spleen cells and myeloma cells were fused after treatment with pronase and dielectrophoresis (Pohl, 1978; Zimmermann *et al.*, 1985). A high yield of lymphocyte–myeloma cell pairs was reported. However, these reports lacked actual

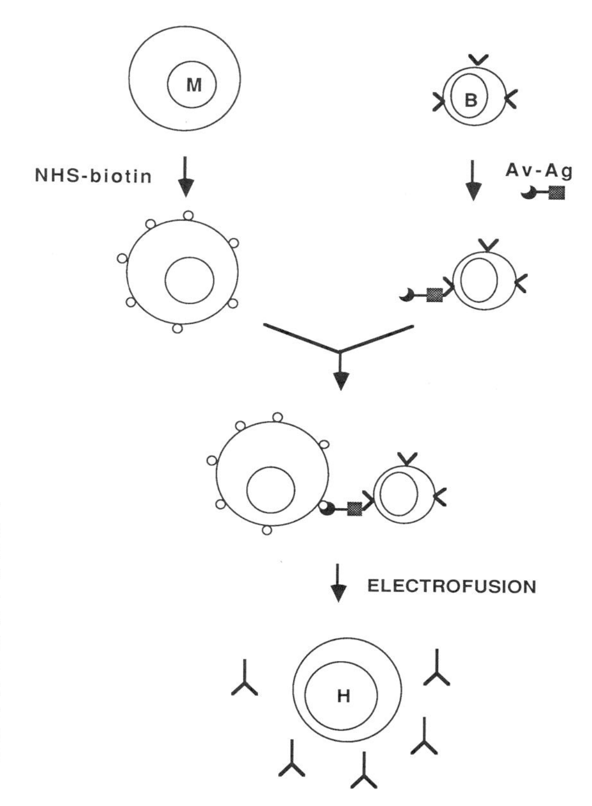

Figure 2. Hybridoma production by bioselective electrofusion. Myeloma cells (M) are biotinylated with NHS-biotin. B cells (B) prepared for immunized spleen, treated with an antigen conjugate (Av-Ag), become effectively coated with avidin. Myeloma and appropriate B cell pair are formed when the two cell types are mixed and incubated together. Cells are fused and hybridized in a high-voltage electric field. Hybrids spontaneously divide and grow in tissue culture and some also secrete a unique monoclonal antibody.

evidence that the hybridoma cells produced were able to proliferate and secrete monoclonal antibodies over any length of time. Subsequently, the same authors reported an improved electrofusion technique using dielectrophoresis alone to achieve cell contact followed by electroporation leading to membrane fusion (Vienken and Zimmermann, 1985). Evidence of stable hybridoma production was again not evident; no results pertaining to either the quality or the quantity of monoclonal antibodies secreted by these hybridomas were provided. Karsten *et al.* (1985) obtained mouse hybridomas by electrofusion and compared them with those obtained by conventional PEG techniques. Electrofusion was about ten times more efficient, producing about one hybrid per 5000 spleen cells used. However, these techniques have in common nonselective methods of promoting cell contact, creating nonspecific cell–cell fusion. Nonspecific electrofusion can potentially result in millions of growing hybrid colonies. For most laboratories, screening this number of samples is an insurmountable task, especially in the development of monoclonal antibodies to rare or hard-to-isolate antigens. Other methods of preselecting antibody-secreting lymphocytes by electrical fields have been proposed (Zimmermann *et al.*, 1985; and see other chapters); however, the practical utilization of such methods has yet to be proven.

1.4. Monoclonal Antibody Production

The method of hybridoma production in current widespread use was originally described by Milstein and co-workers (Kohler and Milstein, 1975; Galfre and Milstein, 1981). Although many modifications have been made by different laboratories, the basic protocol has not changed appreciably. The underlying principles are to maximally stimulate the immune system, and to maintain optimal cell viability in tissue culture. These basic principles are essential for the successful outcome when substituting B-cell preselection and electrofusion for conventional random fusion induced by PEG. It was necessary to establish that the conditions used in these new processes did not interfere with cell function and proliferation.

The humoral immune response to an antigen is often unpredictable even in selected mouse strains. It is, therefore, very difficult to compare results obtained with different antigens. Quite often, hybrid yields vary greatly from one fusion to the next. However, we have attempted to assess bioselective and conventional PEG fusions by comparing monoclonal antibodies secreted from hybridomas produced from fusions using the same antigen and pooled spleens.

2. STRATEGIES FOR BIOSELECTIVE CELL ADHESION

2.1. Biotinylation and Avidinylation of Myeloma Cells

Myeloma cells are most easily biotinylated with an *N*-hydroxysuccinimide (NHS) derivative of *d*-biotin (Fig. 3). We have previously described a procedure for the biotinylation of myeloma cells (Lo *et al.*, 1984). Briefly, myeloma cells are grown in log phase, harvested by centrifugation, and washed with phosphate-buffered saline (PBS). NHS-biotin (0.1 M) is freshly dissolved in dimethylformamide (DMF) and diluted with PBS. DMF, an organic solvent, lyses cells at high concentrations. It is, therefore, important to use a minimal amount of this solvent for dissolving NHS-biotin. Myeloma cells are resuspended in 100 μM NHS-biotin, incubated for 15 min at room temperature without agitation, and centrifuged through 10 ml of serum at 200*g* for 8 min. Free biotin is removed by washing myeloma cells twice by resuspension and centrifugation in Dulbecco's modified Eagle's medium (DME). Biotin, added as a nutrient to most commonly used tissue culture media, interferes with the avidin–biotin bridging system. For this reason we use DME, which is prepared without biotin. Deoxyribonuclease-1 (DNase) added to DME reduces cell aggregation caused by the release of intracellular components. This is a common tissue culture practice, especially used in preparing primary cells from tissues.

We have tested the degree of biotinylation in four different cell lines (P3X63 Ag8.653, SP2/0,

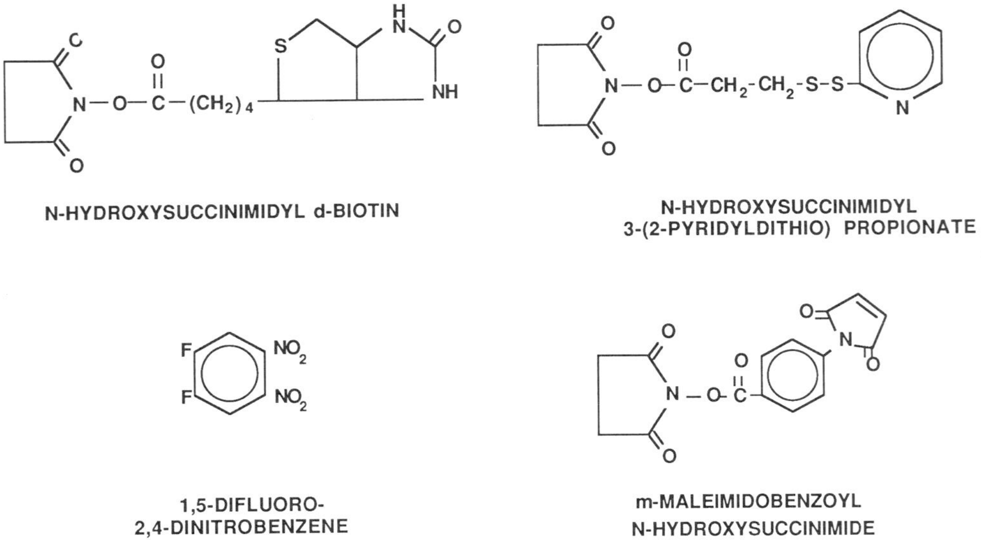

N-HYDROXYSUCCINIMIDYL d-BIOTIN

N-HYDROXYSUCCINIMIDYL
3-(2-PYRIDYLDITHIO) PROPIONATE

1,5-DIFLUORO-
2,4-DINITROBENZENE

m-MALEIMIDOBENZOYL
N-HYDROXYSUCCINIMIDE

Figure 3. Chemical structures of bifunctional cross-linkers.

Table 1. Viability of Biotinylated Cells[a]

Cell fraction	Number of initial cells	Number of clones	Percent
Normal unstained	3000	5	0.17
	100	2	2.0
High density	3000	38	1.23
	100	16	16.0
Low density	3000	23	0.12
	100	1	1.0
Middle fraction	3000	4	0.13
	100	3	3.0

[a]Myeloma cells were biotinylated with NHS–biotin and separated on a fluorescence-activated cell sorter into three fractions. Collected cells were counted and cloned into soft agar. The number of cell colonies was counted after 14 days. Results are from Schwartz and Lo (unpublished).

S194, and FOX-NY). Using the described procedure, cells become maximally labeled with biotin. The viability of these biotinylated cells was determined by cloning after limiting dilution. These experiments also indicate that there is no significant difference between different myeloma cell lines. Cell surface biotin was measured after incubation at 37°C for several hours and overnight. Cells stained with fluorescein-conjugated avidin exhibited no difference in their fluorescence intensity up to 3 hr after biotinylation. However, a significant difference was observed after overnight incubation. P3X63 Ag8.653 retained the greatest fluorescence intensity (approximately 10–30% of the original label), whereas very little staining was detected in the other cell lines.

Another consideration when choosing myeloma cell lines is whether biotinylation interferes with their ability to fuse with spleen cells. Myeloma cells were biotinylated, fused with spleen cells using PEG, and plated into HAT selection medium (containing hypoxanthine, aminopterin, and thymidine) for 2 weeks. No difference in the yield of hybrids was detected between biotinylated and untreated myeloma cells. However, in these experiments we have also routinely found that the P3X63 Ag8.653 cells gave three to five times more hybrids compared with other myeloma cell lines.

The extent of cell surface biotinylation was quantitated with a fluorescence-activated cell sorter. Biotinylated myeloma cells stained with fluorescein-conjugated avidin for 15 min at 4°C and washed were analyzed with a cell sorter. Normal unlabeled cells grown in log phase gave rise to two distinct size populations. Biotinylated cells showed no difference in their size distribution; however, two distinct groups were detected with fluorescent intensities differing by two log units. The large span of fluorescent labeling raised the question of whether heavily biotinylated cells were also affected in their viability. Biotinylated cells were sorted into fractions of (1) high density, (2) low density, and (3) intermediate density. The fluorescence of cells in the high- and low-density fractions differed by approximately one log unit. Cells collected were diluted serially and grown in soft agarose (Table 1). Untreated myeloma cells, serving as controls, were also diluted and cloned. No significant differences were detected after 2 weeks between normal and biotinylated cells. Viable cells were determined in all the above experiments by microscopic examination with trypan blue exclusion, which showed at least 98% cell viability. Together, these experiments indicated that myeloma cell viability, and membrane fusion are not significantly affected by biotinylation with NHS-biotin.

A further consideration in using electrofusion for hybrid fusion is the effect of intense electric fields on myeloma cell viability. In order to assess this factor, myeloma cells were exposed to electrical fields of 2 to 6 kV/cm, and 5 or 20 μsec pulse duration. Cells were cultured for 2 weeks after serial dilution into growth medium. Myeloma cells were completely unaffected by electrical field intensities less than 3 kV/cm, and 5 μsec pulse duration. Early experiments showed that cells exposed to high-voltage fields were more viable if they were suspended in an isosmotic medium (0.3 M sucrose, 5 mM sodium phosphate buffer pH 7.1 adjusted to 300 mosm) instead of isotonic solutions. Due to the high electrical conductance, bulk isotonic cell suspensions may dissipate most of the electrical energy as heat. Experiments conducted with a high-power pulse generator (2 kV, 11 A maximum output) show as much as 50% loss in the potential difference across the cell fusion chamber containing isotonic solutions, whereas very little reduction was experienced with isosmotic sucrose.

However, prolonged exposure of cells to sucrose greatly increases nonspecific cell aggregation, and also reduces their viability. Bearing these problems in mind, isosmotic sucrose may be employed effectively, if cells are handled quickly, electrofused, and returned into isotonic growth medium within a few minutes. Divalent cations or proteins are also excluded from the fusing medium in order to reduce nonspecific cell–cell contact.

Electroporation creates pores in cell membranes. These pores may persist for some period (Sowers, 1986) causing cells to become particularly susceptible to lysis. Following electrofusion, cells become more viable if they are allowed to incubate for at least 30 min at 37°C. Small cell aggregates observed microscopically are most able to survive electrofusion, whereas larger aggregates (typically more than four or five cells) usually lyse shortly after electrofusion. This phenomenon does not apparently depend on the myeloma cell line used.

In order to extend the applicability of this method, we have developed a second method in which avidin is covalently attached to the myeloma cell surface. This procedure, developed in our laboratory by Tomita (in preparation), is based on previous reports (Carlsson *et al.*, 1979; Godfrey *et al.*, 1981) on the use of the heterobifunctional cross-linker *N*-succinimidyl 3-(2-pyridyldithio) propionate (SPDP) (Fig. 3). Avidin is prepared by reaction with a 5 molar excess of SPDP for 1 hr at room temperature, and gel filtered on a Sephadex G-25 column with PBS to remove the free reagent. The degree of conjugation is determined from the absorbance of the avidin–SPDP solution at 253 and 282 nm. The avidin/SPDP ratio is typically 1:4. Myeloma cells (about 10^7) grown in a medium containing 0.1 mM β-mercaptoethanol are reacted with 1 μM avidin–SPDP for 1 hr at room temperature without agitation. Cells are collected by centrifugation through a cushion of solution made from equal volumes of isosmotic sucrose and DME. The degree of labeling is measured by binding of biotinylated fluorescein-conjugated lysozyme. The intensity of fluorescent staining is about 10-fold lower compared to biotinylated myeloma cells stained with fluorescein-conjugated avidin. The percent of cells labeled with avidin by this method is usually greater than 80%.

The cell surface avidin density can be greatly increased by prior reaction of myeloma cells with reduced SPDP. Briefly, reduced SPDP (prepared by dissolving 1 mg SPDP and 0.5 mg dithiothreitol in 30 μl DMF) is diluted into 30 ml PBS and added to myeloma cells. After a 15 min incubation at room temperature, cells are centrifuged through sucrose/DME and resuspended in PBS. Myeloma cells reacted with reduced SPDP are then incubated with 1 μM avidin–SPDP as described above. Myeloma cells avidinylated by prior reaction with reduced SPDP showed a staining intensity equivalent to biotinylated cells stained with fluorescein-conjugated avidin. Further characterization of avidinylated cells indicates that this process is very suited for use with bioselective electrofusion.

Although the procedure for avidinylating myeloma cells is more complex, requiring the preparation of avidin–SPDP, it does permit the use of biotinylated antigen in forming the cell cross-linking between myeloma and desired spleen cells. In many cases, especially when working with large or complex proteins, biotinylation provides the simplest and most convenient method. Bioselection can also be employed using streptavidin to bridge biotinylated myeloma cells and B cells treated with biotinylated antigen. Although this method has been successfully exploited by others (Wojchowski and Sytkowski, 1986), we have found that inappropriate cell pairs (such as myeloma–myeloma and spleen–spleen) are also formed when streptavidin is used. The use of avidinylated myeloma cells and biotinylated antigens ensures a unidirectional cell cross-linking.

2.2. Antigen Conjugation

A critical step in any bioselective fusion is the chemical conjugation of the antigen of interest to avidin or biotin. Conjugation of the antigen to avidin is preferred, and a variety of chemical cross-linking procedures may be employed depending largely on the nature of the antigen. Bifunctional cross-linking reagents are commonly used with proteins or peptides containing amino groups (Tager, 1976; Golds and Braun, 1978).

We have already described the use of a small homobifunctional cross-linker, 1,5-difluoro-2,4-dinitrobenzene (DFDNB) (Fig. 3). Avidin immobilized on iminobiotin–Sepharose is reacted with a 100-fold excess of DFDNB for 10 min and washed with sodium borate buffer (pH 10.5) to remove the unreacted cross-linking reagent. Bound avidin becomes chemically activated, and the subsequent addition of antigen proteins or peptides causes covalent conjugation to avidin. Unreacted DFDNB groups are blocked with either glycine or lysine depending on the desired charge modification of the conjugate. The antigen–avidin conjugate is then eluted at pH 3.5, and gel filtered on a Sephadex G-25 column with PBS. This method of conjugation is particularly effective with larger proteins where modification of basic amino acids does not drastically affect the antigenicity of the protein of interest. However, we have found that acidic proteins or small peptides conjugated in this manner subsequently become insoluble. Furthermore, any free amino groups on the iminobiotin resin must be completely blocked since these groups could also react with DFDNB and antigen. Resins containing high concentrations of iminobiotin can be effectively blocked, whereas commercially available resins, which have much lower ligand densities, are more difficult to derivatize. Conjugation of purified immunoglobulin (about 1 μg) to avidin yielded about 2% of the antibody as an avidin conjugate. Small proteins or peptides (less than 50 amino acids) can be easily derivatized with difluorodinitrobenzene and conjugated to avidin (Tager, 1976). The coupling efficiency of this method is very high (typically 70–80%); however, multiple amino groups on the peptide may all become conjugated to avidin, forming large covalently conjugated complexes or secondary structures, and altering the antigenicity of the peptide.

Another consideration in choosing chemical cross-linkers is whether the cross-linker itself may become an epitope for B-cell antigen receptors and antibodies. Modification of basic residues on proteins and peptides may also adversely affect the solubility of the antigen when conjugated to avidin. Charged residues are also commonly found in antigenic epitopes (Geysen *et al.*, 1987). These can become chemically modified, thereby inactivating the antigen.

We have also devised covalent attachment with another heterobifunctional crosslinker, *m*-maleimidobenzoyl *N*-hydroxysuccinimide (MBS) (Lui *et al.*, 1979; Youle and Neville, 1980) (Fig. 3). Antigen can be conjugated to avidin with MBS in aqueous solution. The conjugate's charged residues are unaffected. Furthermore, sulfhydryl groups are normally paired in disulfide bonds which are not normally accessible to the immune system. Therefore, covalent coupling through disulfhydryl groups on the antigen is less detrimental to the antigenicity of the molecule. Avidin is activated using the same procedure described for SPDP (Tomita, Lo, and Tsong, unpublished). MBS reacts with lysine residues in avidin. This provides a substantial reduction in the isoelectric point of avidin to a more neutral pH value (native avidin has a pI of 10) which also reduces the nonspecific binding usually experienced with native avidin. This reaction is quantitative, the molar ratio of MBS/avidin being 4 or 5 : 1. Avidin MBS prepared in this manner and stored in aliquots at −70°C is stable over a period of 2 years. The antigen is first reduced with 10 mM DTT for 15 min at room temperature and gel filtered. The reduced antigen is then added to avidin–MBS and reacted for 20 hr. Cysteine is added to block any remaining unreacted maleimide groups. The conjugate can be used without further purification. In some cases, the conjugate preparation is purified to remove any unbound antigen which could potentially compete with the avidin–antigen binding to B cells. Purification is accomplished by ion-exchange chromatography, or by gel filtration, depending on the nature of the antigen used.

Most of the conjugation methods described so far are not particularly well suited for haptens or carbohydrates. Haptens or carbohydrates can be chemically activated by periodate oxidation to the

N-hydroxysuccinimide derivative (Tijssen, 1985). These reactions are usually very efficient and the conjugate can be readily purified by ion-exchange chromatography because of the step reduction in the isoelectric point of avidin. Avidin from egg white is more suitable than streptavidin. At least five of the nine lysine residues present in avidin can be conjugated without impairing the conjugate solubility.

Previously, we mentioned antigen biotinylation as a method to produce a myeloma–B cell linkage. Biotinylation is easily accomplished with either NHS–biotin or maleimidyl–biotin. The latter reacts with sulfhydryl groups present on protein antigens. This method is particularly suited for preparing antibodies to crude protein mixtures, or to mixed antigens which are not fully characterized. The extent of biotinylation is easily assessed by Western blot analysis using denaturing polyacrylamide gel electrophoresis and transfer to nitrocellulose membranes. Biotinylated proteins are visualized by incubation with peroxidase-conjugated avidin and reaction with 4-chloronaphthol. Biotinylated protein may be isolated in this manner, or by other chromatographic procedures.

2.3. Characterization of Cell–Cell Cross-Linking

The overall effectiveness of the antigen conjugate preparation in producing specific hybridomas by bioselective electrofusion can be predicted by *in vitro* assays. These assays have been developed to assess the quality of the conjugate, i.e., to determine whether both the avidin and the antigen moieties are functional. The amount of antigen conjugate which is to be used in the actual fusion process may be determined with the same assay. We have devised a number of methods using either enzyme-linked immunoassay (EIA), radioimmunoassay (RIA), or fluorescent immunoassay. In many instances, animals immunized in preparation for hybridoma production exhibit detectable serum titers to the antigen. The antiserum is tested using the native antigen and an EIA or RIA (Mishell and Shiigi, 1980; Tijssen, 1985). Binding activity of the conjugate is measured by incubation of biotinylated myeloma cells, the avidin–antigen conjugate, a dilution of the antiserum, and a goat anti-mouse antibody conjugated to either peroxidase or fluorescent microspheres, or radioiodinated. Cells are then washed and the degree of second antibody binding to myeloma cells is quantitated depending on the reporter group. As with any assay, it is important to determine the nonspecific binding using a negative control such as (1) normal biotinylated myeloma cells, (2) normal mouse serum, (3) replacing the conjugate with avidin or streptavidin. The degree of binding can be quantitated by titration of the avidin–antigen conjugate or by inhibition with a known concentration of the free antigen.

3. HYBRIDOMA PRODUCTION

The procedures described in this section pertain to the immunization of animals, the preparation of the immunogen, the preparation of spleen cells, and the actual fusion procedure. Developing quick and sensitive screening procedures for selecting hybrid colonies is particularly important. These general procedures, although applicable to our research, do by no means constitute a rigid protocol. Many alternatives are possible.

3.1. Immunizations

Initial immunizations are performed by absorption of the protein antibody to bentonite or alumina. Two weekly intraperitoneal injections are given to either C57BL/6 or BALB/c mice between 4 and 6 weeks of age. When minute quantities of the antigen are available, final immunization can be performed by intravenous or intrasplenic injections 3 days prior to the fusion. Otherwise, the final immunization is carried out intraperitoneally. Following three or four immunizations, animals are bled and the serum titer is determined with an EIA or RIA. All peptides or haptens are conjugated to keyhole limpet hemocyanin (KLH), absorbed onto bentonite and injected as described above. In some cases, mice first primed with KLH alone more rapidly developed titers to the peptide or hapten conjugated to KLH.

3.2. Preparation of Spleen Cells

Three days after the final immunization, spleens are removed and dissociated on a wire mesh using a rubber policeman, and collected in DME containing DNase (50 μg/ml). Since several sequential steps during preselection also involve mechanical agitation or disruption of cell pellets, harsh mechanical resuspension, such as pipetting, should be avoided. Instead, cells may be gently disrupted by tapping or "fingerflicking" the tube containers. All subsequent procedures avoid excessive mechanical treatment of cells. Another precaution is to exclude divalent cations which promote nonspecific cell aggregation. Red blood cells are lysed with freshly prepared ammonium chloride solution (0.84%), and cells are spun through serum and washed once with DME/DNase. Spleen cells are then incubated with the antigen–avidin conjugate for a minimum of 30 min, or up to 2 hr, at 4°C to prevent capping and internalization of the conjugate. Treated spleen cells mixed with biotinylated myeloma cells (4 : 1) are centrifuged, and the mixed cell pellet is loosened and spun for 10 sec at 50*g*. The cell pellet is incubated for 30 min and then diluted in DME. An aliquot containing up to 2×10^7 cells is underlaid with sucrose and centrifuged at 200*g* for 6 min. The cell pellet is resuspended in 0.5 ml isosmotic sucrose and placed into the cell fusion chamber. Cells are exposed to two 5 μsec pulses, 2.5 kV/cm field intensity. The field polarity is alternated between the pulses to reduce bulk cell transport. Cells are transferred into isotonic medium, and after 30 min at 37°C are plated into 96-well plates in DME containing 10% FBS and aminopterin, hypoxanthine, and thymidine (Galfre and Milstein, 1981). Typically, cell growth is detected from 1 to 6 weeks after electrofusion. Aliquots of the cell supernatant are screened for antibodies. It is beyond the scope of this chapter to discuss screening strategies. The reader is referred to Mishell and Shiigi (1980).

3.3. Applications to Specific Antigens

Our first successful application of bioselective electrofusion was to make anti-rat lung angiotensin-converting enzyme (ACE). ACE is a large membrane protein of approximately 180,000 daltons composed of 20% (w/w) carbohydrates. Our lack of success in producing hybridomas using PEG fusions prompted the development of bioselective electrofusion. Avidin was cross-linked to ACE with DFDNB using the solid-phase procedure. Four fusions produced 31, 11, 4, and 6 hybrid colonies, respectively; all tested positive by RIA using [^{125}I]-ACE. The RIA used 2 pg of purified ACE, which corresponded to a final concentration of approximately 5 fM. Bound [^{125}I]-ACE was precipitated with *S. aureus* cells (2.5 mg) coated with a rabbit anti-mouse antibody (2 μg). This assay was devised to detect only high-affinity antibodies. Hybrid colonies were cloned by limiting dilution (Mishell and Shiigi, 1980) and expanded in tissue culture for monoclonal antibody production.

Most of the resultant antibodies were either IgG2a or 2b. Only one of the monoclonal antibodies was an IgM. The affinities of the monoclonal antibodies were measured from antibody concentrations required to bind 50% of the added [^{125}I]-ACE. The dissociation constants ranged from 4×10^{-8} to 2×10^{-10} M. The specificity of the anti-ACE monoclonal antibodies was analyzed by Western blot using crude rat lung membrane. All the anti-ACE antibodies bound to only a single protein band corresponding to authentic ACE. The epitope specificity of these monoclonal antibodies was determined by measuring binding of native and denatured [^{125}I]-ACE. The differential binding of native and denatured ACE revealed a large range of epitope specificities in our collection of ACE monoclonal antibodies.

Based on similar experiments using PEG fusions, the results of these experiments were unpredicted. Other laboratories have reported the production of monoclonal antibodies to rat lung ACE using the conventional method (Auerbach *et al.*, 1982; Moore *et al.*, 1984). The hybridoma yields were 1 (Auerbach *et al.*, 1982) and 2 (Moore *et al.*, 1984) anti-ACE monoclonal antibodies produced from a total of 960 and 400 hybrid colonies screened, respectively. Using data presented in those reports, we calculated the affinities of the monoclonal antibodies as 10^{-7} M (Moore *et al.*, 1984) and 7×10^{-9} M (Auerbach *et al.*, 1982). In comparison, our monoclonal antibody affinities ranged between 4×10^{-8} and 2×10^{-10} M, i.e., most of our antibodies had at least 50- to 100-fold higher affinities. The numerous monoclonal antibodies recovered showing diverse antigen specificity of our monoclonal antibodies also indicate that the bioselective fusion method preselects spleen cells on the basis on their surface receptors.

The formation of antibodies to peptide antigens was also assessed using human parathyroid hormone (PTH) and insulin. Both peptide antigens were biotinylated using NHS–biotin. The evaluation of the biotinylated peptide conjugates showed significant nonspecific binding. However, fusions performed using C57BL/6 mice produced 53 hybrid colonies. Hybrid growth was first observed 5 days after fusion. Immunoreactivity to PTH was assayed using an EIA with polystyrene wells coated with 0.5 μg of PTH(1–84) and treated with 1 μg/100 μl 1-ethyl-3-(3-dimethylamino propyl) carbodiimide dissolved in water. The EIA showed positive immunoreactivity in five of the hybrid colonies. Four additional hybrid colonies showed weak immunoreactivity. Isotyping experiments showed that all of these antibodies were IgM. In contrast, anti-insulin antibodies were detected in 4 of 12 growing hybrid colonies. Three of these antibodies showed weak immunoreactivity (10^{-7} to 10^{-8} M); however, one colony was found to exhibit very high binding affinity (10^{-9} M). The subtype of this antibody was an IgG.

4. CONCLUSION

We were unsuccessful in producing monoclonal antibodies with PEG fusions using mice immunized in the same manner. However, antibodies may be preselected by controlling cell–cell contact using a suitable preparation of an appropriate antigen conjugate. Binding activity of the antigen and avidin moieties on the conjugate can be characterized with *in vitro* assays. Chemical coupling reactions can in some cases adversely affect the binding properties of the antigen conjugate. In many instances, especially with peptides, we have observed a high degree of nonspecific binding with the derivatized antigen. This nonspecific binding is a probable reason for nonspecific hybridoma production. Specificity of cell–cell adhesion may be controlled by the amount of antigen conjugate added to the incubation medium and the conditions used. However, conjugates with multiple biotin or avidin binding sites, having greater binding avidity, may select for spleen cells expressing low-affinity receptors present on the cell surface at a very high density.

Appropriate immunization schedules may induce B-cell differentiation and maturation, forming cells with low-density high-affinity receptors. Fusion of these cells with myeloma cells results in the production of hybridomas secreting antibodies with very high binding affinity. Experiments using rat lung ACE immunized with very small amounts of the enzyme protein (typically about 5 to 2 μg), which may preferentially stimulate memory cells with high-affinity receptors to ACE, resulted in a 100% yield of hybridomas producing high-affinity antibodies. Although experiments with PTH and insulin were conducted with 10–20 μg immunizations, the molar concentration of the small peptide antigen was about 100 times greater than for immunizations with ACE. Immunization with low antigen doses may be particularly effective in stimulating high-affinity antibodies. Investment of some effort to synthesize the antigen conjugate for preselected fusions may be worthwhile, as these fusions produce antibodies with very high affinities, thereby reducing costly and time-consuming screening of numerous nonspecific hybrid colonies.

Cells in close membrane contact during or shortly after (Sowers, 1986) electrofusion become fused after pulse treatment. Nonspecific cell–cell contact may be reduced by addition of DNase. Chromosomal DNA binds avidly to the cell surface and this interaction is augmented by divalent cations. Cell aggregation induced by treatment with pronase can be minimized by adding DNase.

Cell lysis and subsequent DNA release also results in cell aggregation. Cells which aggregate nonselectively are difficult to control and result in the formation of very large cell aggregates.

Previously, antigen-targeting alone was reported to promote cell fusion (Bankert *et al.*, 1980). We have never observed hybrids formed by antigen-targeting alone. It is possible that the lipopolysaccharides used in those studies may also actually cause membrane fusion.

We have also investigated whether targeting can be used in conjunction with PEG. We have not detected any differences between normal and antigen-targeted fusions conducted with PEG. A critical step in PEG fusions involves pelleting together B cells and myeloma cells prior to the addition of PEG. Pelleting B cells and myeloma cells diminishes the effectiveness of bioselection with antigen conjugate. It is conceivable that PEG fusion may be performed in dilute cell suspension where specific cell–cell contact can be selectively preserved.

Nonselective electrofusion of cells may be useful when cell types can be isolated in advance. However, when applied to fusing cells in heterogeneous suspension, electrofusion alone does not facilitate monoclonal antibody production. Instead, it raises the impossible scenario of screening millions of hybrid colonies which may only contain a few potentially useful antibodies. This situation may be aggravated when nonselective electrofusion is applied to the production of human monoclonal antibodies. The human immune repertoire contains approximately 10^{12} possible antibody specificities instead of the 10^{8} in the mouse immune system. Human biopsies or peripheral blood cells may represent only a very minor fraction of the total human immune repertoire. When producing mouse monoclonal antibodies, the entire immune system, contained in the spleen, is used in the cell fusion. Targeting promises a great advantage in selecting human lymphocytes which express on the cell surface high-affinity antigen receptors. Likewise, bioselection alone, without the improved fusion efficiency offered by electrofusion, confers no advantage over existing methodology. The combination of both strategies greatly enhances monoclonal antibody production in both mouse and human systems.

ACKNOWLEDGMENTS. This work was supported in part by NIH Grant GM-28795 to T.Y.T. We thank Terrie Pierce for manuscript preparation.

REFERENCES

Auerbach, R., Alby, L., Grieves, J., Joseph, J., Lindgren, C., Morrissey, L. W., Sidkey, Y. A., Tu, M., and Watt, S. L., 1982, Monoclonal antibody against angiotensin-converting enzyme: Its use as a marker for murine, bovine, and human endothelial cells, *Proc. Natl. Acad. Sci. USA* **79:**7891–7895.

Bankert, R. B., DesSoye, D., and Power, L., 1980, Antigen-promoted cell fusion: Antigen-coated myeloma cells fused with antigen-reactive spleen cells, *Transplant. Proc.* **12:**443–448.

Berek, C., Griffiths, G. M., and Milstein, C., 1985, Molecular events during maturation of the immune response to oxazolone, *Nature* **316:**412–418.

Bischoff, R., Eisert, R. M., Schedel, I., Vienken, J., and Zimmermann, U., 1982, Human hybridoma cells produced by electrofusion, *FEBS Lett.* **147:**64–68.

Burnet, M. F., 1959, *The Clonal Selection Theory of Acquired Immunity*, Cambridge University Press, London.

Carlsson, J., Drevin, H., and Axen, R., 1979, Protein thiolation and reversible protein–protein conjugation: N-hydroxysuccinimidyl 3-(2-pyridyldithio) propionate, a new bifunctional reagent, *Biochem. J.* **173:**723–737.

Galfre, G., and Milstein, C., 1981, Production of monoclonal antibodies: Strategies and procedures, *Methods Enzymol.* **73:**3–46.

Galfre, G., Howe, S. C., Milstein, C., Butcher, G. W., and Howard, J. C., 1977, Antibodies to major histocompatibility antigens produced by hybrid cell lines, *Nature* **266:**550–552.

Geysen, H. M., Taimer, J. A., Kodda, S. J., Mason, T. J., Alexander, H., Getzoff, E. D., and Lerner, R. A., 1987, Chemistry of antibody binding to a protein, *Science* **235:**1184–1190.

Godfrey, W., Doe, B., Wallace, E. F., Bredt, B., and Wofsy, L., 1981, Affinity targeting of membrane vesicles to cell surface, *Exp. Cell Res.* **135:**137–145.

Golds, E. E., and Braun, P. E., 1978, Protein association in basic protein conformation in the myelin membrane: The use of difluorodinitrobenzene as a cross-linking reagent, *J. Biol. Chem.* **253:**8162–8170.
Golub, E. B., 1987, Somatic mutation: Diversity in regulation of the immune repertoire, *Cell* **48:**723–724.
Green, N. M., 1975, Avidin, *Adv. Prot. Chem.* **29:**85–133.
Griffiths, G. M., Berek, C., Kaartinen, M., and Milstein, C., 1984, Somatic mutation and maturation of immune response to 2-phenyloxazolone, *Nature* **312:**271–275.
Karsten, U., Papsdorf, G., Roloff, G., Stolley, P., Abel, H., Walther, I., and Weiss, H., 1985, Monoclonal anti-cytokeratin antibody from a hybridoma clone generated by electrofusion, *Eur. J. Cancer Clin. Oncol.* **21:**733–740.
Kohler, G., and Milstein, C., 1975, Continuous cultures of fused cells secreting antibody of predefined specificity, *Nature* **256:**495–497.
Lo, M. M. S., Tsong, Y. T., Conrad, M. K., Strittmatter, S. M., Hester, L. H., and Snyder, S. H., 1984, Monoclonal antibody production by receptor-mediated electrically induced cell fusion, *Nature* **310:**792–794.
Lui, F. T., Finnecker, M., Hamaoka, T., and Katz, D., 1979, New procedures for preparation and isolation of conjugates of proteins and a synthetic copolymer of D-amino acids and immunochemical characterization of such polymer, *Biochemistry* **18:**690–697.
Mishell, B. B., and Shiigi, S. M., 1980, *Selected Methods in Cellular Immunology,* Freeman, San Francisco.
Moore, M. G., Chrzanowski, R. R., McCormick, J. R., Cieplinski, W., and Schwink, A., 1984, Production of monoclonal antibodies to rat lung angiotensin converting enzyme, *Clin. Immunol. Immunopathol.* **33:**301–312.
Neumann, E., Gerisch, G., and Opatz, K., 1980, Cell fusion induced by high electric impulses applied to *Dictyostelium, Naturwissenschaften* **67:**414–415.
Pohl, H. A., 1978, *Dielectrophoresis,* Cambridge University Press, London.
Rudikoff, S., Pawlita, M., Pumphrey, T., and Heller, M., 1984, Somatic diversification of immunoglobins, *Proc. Natl. Acad. Sci. USA* **81:**2162–2166.
Sowers, A., 1986, A long lived fusogenic state is induced in erythrocyte ghost by electric pulse, *J. Cell Biol.* **102:**1358–1362.
Tager, H. S., 1976, Coupling of peptides to albumin with fluorodinitrobenzene, *Anal. Biochem.* **71:**367–375.
Tijssen, P., 1985, *Practice and Theory of Enzyme Immunoassays,* Volume 26, Elsevier, Amsterdam.
Tsong, T. Y., 1983, Voltage modulation of membrane permeability and energy utilization in cells, *Biosci. Rep.* **3:**487–505.
Vienken, J., and Zimmermann, U., 1982, Electric field-induced fusion: Electro-hydraulic procedure for production of heterokaryon cells in high yield, *FEBS Lett.* **137:**11–13.
Vienken, J., and Zimmermann, U., 1985, An improved electrofusion technique for production of mouse hybridoma cells, *FEBS Lett.* **182:**278–280.
Wojchowski, D. M., and Sytkowski, A. J., 1986, Hybridoma production by simplified avidin-mediated electrofusion, *J. Immunol. Methods* **90:**173–197.
Wormmeester, J., Stiekema, F., and Groot, K. D., 1984, A simple method for immunoselective cell separation with avidin-biotin system, *J. Immunol. Methods* **67:**389–394.
Youle, R. J., and Neville, D. M., Jr., 1980, Anti-Thy 1.2 monoclonal antibody linked to ricin is a potent cell-type-specific toxin, *Proc. Natl. Acad. Sci. USA* **77:**5483–5486.
Zimmermann, U., Vienken, J., Halfmann, J., and Emeis, C. C., 1985, Electrofusion: A novel hybridization process, *Adv. Biotech. Processes* **4:**79–150.

Chapter 17

Generation of Human Hybridomas by Electrofusion

Mark C Glassy and Michael Pratt

1. INTRODUCTION

Electrofusion is a technique which uses the application of precise, transient electric fields to induce somatic cell hybridization (Zimmermann, 1982; Knight and Scrutton, 1986). This recent methodology is effective for obtaining viable hybrids of plant (Bates *et al.*, 1983; Saunders, 1985), yeast (Schnettler *et al.*, 1984), and mammalian (Scheurich and Zimmermann, 1981; Tiessié *et al.*, 1982; Vienken *et al.*, 1983; Finaz *et al.*, 1984; Lo *et al.*, 1984; Neumann, 1984; Podesta *et al.*, 1984; Karsten *et al.*, 1985; Pratt *et al.*, 1987) origin, including interkingdom hybridization (Cocking, 1984). Hybridomas of both mouse (Berg *et al.*, 1983; Shimizu and Watarobe, 1985; Wojchowski and Sytkowski, 1986) and human (Bischoff *et al.*, 1982; Shimizu and Watarobe, 1985; Pratt *et al.*, 1987) derivation have been produced at relatively high efficiency, but Ig secretion has been limited (Vienken and Zimmermann, 1985; Hofmann and Evans, 1986).

Electrofusion offers several distinct advantages as an alternative to the more traditional method of using PEG to generate hybridomas. Inasmuch as electrofusion relies only on the biophysical effects of dielectrophoresis (Zimmermann, 1982; Hofmann and Evans, 1986) and membrane dielectric breakdown (Sugar and Neumann, 1984) to induce heterokaryon formation, it is less labor intensive because a chemical incubation period and repeated washings are not required. Also, electrofusion depends on the physical effects of AC (dielectrophoresis) and DC (fusion), which are independent of genetic or biochemical predispositions. Cell types which have been recalcitrant in PEG or other fusogens can potentially be considered for electrofusion experiments. Moreover, PEG is cytotoxic to some fusion partners, including human cell lines (Engleman *et al.*, 1985), and electrofusion would eliminate this problem.

It is expected that with further studies of the mechanics and kinetics of dielectrophoresis and membrane dielectric breakdown efficiencies, fusion protocols will be substantially improved (Zimmermann, 1982; Hoffmann and Evans, 1986). Considering the immense potential for both clinical

---

MARK C GLASSY • Brunswick Biotechnetics, 4116 Sorrento Valley Boulevard, San Diego, California 92121. MICHAEL PRATT • UCSD Cancer Center, University of California at San Diego, La Jolla, California 92093.

and research applications of human monoclonal antibodies (Dorfman, 1985; Olsson, 1985; Glassy *et al.*, 1987a), electrofusion could easily replace PEG fusion protocols as the laboratory standard for the production of Ig-secreting human hybridomas. Here, we summarize some of the work we have done in generating Ig-secreting human–human hybridomas using the UC 729-6 fusion partner (Glassy *et al.*, 1983a) with an electrofusion protocol (Pratt *et al.*, 1987).

2. EXPERIMENTAL PROCEDURES

2.1. Cells and Tissue Culture

UC 729-6 is a 6-thioguanine-resistant human lymphoblastoid B-cell line (Glassy *et al.*, 1983a, 1985), shown to be optimal in generating genetically stable human–human hybridomas (Abrams *et al.*, 1983; Thielmans *et al.*, 1984; Glassy *et al.*, 1987b). UC 729-6 cells were grown and maintained in RPMI-1640 medium supplemented with 10% fetal calf serum (FCS) and glutamine (Glassy *et al.*, 1983a).

Peripheral blood lymphocytes (PBL) were obtained from healthy volunteers and purified by Ficoll centrifugation (Boyum, 1968). Lymph node lymphocytes from cancer patients and lymphoma cells (nodular lymphoma, poorly differentiated; NLPD) were obtained and prepared as previously reported (Glassy *et al.*, 1983a; Shawler *et al.*, 1985).

All cell counts were obtained with a hemocytometer. Cell viability was measured by the exclusion of 0.05% trypan blue; all viabilities were greater than 90% before electrofusion.

2.2. Cell Preparation

Cells to be used for fusion were washed (500*g* for 10 min) 3 times and resuspended in 0.3 M mannitol (Sigma), pH 7.2, at the appropriate cell concentration.

2.3. Electrofusion Equipment

All work was conducted using the BTX model 401A Electro Cell Manipulation System (San Diego, Calif.). The design ideology of this instrumentation has been recently reviewed (Hofmann and Evans, 1986) and described (Pratt *et al.*, 1987). The model 401A Electro Cell Manipulator generator is an instrument which consists of an AC function generator and a high-voltage, square-wave DC pulse generator.

2.4. Fusion Chambers

All fusions were performed in a BTX coaxial fusion chamber. This chamber has divergent field geometry (see Hofmann and Evans, 1986) which is necessary for a maximum dielectrophoretic effect (Pohl, 1978; Zimmermann, 1982). However, a divergent field does create an inhomogeneous field concentration when the DC pulse is applied. Therefore, the field strength across the chamber varies with the radial position. This causes the actual transmembrane voltage potential (Zimmermann, 1982) applied to each cell to depend on the cell's location in the chamber. Inhomogeneous field geometry can have deleterious effects on fusion efficiency (see Section 4).

Optimization of electrofusion parameters was performed on a microslide (20 μl volume) which was equipped with two parallel wires positioned 0.5 mm apart. Since the wires were round, a divergent field was induced which elicits dielectrophoresis in cells. It should be noted, however, that the field geometry is quite different in the microslide electrodes compared to the coaxial chambers.

Lysis analysis was performed utilizing the BTX 470 cuvette electrode. This apparatus fits firmly into a 1.5-ml semidisposable plastic curvette (VWR part no. 58017-847) and has an electrode gap width of 1.9 mm and a volume of 650 μl to hold a cell suspension. The field geometry is uniform so a homogeneous field strength is applied to the entire sample.

2.5. Electrofusion Method

Cells suspended in 0.3 M mannitol were exposed to electrical conditions which were visually determined to be optimal (Glassy and Hofmann, 1985; Pratt *et al.*, 1987). One-milliliter samples of cells (ratio of two lymphocytes to one UC 729-6 cell) were pipetted into the 1-ml coaxial chamber. The chamber was previously autoclaved for sterility and all work was conducted in a laminar flow hood. A fusion sequence consisting of 20 sec of 580 V/cm (maximum field strength) of AC 800 kHz, sinusoidal wave shape, followed by a 3-sec 250 V/cm increase in AC amplitude to 830 V/cm was used. These dielectrophoretic conditions were followed immediately by two DC pulses of 7.69 kV/cm (maximum field strength) each having a duration of 20 μsec.

After completing the electrofusion sequence, the cells were allowed to sit for 10 min in the coaxial chamber before removal. The cells were then washed and resuspended in RPMI-1640 medium (supplemented with 10% FCS and glutamine) and seeded at 1.0×10^5 cells per well in 96-well microtiter plates (Costar). The following day the wells were refed with RPMI-1640 medium plus HAT components (Littlefield, 1964) as described (Glassy *et al.*, 1983a). Hybrids were visible 2–6 weeks postfusion.

2.6. Lysis Curve Determination

Homogeneous cell suspensions of PBLs and UC 729-6 were washed and resuspended in 0.3 M mannitol, pH 7.2. The change in the number of viable cells was observed as a result of exposure of various DC field strengths with a duration of 20 μsec. These values were then converted to yield a percent change in viability and plotted as a function of field strength.

2.7. Enzyme Immunoassay

Human IgG and IgM quantitation and qualitation were determined by an enzyme-linked immunoassay (EIA) as described (Glassy *et al.*, 1983b; Glassy and Cleveland, 1986) and modified (Glassy and Surh, 1985).

2.8. Cell Size Analysis

Cell size and cell cycle determinations were obtained using previously described methodology (Glassy and Furlong, 1981; Sarkar *et al.*, 1980).

3. RESULTS

3.1. Cell Viability

Optimal electrofusion requires very hypotonic media (i.e., low conductivity). We assessed the viability over time of the UC 729-6 fusion partner in the fusion medium, 0.3 M mannitol, pH 7.2. This is shown in Fig. 1. Over a 4 hr period there was very little loss in cell viability. Most fusions can easily be completed within an hour or two.

We also assessed the effect of alignment on the viability of both UC 729-6 and PBL cells. There was less than a 10% cell loss within the 500 kHz to 2 MHz range (data not shown).

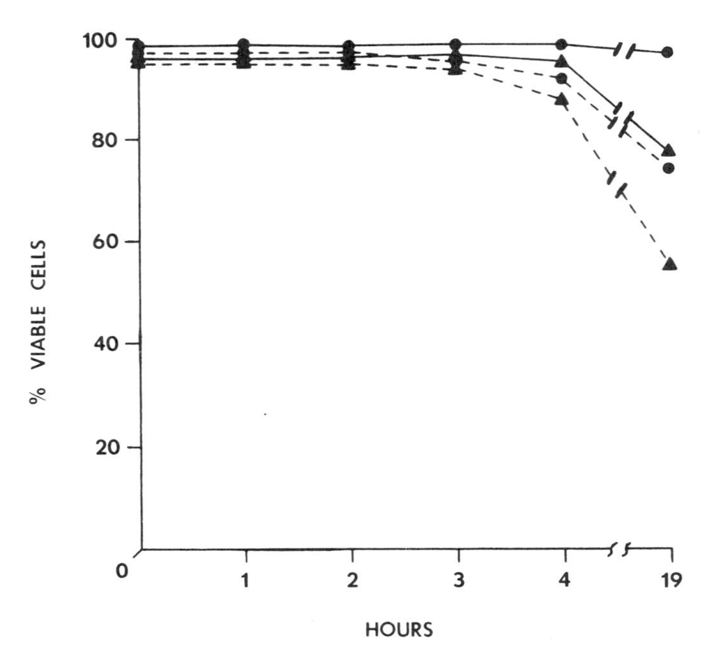

Figure 1. Effect of mannitol on the viabilities of UC 729-6 and PBL lymphocytes. Cells were harvested, washed, and resuspended to 1.0×10^6 cells/ml in 0.3 M mannitol, pH 7.2, either at room temperature or at 4°C. At the indicated time points, aliquots were removed from the cultures and viabilities were assessed by the exclusion of 0.4% trypan blue. ●——●, UC 729-6 at room temperature; ●- - -●, UC 729-6 at 4°C; ▲——▲, PBL at room temperature; ▲- - -▲, PBL at 4°C.

3.2. Cell Density

Analysis of the effects of cell density on electrofusion is shown in Table 1. Over four separate cell concentrations, ranging from 5×10^5/ml to 1×10^7/ml, the yield in hybridomas was assessed. The optimal concentration of cells was 5×10^6/ml, resulting in 12 hybrids, of which 5 secreted IgM at > 1.0 μg/ml per 10^6 cells per day. In this series of experiments no IgG secretors were observed, although they have been detected in unrelated fusions (unpublished observations).

Table 1. The Effect of Cell Density on Electrofusion Efficiency[a]

Cell density	Number of hybrids/ total cells used[b]	Number of Ig secretors
5×10^5/ml	$1/1.5 \times 10^6$	0
1×10^6/ml	$3/3.0 \times 10^6$	1 IgM
5×10^6/ml	$12/1.5 \times 10^7$	5 IgM
1×10^7/ml	$14/3.0 \times 10^7$	5 IgM

[a]Cells for fusion were prepared as described in Materials and Methods. Cell ration was 2:1 (PBL to UC 729-6). Electrofusion parameters were: alignment (AC: 800 kHz) 20 sec, 580 V/cm; compression (AC: 800 kHz) 3 sec, 830 V/cm; fusion (DC pulse) 2 × 20 μsec, 7.69 kV/cm (see Table 3). The fusion was performed in a 1.0-ml chamber with two autopulses (40 μsec total exposure time).

[b]The number of hybrids was scored 3 weeks after fusion.

Table 2. Electrofusion Efficiency[a]

Fusion partners	Fusion experiment	Wells plated[c]	Number of hybrids[d]	Number of Ig secretors	
				IgM	IgG
UC 729-6	1	240	14	6	0
× PBL	2	240	24	9	2
	3	192	30	13	0
	4	240	17	6	1
	5	240	13	4	0
UC 729-6	1	192	121	98	0
× NLPD[b]	2	240	114	0	93
UC 729-6	1	144[e]	57	20	16
×	2	240	77	25	12
lymph node	3	240	47	19	8
lymphocytes	4	192	80	36	20

[a]Cell fusion parameters are as described in Table 1.
[b]NLPD: nodular lymphoma, poorly differentiated. Patient number 1 was an IgM kappa.
[c]All cells plated into 96-well microtiter plates: 1×10^5 cells/well.
[d]Number of hybrids was scored 3 weeks after fusion.
[e]Lymph node lymphocytes were from the following somatic sources: 1 = vulva lymph node; 2 = prostate lymph node; 3 = teratoma lymph lymph node; 4 = lung lymph node. All were from cancer patients.

3.3. Hybridoma Characteristics

Various parameters of human–human hybridomas were analyzed. The hybrids were true tetraploid cells as judged by propidium iodide DNA staining and karyotyping; the hybrids contained antigens from both parental cell types (e.g., HLA); and the $21p^+$ chromosomal marker of UC 729-6 (Glassy *et al.*, 1983a).

3.4. Electrofusion Efficiency

To compare the relative electrocell fusion efficiency of UC 729-6, we used several fusion partners consisting of PBLs, lymphoma cells, and lymph node lymphocytes. These results are shown in Table 2. Five separate fusions of UC 729-6 with PBLs were performed with an average yield of 20 hybrids per experiment, 8 of which secreted Ig.

When UC 729-6 was fused with lymphoma cells (an excellent fusion partner), there was a significant increase in the hybrid yield in each experiment (63 and 48%, respectively). All the hybrids from the patient who was an IgM kappa were of identical class and light chain type. The same was true of the patient who was typed as IgG lambda.

The four separate lymph node experiments resulted in hybrids that secreted IgG and IgM. The range of Ig secretion by all of the hybrids described in this report was from 200 ng to 5 μg/10^6 cells per ml per day. The majority average was approximately 1 μg/10^6 cells per ml per day.

3.5. Viability Analysis

The relationship between cell viability and field strength of UC 729-6 and normal PBLs is shown in Fig. 2. Figure 2A shows the percent viability of cells in an inhomogeneous suspension

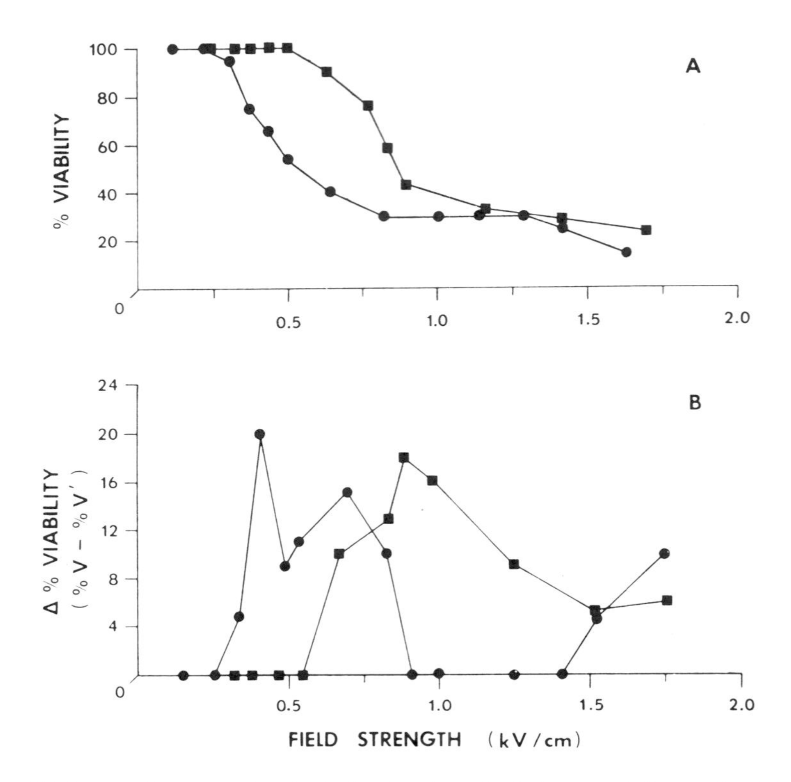

Figure 2. Viability and field strength analysis. Panel A shows the lysis versus field strength relationship for UC 729-6 (●) and human PBL (■). The difference in these curves shows that all lymphoid cells do not have the same dielectric breakdown thresholds. Panel B was derived by graphing the change in cell survival as a function of the periodic increases (x–x′) in the field strength. Periods where larger changes in cell survival occur indicate dielectric breakdown of that percent of the cell population.

exposed to a range of kV/cm pulses at 20 μsec durations. The 50% loss of UC 729-6 occurred at approximately 0.5 kV/cm, whereas the 50% loss of PBLs occurred at approximately 0.85 kV/cm. Figure 2B illustrates the change in cell viability over a periodic (X to X′) increase in field strength. Noteworthy is the fact that the increments in the values of "X" were not held constant; thus, this does not represent a direct relationship to the derivative of the true viability curve. Insight into the dielectric breakdown threshold of the various cell sizes in suspension can, however, be derived from these periodic changes (see below).

3.6. Physical Conditions

Table 3 gives the AC and DC field strength concentrations within 10 concentric radial sectors of the coaxial chamber. The coaxial chamber is 1 mm in diameter and each divided radial sector contains 50 μm of radial distance space. As each sequential radial sector has greater circumference, the value of each sector exposed to a given field passing through this chamber thus causes over 50% of these suspension volumes to be exposed to less than 14% of the maximum field strength applied. This must be considered when designing or optimizing protocol conditions. From the data shown in Table 3, it is apparent that the viability of the cells is considerably diminished when closer than 0.5 mm from the chamber center. When the field strength is the weakest (here, 0.77 kV/cm), the cell

Table 3. Physical Conditions within the Fusion Chamber

Relative r from center	Percent vol. of total	Percent vol. in E[a]	Min.E(DC) in percent vol.[b]	Percent viability at min.E(DC)[c]		Min.E(AC) in percent vol.	
				UC 729-6	PBL	20 sec	3 sec
0.0	0	0	7.69	20	20	578	829
0.1	1	1	7.69	20	20	283	406
0.2	5	4	3.77	20	20	174	250
0.3	9	4	2.31	20	20	127	182
0.4	14	5	1.69	28	25	104	149
0.5	25	11	1.38	30	32	100	133
0.6	37	12	1.23	30	36	100	116
0.7	49	12	1.08	30	46	100	100
0.8	63	14	0.92	30	61	100	100
0.9	80	17	0.85	31	78	100	100
1.0	100	20	0.77	35	85	100	100

[a]This number refers to the percent of the volume which is occupied by the suspension in a radial sector of a given relative distance from the center of the chamber.
[b]This is the min. DC field strength (E=kV/cm) value in each radial sector.
[c]Values taken from Fig. 1 at 20 μsec. The 40 μsec (2 × 20 μsec) pulse duration applied during the electrofusion protocol is assumed to be as or more lethal to the cells than the 20 μsec pulse.

viabilities of UC 729-6 and PBLs are the greatest. One would expect hybrids to form where the cells would be the most viable, the outer distances of the 1-ml chamber.

3.7. Cell Size Analysis

In conjunction with the field strength data shown in Table 3, we also analyzed the relative cell sizes of both UC 729-6 and PBLs. These data are shown in Table 4. The cell cycle was divided into four divisions represented as G_0, G1, S, and G2/M. The UC 729-6 cells were analyzed while in midlog growth and the PBLs immediately after harvesting. A total of 93% of the UC 729-6 cells were in log phase (G1, S, G2/M) and, therefore, in the optimal "fusogenic state" (see Sowers, this volume), whereas only 56% of the PBLs were optimally "fusogenic".

Table 4. Cell Size Analysis[a]

	Cell cycle phase	Cell volume (μm^3)		Cell radius (μm)		Percent cells	Number of cells
UC 729-6 (1.65×10^6 cells)	G_0	623	273	5.3	4.0	6.7	1.1×10^5
	G1	720	198	5.6	3.6	35.5	5.9×10^5
	S	1092	372	6.4	4.5	35.6	5.9×10^5
	G2/M	1248	423	6.7	4.6	22.2	3.7×10^5
PBL (3.35×10^6 cells)	G_0	582	219	5.2	3.9	44.1	1.47×10^6
	G1	707	220	5.5	3.5	32.6	1.09×10^6
	S	983	349	6.2	4.0	14.5	4.85×10^5
	G2/M	1192	397	6.6	4.3	8.8	2.94×10^5

[a]Cell size analysis was performed as described (Sarkar *et al.*, 1980; Glassy and Furlong, 1981). UC 729-6 cells used here were from a mid-log phase culture as described (Sarkar *et al.*, 1980; Glassy and Furlong, 1981). PBLs were from a healthy volunteer and Ficolled to enrich for lymphocytes.

4. DISCUSSION

A flow scheme illustrating the general steps in the generation of Ig-secreting human–human hybridomas by electrofusion and system variables is shown in Fig. 3. The immunization and lymphocyte preparation steps are the same whether done for PEG-mediated fusion or electrofusion. In preparation for the actual fusion, there are several steps that can be used. The fusion can be done either at room temperature or at 4°C; the cells can be treated with facilitators such as pronase which enzymatically clips the membrane surface and may potentially yield more hybrids; fusion media composed of nonelectrolytes other than mannitol at different pH conditions; and optimal cell concentrations, ratios, and physiology will have to be coordinated so the highest number of hybridomas will be obtained. For the actual fusion itself, the type, style, and geometry of the fusion chamber are critical; the optimal fusion parameters of the instrumentation and the number of repetitions of the fusion sequence all must be taken into account. Postfusion care in HAT selection and the various immunoassays available (Gaffar and Glassy, 1988) are conventional. This is based on the philosophy that a hybrid is a hybrid, whether derived with PEG or by electrofusion.

Teissié and Tsong (1981) and Knight and Baker (1982) have shown that electric-field-induced permeability changes are strongly dependent on the temperatures of the medium and partially due to high osmotic potentials which are induced by cytosolic exposure to the medium. Our experiments were done in hypotonic medium and cell survival after a significantly porative state is initiated is unlikely. The low overall efficiency of hybridoma formation reported suggests that the fusogenic state was either relatively transient or the conditions used superseded the optimal fusion (i.e., DC) field strengths.

Dielectrophoresis is an electrical phenomenon which can exhibit physical forces on polarizable objects within an alternating field (i.e., AC). For lymphoid cells in an inhomogeneous field, AC frequencies of 800 kHz to 1.2 MHz, the net attraction of the oppositely charged ends of neighboring cells in suspension, the field brings about pearl chaining (cell–cell contact aligned with the vectorial polarity of the AC). The AC field also imparts a force on the cell which will move the aligned cells toward higher field strengths. The duration of exposure to AC was 30 sec with a 3 sec increase. It can be assumed, based on the divergent nature of the coaxial chamber, that, over this period, there is a significant migration of the aligned cells toward the center electrode. Any attempt to compute the cell concentrations in the various field microenvironments treated by the field divergence must include this factor. Additionally, the probability of adjacent fusion partner cell type must also be considered (see below).

The field divergence present in the chamber used is necessary to effect cell alignment. This divergence also creates a decreasing cross-sectional area at radial locations closer to the center of the chamber (see Table 3). This becomes more complex than a simple statistical treatment because there are only certain cell subpopulations (G1, S, G2/M), based on size, which are going to be fusogenic in a given field strength. These size subpopulations will be very different in the two different cell types used. Combining the data in Tables 3 and 4, we can see that the optimal fusogenic partners, the G1, S, and G2/M cells, can be randomly distributed throughout the radial sectors within the fusion chamber and away from the sectors that yield the highest cell viability. This would then diminish the probability of the "right" two cells coming in contact, fusing, and surviving. It can easily be seen that the number of hybrids we obtained were nowhere near what could be obtained under truly optimal conditions. It should also be noted that different optimal conditions would be required for each type of chamber used.

It is tempting to speculate that for most of the cells in the chamber there was complete lysis in a cell size distribution of less than 8 μm (at 1 kV/cm, $V_0 = 1.2$), and that only a fraction of the cell suspension was actually fusogenic at the field conditions applied here. Judging by the efficiencies seen here, the actual fusogenic state must either characteristically occur in a small percent of the suspension at any given field strength, or occur to a greater degree in other electrical or medium conditions.

Using the protocol described here, we demonstrated that electrofusion can be used to produce

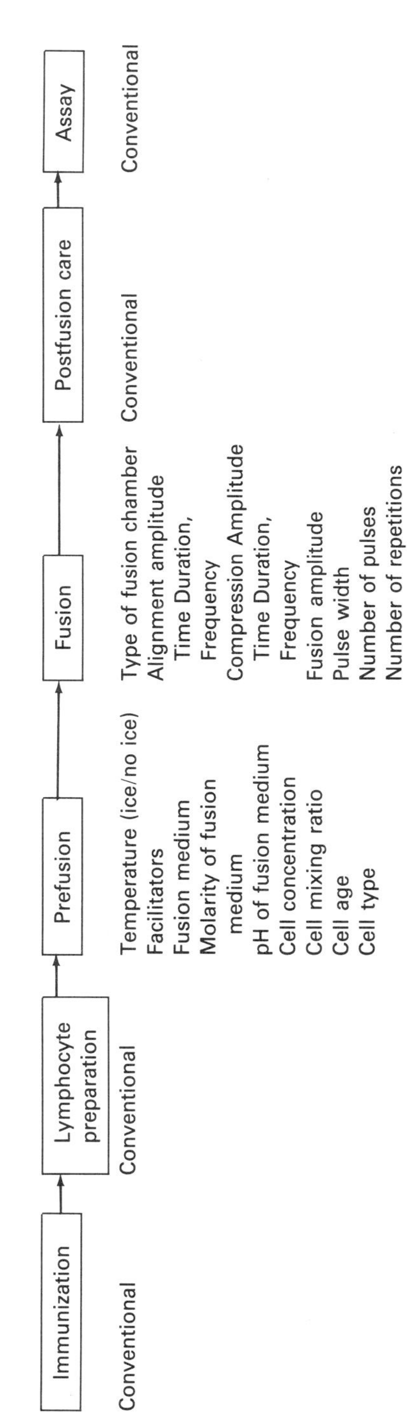

Figure 3. Flow scheme illustrating the steps and variables in electrocell fusion. See Section 4 for details.

heterokaryons and Ig-secreting human hybridomas, thus confirming and extending previous work. The priority of the electrofusion efficiencies of the lymphocytes with UC 729-6 is in agreement with previous published data (Glassy *et al.*, 1983a), namely, lymphoma cells fuse the best, followed by lymph node lymphocytes, and PBLs fuse poorly. Essentially, transformed or malignant (i.e., lymphoma cells) lymphocytes are relatively easy to "immortalize" by fusion since they are "immortal" already and are a clonal cell population which would behave uniformly. Lymph node lymphocytes are essentially "activated" lymphocytes (Glassy, 1987), and therefore, within the cell cycle and would yield a significant number of hybrids. PBL cells are predominantly in the G_0, or stationary phase of growth (see Table 4). These cells are traditionally the most difficult to fuse.

The human hybridomas described here have behaved as "typical" hybridomas, as judged by PEG standards. Also, in comparison with published data (Bischoff *et al.*, 1982; Vienken *et al.*, 1983; Berg *et al.*, 1983; Lo *et al.*, 1984; Karsten *et al.*, 1985; Wojchowski and Sytkowski, 1986), our hybrids, though human, are also typical, and within these parameters.

We have also shown that a pretreatment with calcium or proteases is not essential to electrofusion, at least with the UC 729-6 cell line. However, further work with ionic environments and other electrical conditions may yield significantly improved results. Also, fusion facilitators such as enzymes and other agents active within plasma membranes (such as anesthetics) on V_c can be assessed for their effect on cell survival and longevity of the fusogenic (or porative) state.

It can be concluded here that the potential of the technology of electrofusion (and electroporation) for immunology as well as other disciplines has only been initially explored. There are many other aspects of electrofusion which must be experimentally addressed before truly "optimal" protocol conditions for human hybridoma generation can be obtained.

5. FUTURE DIRECTIONS

We have established that hybridomas can be generated by electrofusion. The major question to be answered is whether PEG is the easiest and most practical method to produce hybridomas. The majority of biologists doing fusions do not have the expertise nor desire to construct their own electrofusion systems and those currently on the market are deemed too expensive (at least compared to the cost of PEG). To make electrocell fusion acceptable will require that the cost of the systems be lower than current prices and that protocols be available that yield results superior to PEG.

Areas worth exploring consist of, though not limited to, use of fusion facilitators, manipulations of the fusion medium (e.g., pH, tonicity, nonelectrolyte composition), and optimal chamber design that maximizes the nearest neighbor orientation of usable cells.

REFERENCES

Abrams, P. G., Knost, J. A., Clark, G., Wilburn, S., Oldham, R. K., and Foon, K. A., 1983, Determination of the optimal human cell lines for development of human hybridomas, *J. Immunol.* **131:**1201–1205.

Bates, G. W., Gaynor, J. J., and Shekhawat, H. S., 1983, Fusion of plant protoplasts by electric fields, *Plant Physiol.* **72:**1110–1113.

Berg, D., Schumann, I., and Stelzner, A., 1983, Electrically stimulated fusion between myeloma cells and spleen cells, *Stud. Biophys.* **94:**101–102.

Bischoff, R., Eisert, R. M., Schnedek, I., Vienken, J., and Zimmermann, U., 1982, Human hybridoma cells produced by electrofusion, *FEBS Lett.* **147:**64–68.

Boyum, A., 1968, Separation of leucocytes for blood and bone marrow, *Scand. J. Clin. Lab. Invest.* **21**(Suppl. 97)**:**77–89.

Cocking, E. C., 1984, Plant–animal cell fusion, *Ciba Found. Symp.* **103:**119–128.

Dorfman, H. A., 1985, The optimal technological approach to the development of human hybridomas, *J. Biol. Respir. Mod.* **4:**213–239.

Engleman, E. G., Foung, S., Larrick, J., and Raubitschek, A. (eds.), 1985, *Human Hybridomas and Monoclonal Antibodies,* Plenum Press, New York.

Finaz, C., Lefevre, A., and Teissié, J., 1984, A new highly efficient technique for generating somatic cell hybrids, *Exp. Cell Res.* **150:**477–482.

Gaffar, S. A., and Glassy, M. C., 1988, Applications of mouse and human monoclonal antibodies in non-isotopic enzyme immunoassays, in: *Reviews on Immunoassay Technology,* Volume 1 (S. B. Pal, ed.), Macmillan Press, Basingstoke, U.K., pp. 123–146.

Glassy, M. C., 1987, Immortalization of human lymphocytes from a tumor involved lymph node, *Cancer Res.* **47:**5181–5188.

Glassy, M. C., and Cleveland, P. H., 1986, Use of mouse and human monoclonal antibodies in enzyme immunofiltration, *Methods Enzymol.* **121:**525–541.

Glassy, M. C, and Furlong, C. E., 1981, Neutral amino acid transport during the cell cycle in human lymphocytes, *J. Cell. Physiol.* **107:**69–74.

Glassy, M. C., and Hofmann, G., 1985, Optimization of electrocell fusion parameters in generating human–human hybrids with UC 729-6, *Hybridoma* **4:**61.

Glassy, M. C., and Surh, C., 1985, Immunodetection of cell-bound antigens using both mouse and human monoclonal antibodies, *J. Immunol. Methods* **81:**115–122.

Glassy, M. C., Handley, H. H., Hagiwara, H., and Royston, I., 1983a, UC 729-6, a human lymphoblastoid B cell line useful for generating antibody secreting human–human hybridomas, *Proc. Natl. Acad. Sci. USA* **80:**6327–6331.

Glassy, M. C., Handley, H. H., Cleveland, P. H., and Royston, I., 1983b, An enzyme immunofiltration assay useful for detecting human monoclonal antibody, *J. Immunol. Methods* **58:**119–126.

Glassy, M. C., Handley, H., and Royston, I., 1985, Design and production of human monoclonal antibodies to human cancers, in: *Human Hybridomas and Monoclonal Antibodies* (E. G. Engleman, S. Foung, J. Larrick, and A. Raubitschek, eds.), Plenum Press, New York, pp. 211–225.

Glassy, M. C., Gaffar, S. A., Peters, R. E., and Royston, I., 1987a, Immortalization of the human immune response to human cancers, in: *Human Hybridomas: Diagnostic and Therapeutic Applications* (A. J. Strelkauskas, ed.), Dekker, New York, pp. 205–225.

Glassy, M. C., Handley, H. H., Surh, C., and Royston, I., 1987b, Genetically stable human hybridomas secreting tumor reactive human monoclonal IgM, *Cancer Invest.* **5:**449–457.

Hoffmann, G. A., and Evans, G. A., 1986, Physical and biological aspects of cellular electromanipulation, *IEEE Eng. Med. Biol.* **5:**6–25.

Karsten, U., Papsdorf, G., Roloff, G., Stolley, P., Abel, H., Walther, I., and Weiss, H., 1985, Monoclonal anti-cytokeratin antibody from a hybridoma clone generated by electrofusion, *Eur. J. Cancer Clin. Oncol.* **21:**733–740.

Knight, D. E., and Baker, P. F., 1982, Calcium-dependence of catecholamine release from bovine adrenal medullary cells after exposure to intense electric fields, *J. Membr. Biol.* **68:**107–140.

Knight, D. E., and Scrutton, M. C., 1986, Gaining access to the cytosol: The technique and some applications of electropermeabilization, *Biochem. J.* **234:**497–506.

Littlefield, J. W., 1964, Selection of hybrids from matings of fibroblasts in vitro and their presumed recombinants, *Science* **145:**709–711.

Lo, M. M. S., Tsong, T. Y., Conrad, M. K., Strittmatter, S. M., Hester, L. D., and Snyder, S. H., 1984, Monoclonal antibody production by receptor-mediated electrically induced cell fusion, *Nature* **310:**792–794.

Neumann, E., 1984, Electric gene transfer into culture cells, *Bioelectrochem. Bioenerg.* **13:**219–223.

Olsson, L., 1985, Human monoclonal antibodies in experimental cancer research, *J. Natl. Cancer Inst.* **75:**397–403.

Podesta, E. J., Solano, A. R., Molina Y Vedia, L., Paladini, A., Jr., and Sanchez, M. L., 1984, Production of steroid hormone and cyclic AMP in hybrids of adrenal and Leydig cells generated by electrofusion, *Eur. J. Biochem.* **145:**329–332.

Pohl, H. A., 1978, *Dielectrophoresis,* Cambridge University Press, London.

Pratt, M., Mikhalev, A., and Glassy, M. C., 1987, The generation of Ig-secreting UC 729-6 derived human hybridomas by electrofusion, *Hybridoma* **6:**469–477.

Sarkar, S., Glassy, M. C., Ferrone, S., and Jones, O. W., 1980, Cell surface HLA antigen density in the cell cycle of a human B lymphoid cell line, *Proc. Natl. Acad Sci. USA* **77:**7297–7301.

Saunders, J. A., 1985, Electrically induced fusion of cells and protoplasts, *Front. Membr. Res. Agr.* **9:**147–156.

Scheurich, P., and Zimmermann, U., 1981, Giant human erythrocytes by electric-field induced cell to cell fusion, *Naturwissenschaften* **68:**45–46.

Schnettler, R., Zimmermann, U., and Emeis, C. C., 1984, Large-scale production of yeast saccharomyces-cerevisiae hybrids by electrofusion, *FEMS Microbiol. Lett.* **24:**81–85.

Shawler, D. L., Wormsley, S. B., Dillman, R. O., Frisman, D., Baird, S. M., Glassy, M. C., and Royston, I., 1985, Diffuse, poorly differentiated lymphoma with peripheral blood and bone marrow involvement: Phenotypic characterization and serotherapy with monoclonal antibodies, *Int. J. Immunopharmacol.* **7:** 423–432.

Shimizu, H., and Watarobe, T., 1985, Application of electrofusion on the gene transfer and hybridoma production, *Med. Immunol.* **2:**105–110.

Sowers, A. E., and Lieber, M. R., 1986, Electropore diameters, lifetimes, numbers, and locations in individual erythrocyte ghosts, *FEBS Lett.* **205:**179–184.

Sugar, I. P., and Neumann, E., 1984, Stochastic model for electric field induced membrane pores electroporation, *Biophys. Chem.* **19:**211–225.

Teissié. J., and Tsong, T. Y., 1981, Electric field induced transient pores in phospholipid bilayer vesicles, *Biochemistry* **20:**1548–1554.

Teissié, J., Knutson, V. P., Tsong, T. Y., and Lane, M. D., 1982, Electric pulse induced fusion of 3T3 cells in monolayer culture, *Science* **216:**537–538.

Thielmans, K., Maloney, D. G., Meeker, T., Fujimoto, J., Doss, C., Warnke, R. A., Bindle, J., Gralow, J., Miller, R. A., and Levy, R., 1984, Strategies for production of monoclonal anti-idiotype antibodies against human B cell lymphomas, *J. Immunol.* **133:**495–502.

Vienken, J., and Zimmermann, U., 1985, An improved electrofusion technique for production of mouse hybridoma cells, *FEBS Lett.* **182:**278–280.

Veinken, J., Zimmermann, U., Fouchard, M., and Zagury, D., 1983, Electrofusion of myeloma cells on the single cell level: Fusion under sterile conditions without proteolytic enzyme treatment, *FEBS Lett.* **163:**54–56.

Wojchowski, D. M., and Sytkowski, A. J., 1986, Hybridoma production by simplified avidin-mediated electrofusion, *J. Immunol. Methods* **90:**173–177.

Zimmermann, U., 1982, Electric field-mediated fusion and related electric phenomena, *Biochim. Biophys. Acta* **694:**227–277.

Chapter 18

Gaining Access to the Cytosol

Clues to the Control and Mechanisms of Exocytosis and Signal Transduction Coupling

D. E. Knight

1. INTRODUCTION

The generally accepted sequence of events following the arrival of an agonist at the surface of a cell is (1) the transduction of the signal across the plasma membrane leading to (2) the generation of a second messenger which (3) controls the mechanism underlying the cellular response. An understanding of any of these processes requires access to the cytosol, either to sample the chemical changes associated with cell activation, or to experimentally manipulate the chemical environment of the cell. Such access is hampered by the effective barrier of the plasma membrane. Techniques such as membrane-permeant dyes, microinjection, and using broken cell preparations have been employed to overcome this problem. Another technique is to breach the plasma membrane in localized regions and thus allow either chemicals generated within the cytosol to leak out and so be measured, or to introduce chemicals into the cytosol by diffusion through the breaches from the extracellular fluid. Such preparations can be regarded as midway between the intact cell and the broken cell preparations.

As it is necessary to preserve the functional integrity of the intracellular organelles and their surrounding protein matrices for various cellular processes to operate, it follows that it is necessary to breach the plasma membranes of cells such that (1) their plasma membranes remain impermeable to the essential proteins/enzymes within the cell, (2) the permeation process does not alter the bulk of the plasma membrane such that it is unable to participate in the response mechanism, and (3) the permeation process is selective for the plasma membrane and does not affect the intracellular organelles. Although experiments involving the solubilization of plasma membranes by detergents have yielded significant results (Dunn and Holz, 1983), there is some uncertainty that pore sizes are not predictable or reproducible from experiment to experiment. The site of action of the detergents

D. E. Knight • M. R. C. Secretory Mechanisms Group, Department of Physiology, Kings College London, London WC2R2LS, United Kingdom.

is not restricted to the plasma membrane, and the bulk of the plasma membrane is not affected such that some important membrane reactions are inhibited. Other techniques of permeation have been successful (Gomperts and Fernandez, 1985), not least of all the electropermeabilization method which is chemically clean and where the above three criteria have been shown to be fulfilled (Baker and Knight, 1983; Knight, 1981, 1986; Knight and Scrutton, 1986a).

This chapter will concentrate in the main on one area of cell physiology, i.e., exocytosis, as such a mechanism has been extensively studied in a variety of preparations using this technique. Studies of signal transduction mechanisms will be described briefly at the end of the chapter, and other areas of study, e.g., enzyme activity and phosphorylation *in situ,* and calcium homeostasis will be referred to even more briefly.

2. THE EXPERIMENTAL APPROACH TO STUDYING EXOCYTOSIS

Exocytosis—the fusion of intracellular vesicles with the inner surface of the plasma membrane—is thought to be not only the mechanism underlying secretion of many hormones, transmitter substances, and enzymes, but also a means of altering the composition of the plasma membrane by insertion of the vesicle membrane. In addition to storing the substances to be secreted, the vesicles are packed with osmotically active solutes and any perturbation of the membrane could lead to their lysis within the cytosol. As the influence of the electric field on the limiting membrane of a cell or vesicle depends approximately upon its size, it is relatively easy to choose an intensity of electric field that will impose a voltage across the plasma membrane sufficient to cause membrane breakdown and yet be too weak to disturb the much smaller intracellular organelles. This was first appreciated and applied by Baker and Knight (1978) to bovine adrenal medullary cells and has since been extended to many other cell types ranging in size from platelets and synaptosomes, up to sea urchin eggs and paramecium. By applying a series of electric discharges through a cell suspension, each of ~100 μs duration sufficient to induce a maximum of 3 V across the plasma membrane, it is possible to effectively "pepper" cells with small localized pores—the number of pores being equal to twice the number of discharges. This imposed maximum voltage is achieved, for example, by exposing 20-μm-diameter cells (e.g., chromaffin cells) to 2 kV/cm or 2-μm cells (e.g., platelets or synaptosomes) to 20 kV/cm. The experimental setup to electropermeabilize this range of cell sizes is shown in Fig. 1. Very simply, cells are exposed to the electric field by discharging a capacitor through the cell suspension. Figure 1A shows the setup suitable for the 20-μm chromaffin cell where 2 kV is discharged across electrodes placed 1 cm apart. For the smaller cells, 20 kV/cm is achieved by discharging 2 kV across electrodes placed 1 mm apart. In this case, where the field strength is very high, the equipment must not only be easily dismantled, but of a design that avoids any part of the electrodes being exposed to an air–water interface. Figure 1B shows an exploded view of such a discharge chamber. It is necessary in this case for the electrodes to be completely covered by fluid rather than being partly immersed as in Fig. 1A. In the case of the 20-μm chromaffin cell, the approximate size of the pores induced in the plasma membrane by such electric fields has been calculated (Knight and Baker, 1982) by measuring the efflux rates of endogenous enzymes, e.g., 120,000-dalton lactate dehydrogenase, or preloaded markers, e.g., ^{86}Rb or ^{14}C-3-O-methylglucose, or by measuring the influx rates of extracellular solutes, e.g., ^{45}CaEGTA or ^{14}C-ATP, and so forth. Taken together, these data suggest that each electric discharge of 2 kV/cm across a cell suspension induces two pores at opposite ends of a cell (Knight, 1981) that are effectively 4 nm in diameter. Such an effective pore size would allow rapid equilibration of extracellular solutes of molecular weight about 1000, but retain the large cytosolic proteins.

3. CHOICE OF INTRACELLULAR SOLUTION

Once the plasma membrane has been permeabilized, the diffusible cytosolic components of low molecular weight will exchange with those in the extracellular fluid. The composition of the

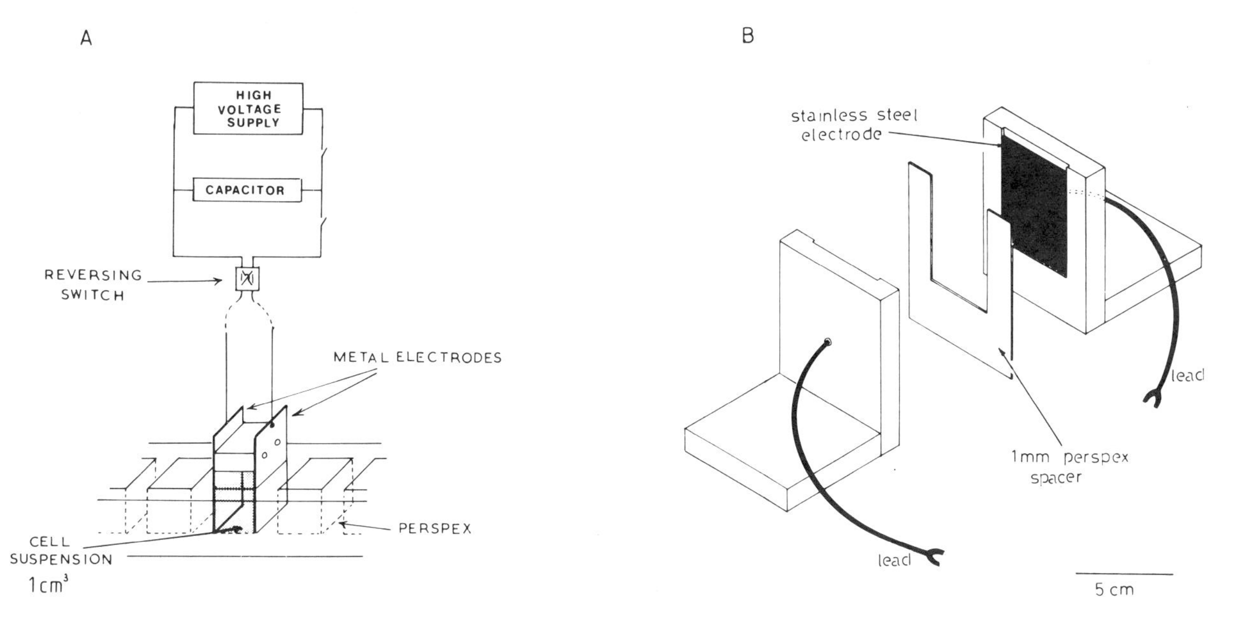

Figure 1. Experimental setup to electropermeabilize cells in suspension. (A) A 2-μF capacitor charged to 2 kV is discharged several times through 1 ml of a cell suspension (resistance ~ 100 ohms) to give a field strength of 2 kV/cm decaying exponentially with a time constant of 200 μsec. This field strength is suitable for 20-μm-diameter cells. (B) A cell suspension chamber for cells of 2-μm diameter. The electrodes, spaced 1 mm apart by the spacer, are charged to 2 kV. The field strength is 20 kV/cm.

Table 1. Composition of Medium in Which Adrenal Chromaffin Cells are Rendered Permeable and Still Able to Participate in Ca^{2+}-Dependent Exocytosis

Component	Concentration	Comments
Potassium glutamate	150 mM	K^+ can be replaced by Na^+. The nature of the anion is important. 150 mM Cl^- is inhibitory.
MgATP	5 mM	Essential EC_{50} normally around 1 mM. In a sucrose-based medium, however, the EC_{50} can reduce to 50 μM.
Pipes pH 6.6 or Hepes pH 7.4	20 mM	pH can be varied from 6 to 8. The optimal range for exocytosis is 6.5–7.0.
EGTA	0.4 mM	For very low free Ca^{2+}, it is preferable to use BAPTA in place of EGTA. Ca^{2+} is altered by adding Ca^{2+}-EGTA if the pH is between 6 and 6.9, or Ca^{2+}/Mg^{2+}/EGTA/EDTA buffer if the pH is above 6.9. The buffer concentration can be raised to 50 mM without altering the result.
Free Mg^{2+}	2–4 mM	Higher levels inhibit
Free ionized Ca^{2+}	<0.1 μM	

fluid is critical if the secretory mechanism is to be preserved (Baker and Knight, 1981; Knight and Baker, 1982). Table 1 lists the composition of the medium that seems effective in preserving the ability of the leaky cells to participate in exocytosis. Secretion seems not to depend critically on the nature of the major cation, as K^+ can be replaced by Na^+ or even by sucrose without much effect. The choice of the anion is of importance, however. Exocytosis from leaky cells is inhibited if the cytosol is exposed (at room temperature or above) to physiological levels of extracellular Cl^-. This inhibitory effect is not peculiar to Cl^- alone. The order of potency of the various anions follows the Hoffmeister or lyotropic series. Thus, low levels of thiocyanate are inhibitory whereas higher levels of Cl^-, NO_3^-, and Br^- are needed to have similar inhibitory effects. Anions such as acetate or glutamate have little effect at concentrations of 100 mM (but see Section 5). It is possible, therefore, that these anions interact with and perhaps solubilize some protein or protein matrix essential for secretion. The inhibitory effects on leaky cells are not readily reversible after a few minutes.

4. CALCIUM DEPENDENCE OF SECRETION

Many physiological responses are thought to be triggered by low levels of Ca^{2+} (Douglas, 1968). Cytosolic Ca^{2+} levels can be defined by exposing leaky cells to extracellular Ca^{2+} buffers. At pH 6 to 6.9, CaEGTA buffers are suitable, but above this pH range, i.e., at physiological pH, a mixture of Ca^{2+}, Mg^{2+}, EGTA, and EDTA buffers is more appropriate. Figure 2A shows the free Ca^{2+} obtained by these buffers at different pH levels when various concentrations of calcium are added to 5 mM of either EGTA or EGTA/EDTA buffers. Figure 2B shows that the secretory responses of many different electropermeabilized preparations are triggered when they are exposed to Ca^{2+} buffers corresponding to micromolar levels of Ca^{2+}. Results using unbuffered Ca^{2+} are unreliable as the cytosolic concentration may be considerably lower than the extracellular levels because of the readiness of the mitochondria and endoplasmic reticulum to sequester large amounts of Ca^{2+} (Baker and Knight, 1981; Knight and Baker, 1982). It could be argued that the mechanism of release into the extracellular fluid of secretory products from "leaky" cell preparations might not be the same as that underlying exocytosis. Instead, it could be a consequence of secretory granules lysing *in situ* and the products diffusing out from the cell through the electrically induced pores in the membrane. In the case of the adrenal medullary cell, however, this is unlikely as two components of the secretory granules, catecholamines (mol wt 300) and dopamine-β-hydroxylase (DBH; mol wt ~300,000), appear in the extracellular fluid in response to a Ca^{2+} challenge at the same

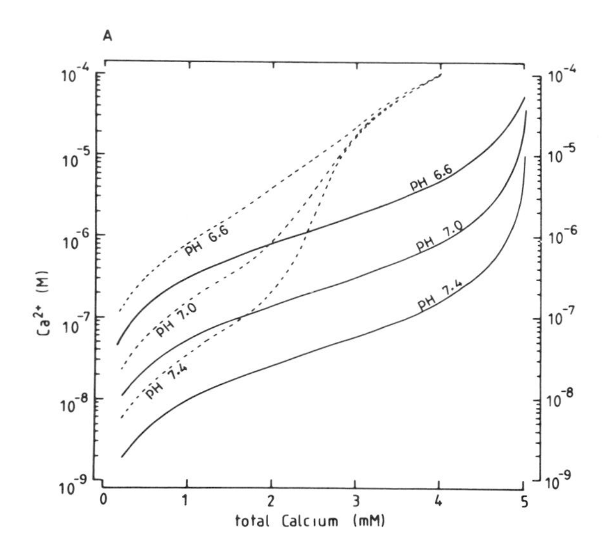

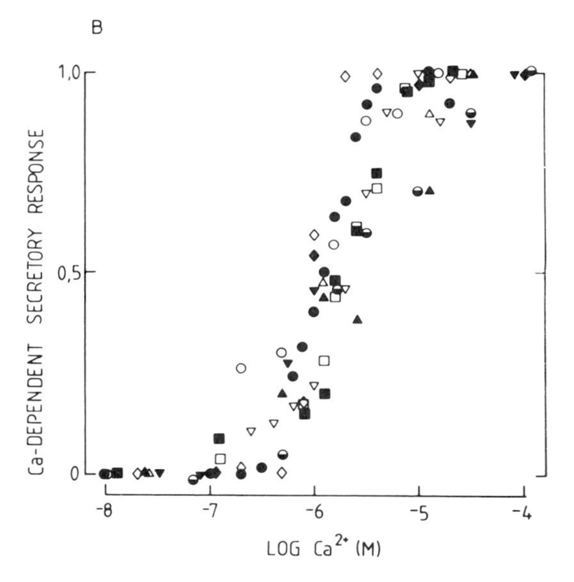

Figure 2. (A) Calculated Ca^{2+} concentration resulting from a total Ca, shown on the abscissa, being added to either 5 mM EGTA (solid lines) or 2.5 mM EDTA, 2.5 mM EGTA, and 12.5 mM Mg^{2+} (dashed lines). (B) Calcium dependence of secretion in various electropermeabilized preparations. Catecholamine (●) and Met-enkephalin (◇) from bovine chromaffin cells (Baker and Knight, 1981). Cortical granule discharge (○) from sea urchin eggs (Baker *et al.*, 1980). Insulin (◆) from pancreatic B cells (Jones *et al.*, 1985). Serotonin (■) and *N*-acetyl-β-glucosaminidase (□) from platelets (Knight *et al.*, 1982). Acetycholine (△) and ATP (▲) from *Torpedo* synaptosomes (D. E. Knight and H. Zimmermann, unpublished data). Amylase (▽) from pancreatic acinar cells (Knight and Koh, 1984). Catecholamine secretion from bovine chromaffin cells at pH 7.0 (▼) and 7.4 (⊖).

time and in the same relative proportions as found in the soluble granule pool. If the secretory products were released from the leaky cell by diffusion from the cytosol, the DBH would be expected to leave the cell at a much slower rate than either the catecholamine or the cytosolic protein lactate dehydrogenase (LDH; mol wt 100,000). This is not the case as the cytosolic enzyme effluxes from the cell through the pores at the slowest rate and in a Ca^{2+}-independent manner. A similar argument may be made for secretion of the lysosomal enzyme *N*-acetyl-β-glucosaminidase from leaky platelets. This enzyme is larger than the cytosolic enzyme LDH.

The Ca^{2+} activation curve for most of these preparations can best be fitted by a model involving two calcium ions per exocytotic event (Knight and Baker, 1982). For the adrenal medullary cell and the platelet, at least, the suboptimal levels of Ca^{2+} affect mainly the extent of secretion rather than simply the rate. The kinetics of release from leaky cells resembles that from intact cells.

The physiological significance of this observation is that a sustained rise in the intracellular level of Ca^{2+} would trigger not a sustained rate of secretion but rather a transient one. In many studies on secretion from electropermeabilized cells, levels of Ca^{2+} above 100 μM result in less secretion than if evoked by 10 μM Ca^{2+}. If the inhibitory action of high Ca^{2+} is intracellular, then the data suggest that the rate of exocytosis is not fast enough to be triggered as the Ca^{2+} level rises from the resting value of 0.1 μM through the 10 μM level, which is optimal for secretion, and toward the inhibitory level of 100 μM. This rate of secretion cannot easily be measured by simply adding Ca^{2+} to a leaky cell preparation as the rate-limiting step might be the rate at which Ca^{2+} enters the cell. The problem can be overcome by equilibrating the cells with Ca^{2+} at 0°C and then triggering exocytosis by a temperature jump. Such experiments reveal that the time taken to reach half-maximal secretion is close to 1 min. This slow rate of secretion, therefore, accounts for the inhibitory action seen by exposing leaky cells to 100 μM Ca^{2+}. What mechanisms are involved are not clear, but the inhibition by high Ca^{2+} is unlikely to arise from simple Ca^{2+}-activated proteases as a cocktail of protease inhibitors (TLCK, PMSF, and leupeptin) fail to neutralize the high Ca^{2+} effect. This phenomenon might, however, also reflect a physiological mechanism whereby a rise in intracellular Ca^{2+} level inhibits secretion, as is seen with the electropermeabilized parathyroid cell (Oetting *et al.*, 1986).

5. NUCLEOTIDE DEPENDENCE OF SECRETION

When bovine adrenal medullary cells are rendered leaky in a medium as described in Table 1 but lacking MgATP, they very rapidly (minutes) lose their responsiveness to a subsequent Ca^{2+} challenge. Adding back MgATP recovers the secretory response to some extent but the degree of recovery gets smaller the longer the cells are left without MgATP. The actual levels of nucleotide in the cytosol of the leaky cell can be measured *in situ* by measuring the activity of the endogenous ATPase pump on the granules. Such measurements reveal that the cytosolic MgATP level is very close to the extracellular nucleotide concentration–milli-molar levels being needed for secretion. As ATP^{4-} and nonhydrolyzable analogues of ATP cannot substitute for MgATP, a possible role for the nucleotide is one of phosphorylation. The nucleotide specificity and extent of phosphorylation have been studied in the electropermeabilized cell preparation. But as there are at least ten protein bands obviously phosphorylated under conditions that trigger Ca^{2+}-dependent exocytosis, it is not possible to select any one protein as being intimately associated with exocytosis (Niggli *et al.*, 1984). The results showing the dependence of secretion on MgATP can best be fitted by a model in which a chemical reaction directly related to exocytosis involves 1 MgATP molecule. If this is in fact the case, then the amount of phosphorylated protein associated with normal levels of secretion would be below detection limits of simple experiments. Furthermore, a model to accommodate the finding that suboptimal levels of either Ca^{2+} or MgATP affect mainly the extent of secretion rather than simply the rate might suggest that the rate of phosphorylation of some protein was of importance in the control of exocytosis rather than the extent. If this was so, then the chances of isolating and identifying the phosphorylation step would also be slight. The decline over minutes in the responsiveness of chromaffin cells rendered leaky in a medium lacking MgATP can be correlated with the "expected" loss of endogenous nucleotide from the cytosol. With platelets, however, the decline still occurs over minutes even though the nucleotide effluxes from the cell in less than 1 min. This might suggest at first sight that either Ca^{2+}-dependent exocytosis can occur in the absence of MgATP, that there is a bound pool of nucleotide, a stable Ca^{2+}-independent phosphorylated intermediary, or perhaps that one of the roles of MgATP is simply to give stability to the system. This latter possibility is supported by the finding that (1) in platelets the nucleotide requirement is not specific for MgATP and (2) the cortical granule reaction of the sea urchin egg can proceed in the absence of MgATP. An interesting observation is that the amount of MgATP needed to protect Ca^{2+}-dependent secretion can be reduced by over an order of magnitude simply by lowering the ionic strength of the suspending medium. This is shown in Fig. 3B where chromaffin cells were

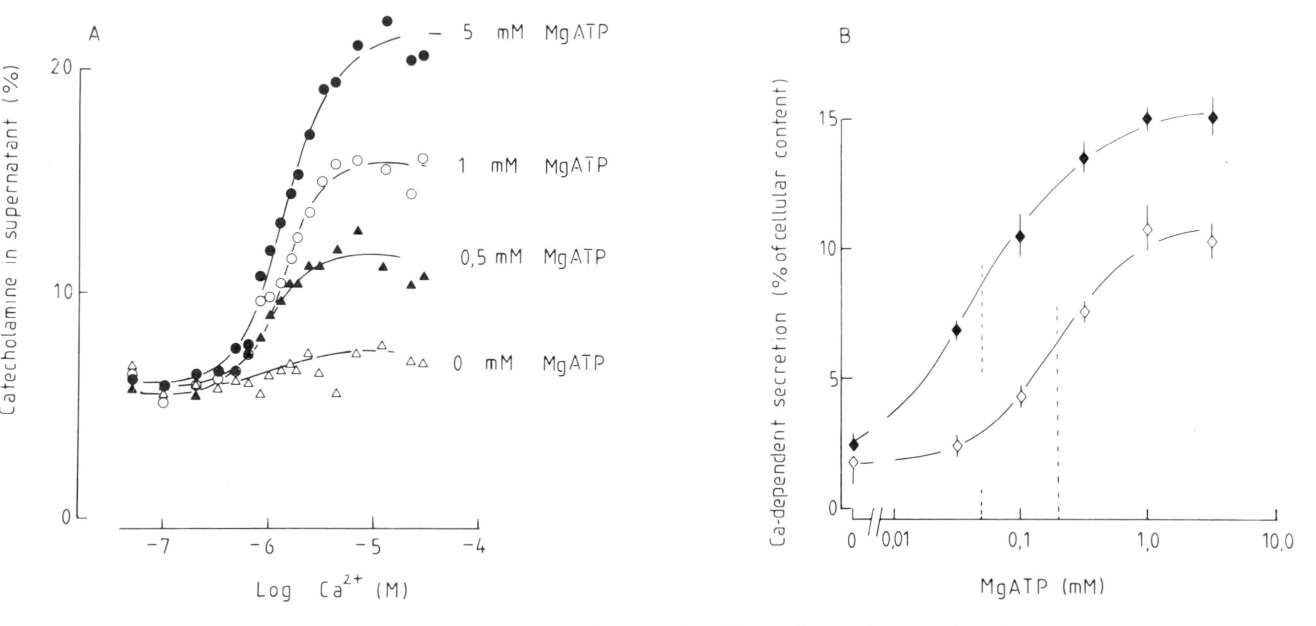

Figure 3. MgATP dependence of catecholamine secretion. (A) Bovine adrenal medullary cells were incubated in various concentrations of MgATP and then challenged with Ca^{2+}-EGTA buffers corresponding to the Ca^{2+} shown. (From Knight and Baker, 1982). (B) Cells initially rendered leaky at 0°C and 0.01 μM Ca^{2+} in a potassium glutamate-based medium were washed extensively in either a similar medium (◇), or in one in which the potassium glutamate was replaced isosmotically with sucrose (◆). MgATP was added and the cells challenged with 10 μM buffered Ca^{2+}. The dashed lines refer to the EC_{50} values of MgATP.

electropermeabilized at 0°C in a K glutamate-based medium lacking MgATP (see Table 1) and then washed extensively either in a similar buffer, or in one in which the K glutamate had been isosmotically replaced by sucrose, before various levels of MgATP were added back. Although there may be several interpretations to these data, e.g., effect of ionic strength on ATP binding, another possibility is that the secretory system is unstable in K glutamate medium unless high MgATP levels are present. The system is more stable at low ionic strength.

6. ENDOCYTOSIS ACCOMPANIES EXOCYTOSIS

In order that the surface area of the cell remain constant, the events surrounding secretion involve not only exocytosis but also endocytosis of an equivalent area of membrane. One way to quantify endocytosis is to measure the uptake of extracellular marker into the endocytosed organelles during secretion. With intact adrenal medullary cells, for example, [^{3}H]sucrose or horseradish peroxidase uptake into membrane-bound organelles can be visualized and measured (Phillips *et al.*, 1983; von Grafenstein *et al.*, 1986). In electropermeabilized cells, the same phenomenon occurs. For example, extracellular markers are trapped with the cell when the latter is triggered to secrete by a Ca^{2+} challenge, and cannot be removed by extensive washing (Knight and Baker, 1982; Baker *et al.*, 1982). Figure 4 shows that [^{3}H]sucrose uptake into leaky cells occurs at approximately the same time as exocytosis and at the same Ca^{2+} levels that trigger exocytosis. If this uptake does reflect endocytosis, and is not an artifact such as Ca^{2+}-induced vacuolization, then the finding that Ca^{2+} with MgATP triggers a cycle involving both exocytosis and endocytosis in leaky cells greatly complicates interpretation of the experimental results. It may be, for example, that the requirement for Ca^{2+} satisfies exocytosis, whereas MgATP is needed for the endocytosis part of the cycle of events. A clearer understanding must await further experiments with electropermeabilized cells designed to dissect these two events and hence define their separate ionic requirements.

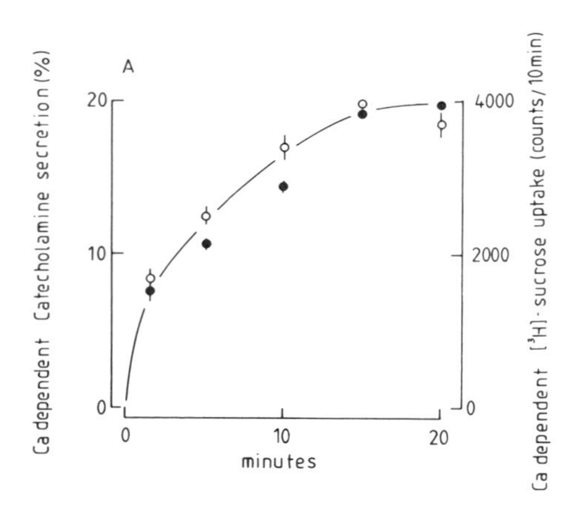

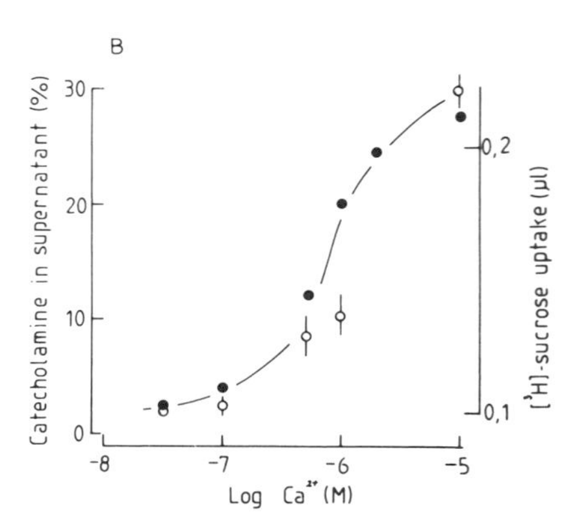

Figure 4. Endocytosis accompanies exocytosis in electropermeabilized cells. (A) Bovine adrenal medullary cells were rendered permeable in the presence of [^{3}H]sucrose and challenged with 0.01 and 10.0 μM buffered Ca^{2+}. At the various times shown, the cell suspension was diluted into 50 volumes of ice cold buffer, and the cells washed three times before being separated from the extracellular fluid by centrifuging through oil. The Ca^{2+}-dependent amounts of catecholamine secreted (●) and ^{3}H associated with the pellets (○) are shown. (Mean ± S.E.M., N = 4.) (B) As (A) except the leaky cells were challenged with Ca^{2+}-EGTA buffers corresponding to the various Ca^{2+} concentrations shown for 15 min.

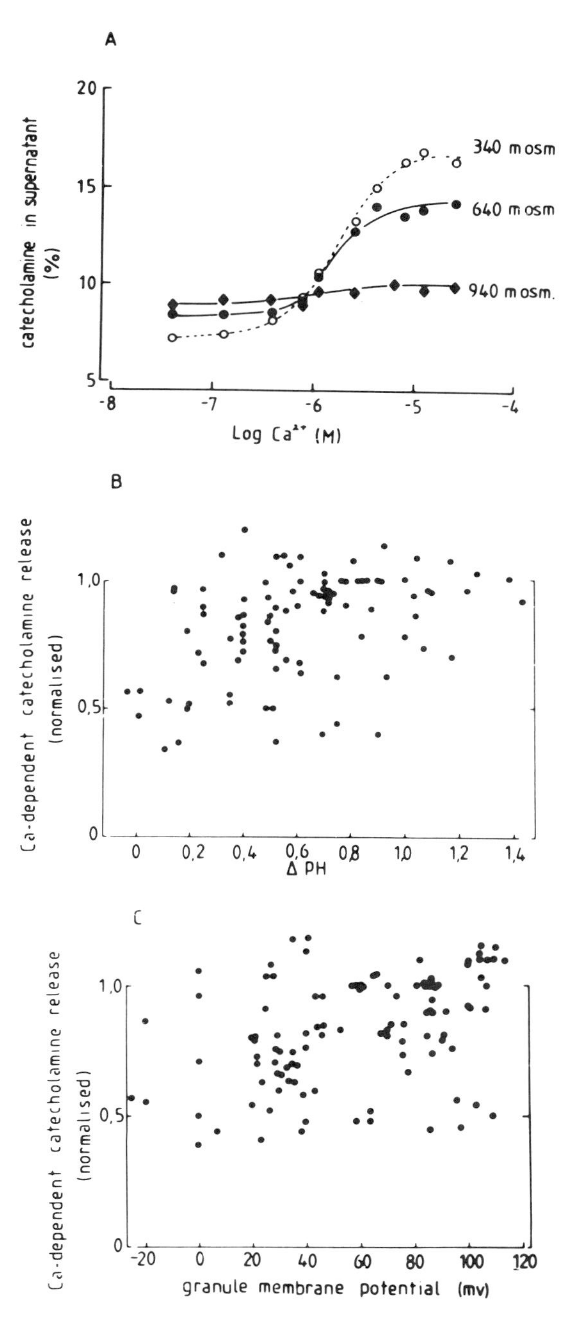

Figure 5. Secretion is inhibited by a rise in osmotic pressure but not by varying the acidity and potential of the granule interior. (A) Increasing the osmotic pressure with sucrose decreases the Ca^{2+}-dependent secretory response from electropermeabilized chromaffin cells without altering the Ca^{2+} sensitivity. (B, C) The lack of correlation between Ca^{2+}-evoked secretion from electropermeabilized cells and the pH gradient across the granule membrane (inside positive). The pH and the potential gradients across the membrane of the granules were measured *in situ* in the leaky cell. From Knight and Baker (1985b).

7. OSMOTIC PRESSURE AND THE CHEMIOSMOTIC THEORY

One possible site of action of MgATP is the ATPase associated with the proton pump on the secretory granule. This pump establishes both a positive potential and an acid interior of the granule, and these could act as a source of potential energy necessary for secretion. The chemiosmotic theory (Pollard *et al.*, 1977) suggests that the electric potential and acid interior of the granules could act as a source of energy which would be associated with an increase in intragranular osmotic pressure. The resultant swelling of the granule could be an essential event promoting membrane fusion and hence secretion. This theory is supported in part by the finding that increasing the osmotic pressure difference across the granule membrane by increasing the osmotic pressure within leaky cells (using either sucrose, glucose, or glycine) inhibits secretion (Fig. 5A). The involvement of this source of potential energy has been experimentally tested by measuring the potential and pH of granules *in situ* in leaky cells together with their Ca^{2+}-dependent secretory response (Knight and Baker, 1985b). Over a range of experimental conditions that either enhance or collapse the pH or potential gradients, there appears to be little correlation between these parameters and secretion and as such provide no evidence to support the idea that the intragranular potential and pH play an essential role in exocytosis (Fig. 5B).

8. FACTORS THAT ALTER Ca^{2+}-DEPENDENT SECRETION

8.1. Activators—A Role for Protein Kinase C

The phorbol ester 12-O-tetradecanoylphorbol-13-acetate (TPA) enhances Ca^{2+}-dependent secretion from leaky cells (Knight and Baker, 1983; Knight and Scrutton, 1984b). Two possible sites of action of TPA are protein kinase C and phospholipases. The finding that the effect of TPA can be mimicked by mezerein, diacylglycerols, and some other phorbol esters strongly suggests that the site of action of TPA in exocytosis is the kinase C (Nishizuka, 1984). Additional evidence implicating a role of protein kinase C in exocytosis comes from the finding that agents that inhibit the kinase also inhibit Ca^{2+}-dependent secretion, and the divalent cations that activate secretion also activate, in the same order of potency, protein kinase C. Thus, Ba, Sr, and Ca are equipotent, Mn is 20% effective, whereas Zn and Co are without effect.

Secretory systems do not all respond to phorbol esters or diacylglycerol in the same way, and Fig. 6 summarizes the three main types of response that have been observed from electropermeabilized cells. Figure 6A illustrates the first type where activation of protein kinase C causes a rather small leftward shift in the Ca^{2+} activation curve. Figure 6B is of the second type where a much larger leftward shift occurs at high concentrations of TPA or diacylglycerol, and the third type of response is shown in Figure 6C where activation of protein kinase C does not alter the affinity for Ca^{2+} but increases the extent of secretion. In the platelet there are at least two distinct populations of secretory granules, one of which exhibits the second type of response and the other the third type. The coexistence of different types of exocytosis in the same cell may provide a mechanism for effecting differential release of secretory products (Knight *et al.*, 1982, 1984). Thus, at low cytosolic Ca^{2+}, activation of protein kinase C by diacylglycerol would lead to the secretion of serotonin (type 2 response) but to no lysosomal release (type 3 response).

8.2. Inhibitors—A Role for Guanine Nucleotide Binding Proteins

In addition to the inhibitors of protein kinase C and some anions, the nonhydrolyzable analogue of GTP (GTPγS) and also botulinum toxin have been shown to inhibit Ca^{2+}-dependent catecholamine secretion from electropermeabilized chromaffin cells (Fig. 6A). The mechanism by which these two agents operate is not clear but as their effects cannot be overcome by activators of protein kinase C, it would appear that they act downstream of the Ca^{2+} transient, either at or near

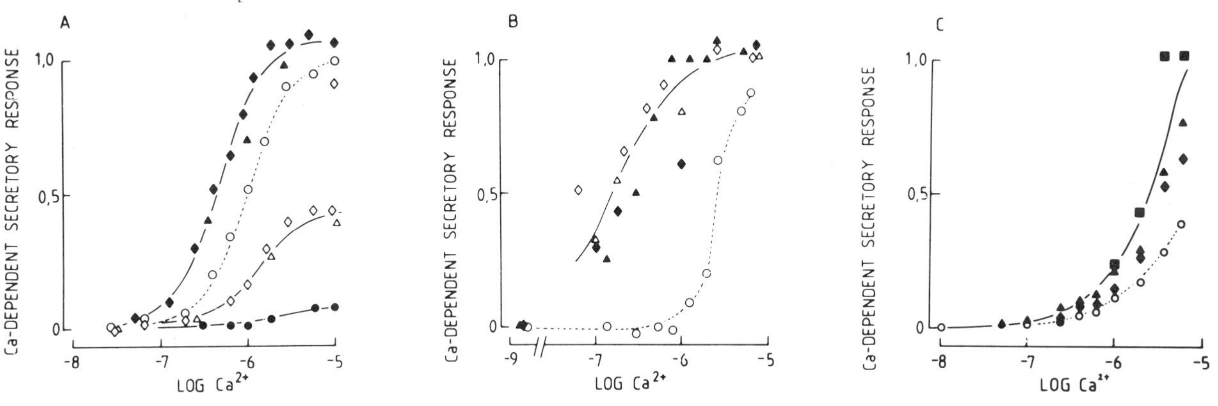

Figure 6. Modulation of Ca^{2+}-dependent secretion from electropermeabilized cells. (A) Bovine adrenal medullary cells; (B, C) human platelets. (A) Catecholamine secretion. Cells at about 0.01 μM Ca^{2+} concentration were incubated with 30 nM TPA (◆), 100 μM dioctanoylglycerol (▲) before being challenged with a range of Ca^{2+} concentrations. Cells with no additions and challenged with Ca^{2+} alone (○). The small shift to the left of the Ca^{2+} activation curve by these activators of protein kinase C is representative of the first type of secretory response. Incubation of leaky cells with 50 μM GTPγS (◇), or 10 mM fluoride ions (△), or preincubation of intact cells with botulinum toxin type D (●) all inhibit secretion. (B) Serotonin secretion. Electropermeabilized platelets, under the same experimental conditions as (A), were incubated with either 30 nM TPA (◆), 20 μM 1-oleoyl-2-acetylglycerol (OAG; ▲), 50 μM GTPγS (◇), or 10 mM fluoride (△) before being challenged with Ca^{2+}. The large shift to the left is representative of the second type of secretory response. (C) *N*-Acetylglucosaminidase secretion. Electropermeabilized platelets were incubated with either 30 μM OAG (▲), 10 nM TPA (◆), or 0.5 U/ml thrombin (■). Control cells (○). The increase in the extent of release rather than a shift in the activation curve represents the third type of secretory response to activators of protein kinase C.

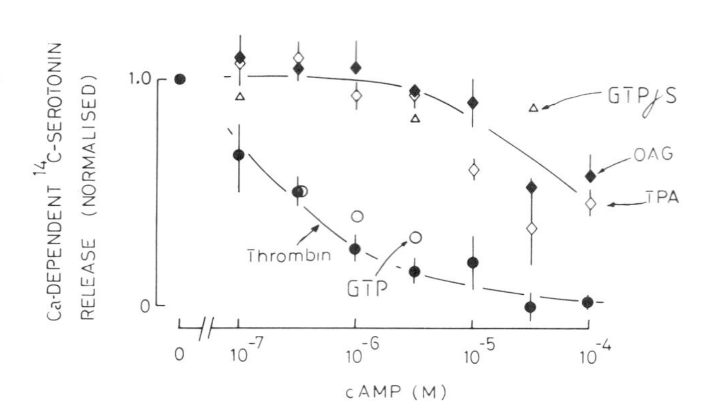

B

Figure 7. (A) The effect of cAMP on serotonin secretion from electropermeabilized platelets. Platelets were rendered leaky in a solution containing about 0.001 μM Ca^{2+}, incubated in various concentrations of cAMP before being challenged with 0.6 μM Ca^{2+} together with 0.6 U/ml thrombin (●), 10 μg/ml OAG (◆), 8 nM TPA (◇), 10 μM GTPyS (△), or 10 μM GTP (○). The amounts secreted as a result of raising the Ca^{2+} to 0.6 μM are shown and are normalized in terms of the amounts secreted in the absence of cAMP. (B) Some of the events linking the arrival of an agonist at the surface of a cell and secretion that have been studied by gaining access to the cytosol using the electropermeabilized cell. Receptor coupling to phospholipase C is mediated by a guanine nucleotide binding protein (i, Haslam and Davidson, 1984a,b,c; Knight and Scrutton, 1986b; ii, Merritt *et al.*, 1986). IP_3 or diacylglycerol are released (iv, Merritt *et al.*, 1986; vi, Haslam and Davidson, 1984b). IP_3 mobilizes Ca^{2+} from intracellular stores (v, Moos *et al.*, 1985). Diacylglycerol modulates the secretory response probably via protein kinase C (vii, Knight and Baker, 1983; viii, Knight and Scrutton, 1984b). The secretory response itself may be a complex event involving not only exocytosis but also endocytosis. The exocytotic site may be regulated by guanine nucleotide binding proteins (ix, Knight and Baker, 1985a; Barrowman *et al.*, 1986). An inhibitory effect of cAMP may be expressed at the level of phospholipase C coupling (iii, Knight and Scrutton, 1986b) and on Ca^{2+} mobilization (v, Moos and Goldberg, 1985). A rise in intracellular Ca^{2+} is buffered by intracellular organelles (Baker and Knight, 1981).

the site of exocytosis. The action of GTPγS (and fluoride$^-$) suggests an involvement of a guanine nucleotide binding protein and this may be very relevant to the mechanism of action of botulinum toxin. Other toxins such as cholera or pertussis toxin have been shown to express their potencies by activating guanine nucleotide binding proteins via ADP ribosylation. It may be significant that in electropermeabilized chromaffin cells, botulinum toxin type D ADP ribosylates a protein near 21k mol wt and a doublet close to 24k mol wt. The onset of ADP ribosylation is paralleled by the onset of inhibition of exocytosis (Adam-Vizi and Knight, unpublished). It is not improbable, therefore, that these proteins are guanylate binding proteins and that one or more of them are intimately associated with the control of exocytosis (Knight and Baker, 1985a; Knight *et al.*, 1985; Barrowman *et al.*, 1986).

9. SIGNAL TRANSDUCTION COUPLING

Measurement of intracellular Ca^{2+} in platelets or chromaffin cells suggests that the Ca^{2+} level necessary to trigger secretion may be much lower than that needed in leaky cells (Baker and Knight, 1986). The addition of diacylglycerol to the leaky cell preparations increases the Ca^{2+} affinity of the secretory process (presumably via protein kinase C) and brings closer together the Ca^{2+} requirements for secretion from leaky and intact cells. Endogenous diacylglycerol is produced as a result of an extracellular agonist activating phospholipase C. By manipulating the chemical environment of the cytosol it has been shown that a guanine nucleotide binding protein couples a receptor to phospholipase C activity. In the leaky platelet for example, thrombin produces diacylglycerol. The production of diacylglycerol by thrombin is further enhanced by GTP, or in the absence of thrombin by the nonhydrolyzable analogue GTPγS (Haslam and Davidson, 1984a,b). Similar effects are also seen on secretion. Thus, both diacylglycerol and thrombin increase the Ca^{2+} sensitivity of secretion (Knight and Scrutton, 1983, 1984a,b; Haslam and Davidson, 1984c). The effect of thrombin is further enhanced by micromolar levels of GTP, or as Fig. 6C shows by GTPγS or fluoride$^-$ in the absence of the agonist (Haslam and Davidson, 1984a; Knight and Scrutton, 1986b).

Agents that elevate intracellular cAMP, e.g., forskolin, also block secretion. Using the electropermeabilized platelet, it is possible to identify the main site of action and to propose a mechanism by which cAMP inhibits secretion (Knight and Scrutton, 1986b). Secretion triggered by Ca^{2+}, in the absence of other additives, is not affected by micromolar levels of cAMP. The increase of serotonin and lysosomal secretion induced by either diacylglycerol or TPA is also not affected by cAMP. However, the effect of thrombin and of GTP on Ca^{2+}-dependent secretion is inhibited by micromolar levels of cAMP (Fig. 7A). This suggests that cAMP acts at the level of phospholipase C and the finding that GTPγS-induced secretory response is unaffected by cAMP might suggest that cAMP operates by increasing the GTPase activity of the guanine nucleotide binding protein.

10. A SUMMARY OF SOME INTRACELLULAR EVENTS STUDIED BY GAINING ACCESS TO THE CYTOSOL USING THE ELECTROPERMEABILIZATION TECHNIQUE

Figure 7B gives an overall picture of part of the cascade of events involving stimulus–secretion coupling that have been investigated by gaining access to the cytosol using the electropermeabilization technique. Essentially an agonist interacts with its receptor and by means of guanine nucleotide binding proteins activates phospholipase C. The two products IP_3 and diacylglycerol release Ca^{2+} from intracellular stores, and increase the affinity of protein kinase C for Ca^{2+}, respectively. Elevated intracellular Ca^{2+} is buffered by the intracellular organelles. Activation of protein kinase C triggers secretion, which may involve both exocytosis and endocytosis, part of the secretory process possibly being regulated by guanine nucleotide binding proteins.

REFERENCES

Baker, P. F., and Knight, D. E., 1978, Calcium dependent exocytosis in bovine adrenal medullary cells with leaky plasma membranes, *Nature* **276:**620–622.

Baker, P. F., and Knight, D. E., 1981, Calcium control of exocytosis and endocytosis in bovine adrenal medullary cells, *Philos. Trans. R. Soc. London Ser. B* **296:**83–103.

Baker, P. F., and Knight, D. E., 1983, High voltage techniques for gaining access to the interior of cells: Application to the study of exocytosis and membrane turnover, *Methods Enzymol.* **98:**28–37.

Baker, P. F., and Knight, D. E., 1986, Exocytosis: Control by calcium and other factors, *Br. Med. Bull,* **42:** 399–404.

Baker, P. F., Knight, D. E., and Whitaker, M. J., 1980, The relation between ionised calcium and cortical granule exocytosis in eggs of the sea urchin Echinus esculentus, *Proc. R. Soc. London Ser. B* **207:**149–161.

Baker, P. F., Knight, D. E., and Roberts, C. S., 1982, Morphology of bovine adrenal medullary cells after exposure to brief intense electric fields, *Proc. Phys. Soc. London* **326:**6–7P.

Barrowman, M. M., Cockcroft, S., and Gomperts, B. D., 1986, Two roles for guanine nucleotides in the stimulus secretion sequence of neutrophils, *Nature* **319:**504–507.

Douglas, W. W., 1968, Stimulus secretion coupling: The concept and clues from chromaffin and other cells, *Br. J. Pharmacol.* **34:**451–474.

Dunn, L. A., and Holz, R. W., 1983, Catecholamine secretion by digitonin treated adrenal medullary chromaffin cells, *J. Biol. Chem.* **258:**4989–4993.

Gomperts, B. D., and Fernandez, J. M., 1985, Techniques for membrane permeabilisation, *Trends Biochem. Sci.* **10:**414–417.

Haslam, R. J., and Davidson, M. M. L., 1984a, Potentiation by thrombin of the secretion of serotonin from permeabilised platelets equilibrated with Ca^{2+} buffers, *Biochem. J.* **222:**351–361.

Haslam, R. J., and Davidson, M. M. L., 1984b, Receptor induced diacylglycerol formation in permeabilised platelets; possible role for a GTP binding protein, *J. Rec. Res.* **4:**605–629.

Haslam, R. J., and Davidson, M. M. L., 1984c, Guanine nucleotides decrease the free Ca^{2+} required for secretion of serotonin from permeabilised blood platelets, *FEBS Lett.* **174:**90–95.

Jones, P. M., Stutchfield, J., and Howell, S. L., 1985, Effects of Ca^{2+} and phorbol esters on insulin secretion from islets of Langerhans permeabilised by high voltage discharge, *FEBS Lett.* **191:**102–106.

Knight, D. E., 1981, Rendering cells permeable by exposure to electric fields, in: *Techniques in Cellular Physiology* (P. F. Baker, ed.), Elsevier, Amsterdam, pp. 1–20.

Knight, D. E., 1986, Calcium and exocytosis, *Ciba Found. Symp.* **122:**250–265.

Knight, D. E., and Baker, P. F., 1982, Calcium dependence of catecholamine release from bovine adrenal medullary cells after exposure to intense electric fields, *J. Membr. Biol.* **68:**107–140.

Knight, D. E., and Baker, P. F., 1983, The phorbol ester TPA increases the affinity of exocytosis for Ca in leaky adrenal medullary cells, *FEBS Lett.* **160:**98–100.

Knight, D. E., and Baker, P. F., 1985, Guanine nucleotides and Ca-dependent exocytosis: Studies on two adrenal preparations, *FEBS Lett.* **189:**345–349.

Knight, D. E., and Baker, P. F., 1985b, The chromaffin granule proton pump and Ca dependent exocytosis in bovine adrenal medullary cells, *J. Membr. Biol.* **83:**147–156.

Knight, D. E., and Koh, E., 1984, Ca^{2+} and cyclic nucleotide dependence of amylase release from isolated rat pancreatic acinar cells rendered permeable by intense electric fields, *Cell Calcium* **5:**401–418.

Knight, D. E., and Scrutton, M. C., 1983, Secretion induced by Ca^{2+} and thrombin in platelets subjected to dielectric membrane breakdown, *Thromb. Haemost.* **50:**93.

Knight, D. E., and Scrutton, M. C. 1984a, The relationship between intracellular second messengers and platelet secretion, *Biochem. Trans.* **12:**969–972.

Knight, D. E., and Scrutton, M. C., 1984b, Cyclic nucleotides control a system which regulates the Ca sensitivity of platelet secretion, *Nature* **309:**66–68.

Knight, D. E., and Scrutton, M. C., 1986a, Gaining access to the cytosol: The technique and some applications, *Biochem J.* **234:**497–506.

Knight, D. E., and Scrutton, M. C., 1986b, Effects of guanine nucleotides on the properties of 5-hydroxytryptamine secretion from electropermeabilised human platelets, *Eur. J. Biochem.* **160:**183–190.

Knight, D. E., Hallam, T. J., and Scrutton, M. C., 1982, Agonist selectivity and second messenger concentrations in Ca^{2+} mediated secretion, *Nature* **296:**256–257.

Knight, D. E., Niggli, V., and Scrutton, M. C., 1984, Thrombin and activators of protein kinase C modulate the secretory response of permeabilised human platelets induced by Ca^{2+}, *Eur. J. Biochem.* **143:**437–446.

Knight, D. E., Tonge, D. A., and Baker, P. F., 1985, Inhibition of exocytosis in bovine adrenal medullary cells by botulinum toxin type D, *Nature* **317:**719–721.

Merritt, J. E., Taylor, C. W., Rubin, R. P., and Putney, J. W., 1986, A G protein couples receptor to phospholipase C in exocrine pancreas, *Biochem. J.*

Moos, M., and Goldberg, N. D., 1985, Inositol trisphosphate promotes calcium release from internal platelet stores and cyclic AMP counteracts this effect, *Fed. Proc.* **44:**729.

Niggli, V., Knight, D. E., Baker, P. F., Vigny, A., and Henry, J.-P., 1984, Tyrosine hydroxylase in leaky adrenal medullary cells: Evidence for in situ phosphorylation by separate Ca^{2+} and cAMP dependent systems, *J. Neurochem.* **43:**646–658.

Nishizuka, Y., 1984, The role of protein kinase C in cell surface signal transduction and tumour promotion, *Nature* **308:**693–698.

Oetting, M., LeBoff, M., Swiston, L., Preston, J., and Brown, E., 1986, Guanine nucleotides are potent secretagogues in permeabilised parathyroid cells, *FEBS Lett.* **208:**99–104.

Phillips, J. H., Burridge, K. E., Wilson, S. P., and Kirshner, N., 1983, Visualisation of the exocytosis and endocytosis in secretory cycle in cultured adrenal chromaffin cells, *J. Cell Biol.* **97:**1906–1917.

Pollard, H. B., Pazoles, C. J., and Creutz, C. E., 1977, A role for anion transport in the regulation and release from chromaffin granules and exocytosis, *J. Supramol. Struct.* **7:**277–285.

von Grafenstein, H., Roberts, C. S., & Baker, P. F., 1986, The kinetics of the exocytosis endocytosis secretory cycle in bovine adrenal medullary cells, *J. Cell Biol.* **103:**2343–2352.

Chapter 19

Gene Transfer by Electroporation
A Practical Guide

Walter Förster and Eberhard Neumann

1. INTRODUCTION

Electric field effects on living cells have found considerable interest in cell biology, genetics, and biotechnology. In particular, electroporative gene uptake and electrofusion have become powerful new techniques based on the reversible electropermeabilization of cell membranes by high electric field pulses (Neumann and Rosenheck, 1972, 1973). Electropermeabilization constitutes a simple method for a controlled release of internal material from organelles and cells as well as, on the other hand, the introduction of foreign substances into cells. Of special interest for gene engineering is the direct transfer of DNA into recipient cells. Electric field pulse-mediated uptake of DNA into cells followed by the expression of the foreign gene was first demonstrated in 1982 (Wong and Neumann, 1982; Neumann *et al.*, 1982). The operative term "electroporation" (electric pore formation) was introduced in order to describe the electric field effect. The general applicability of the electroporation method has since been confirmed by numerous reports on electroporative gene transfer. Successful applications of this new technique include bacteria and yeast as well as plant and mammalian cells. Thus, direct gene transfer by electroporation, in addition to cell electrofusion, is an impressive new possibility offered by electric techniques of cell manipulation. In this chapter we discuss the new technique and some of the main aspects important for a proper understanding and practical application of electroporative gene transfer. This new technique was developed between 1980 and 1982 in our laboratory.

2. REVERSIBLE MEMBRANE ELECTROPERMEABILIZATION

Both electroporative gene transfer and electrofusion are apparently based on the same fundamental event: the reversible, nondestructive electropermeabilization of cell membranes by the action

Walter Förster and Eberhard Neumann • Faculty of Chemistry, Department of Physical and Biophysical Chemistry, University of Bielefeld, D-4800 Bielefeld 1, Federal Republic of Germany.

of short-duration high electric field pulses. This phenomenon was discovered in 1972 (Neumann and Rosenheck, 1972, 1973). It was demonstrated that, above a threshold field strength, a controlled release of internal material from suspended membranous vesicles (concomitant with uptake of solvent molecules) occurs in response to a field pulse of several microseconds' duration. There was no irreversible membrane damage (no dielectric breakdown). Apparently reversible changes in the membrane structure, of long-lived nature compared to the much shorter pulse duration, had been indicated by electrooptic techniques (Rosenheck *et al.*, 1975).

In contrast to reversible electroporation, irreversible and lethal effects on microorganisms caused by a whole series of similar electric field pulses were reported in 1967 (Sale and Hamilton, 1967, 1968). An inverse relationship has been found between cell size and a critical external electric field strength, E_c. Only if $E > E_c$ is there lysis of bacterial protoplasts. These results indicated that the cell membranes are the field-sensitive targets and that a critical transmembrane voltage V_c (of about 1 volt, independent of cell size) has to be created by the external field pulses. Sale and Hamilton (1968) suggested that the irreversible loss of the semipermeable membrane properties and ultimate lysis and cell death might be caused by conformational changes in the membrane structure.

In summary, short electric field pulses may either cause reversible or irreversible effects on membrane systems, depending on parameters such as, e.g., strength, duration, shape, and number of the pulses. These early results remain fundamental also for the application of electric field pulses in electroporative gene transfer. Details and current views on membrane polarization, generation of the critical transmembrane voltage by the external electric field pulse, and the electroporation process itself are discussed in the various chapters of this book as well as possible mechanisms of electropore formation. This chapter concentrates on practical, experimental features important for electroporative gene transfer.

3. ELECTROPORATIVE GENE TRANSFER, ELECTROTRANSFECTION, ELECTROTRANSFORMATION

3.1. Instrumentation

High electric field pulses are conveniently applied to suspended biological cells and other membrane particles by discharge of electric energy (current flow) through the suspension enclosed in a discharge chamber. This is easily achieved by the capacitor discharge (CD) technique adapted from fast relaxation kinetics (Eigen and DeMaeyer, 1963). The CD technique was originally applied in the experiments on reversible electropermeabilization (Neumann and Rosenheck, 1972, 1973). Obviously, because of its ease and fairly low costs, the CD technique has been preferred in most of the papers on electroporative gene transfer (see Appendix pp. 308–315).

With respect to the time characteristics, mainly two different types (shapes) of electric field pulses are currently used for membrane electropermeabilization. These types are either exponentially decaying pulses obtained by capacitor discharge, or rectangular (square) pulses delivered by, e.g., a cable discharge (Ilgenfritz, 1966; Grünhagen, 1973; Lindner *et al.*, 1977) or by special high-voltage pulsers (Sale and Hamilton, 1967; Kinosita and Tsong, 1977). Depending on the maximum voltage and power requirements in the discharge circuit, the technical realization of the pulsing equipment for given pulse shapes may vary considerably. These requirements are defined mainly by the geometry (electrode distance and surface, volume) of the discharge chamber, by the size and nature of the cells to be electroporated as well as by the electric conductance of the electroporation medium.

The discharge of a storage capacitor of capacity C, charged to the desired initial voltage, V_0, by a power supply, ensures a defined electric field pulse. If the inductance of the discharge circuit is negligibly small, the time characteristic of the field pulse within the discharge chamber of electrode distance d is given by

$$E(t) = E_0 \exp(-t/\tau)$$

with $E_0 = V_0/d$ and $\tau = RC$. The initial electric field strength, E_0, as well as the time constant of the electrical discharge, τ (i.e., the time of the field to decrease to E_0/e), are given to characterize the applied electric field pulse (see Appendix). The resistance, R, of the discharge chamber filled with the cell suspension together with the chosen condensor capacity, C, determines the pulse duration (τ). For details of the CD technique, see Eigen and DeMaeyer (1963).

The power supplies for electrophoresis, directly connected to a discharge chamber as described by Potter *et al.* (1984), are not well-suited to deliver a defined exponential pulse. The voltage knob settings are usually not accurate (Fromm *et al.*, 1985), and direct measurement of the pulse characteristics by an oscilloscope is necessary.

A broad variety of discharge chambers differing in geometry, volume, electrode material, and other characteristics have been designed and used in practice for the application of nearly uniform (homogeneous) electric field pulses to cell suspensions. Closed and open chambers, mostly of cubic or cylindrical shape and the filling volume between parallel, rectangular, or disc electrodes, have been used (Sale and Hamilton, 1967; Neumann and Rosenheck, 1972; Zimmermann *et al.*, 1974; Kinosita and Tsong, 1977; Berg *et al.*, 1984). The handling of fixed or variable volume is possible from several milliliters to a few microliters. Concentric electrode arrangements providing a moderate inhomogeneous field distribution have also been chosen. Commercially, a special "pipetting" discharge chamber based on this geometry has been designed. There are means for temperature control as well as for sterile and easy handling. Even a combination of controlled centrifugation with *in situ* electrical pulsing was reported. Among the easiest and most popular designs are: (1) disposable plastic spectrophotometer cuvettes ("cuvette chamber," semimicro, 1 cm pathlength, 0.4 to 0.5 cm electrode distance, up to 1 ml filling volume) with either platinum or stainless steel paddle electrodes inserted (Evans *et al.*, 1984) or (2) aluminum foil electrodes directly glued to the cuvette walls (Potter *et al.*, 1984). Most of the discharge chambers are now available as commercial equipment. A commercial instrument for electroporation became available in 1984 (by W. Rauhaus, Fa. Dialog, Düsseldorf); the number of manufacturers and models is still growing. For "historical" reasons the first version of the original Elektroporator® is shown in Fig. 1.

3.2. Early Experiments

Soon after studies were completed on electric hemolysis of red blood cells (Zimmermann *et al.*, 1974), Auer *et al.* (1976) demonstrated the possibility of uptake of genetic material into human red blood cells permeabilized by electric field pulses. Applying the CD technique (E_0 = 11 kV/cm; τ = 100 μsec), Auer *et al.* succeeded in trapping ^{3}H-labeled SV40 DNA in the cells. Hemolysis was the first measurable consequence of the electric pulsing. Subsequently, a defined resealing procedure (90 min incubation at 37°C) in the presence of the DNA or RNA was applied. Up to 35% of the added linearized SV40 DNA was finally sequestered within the cells. The entrapped DNA turned out to be DNase resistant due to the package. The authors noted that DNA trapping was much more effective with the linearized DNA than with the supercoiled form of SV40 DNA.

In 1982, the first electroporative gene transfer with subsequent actual gene expression was reported (Wong and Neumann, 1982; Neumann *et al.*, 1982). A plasmid DNA carrying the thymidine kinase (tk) gene was introduced into tk-deficient mammalian culture cells (mutant L tk$^-$ mouse cells) by the application of short electric field pulses. Stable transformants surviving in the HAT selection medium proved the direct gene transfer and expression. Several main features of the procedure, already worked out in the early paper (Neumann *et al.*, 1982), became a standard of application: (1) the highly conductive electroporation medium (HBS: Hepes-buffered saline, with 140 mM NaCl); (2) the fairly easy CD technique for pulse application; (3) the use of a few successive electric field pulses (e.g., three pulses at time intervals of several seconds); (4) the necessity of a postincubation of the cell–DNA mixture for about 10 min after pulsing, suggesting that the actual gene transfer is an afterfield effect; (5) the existence of a fairly sharp optimum in electric field strength, above a threshold value, at a given pulse duration in the microsecond range [e.g., E_0 = (8 ± 0.5) kV/cm for τ = 5 μsec]; (6) the limitation of transfer efficiency at higher field

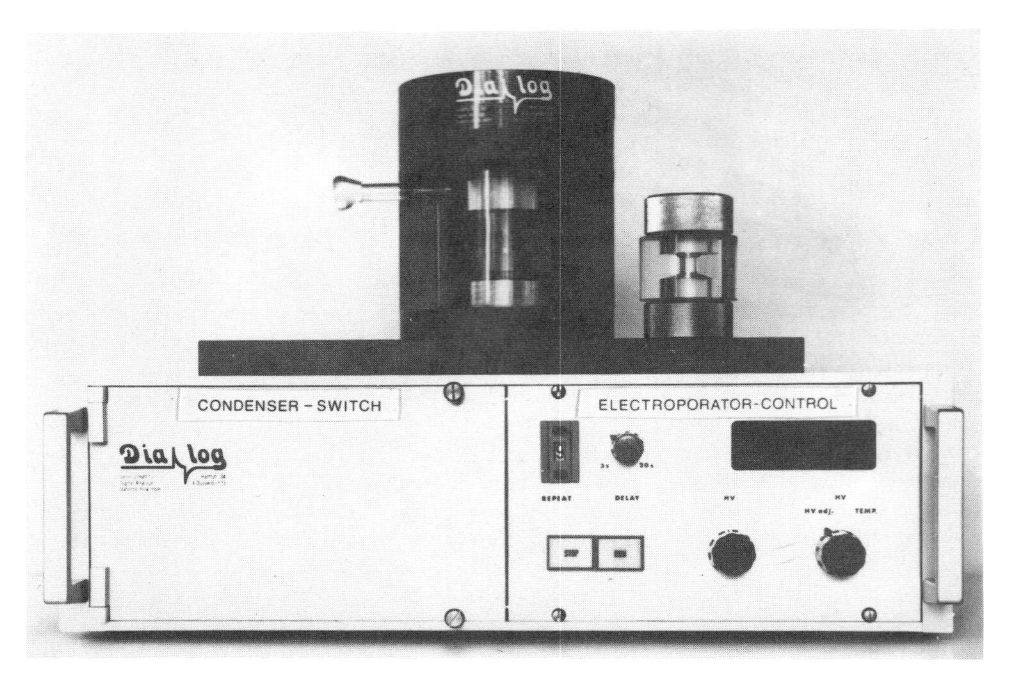

Figure 1. The first commercial electroporation apparatus: Elektroporator® of the Fa. Dialog, Düsseldorf, developed by W. Rauhaus.

strengths by lethal effects; (7) the increase in the number of transfected colonies with an increase in cell density and plasmid DNA concentration; and (8) the linearization of the plasmid resulting in significantly higher transformation yield compared to the circular form.

In fact, the first direct gene transfer into plant protoplasts by electroporation was achieved in 1983 in our laboratory (by Shillito, Bierth, and Neumann) along the lines of experience with animal cells. Subsequently, a combination of the electroporation technique with PEG treatment greatly enhanced the transformation efficiency (Shillito *et al.*, 1985). A similar combination of PEG treatment and electric field pulse application was used to transform bacterial protoplasts (Shivarova *et al.*, 1983) and even intact yeast cells (Hashimoto *et al.*, 1985; see Appendix).

The rapidly growing number of reports on electroporative gene transfer, especially since 1985, demonstrates the general applicability of this technique not only for mammalian, but also for plant, bacterial, and yeast cells. Several experimental parameters of the procedure have since been investigated in more detail, including: (1) the temperature (pulsing and postincubation at 0°C recommended); (2) the pulse shape (exponential versus rectangular); (3) the pulse duration (microsecond versus millisecond range); and (4) the addition of divalent ions to the electroporation medium (Ca^{2+}, Mg^{2+}).

In the Appendix, we have extracted the main features from representative papers in order to give an overview as well as to provide data for comparison and generalization.

3.3. Guidelines for Gene Transfer by Electroporation—Important Experimental Parameters

3.3.1. Electric Field Pulse Duration and Shape

3.3.1a. Microsecond Time Range. The reversibility of membrane electropermeabilization has been linked to a sufficiently short duration of the external electric field pulses applied. Using the experience accummulated until 1981, the first electroporative gene transfer experiments were performed with electric field pulse durations in the microsecond range in order to avoid irreversible destructive effects on the cell membranes. In fact, pulse times between 5 and 200 μsec of exponential and square pulses have been chosen in systematic studies on "electrotransfection" and "electrotransformation" (see Appendix; *exponential pulses:* Neumann *et al.*, 1982; Shivarova *et al.*, 1983; Evans *et al.*, 1984; Karube *et al.*, 1985; Zerbib *et al.*, 1985; Stopper *et al.*, 1985; Shillito *et al.*, 1985; Riggs and Bates, 1986; *square pulses:* Zerbib *et al.*, 1985; Langridge *et al.*, 1985; Hama-Inaba *et al.*, 1986; Ou-Lee *et al.*, 1986; Hibi *et al.*, 1986). Equipment originally developed for cell electrofusion (square pulse generators and chambers) has been applied in some of the studies on electroporative gene transfer into plant protoplasts (Langridge *et al.*, 1985; Nishiguchi *et al.*, 1986; Ou-Lee *et al.*, 1986; Hibi *et al.*, 1986). The pulse times were also adapted from typical electrofusion conditions (1 to 100 μsec) as well as the fairly high repetition rate in multipulse application. Because of power and field strength limitations, however, such equipment including the dielectrophoresis setup is not advised for systematic studies on electroporative gene transfer.

In the case of the square pulse experiments with pulse times of 10 to 100 μsec, it is necessary to use an electroporation medium of sufficiently low conductance. This limitation was circumvented by Hama-Inaba *et al.* (1986) using a "semirectangular" pulse shape generated by a time-controlled cutoff of a fairly slow CD. In this manner, pulse times of up to 50 μsec can be realized even in high-conductance phosphate-buffered saline (PBS) medium.

A systematic comparison of exponential versus rectangular pulse shapes with respect to electrotransfection efficiency is still lacking. In a paper dealing with this question (Zerbib *et al.*, 1985), no clear-cut conclusion could be drawn because, for technical reasons, different electroporation media for both types of experiments had to be used (Appendix). Nevertheless, in these experiments with tk^- Chinese hamster ovary (CHO) cells, exponential pulses (1) resulted in a higher yield and (2) revealed more specifically the threshold effect with respect to electric field strength that is typical for the electropermeabilization phenomenon.

3.3.1b. Millisecond Time Range. The alternative pulse duration range used in practice is between 5 and 20 msec: it obviously was obtained by chance by Potter *et al.* (1984) when a standard electrophoresis power supply was directly discharged through a cuvette chamber (see Section 3.1); PBS was used as the electroporation medium. Fromm *et al.* (1985) showed that, in this case, instead of the quoted 2000 V setting, only some 120 to 150 V is really delivered to the cell suspension. This is due to the load mismatch within the discharge circuit (see Hofmann, this volume). This creates an initial electric field strength of about 300 V/cm and yields a nearly exponential pulse with a time constant (τ) of about 15 to 20 msec (Shigekawa, 1987). Nevertheless, electrotransfection was observed and because of the widespread availability of electrophoresis equipment, the experimental procedure described by Potter *et al.* (1984) was subsequently applied by many groups to a large variety of mammalian cell lines (see Appendix). Of course, no optimization with respect to electric field strength and pulse duration can be expected with only one "standard" parameter setting. Chu *et al.* (1987) reported on systematic reinvestigations of electrotransfections of a variety of mammalian cell lines (millisecond pulses by CD) and on optimizations with respect to various electroporation parameters.

Optimum electroporation conditions for transient expression and stable transformation of plant protoplasts were investigated by Fromm *et al.* (1985, 1986) using millisecond pulses delivered by CD (see Appendix). Fairly long electric field pulses were also applied to introduce TMV and CMV RNA and even whole TMV and CMV particles into tobacco protoplasts (Okada *et al.*, 1986). The transformation of *intact* yeast cells with plasmid DNA was reported by Hashimoto *et al.* (1985) who applied one or several millisecond pulses in combination with PEG treatment and long incubation times. Thus, millisecond pulse durations have turned out to be useful in electroporative gene transfer. The popularity of these conditions was due to the reduced electric field strength. Charging voltages of only 200 to 400 V are sufficient for the experiments (standard 0.4 cm electrode distance in cuvette chambers, pulse times between 2 and 20 msec) as reported for mammalian cells (Chu *et al.*, 1987; Shigekawa, 1987) and for plant protoplasts (Fromm *et al.*, 1985, 1986; Okada *et al.*, 1986).

A systematic comparison of microsecond versus millisecond pulses with respect to transfection efficiency using the same transfection system has been undertaken by Shigekawa (1987). CHO cells were transfected with the pSVneo vector and pulses of 500 μsec were compared with pulses of 2 to 10 msec. As already mentioned for the comparison between exponential and square pulse shapes, no clear-cut result was obtained owing to technical limitations. For millisecond pulses, a fairly narrow peak was observed in the voltage dependence of transfection efficiency (1 kV/cm, at about 60 to 80% cell survival). The survival rate at higher fields very quickly decreased, thus limiting the useful range of field strength. In contrast, a much broader dependence is seen at the shorter pulses (500 μsec), but in this case the maximum transfection efficiency unfortunately could not be reached because of the voltage limitation of the equipment.

3.3.2. Electric Field Strength

One of the fundamental characteristics of membrane electropermeabilization by short electric field pulses in the microsecond range is the existence of a threshold electric field strength for the onset of reversible permeability changes. A corresponding critical transmembrane voltage of $V_c \approx$ 0.5 to 1 V has been established in many cases. For practical purposes, the necessary critical external electric field strength, E_c, can be estimated:

$$E_c \approx \frac{4\ V_c}{3\ D} \quad \text{and} \quad E_c\ (\text{kV/cm}) \approx \frac{13.1 \text{ to } 6.7}{D\ (\mu\text{m})}$$

where D is the diameter of the suspended spherical cells. Such estimates are, however, only crude approximations. A variety of parameters such as the cell membrane composition and the physiological state of the cells will affect the value of E_c. Of course, the inverse relationship between E_c and the dimensions of the cells is fundamental; this means that electroporation of small bacteria requires much higher field strengths than that of big plant protoplasts (at the same pulse duration).

In this connection the relationship between pulse duration and critical electric field strength is practically important, but there is no quantitative general expression for the functional dependence between these parameters. Therefore, it became common practice to study the survival of cells after treatment with electric pulses of increasing field strength, as proposed in the original electric gene transfer experiments (Wong and Neumann, 1982). The resulting "survival curves" for given conditions (pulse duration and shape, medium, temperature) are the basis for selecting electroporation conditions for the actual gene transfer experiments. A typical example of a survival curve and of the field strength dependence of electrotransformation is shown in Fig. 2. It should be noted that there are noticeable differences between short-term survival (judged, e.g., within 2 hr after pulsing by dye exclusion test) and long-term growth ability of electrically treated cells (Riggs and Bates, 1986; Hama-Inaba *et al.*, 1986). Long-term stability is needed for successful electrotransfection.

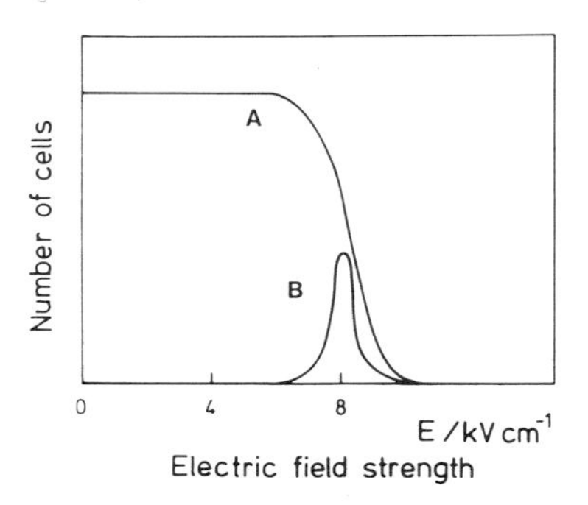

Figure 2. Scheme (A) of survival curve and (B) of the number of electrotransformed cells.

Numerous reports have established that a relative optimum in electrotransfection efficiency exists at field strengths yielding survival rates between 80 and 30% (Hashimoto *et al.*, 1985; Karube *et al.*, 1985; Hama-Inaba *et al.*, 1986; Chu *et al.*, 1987; Fromm *et al.*, 1985; Riggs and Bates, 1986). These results can be qualitatively rationalized. At increased field strengths of the electric pulses, reversible electroporation precedes the irreversible membrane permeability changes, leading to cell death. Successful electroporative gene transfer only occurs in the range of reversible electric field effects (reversible electroporation). Cell populations typically are nonsynchronized and often have a fairly broad size distribution. A field pulse of given strength is critical for only the subpopulation of cells of a proper size; these cells are then electrotransfected. The lethal effects indicate the upper limit of the range of reversibility of the electric field pulse effects.

The electric field strength limitation was critical in some experiments on the electrotransformation of small objects like bacteria or yeast (Shivarova *et al.*, 1983; Karube *et al.*, 1985; Miller *et al.*, 1988). The optimum field strength for maximum transformation efficiency has not been discovered, partly because of technical problems. Early attempts on the electrotransformation of *Bacillus subtilis* were not successful with short microsecond pulses in the field strength range (up to about 30 to 40 kV/cm).

A recent example from our laboratory illustrates the typical approach to elaborate on suitable electroporation conditions. It refers to the attempts at electrotransformation of gram-positive amino acid-producing Corynebacteria (Wolf *et al.*, 1988). At first the survival of osmotically sensitized cells (spheroplasts) was investigated at pulse durations of 5 μsec, 300 μsec, and 5 msec; different media as well as different equipment are necessary. It was noticed that 5-μsec pulses up to $E_0 = 40$ kV/cm did not affect the viability of the bacterial spheroplasts. In contrast, using the longer pulse durations, lethal effects were observed within the field strength ranges available (up to 25 kV/cm for the 300-μsec pulses and up to 6 kV/cm for the 5-msec pulses). At field strengths of 20 kV/cm and 300-μsec pulses, 60% survival and stable transformants were obtained after regeneration and selection.

3.3.3. Electroporation Medium

No serious problems arose concerning the proper choice of the electroporation medium for mammalian cell lines. In the very first experiments, highly conductive media, namely HBS (Neumann *et al.*, 1982) or PBS (Wong and Neumann, 1982; Potter *et al.*, 1984), were used. They are now standard media (without divalent ions added; see Appendix). Usually, the cells are washed and resuspended in one of these media just before electrotransfection. After mixing with the plasmid DNA, the suspension is pulsed (''zapped'') and, after an incubation period (see below), transferred to the growth medium.

The influence of divalent cations (Mg^{2+}, Ca^{2+}) on the transfection efficiency was also investigated (Wong and Neumann, 1982; Neumann *et al.*, 1982; Fromm *et al.*, 1985; Hibi *et al.*, 1986). While Mg^{2+} ions increase DNA adsorption on the cell surfaces, the actual overall electrotransfection efficiency is largest when these ions are not added (Wong and Neumann, 1982; Neumann *et al.*, 1982). Therefore, divalent ions are not usually added to mammalian cells.

In the case of plant protoplasts, on the other hand, it has been found that besides the necessary osmotic stabilization by sugars such as mannitol (see Appendix), 4 mM $CaCl_2$ is favorable (Fromm *et al.*, 1985). Hibi *et al.* (1986) sought to systematically study the effect of Mg^{2+} and Ca^{2+} ions on the electrotransfection of tobacco protoplasts by TMV RNA. Because of technical limitations, they could only check the concentration range up to 100 μM. There was no effect of Ca^{2+} ions, thus confirming the results of Nishiguchi *et al.* (1986) who found no dependence on Ca^{2+} ions up to 8 mM. On the other hand, Mg^{2+} ions at 100 μM had a small promotive effect. The optimized electrotransformation protocol of Shillito *et al.* (1985) for tobacco leaf protoplasts applies 6 mM $MgCl_2$. It has been reported that Mg^{2+} ions are essential for electric field-mediated transformation of carrot protoplasts (Langridge *et al.*, 1985). In summary, with plant protoplasts usually some divalent ions up to about 5 mM are added to the electroporation medium (see Appendix).

The nature of the buffer used (e.g., PBS, HBS, or MES) has no marked influence on the infection rate of tobacco protoplasts if TMV RNA is electrically introduced. The yield is also independent of the pH between pH 5 and 8 (Okada *et al.*, 1986). Fromm *et al.* (1985) found a dependence of the efficiency of transient expression (CAT activity) in electrotransformed carrot protoplasts on PBS concentration. However, this dependence most probably reflects the different pulse times. Thus, the different survival rates result from the increase in the electric resistance when the PBS buffer is diluted. Chu *et al.* (1987) studied the dependence on the salt concentration of the CAT activity of mammalian cells after electroporative gene transfer in a Hepes buffer. The salt concentration was varied between 25 and 100 mM NaCl keeping the osmolarity constant by addition of sucrose. The highest salt concentration (100 mM) gave the strongest CAT signal. The standard HBS buffer with 137 mM NaCl was even more suitable for the gene transfer. The role of different pulse time constants due to changing salt concentrations remains unclear.

Besides the choice of electroporation medium for pulsing, special medium conditions may also be favorable for a definite resealing process after pulsing, as claimed by Stopper *et al.* (1985). This study confirms the early electroporation data of mouse L cells by Wong and Neumann (1982) and Neumann *et al.* (1982).

3.3.4. Incubation Times and Temperature

It was noticed in the first electroporative gene transfer experiments (Neumann *et al.*, 1982) that a postincubation period of up to 10 min (after the electric field pulse application) is necessary for the cell–DNA mixture. This clearly indicates that the DNA uptake resulting in gene expression is an "afterfield" effect which is triggered by the short-duration electric field pulse.

The longevity of the reversibly electropermeabilized membrane state is well known (Neumann *et al.*, 1982). It has been studied systematically by measuring the uptake of normally impermeable molecules (drugs, dyes) added to the cells at different times after pulse application (e.g., Jacob *et al.*, 1981; Teissié and Rols, 1986). A strong temperature dependence of the resealing process has been documented in these studies. It has been repeatedly found that at low temperatures, the electropermeabilized state of the cell membrane can be maintained for minutes and even up to an hour (Knight and Baker, 1982). The slow resealing at low temperatures fits the observation made in electrotransfection experiments (Potter *et al.*, 1984) that the efficiency of gene expression can be increased by pulsing and postincubation at low temperatures. Thus, electroporative gene transfer experiments are usually performed at 0 to 4°C. According to an often used standard protocol (see Table 1), the washed cells are resuspended in an ice-cold electroporation buffer, the plasmid DNA is added, and this mixture is incubated for 5 to 10 min. Then one or several pulses are applied. After a postincubation of about 10 min (all at the low temperature), the cells are transferred to a growth medium at the necessary elevated temperature.

This protocol also reflects some of the supposed steps in the overall electrotransfection and electrotransformation procedures: (1) The necessary DNA–cell contact is established within the preincubation period by DNA adsorption on the cell membrane surface. Adsorption may be enhanced either by charge neutralization through the binding of divalent cations or by pretreatment with enzymes partially removing the extracellular matrix or cell walls. (2) DNA transport across the cell membrane is initiated by one or several subsequent short electric field pulses. (3) The actual permeation is a sequence of fairly slow (millisecond to minute range), but still unknown, "transfer steps" within the postincubation period. (4) The DNA transport is terminated by membrane resealing at elevated temperature. (5) If the DNA is inside the cell, intracellular events lead to actual gene expression.

3.3.5. Miscellaneous

3.3.5a. Number of Electric Field Pulses. The possible advantage of the application of several subsequent field pulses (e.g., three to five pulses at time intervals of several seconds; see Appendix) is not clear. Two simple effects might be envisaged. First, a mere summation of the favorable events might occur if, after a second pulse, other independent sites of DNA–cell contacts become electroporated and thus also these sites become involved in DNA-transfer steps. Second, if the time between two pulses is small compared to the rotational diffusion of the cells, a second pulse could enhance or accelerate the sequence of transfer steps at the same electropermeabilized site following an event triggered by the first pulse. Most probably this was the case in the study by Hama-Inaba *et al.* (1986), where the time interval between two pulses was only 50 msec. Here, the efficiency in the double-pulse experiment was higher than that in an optimized single-pulse experiment.

3.3.5b. Linearization of Plasmid DNA. If tolerable, plasmid DNA should be linearized for electroporative gene transfer experiments. An enhancement of the gene expression yield between 3- and 20-fold was observed with linearized plasmid DNA compared to the yield of the circular or supercoiled form for transient expression as well as for stable transformation of different kinds of cells (see Appendix) (Neumann *et al.*, 1982; Falkner *et al.*, 1984; Potter *et al.*, 1984; Shillito *et al.*, 1985; Stopper *et al.*, 1985; Toneguzzo *et al.*, 1986). These observations point to a possible mechanism of electroporative DNA transfer and might be correlated with the size of the electropores or "electrocracks" (Sugar *et al.*, 1987) necessary for transfer steps. It is expected that the permeation sites for the linear DNA molecules fitting the diameter of the hydrated double-helical DNA "rod" can be smaller than those necessary for a supercoiled, i.e., more "globular," tertiary structure.

3.3.5c. Addition of Carrier DNA. It is well known from classical transfection, transformation, and microinjection procedures that the addition of "carrier" DNA (i.e., nonspecific high-molecular-weight DNA) leads to an increase in the level of gene expression of the specific plasmid transferred (Krens *et al.*, 1982). A similar positive effect of the addition of carrier DNA to the cell–plasmid DNA mixture was also established in electrotransfection experiments (Shillito *et al.*, 1985; Langridge *et al.*, 1985; Chu *et al.*, 1987). The nature of the carrier DNA was found to be important (Chu *et al.*, 1987). This is in line with the results of other, nonelectroporative cell transfection and transformation methods.

The mechanism of the effect of carrier DNA is still unknown. It is argued that it is linked to the activity of intracellular nucleases. In the presence of excess carrier DNA, a sufficient amount of the specific foreign plasmid DNA is preserved from cleavage.

4. PERSPECTIVES

Recent progress in molecular genetics and in biotechnology demonstrates that efficient and reliable gene transfer is an important prerequisite. It appears that direct transfer of DNA or mRNA by electroporation is not only simple and readily controlled but also generally applicable. The efficiency of the electric technique in many cases compares well with the chemical methods. In several cases the electric method is clearly superior (see Appendix).

One of the main aims of further improvements of DNA transfer techniques is the development of conditions such that cells can be transformed without excessive pretreatment, for instance, avoiding protoplast formation. Electroporative gene transfer, in our opinion, is a likely candidate for realization of these goals. In addition, easy and reliable handling of the electroporation method together with further technical development of equipment will make this method even more attractive to a broad range of cell biologists and biotechnologists.

5. APPENDIX

The following pages present information regarding research on the use of electroporation for direct gene transfer.

Appendix. Direct Gene Transfer by Electroporation

	Authors	Recipient cells	Plasmid/gene	Pulsing CD: E_0; τ	Results
			Mammalian cells		
1982	Neumann *et al.* (München)	Mouse L tk^- fibroblast cells	pAGO with tk gene from herpes virus (HAT select.)	CD: 3×8 kV/cm; 5 μsec 20°C, 10 min postincub., HBS (without Mg^{2+})	Sharp optimum in field strength, incubation after pulse necessary, linear plasmid better than circular, 100 col./10^6 cells/μg DNA
1984	Falkner *et al.* (München, Bielefeld)	Mouse lymphoid cell lines	Plasmids with Ig κ gene	CD: 3×8 kV/cm; 5 μsec 20°C, 10 min incub., HBS	Linearized plasmid better, but κ gene not intact, few copy number integrated
	Evans *et al.* (La Jolla)	Rat neuronal cells (B50 neuroblastoma)	Thy-1 glycoprotein encoding gene	CD: 3×8 kV/cm; 5 μsec 20°C, 10 min incub., DME medium + 20 mM $MgCl_2$ (plastic cuvette)	Two to five copies of plasmid per genome integrated in transformed clones
	Potter *et al.* (Boston)	Mouse B and T lymphocytes and fibroblasts	Mouse and human Ig κ gene	Pulse: ISCO 494 power supply directly discharged through cuvette, no definite pulse parameters given, estimated: 320 V/cm; 17 msec (CD also tried) 0°C, 5 min preincub., pulse, 10 min postincub., PBS	Up to 300 transf./10^6 cells, linear > supercoiled, low temperature favorable, few copy number (1–15) integrated, mitotic arrest by colcemid favorable
1985	Sugden *et al.* (Madison)	B lymphoblasts (transformed by Epstein–Barr virus)	pHEBo (hygromycin B resistance)	Pulse: ISCO power supply (acc. to Potter) (no definite pulse amplitude given; $\tau \sim 30$ msec) 0°C, PBS, cuvette	Stable transformation of up to 3% of survivors, 200- to 600-fold more effective than protoplast fusion
	Prochownik (Ann Arbor)	Mouse M12 myeloma cells (transient expression)	Plasmid carrying CAT gene	Pulse: ISCO power supply (acc. to Potter) 0°C. PBS, cuvette	Successful transformation, CAT activity after 48 hr observed
	Zerbib *et al.* (Toulouse)	Hamster CHO tk^- cells in suspension or monolayer	pAGO with tk gene from herpes virus	CD: 3×6 kV/cm; $\tau \sim 10$ μsec (20°C, HBS)	150 transf./10^6 cells/μg DNA, threshold: > 4 kV/cm
				Square pulses: 3×1.5 kV/cm; 50 μsec (low ionic strength)	70 transf./10^6 cells/μg DNA, 4 plasmids/transformed cell in monolayer
	Smithies *et al.* (Madison)	Mouse/human hybrid cell line HU 11	Plasmid with human β-globulin (neo resistance)	CD: 2 kV/cm; τ not given, estimated 700 μsec (14 μF) 0°C, 10 min incub., HBS	Insertion of plasmid in homologous sequences at β-globulin locus detected in one of 10^3 transformed cells
	Stopper *et al.*	Mouse L cells	pSV2-neo, lin-	CD: 10kV/cm; 5 μsec	Linearized DNA 20-fold better, low T:

Year	Author	Cells	DNA	Conditions	Results
	(Würzburg)		earized (neo resistance)	4°C, 2 min incub., then 37°C, 20 min incub., 1mM phosphate buffer + 30 mM KCl + 0.2 M inositol	one pulse sufficient, optimum at low DNA concentration (0.1 μg/ml), resealing important
1986	Weir and Leder (Boston)	Mouse B and pre-B cell lines	Functionally rearranged VκII gene	Pulse: ISCO power supply (acc. to Potter)	Gene successfully introduced both transiently and permanently by electroporation
	Tur-Kaspa *et al.* (Bronx)	Primary rat hepatocytes	pSV2CAT (linearized) (transient expression)	Pulse: ISCO power supply (acc. to Potter) 0°C, 15 min pre- and 15 min postincub., PBS	Hepatocytes not transfected by common methods because of cytotoxicity, electroporation successful
	Toneguzzo *et al.* (North Billerica)	Human and mouse lymphoid cell lines	pSV2CAT or pSV7neo: transient/stable expression (circ./lin.)	Pulse: ISCO power supply two-phase pulse characteristics, 30 msec tail 0°C, 10 min incub., PBS	Transient expression and stable transformation: no correlation, linear DNA better in both cases, mitotic arrest not improving yield
	Toneguzzo and Keating (North Billerica, Toronto)	Human bone marrow cells	pSV2cat, pSV2dhfr, pSV7neogpt; transient and stable expression	Pulse: ISCO power supply (acc. to Potter) 0°C, 10 min incub., PBS	Linearized plasmid DNA successfully transferred into human hematopoietic stem cells, DNA stably maintained and expressed in differentiated progeny
	Shimizu *et al.* (Madison)	Human HLA mutant lymphoblastoid cells	Plasmid vector pHP-Te with class I HLA genes	Pulsing: acc. to Potter, Sugden 0°C, PBS	Successful introduction of human MHC genes into chromosomal DNA of EBV-transformed LCLs by electroporation
	Igarashi *et al.* (Boston)	Rat pituitary GH_4C_1 cells, human HeLa cells	Plasmid pPTH-neo101 (encoding parathyroid hormone)	Pulsing: acc. to Potter	One transformant per 10^4 to 10^5 cells for both cell lines
	Yancopoulos *et al.* (New York)	tk^- derivative of 38B9 A-MuLV-transformed pre-B cell line	T cell receptor variable region gene segments on special plasmid construct	Pulse: ISCO power supply (acc. to Potter) 0°C, PBS	Linearized plasmid successfully transfected
	Drinkwater and Klinedinst (Madison)	EBV-transformed lymphoblastoid cell line (LCL-721)	Shuttle vector pHET	Pulse: ISCO power supply (acc. to Potter, Sugden)	Supercoiled plasmid DNA introduced, about 2% of transfected cells stably expressed the gene

(*continued*)

Appendix. *(Continued)*

Authors	Recipient cells	Plasmid/gene	Pulsing CD: E_0; τ	Results
Atchison and Perry (Philadelphia)	Plasmacytoma S 194 cells	Ig κ constructs containing single or tandem V_j promoters	Pulsing: acc. to Potter	Stable integration of linearized plasmid, copy number: one to five transfected κ genes in stably transformed S 194 cells detected
Zamecnik *et al.* (Shrewsbury)	HeLa cells	Synthetic oligodeoxynucleotides (10 to 30 bp)	Pulsing: acc. to Potter (modified) 0°C, DME medium without serum, 15 min incub.	Introduction of oligonucleotides by electroporation; approach to inhibit HTLV-III virus replication and gene expression in infected cells by the introduction of oligonucleotides complementary to viral RNA or proviral DNA
Reiss *et al.* (New Haven)	Epidermal cells (MK-1 murine keratinocytes)	pSV2CAT: transient expression, pSV2neo: stable integration (geneticin resistance)	Pulse: ISCO power supply (acc. to Potter) 0°C, 10min pre- and 10 min postincub., PBS	Eight resistant clones/10^6 cells/μg DNA, linearized plasmids
Hama-Inaba *et al.* (Chiba, Osaka)	Mouse mammary carcinoma cells	pSV2neo (geneticin resistance)	Semirectangular pulses (CD pulses cut-off, times 2 to 50 μsec): typically 3 kV/cm, 50 μsec 20°C, 10 min incub., PBS	Systematic study of pulsing parameters, optimum: double pulses each of 50 μsec at 50 msec time interval 80 times better than calcium phosphate coprecipitation
Sahagan *et al.* (North Billerica)	Murine myeloma cells (non-antibody producing)	Chimeric immunoglobulin genes	Pulse: ISCO power supply (acc. to Potter) 4°C, PBS	Transfectomas producing murine/human chimeric antibody obtained with high frequency, chimeric Ig able to recognize human tumor-specific antigen
Lachman *et al.* (Bronx)	Mouse erythroleukemia cells	pSV2neo and pMT-myc (linearized, cotransfected)	Pulsing: acc. to Potter	Within G418-resistant clones cotransfectants identified which contained 1–2 copies of pMT-myc
Allen *et al.* (Cambridge, Mass.)	Murine R1E cell line	Plasmid construct containing H-2D gene	Pulsing: acc. to Potter	Clones successfully selected which gave a high level of binding of specific monoclonal antibody
Chung *et al.*	Ly65 Burkitt lym-	Human c-myc gene	Pulsing: acc. to Potter	Using transient transfection the role of c-

	(Boston)	phoma cells	joined to bacterial CAT gene	0°C, 10–20 min preincub.	myc regulatory sequences in activation is studied in detail
	Khazaie *et al.* (London)	Putko cells which express embryonic but no adult globin genes	Hybrid plasmid constructs with AGPT gene, mouse H2 gene, and human β-globin promoter	Pulses: ISCO power supply (acc. to Potter), two pulses, 4°C, 30 min postincub., PBS	Stable introduction of hybrid constructs into Putko and K562 cells, expression of the Hβ hybrid gene
	Narayanan *et al.* (New Haven)	Mouse bone marrow cells	RSVCAT and SV_2NEO plasmids, linearized	Pulse: ISCO power supply (acc. to Potter) 4°C, 10 min pre- and 10 min postincub., PBS	Transient expression, cotransfection of both plasmids successful, *in vivo* expression after injection of RSVCAT transfected bone marrow cells into irradiated mice
1987	Chu *et al.* (Stanford)	Human fibroblasts, HeLa cells, CV-1 monkey kidney cells, mouse embryo fibroblasts (NIH3T3)	pRSV-gpt, pRSV-neo, pRSV-cat, pMT-hGH-SV2 plasmids (supercoiled and linearized)	CD: optimal 500–700 V/cm, 7 msec (3.5 msec tested) 20°C, 10 min incub., HBS, carrier DNA	Successful transient and stable transfection in all cell lines, sharp optima in field strength at given $\tau \approx 7$ msec, systematic study of different electroporation conditions 1 to 2 orders of magnitude better than Ca phosphate coprecipitation
	Shigekawa (Richmond)	CHO cells, L cells, Cos cells, Human 293, several suspension cells	pSVneo	CD: 500–750 V/cm; 5 msec (range 2 to 15 msec studied) 0°C, 10 min pre- and 10 min postincub., PBS or phosphate-buffered sucrose, disposable cuvette with Al foil electrodes	Linearized plasmid for stable transfection, circular form for transient expression, usually sharp optimum in field strength at given τ around 5 msec
	Spandidos (Glasgow)	Mouse Friend, human k562 erythroleukemic cells	Plasmid Homer 6 with aph gene (stable transformation), pLW4 carries CAT gene (transient expression)	CD: up to 4 kV/cm tested; no definite τ 5 min preincub. on ice, HBS	Optimum conditions for Homer 6: 80 μg plasmid DNA, 1.5 kV/cm, 1×10^7 recipient cells, 24 hr preselecting time; comparable results with transient expression; 3–5-fold increase in transf. frequency with linearized DNA; 5-fold higher transf. frequency compared to calcium phosphate technique with same DNA concentration

(*continued*)

Appendix. *(Continued)*

	Authors	Recipient cells	Plasmid/gene	Pulsing CD: E_0; τ	Results
				Plant cells	
1985	Shillito *et al.* (Basel)	Tobacco leaf mesophyll protoplasts	pABDI conferring kn resistance	CD: 3 × 1.5 kV/cm; 10 μsec Optimized protocol by strict sequence of steps including carrier DNA, heat shock, 8% PEG, pulsing, incubations; 20°C, 0.4 M mannitol, 6mM $MgCl_2$, ''DIA-LOG'' electroporator	Up to 2% transformants, linear plasmid more effective (10 times)
	Fromm *et al.* (Stanford)	Protoplasts from carrot, tobacco, maize (transient expression) mono- and dicots!	pNOSCAT carrying CAT gene (supercoiled)	CD: 875 V/cm; 15 msec (ISCO power supply also tried, acc. to Potter) 0°C, 10 min incub., HBS + 0.2 M mannitol + 4 mM $CaCl_2$; cuvette	Optimum: if 4 mM $CaCl_2$ added, msec pulses used successfully, ~50% survival, electroporated protoplasts re-form cell wall, divide and form callus
	Langridge *et al.* (Ithaca)	Carrot protoplasts	pTiC58 enabling hormone-independent protoplast regeneration	Square pulses: 6 × 0.5 to 3.8 kV/cm; 6 μsec (0.2 sec. intervals) 0°C, GCA vessel: 10 μl between Pt wire electrodes	200 mM $MgCl_2$ and low temperature essential, carrier DNA added
1986	Fromm *et al.* (Stanford)	Maize protoplasts	pCaMVNEO conferring kn resistance	CD: 500 V/cm; 2 msec 0°C, 10 min incub., medium: see above (Fromm)	Up to 1% of dividing protoplasts become kanamycin resistant, stable transformation of monocot plant cells
	Ecker and Davis (Stanford)	Carrot protoplasts (transient expression)	Plasmids with CAT gene in sense or antisense orientation	CD: 875 V/cm; 10 msec 0°C, 10 min incub., HBS + 0.2 M mannitol + 4 mM $CaCl_2$	''Standard medium'' acc. to Fromm used, supercoiled plasmid DNA, transient expression of CAT gene inhibited by antisense RNA expression
	Riggs and Bates (Tallahassee)	Tobacco leaf mesophyll protoplasts	pMON200 conferring kn resistance	CD: 2 kV/cm; 250 μsec 0°C, 0.5 M mannitol	Stable transformation by linearized plasmid, optimum at about 50% survivors, transformation frequency 2 × 10^{-4}/pulsed protoplasts, single intact plasmid copy or only low copy number detected in regenerated transgenic plants, Mendelian inheritance of transfected DNA shown
	Nishiguchi *et al.*	Tobacco leaf pro-	TMV and CMV	Square pulses: 9 × 0.5 to 11 kV/cm; 1	Optimum: 4 × 10 kV/cm, 50 to 90

(Ithaca)	toplasts	RNA	to 90 μsec (0.1 sec intervals) 0°C, 10 min incub., 0.5 M mannitol (without ions), GCA helix chamber	μsec, position alteration between pulses favorable, up to 75% infection
Okada *et al.* (Nagoya)	Tobacco protoplasts (infection detected by fluorescent antibody to virus protein)	TMV and CMV RNA TMV and CMV particles	CD: 750 V/cm; 6 msec 0°C, 10 min incub., different buffers suitable: PBS, MES, Hepes, all with 0.3 M mannitol, cuvette	Up to 80% of protoplasts infected, RNA addition after pulsing: uptake even after 10 min interval; TMV particles (18 × 300 nm) also introduced as well as CMV particles (diameter 30 nm)
Morikawa *et al.* (Kyoto)	*Intact* tobacco mesophyll cells (infection detected by immunofluorescence)	TMV RNA	CD: 650 V/cm; τ not given 20°C, medium: ≈ 1 mM $CaCl_2$, $ZnSO_4$, $MgSO_4$ + 0.4 M glucose	Infection proved in principle, method not yet optimized
Ou-Lee *et al.* (Ithaca)	Rice, wheat, sorghum protoplasts from leaves and suspension cultures (monocots!, transient expression)	CAT gene with promoters from CMV and *Drosophila:* p35S-CAT and pCopia-CAT	Square pulses: 2 × 2.5 kV/cm; 100 μsec (GCA helix chamber) or 20 pulses, each 62 μsec; 10 kV in "non-contact mode BAEKON 2000 receptacle" (field strength not given), 0°C, 10 min incub., PBS + 5 mM $CaCl_2$ + 0.4 M mannitol	Successful uptake and expression of plasmids in three graminaceous species, *Drosophila* promoter highly effective in plants, short pulses (μsec range) suitable and avoiding excessive heating, optimum pulse intensities inversely related to protoplast size
Hibi *et al.* (Ibaraki)	Tobacco leaf mesophyll protoplasts (infection detected by fluorescent anti-TMV serum)	TMV RNA	Square pulses: 500–900 V/cm; 50 μsec,5 to 10 pulses (1 sec intervals), sandwich chamber, gold electrodes, low conductivity 0°C, 5 min incub., 0.5 M mannitol without salts or with up to 100 μM $CaCl_2$ or $MgCl_2$	Maximum transfection (80%) with 5 to 10 pulses of 800 V/cm, 90% survival; effects of ions (100 μM): Ca^{2+} without effect, Mg^{2+} stimulating by 25%, comparison with electrofusion: same critical (600 V/cm) and optimum pulse voltages
Schocher *et al.* (Basel)	Tobacco mesophyll protoplasts	Cotransformation of selectable (pABDI) and non-selectable genes pMSI (with zein gene) and pGV0422 (with nopaline synthetase gene)	CD: acc. to Shillito	Plasmids integrated into plant genomic DNA and stably maintained in tissue culture for hundreds of cell generations, high efficiency of cotransformation

(*continued*)

Appendix. (*Continued*)

	Authors	Recipient cells	Plasmid/gene	Pulsing CD: E_0; τ	Results
	Hauptmann *et al.* (St. Louis/Gainesville)	Protoplasts of dicot and monocot plants (transient expression)	Plasmid with CAT gene and CaMV35S promoter	CD, no parameters given	CAT activity detected in carrot, *Glycine max* and *Petunia hybrida*, monocots: *Triticum monococcum*, *Pennisetum purpureum*, *Panicum maximum*, not successful with *Zea mays* and *Saccharum officanum*
1987	Watts *et al.* (Norwich)	*Nicotiana tabacum*, *N. plumbaginifolia* protoplasts, *Beta vulgaris* protoplasts (infection detected by immunofluorescence)	CCMV RNA, BMV and CCMV viruses	CD: optimum 1.5/2.5 kV/cm; τ = 1 μsec 0°C, postincub. 10 min + 10 min at RT after dilution with mannitol; 0.7 M mannitol; plastic cuvettes (acc. to Potter)	Long pulses and presence of electrolytes decrease cell survival; no infection with virus RNA when given after pulse only in presence during electroporation; negatively charged CCMV do not enter protoplasts during inoculation
	Boston *et al.* (West Lafayette)	*Daucus carota* protoplasts (W001C)	pCATTi, pCATZ2 (supercoiled)	Pulse: ISCO power supply (acc. to Potter) Preincub. 5 min 45°C + 5 min at RT after plasmid is added + 5 min on ice with PEG; postincub. 10 min at RT; PCM; cuvette with Al foil electrodes (acc. to Potter)	2.0 kV setting results in 40% intact viable cells and maximum CAT activity; presence of PEG is necessary (no sharp optimum related to concentration); no effect of heat-shock treatment; linear DNA and presence of carrier DNA decreases CAT expression

				Bacteria	
1983	Shivarova *et al.* (Sofia, Jena)	*Bacillus cereus* protoplasts	pUB110 from *B. thuringiensis* (kn resistance)	CD: 3 × 14 kV/cm; 5 μsec 40% PEG present, 20°C, 10 min incub.	Small objects, high electric field strength necessary, 10-fold increase in stable transformation
1987	Shigekawa (Richmond)	*E. coli* strains LE392, HB101, JM107, MV1190	pBR322 and linear lambda phage DNA (50 kb)	CD: 6 kV/cm; 5 msec 0°C, 20 min pre- and 20 min postincub., phosphate-buffered sucrose	Mid-log growth phase of *E. coli*, 40% viability at 6 kV/cm: transformation efficiency not yet optimal, 10^5 to 10^6 transformants/μg DNA, independent of plasmid concentration, 22°C: 100-fold less efficiency, multiple pulses not helpful, lambda DNA transforms 1/100 less than pBR322
				Yeast	
1985	Hasimoto *et al.* (Kyoto)	Intact yeast cells (*S. cerevisiae* mutants)	YEp13 complementing Leu deficiency	CD: 3 × 5 kV/cm; 10 msec 35% PEG present, long incub. times	Optimum: 100 transf./5 × 10^7 cells/μg DNA at about 40% survival
	Karube *et al.* (Yokohama)	Yeast spheroplasts (*S. cerevisiae* mutant)	YRp7 shuttle vector complementing tryptophan deficiency	CD: 10 kV/cm; 50 μsec 10 mM $CaCl_2$ + 1.2 M sorbitol + 10 mM Tris	10^3 transf./μg DNA, optimum: one pulse at 10 kV/cm

ACKNOWLEDGMENTS. We thank Elvira Boldt for help in preparing the Appendix and the reference list. We gratefully acknowledge financial support by the DFG (Grant NE 227/4 to E.N.).

REFERENCES

Allen, H., Fraser, J., Flyer, D., Calvin, S., and Flavell, R., 1986, β_2-Microglobulin is not required for cell surface expression of the murine class I histocompatibility antigen H-2D^b or of a truncated H-2D^b, *Proc. Natl. Acad. Sci. USA* **83:**7447–7451.

Atchison, M. L., and Perry, R. P., 1986, Tandem κ immunoglobulin promoters are equally active in the presence of the κ enhancer: Implications for models of enhancer function, *Cell* **46:**253–262.

Auer, D., Brandner, G., and Bodemer, W., 1976, Dielectric breakdown of the red blood cell membrane and uptake of SV 40 DNA and mammalian cell RNA, *Naturwissenschaften* **63:**391.

Berg, H., Augsten, K., Bauer, E., Förster, W., Jacob, H.-E., Mühlig, P., Weber, H., and Kurischko, A., 1984, Possibilities of cell fusion and transformation by electrostimulation, *Bioelectrochem. Bioenerg.* **12:** 119–133.

Boston, R. S., Becwar, M. R., Ryan, R. D., Goldsbrough, P. B., Larkins, B. A., and Hodges, T. K., 1987, Expression from heterologous promoters in electroporated carrot protoplasts, *Plant Physiol.* **83:**742–746.

Chu, G., Hayakawa, H., and Berg, P., 1987, Electroporation for the efficient transfection of mammalian cells with DNA, *Nucleic Acids Res.* **15:**1311–1326.

Chung, J., Sinn, E., Reed, R. R., and Leder, P., 1986, Trans-acting elements modulate expression of the human c-myc gene in Burkitt lymphoma cells, *Proc. Natl. Acad. Sci. USA* **83:**7918–7922.

Drinkwater, N. R., and Klinedinst, D. K., 1986, Chemically induced mutagenesis in a shuttle vector with a low background mutant frequency, *Proc. Natl. Acad. Sci. USA* **83:**3402–3406.

Ecker, J. R., and Davis, R. W., 1986, Inhibition of gene expression in plant cells by expression of antisense RNA, *Proc. Natl. Acad. Sci. USA* **83:**5372–5376.

Eigen, M., and DeMaeyer, L., 1963, Relaxation methods, in: *Techniques of Organic Chemistry,* Volume 8(2) (S. L. Friess, E. S. Lewis, and A. Weissberger, eds.), Wiley, New York, pp. 895–1054.

Evans, G. A., Ingraham, H. A., Lewis, K., Cunningham, K., Seki, T., Moriuchi, T., Chang, H. C., Silver, J., and Hyman, R., 1984, Expression of the Thy-1 glycoprotein gene in DNA-mediated gene transfer, *Proc. Natl. Acad. Sci. USA* **81:**5532–5536.

Falkner, F. G., Neumann, E., and Zachau, H. G., 1984, Tissue specificity of the initiation of immunoglobulin κ gene transcription, *Hoppe-Seyler's Z. Physiol. Chem.* **365:**1331–1343.

Fromm, M., Taylor, L. P., and Walbot, V., 1985, Expression of genes transferred into monocot and dicot plant cells by electroporation, *Proc. Natl. Acad. Sci. USA* **82:**5824–5828.

Fromm, M. E., Taylor, L. P., and Walbot, V., 1986, Stable transformation of maize after gene transfer by electroporation, *Nature* **319:**791–793.

Grünhagen, H. H., 1973, Fast spectrophotometric detection system for coupled physical and chemical electric field effects in solution, *Biophysik* **10:**347–354.

Hama-Inaba, H., Shiomi, T., Sato, K., Ito, A., and Kasai, M., 1986, Electric pulse-mediated gene transfer in mammalian cells grown in suspension culture, *Cell Struct. Funct.* **11:**191–198.

Hashimoto, H., Morikawa, H., Yamada, Y., and Kimura, A., 1985, A novel method for transformation of intact yeast cells by electroinjection of plasmid DNA, *Appl. Microbiol. Biotechnol.* **21:**336–339.

Hauptmann, R., Ozias-Akins, P., Vasil, V., Tabaeizadeh, Z., and Vasil, N., 1986, Expression of electroporated DNA in several mono- and dicotyledonous plant species, *Plant Physiol.* **80**(4/Suppl.)**:**102.

Hibi, T., Kano, H., Sugiura, M., Kazami, T., and Kimura, S., 1986, High efficiency electro-transfection of tobacco mesophyll protoplasts with TMV RNA, *J. Gen. Virol.* **67:**2037–2042.

Igarashi, T., Okazaki, T., Potter, H., Gaz, R., and Kronenberg, H. M., 1986, Cell-specific expression of the human parathyroid hormone gene in rat pituitary cells, *Mol. Cell. Biol.* **6:**1830–1833.

Ilgenfritz, G., 1966, Chemische Relaxation in starken elektrischen Feldern, Dissertation, Universität Göttingen.

Jacob, H.-E., Förster, W., and Berg, H., 1981, Microbiological implications of electric field effects. II. Inactivation of yeast cells and repair of their cell envelope, *Z. Allg. Mikrobiol.* **21:**225–233.

Karube, I., Tamiya, E., and Matsuoka, H., 1985, Transformation of Saccharomyces cerevisiae spheroplasts by high electric pulse, *FEBS Lett.* **182:**90–94.

Khazaie, K., Gounari, F., Antoniou, M., deBoer, E., and Grosveld, F., 1986, β-globin gene promoter generates 5′ truncated transcripts in the embryonic/fetal erythroid environment, *Nucleic Acids Res.* **14:** 7199–7212.

Kinosita, K., Jr., and Tsong, T. Y., 1977, Hemolysis of human erythrocytes by a transient electric field, *Proc. Natl. Acad. Sci. USA* **74:**1923–1927.
Knight, D. E., and Baker, P. F., 1982, Calcium-dependence of catecholamine release from bovine adrenal medullary cells after exposure to intense electric fields, *J. Membr. Biol.* **68:**107–140.
Krens, F. A., Molendijk, L., Wullems, G. J., and Schilperoort, R. A., 1982, In vitro transformation of plant protoplasts with Ti-plasmid DNA, *Nature* **296:**72–74.
Lachman, H. M., Cheng, G., and Skoultchi, A. I., 1986, Transfection of mouse erythroleukemia cells with myc sequences changes the rate of induced commitment to differentiate, *Proc. Natl. Acad. Sci. USA* **83:**6480–6484.
Langridge, W. H. R., Li, B. J., and Szalay, A. A., 1985, Electric field mediated stable transformation of carrot protoplasts with naked DNA, *Plant Cell Rep.* **4:**355–359.
Lindner, P., Neumann, E., and Rosenheck, K., 1977, Kinetics of permeability changes induced by electric impulses in chromaffin granules, *J. Membr. Biol.* **32:**231–254.
Miller, J. F., Dower, W. J., and Tompkins, L. S., 1988, High voltage electroporation of bacteria: Genetic transformation of *Campylobacter jejuni* with plasmid DNA, *Proc. Natl. Acad. Sci. USA* **85:**856–860.
Morikawa, H., Iida, A., Matsui, C., Ikegami, M., and Yamada, Y., 1986, Gene transfer into intact plant cells by electroinjection through cell walls and membranes, *Gene* **41:**121–124.
Narayanan, R., Jastreboff, M. M., Chiu, C. F., and Bertino, J. R., 1986, In vivo expression of a nonselected gene transferred into murine hematopoietic stem cells by electroporation, *Biochem. Biophys. Res. Commun.* **141:**1018–1024.
Neumann, E., and Rosenheck, K., 1972, Permeability changes induced by electric impulses in vesicular membranes, *J. Membr. Biol.* **10:**279–290.
Neumann, E., and Rosenheck, K., 1973, Potential differences across vesicular membranes, *J. Membr. Biol.* **14:**194–196.
Neumann, E., Schäfer-Ridder, M., Wang, Y., and Hofschneider, P. H., 1982, Gene transfer into mouse lyoma cells by electroporation in high electric fields, *EMBO J.* **1:**841–845.
Nishiguchi, M., Langridge, W. H. R., Szalay, A. A., and Zaitlin, M., 1986, Electroporation-mediated infection of tobacco leaf protoplasts with tobacco mosaic virus RNA and cucumber mosaic virus RNA, *Plant Cell Rep.* **5:**57–60.
Okada, K., Nagata, T., and Takebe, I., 1986, Introduction of functional RNA into plant protoplasts by electroporation, *Plant Cell Physiol.* **27:**619–626.
Ou-Lee, T.-M., Turgeon, R., and Wu, R., 1986, Expression of a foreign gene linked to either a plant-virus or a Drosophila promoter after electroporation of protoplasts of rice, wheat, and sorghum, *Proc. Natl. Acad. Sci. USA* **83:**6815–6819.
Potter, H., Weir, L., and Leder, P., 1984, Enhancer-dependent expression of human κ immunoglobulin genes introduced into mouse pre-B lymphocytes by electroporation, *Proc. Natl. Acad. Sci. USA* **81:**7161–7165.
Prochownik, E. V., 1985, Relationship between an enhancer element in the human antithrombin III gene and an immunoglobulin light-chain gene enhancer, *Nature* **316:**845–848.
Reiss, M., Jastreboff, M. M., Bertino, J. R., and Narayanan, R., 1986, DNA-mediated gene transfer into epidermal cells using electroporation, *Biochem. Biophys. Res. Commun.* **137:**244–249.
Riggs, C. D., and Bates, G. W., 1986, Stable transformation of tobacco by electroporation: Evidence for plasmid concatenation, *Proc. Natl. Acad. Sci. USA* **83:**5602–5606.
Rosenheck, K., Lindner, P., and Pecht, I., 1975, Effect of electric fields on light scattering and fluorescence of chromaffin granules, *J. Membr. Biol.* **20:**1–12.
Sahagan, B. G., Dorai, H., Saltzgaber-Muller, J., Toneguzzo, F., Guindon, C. A., Lilly, S. P., McDonald, K. W., Morrissey, D. V., Stone, B. A., Davis, G. L., McIntosh, P. K., and Moore, G. P., 1986, A genetically engineered murine/human chimeric antibody retains specificity for human tumor-associated antigen, *J. Immunol.* **137:**1066–1074.
Sale, A. J. H., and Hamilton, W. A., 1967, Effects of high electric fields on microorganisms. I. Killing of bacteria and yeasts. II. Mechanism of action of the lethal effect, *Biochim. Biophys. Acta* **148:**781–800.
Sale, A. J. H., and Hamilton, W. A., 1968, Effects of high electric fields on microorganisms. III. Lysis of erythrocytes and protoplasts, *Biochim. Biophys. Acta* **163:**37–43.
Schocher, R. J., Shillito, R. D., Saul, M. W., Paszkowski, J., and Potrykus, I., 1986, Co-transformation of unlinked foreign genes into plants by direct gene transfer, *Bio/Technology* **4:**1093–1096.
Shillito, R. D., Saul, M. W., Paszkowski, J., Müller, M., and Potrykus, I., 1985, High efficiency direct gene transfer to plants, *Bio/Technology* **3:**1099–1103.
Shimizu, Y., Koller, B., Geraghty, D., Orr, H., Shaw, S., Kavathas, P., and DeMars, R., 1986, Transfer of

cloned human class I major histocompatibility complex genes into HLA mutant human lymphoblastoid cells, *Mol. Cell. Biol.* **6:**1074–1087.

Shivarova, N., Förster, W., Jacob, H.-E., and Grigorova, R., 1983, Microbiological implications of electric field effects. VII. Stimulation of plasmid transformation of Bacillus cereus protoplasts by electric field pulses, *Z. Allg. Mikrobiol.* **23:**595–599.

Smithies, O., Gregg, R. G., Boggs, S. S., Koralewski, M. A., and Kucherlapati, R. S., 1985, Insertion of DNA sequences into the human chromosomal β-globin locus by homologous recombination, *Nature* **317:** 230–234.

Spandidos, D. A., 1987, Electric field-mediated gene transfer (electroporation) into mouse friend and human K562 erythroleukemic cells, *Gene Anal. Tech.* **4:**50–56.

Stopper, H., Zimmermann, U., and Wecker, E., 1985, High yield of DNA-transfer into mouse L-cells by electropermeabilization, *BioScience* **40:**929–932.

Sugar, I. P., Förster, W., and Neumann, E., 1987, Model of cell electrofusion, membrane electroporation, pore coalescence and percolation, *Biophys. Chem.* **26:**321–335.

Sugden, B., Marsh, K., and Yates, J.-A., 1985, A vector that replicates as a plasmid and can be efficiently selected in lymphoblasts-B transformed by Epstein–Barr virus, *Mol. Cell. Biol.* **5:**410–413.

Teissié, J., and Rols, M. P., 1986, Fusion of mammalian cells in culture is obtained by creating the contact between cells after their electropermeabilization, *Biochem. Biophys. Res. Commun.* **140:**258–266.

Toneguzzo, F., and Keating, A., 1986, Stable expression of selectable genes introduced into human hematopoietic stem cells by electric field-mediated DNA transfer, *Proc. Natl. Acad. Sci. USA* **83:**3496–3499.

Toneguzzo, F., Hayday, A. C., and Keating, A., 1986, Electric field-mediated DNA transfer: Transient and stable gene expression in human and mouse lymphoid cells, *Mol. Cell. Biol.* **6:**703–706.

Tur-Kaspa, R., Teicher, L., Levine, B. J., Skoultchi, A. I., and Shafritz, D. A., 1986, Use of electroporation to introduce biologically active foreign genes into primary rat hepatocytes, *Mol. Cell. Biol.* **6:**716–718.

Watts, J. W., King, J. M., and Stacey, N. J., 1987, Inoculation of protoplasts with viruses by electroporation, *Virology* **157:**40–46.

Weir, L., and Leder, P., 1986, Structure and expression of a human subgroup II immunoglobulin κ gene, *Nucleic Acids Res.* **14:**3957–3970.

Wolf, H., Puhler, A., and Neumann, E., 1989, Electrotransformation of intact and osmotically sensitive cells of *Corynebacterium glutamicum, Appl. Microbiol. Biotechnol.* **30** (in press).

Wong, T.-K., and Neumann, E., 1982, Electric field mediated gene transfer, *Biochem. Biophys. Res. Commun.* **107:**584–587.

Yancopoulos, G. D., Blackwell, T. K., Suh, H., Hood, L., and Alt, F. W., 1986, Introduced T cell receptor variable region gene segments recombine in pre-B cells: Evidence that B and T cells use a common recombinase, *Cell* **44:**251–259.

Zamecnik, P. C., Goodchild, J., Taguchi, Y., and Sarin, P. S., 1986, Inhibition of replication and expression of human T-cell lymphotropic virus type III in cultured cells by exogenous synthetic oligonucleotides complementary to viral RNA, *Proc. Natl. Acad. Sci. USA* **83:**4143–4146.

Zerbib, D., Amalric, F., and Teissié, J., 1985, Electric field-mediated transformation: Isolation and characterization of a TK^+ subclone, *Biochem. Biophys. Res. Commun.* **129:**611–618.

Zimmermann, U., Pilwat, G., and Riemann, F., 1974, Dielectric breakdown of cell membranes, *Biophys. J.* **14:** 881–889.

Chapter 20

Electropermeabilization and Electrosensitivity of Different Types of Mammalian Cells

M. J. O'Hare, M. G. Ormerod, P. R. Imrie, J. H. Peacock, and W. Asche

1. INTRODUCTION

Electrotransfection (Neumann *et al.*, 1982; Chu *et al.*, 1987) and electrofusion (Zimmermann *et al.*, 1985) are both potentially powerful techniques for creating new cell types. Little is presently known, however, about the individual responses of different mammalian cell types to the microsecond electroshocks necessary to create transient permeability and cause fusion of adjacent cell membranes.

In a recent study we have shown that transfection of cultured breast epithelial cells by electroporation can be optimized with respect to field strength, pulse duration, and pulse number using a dedicated electroporator with which all three parameters can be independently controlled. The nature of the suspending medium as well as the total number of pulses given proved to be important determinants of the absolute yield of transfectants (O'Hare *et al.*, 1987). Low-conductivity medium with multiple short pulses proved the most effective combination. Under fully optimized conditions the electroporator was superior to conventional methods using calcium phosphate applied to the same cells; a homemade electroporator based on an electrophoresis power pack was much less efficient. Although this type of comparison has not often been made in published studies of electrotransfection (see O'Hare *et al.*, 1987, for references), others have also observed improvements over conventional techniques (e.g., Hama-Inaba *et al.*, 1986). In our experiments, optimization of precise electrical conditions was facilitated by the use of flow cytometry to measure electropermeabilization directly with permeant and impermeant dyes.

During the course of these experiments, which were carried out with both rat and human breast epithelial cells, it became apparent that different types of cells showed markedly different electrosensitivities, as measured by viability of treated cells 12 hr after shocking. Furthermore, these

---

M. J. O'Hare, M. G. Ormerod, P. R. Imrie, and J. H. Peacock • Institute of Cancer Research, Royal Cancer Hospital, Sutton SM2 5PX, United Kingdom. W. Asche • Krüss GmbH, 2000 Hamburg 61, Federal Republic of Germany.

Table 1. Cells Used for Electroporation

Name	Source	Species	Type	Reference
NSO	Ascites	Mouse	Myeloma	Glafré and Milstein (1981)
HMy2	Blood	Human	Lymphoblastoid	Edwards *et al.* (1982)
UC729-6	Blood	Human	Lymphoblastoid	Glassy *et al.* (1983)
Nalm-6	Blood	Human	Lymphoid	Hurwitz *et al.* (1979)
Comma 1D	Breast	Mouse	Epithelial	Danielson *et al.* (1984)
Rama 37	Breast	Rat	Epithelial	Dunnington *et al.* (1983)
SVep	Breast	Human	Epithelial	O'Hare *et al.* (1987)
ZR75-1	Breast	Human	Epithelial	Engel *et al.* (1978)
SVK14	Skin	Human	Epithelial	Taylor-Papadimitriou *et al.* (1982)
RCC2	Colon	Rat	Epithelial	Borman *et al.* (1972)
NRK	Kidney	Rat	Epithelioid	DeLarco and Todaro (1978)
3T3	Mesenchyme	Mouse	Fibroblastic	Todaro and Green (1963)
3T6	Mesenchyme	Mouse	Fibroblastic	Todaro and Green (1963)
WI-38	Lung	Human	Fibroblastic	Hayflick (1965)
BCE	Cornea	Bovine	Endothelial	Gospodarowicz *et al.* (1977)

differences could not be accounted for by measurable differences in parameters known to influence electropermeabilization, such as cell diameter (Zimmermann, 1982).

To explore further these differences and their possible causes, we have extended these observations by visualizing electropermeabilization using flow cytometry and by measuring electrosensitivity using cell viability. A variety of cultured mammalian cell types from different species and from different tissues have been studied, including cells of lymphoid, epithelial, fibroblastic, and endothelial origin.

2. TECHNIQUES

2.1. Preparation of Cells

Table 1 details the origins and characteristics of the different types of cells used in this study. With the exception of WI-38 and BCE cells, which were cell strains with a finite life span *in vitro,* all cells tested were continuous cell lines derived from both normal and malignant tissues, growing either as suspensions (ascites and blood derived) or as attached monolayers (epithelial, fibroblastic, and endothelial cells). All cells were grown in DMEM medium supplemented with fetal calf serum (10% v/v).

Cells grown as suspensions were harvested directly and pipette flushed, if necessary, to ensure a uniform cell suspension. Cell suspensions were prepared from monolayer cultures by washing them in 0.25% (w/v) EDTA (versene) solution in Ca/Mg-free phosphate-buffered saline (PBS), and then incubating them at 37°C for 3 min in trypsin–versene (0.5 mg/ml). Monolayers were then flushed with complete growth medium (DMEM plus 10% fetal calf serum) to inactivate residual trypsin and generate a uniform single cell suspension. All cells were mycoplasma free and $> 95\%$ viable by trypan blue exclusion when used for electrosensitivity studies.

2.2. Determination of Cell Size

The modal cell diameter for each cell line was calculated from volumes determined using a Coulter counter (Coulter Electronics) with a Mevway multichannel analyzer, on the assumption that the cells in suspension were spherical. The complete distribution of cell sizes was also determined from multichannel analysis of Coulter data to give means, medians, and skew as well as model diameters for selected lines.

2.3. Electroporation

Cells were electroporated using the TA 750 (Krüss GmbH, Hamburg, West Germany) apparatus which delivers a square wave pulse of predefined duration and field strength. Cells were shocked in the TA 756 stainless steel pipetting annular electrode chamber with an electrode gap of 0.5 mm. They were subjected to a train of preset pulses of different field strengths and durations, in varying numbers, at intervals of 1 sec unless otherwise specified.

The electroporation medium was isotonic (0.3 M) mannitol containing 0.5 mM $CaCl_2$ and 0.2 mM $MgCl_2$ plus 200 μg/ml bovine serum albumin (O'Hare *et al.*, 1987). Up to 250 μl of low-conductivity medium could be shocked using the TA 750/756 without exceeding the current limits of the apparatus (< 20A); by contrast, only < 50 μl of high-conductivity medium (e.g., saline) could be shocked without exceeding these limits. As shown elsewhere (O'Hare *et al.*, 1987), the low-conductivity medium also results in a more effective electropermeabilization.

After harvesting and prior to shocking, cells were washed in PBS followed by isotonic mannitol, and then resuspended in the shocking medium at a concentration of 5×10^6–2×10^7 cells/ml. Duplicate aliquots of 250 μl each were shocked at ice temperature (unless otherwise specified) under each set of conditions. Separate determinations of electropermeabilization and electrosensitivity were made for each cell line using samples shocked in this manner.

2.4. Determination of Electropermeabilization

For studies of electropermeabilization by flow cytometry, cells were expelled from the pipetting chamber into 25 μl of a solution of 10μg/ml propidium iodide, in which they were incubated at ice temperature for 20 min. They were then resuspended in warm (37°C) complete culture medium for 5 min to permit resealing, resuspended in PBS, and stained with 10 ng/ml fluorescein diacetate for 10 min at room temperature. Flow cytometric parameters to enable discrimination of living, permeabilized (FDA^+/PI^+) cells from their living nonpermeabilized (FDA^+/PI^-) and dead (FDA^-/PI^+) counterparts have been described in detail elsewhere (O'Hare *et al.*, 1987). Briefly, cells were analyzed using an Ortho 50H flow cytometer with an argon-ion laser tuned to 488 nm and an associated 2150 computer system. Parameters recorded were forward and orthogonal light scatter together with green (fluorescein) and red (propidium iodide) fluorescence. The two parameters of scattered light were displayed and a region set on the cytogram to include single cells and exclude any clumps and debris. Green versus red fluorescence was then displayed. To quantitate uptake of propidium iodide, the median red fluorescence was also recorded.

2.5. Determination of Electrosensitivity

Immediately following shocking, cells were expelled from the pipetting chamber into 25-cm^2 culture flasks containing 10 ml bicarbonate-buffered Dulbecco's modified Eagle's medium plus 10% fetal calf serum, and incubated at 37°C in an atmosphere of 5% CO_2 in air for 12 hr. The total number of viable cells remaining after this time was then determined. In the case of cells in suspension, this was done by counting after staining with 0.1% trypan blue to exclude dead cells; in the case of monolayers, the dead cells failed to attach to the culture flask and were washed off prior to trypsinizing the viable attached cells and counting them. Controls consisted of samples which had been processed in an identical manner, except that the cells were not shocked. The viability of the shocked cells was then expressed as a percentage of control samples, calculations being based on duplicate samples with counting to a precision of ±3% (S.D.).

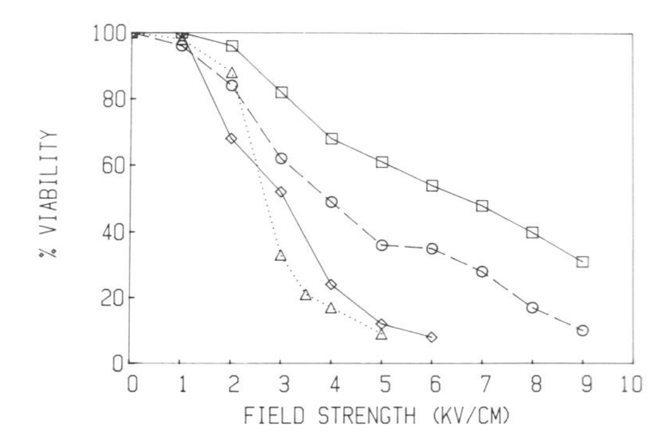

Figure 1. Electrosensitivity of different cell types. The 12-hr postshock viability is shown as a percentage of unshocked controls for Nalm-6 human lymphoid cells (△), SVep human breast cells (◇), 3T3 mouse fibroblasts (○), and bovine endothelial cells (□) as a function of field strength. Cells were shocked in mannitol-based low-conductivity medium (see Techniques) with nine pulses of 10 μsec duration at 1 sec intervals.

3. EXPERIMENTAL RESULTS

3.1. Effect of Cell Type

All cell lines and strains listed in Table 1 were tested for electrosensitivity as described in Section 2. Field strengths from 1 to 9 kV/cm were used with a train of nine pulses of 10 μsec duration at 1 sec intervals. From the resulting curves, some of which are illustrated in Fig. 1, the voltage required to kill 50% of the cells was determined (LV50). Results, together with the modal cell diameters, are given in Table 2.

From these data it is evident that wide differences exist between different types of cells in their response to the same electrical pulses. Similar differences were also seen when pulse number of pulse duration was varied at a constant voltage (unpublished data). Their reproducibility is evident from the fact that the results illustrated in Fig. 1 and summarized in Table 2 have been repeated with a precision of ±10% on a minimum of two separate occasions with all the cell lines and strains.

Table 2. Electrosensitivity of Mammalian Cells[a]

Cell type	LV50 (kV/cm)	Cell diameter (μm)[b]	EI[c]
NSO	2.0	13.3	3.7 (2)
UC729-6	2.7	12.1	3.1 (3)
HMy2	2.7	11.9	3.1 (3)
Nalm-6	2.7	9.5	3.9 (1)
ZR-75-1	2.7	16.6	2.2 (6)
SVep	3.1	16.4	2.0 (7)
SVK-14	3.4	15.0	2.0 (7)
RCC2	3.6	10.7	2.6 (5)
3T3	3.9	16.8	1.5 (11)
Comma 1D	4.3	13.6	1.7 (9)
Rama 37	4.6	12.8	1.7 (9)
3T6	5.6	15.4	1.2 (12)
WI-38	4.6	16.1	1.3 (13)
NRK	5.3	11.8	1.6 (14)
BCE	6.7	17.7	0.9 (15)

[a]For sources and types, see Table 1.
[b]Modal (peak) diameter (MD).
[c]Electrosensitivity index (EI) = (1/LV50 × MD)0.100. Ranking order is shown in parentheses.

As in our previous study, some cells of epithelial type survived better when subjected to a long train of short pulses, compared with the same current applied in a smaller number of longer pulses. Other cells (e.g., lymphoid cells) showed no significant difference when the applied current was fractionated in this manner, with overall survival directly related to the total integrated voltage. Some differences were also seen between cells of like type but from different species (e.g., rodent and human breast epithelial cells) in accord with our original observations (O'Hare *et al.*, 1987). While these could often be correlated with differences in cell diameter, i.e., with smaller cells having the higher LV50, they were usually greater than could be accounted for by a simple direct proportionality to cell size (Table 2). Tissue rather than species differences were, however, even more pronounced and showed no correlation with cell size.

3.2. Effect of Cell Size Distribution

In Table 2, the effect of cell size on the response to electroshocks has been monitored by measuring the peak (modal) cell diameter of each type. Modal diameter does not take into account any marked deviation of the mean and median from the mode, or skewing of the size distribution that might influence the net effect of the shocks. These parameters have, therefore, been examined in detail in four lines covering the range of electrosensitivities and cell types, viz., Nalm-6, a lymphoid (ALL) line which is small but particularly sensitive, and the larger breast epithelial (SVep), fibroblastic (3T3), and endothelial (BCE) cells. The latter are the most electroresistant cells we have thus far discovered. As is evident from Fig. 2, some of these lines showed wide variations in size distributions. These did not, however, correlate with resistance or sensitivity. For example, two lines of differing sensitivity (SVep and BCE) not only had similar mean diameters (16.8 and 17.7 μm) but also similar distributions (median/mode = 1.6 for SVep and 1.24 for BCE) while the cells most heterogeneous in size (WI-38, median/mode = 2.2) were neither the most resistant nor the most sensitive (Table 2). The line least heterogeneous in size was Nalm-6 with a median/mode ratio of 1.11.

3.3. Thresholds for Electropermeabilization

The electropermeabilization of four of the cell types was measured by flow cytometry using the impermeant dye propidium iodide and the permeant vital stain fluorescein diacetate. As described above, propidium iodide was added after shocking the cells but prior to incubation at 37°C (which allowed their membranes to reseal); fluorescein diacetate was added after resealing.

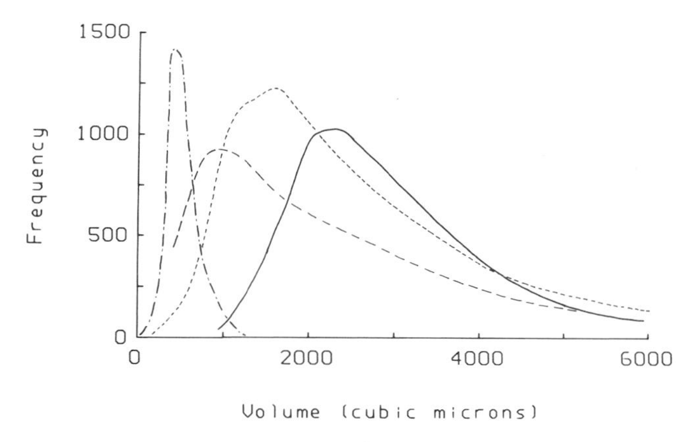

Figure 2. Size distribution of different cell types. Volume distributions from multichannel analysis of Coulter data are shown for Nalm-6 (—-—-—), SVep (-----), WI-38 (—–—), and bovine endothelial cells (——).

Using this method, three classes of cells could be distinguished: nonpermeabilized cells which fluoresced green (fluorescein) but not red (propidium iodide), permeabilized resealed cells which fluoresced both green and red, and cells which had failed to reseal the plasma membrane and consequently fluoresced red but not green.

Typical results using bovine endothelial cells are shown in Fig. 3. The amount of propidium iodide taken up by the cells increased with increasing field strength. This can be represented

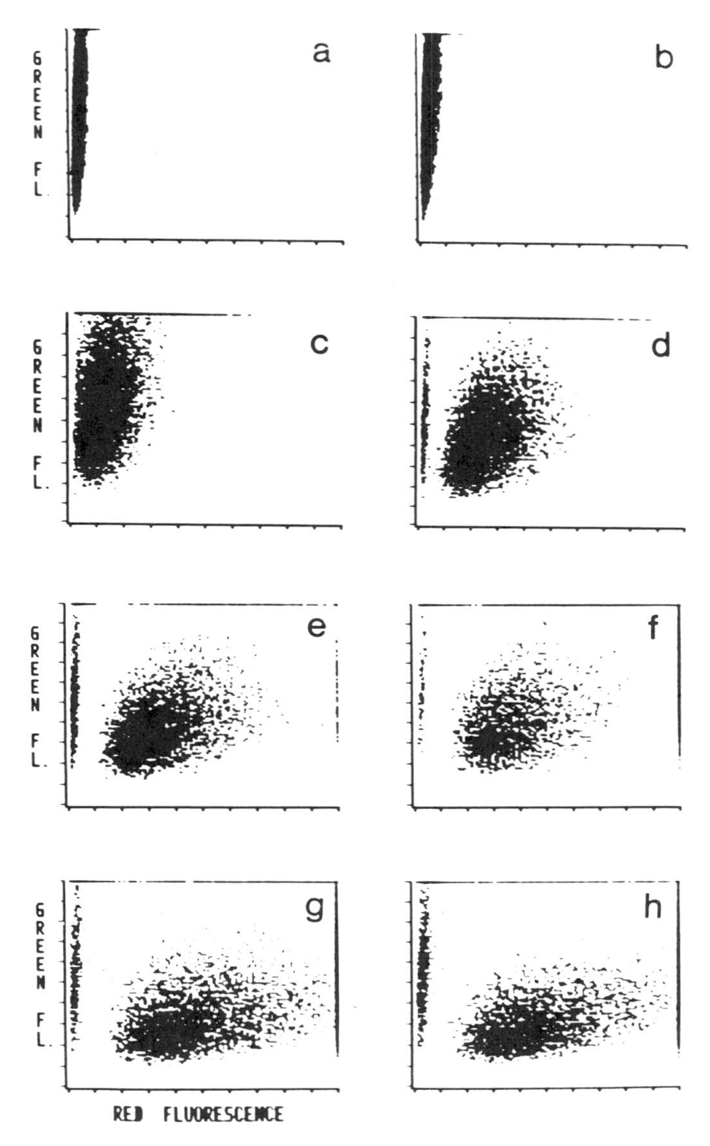

Figure 3. Electropermeabilization of cells visualized by flow cytometry. Cytograms show green (fluorescein) versus red (propidium iodide) fluorescence of bovine endothelial cells. Unshocked cells (a) are shown in comparison with shocked cells (nine pulses of 10 μsec) at field strengths of 1 (b), 2 (c), 3 (d), 4 (e), 5 (f), 6 (g), and 7 kV/cm (h) illustrating the progressive increase in propidium iodide uptake at 2 kV/cm and above. The nonpermeabilized (green only) cells seen in all shocked samples correspond to unshocked cells (~ 5%) remaining in the plastic tip of the pipetting electrode assembly.

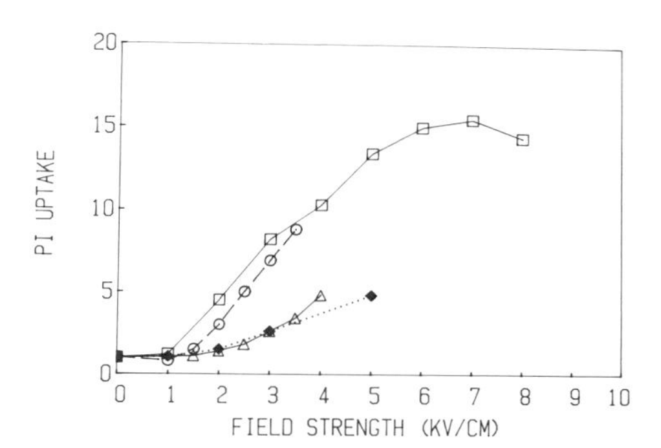

Figure 4. Propidium iodide (PI) fluorescence in different cell types. Median levels from cytograms of shocked cells (nine pulses of 10 μsec at the field strengths indicated) are shown relative to unshocked cells for UC729-6 (♦), Rama 37 (○), 3T3 (△), and bovine endothelial cells (□).

graphically by plotting the ratio of median red fluorescence to that in unshocked cells (Fig. 4). From this graph, it can be seen that the threshold for electropermeabilization for the four cell types lay between 1 and 2 kV/cm. There was no correlation between electrosensitivity (measured by viability 12 hr after shocking) and the measured extent of permeabilization; the endothelial cells were both the most resistant and the most easily permeabilized. The human lymphoblastoid cell line, UC729-6, gave a pattern different from that of the epithelial, fibroblastic, and endothelial cells. At high field strengths, a high proportion of cells failed to reseal (Fig. 5) in contrast to the other cell types in which most shocked cells showed some fluorescein fluorescence over a wide range of field strengths. A similar effect was seen with normal human lymphocytes (data not shown).

3.4. Effect of Culture Conditions

The LV50s were not perceptibly modified by the condition of the cells at the time of harvesting and identical results have been obtained from both dense and dilute cultures of the same cell type. To generate the data in Table 2, all cells were grown in identical medium prior to test, using the same batch of fetal calf serum. When UC729-6 cells adapted to long-term growth in serum-free conditions were tested against their serum-adapted counterparts, they showed similar electrosensitivities, with serum-free cells if anything slightly more resistant (Fig. 6) despite the fact that their modal diameter was larger (13.4 μm). Cells which are notoriously susceptible to adverse culture conditions, such as mouse hybridoma cells, were not exceptionally electrosensitive (unpublished observations).

One obvious difference in the handling of the cells is the exposure of monolayers to trypsin. However, treatment of the electrosensitive lymphoid cells in suspension culture with trypsin did not enhance their resistance to shock to the levels seen in epithelial, fibroblastic, or endothelial cells. A

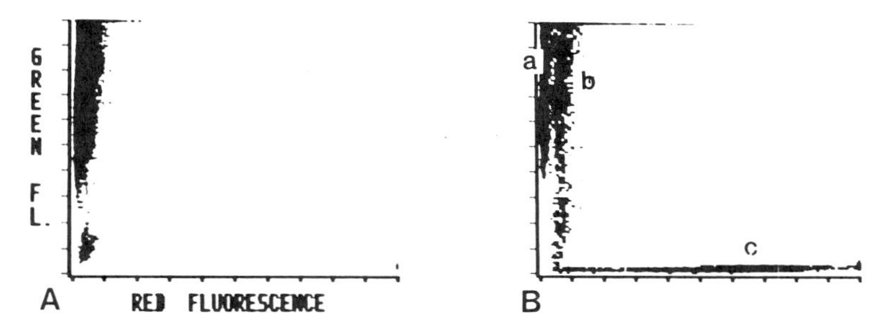

Figure 5. Electropermeabilization of lymphoblastoid cells visualized by flow cytometry. Cytograms show green versus red fluorescence for UC729-6 cells shocked (A) at 3 kV/cm and (B) at 5 kV/cm (nine pulses of 10 μsec). The latter shows nonpermeabilized cells (region a), permeabilized resealed cells (b), and cells which have failed to reseal at the higher voltage (c).

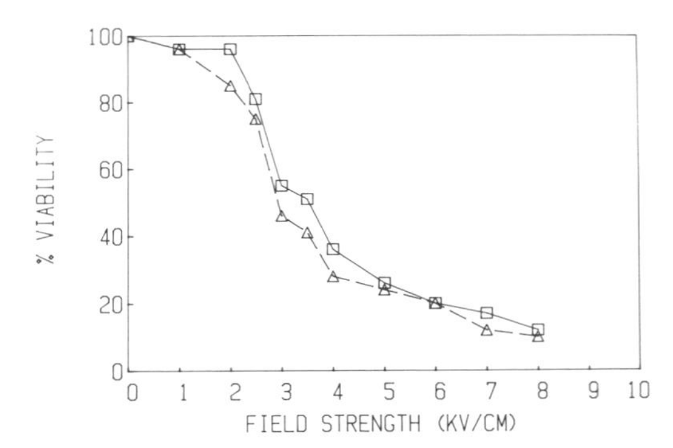

Figure 6. Effect of culture conditions on 12-hr postshock viability. UC729-6 cells from serum-free (□) and serum-containing medium (△) were shocked with nine pulses of 10 μsec at the indicated field strengths.

small effect was observed (LV50 increased < 10%) which was due probably to the removal of a small number of subviable cells present in any cultured population; these cells are usually rapidly lysed by the enzyme. Differences between individual cell types grown in monolayer were not due to differences in trypsin exposure, as this was carefully equalized in all such experiments.

3.5. Effect of Shocking Protocol

In an attempt to exclude other extrinsic causes of the wide differences in electroshock sensitivity, various parameters of the shocking protocol itself were systematically varied. The lethal effects of the shocks were not detectably influenced by the cell density in the shocking medium over at least two orders of magnitude (10^6–10^8 cells/ml); under the conditions used to generate Table 2, densities did not vary by more than a factor of four between any two experiments.

One potential variable in the procedure did have a marked influence on cell survival. This was the length of time that the cells were exposed to the isotonic mannitol-based shocking medium after they had been shocked (Fig. 7). At ice temperature the decline in resultant long-term viability was slow, but at room temperature it was relatively rapid. In contrast, long-term exposure of cells to mannitol medium without shocking had only marginal effects on viability, even at room temperature. No effect was detected when medium was first subject to shocks and then immediately added to the cells, even when voltages of up to 14 kV/cm were used (Fig. 7). The low-conductivity medium used in this study is therefore not itself rendered toxic by the shocks as a result of, for example, electrolysis.

In our experiments, cells were shocked at ice temperature and transferred immediately to warm growth medium. Under these conditions, contingent loss of viability is minimal. Virtually superimposable viability curves are obtained at 37°C, ice and room temperatures provided that cells are immediately transferred to warm growth medium (Fig. 8). When cells were left within the electrode

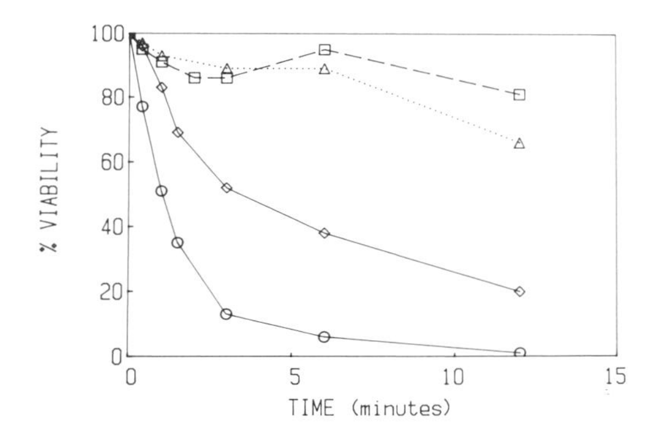

Figure 7. Effect of exposure to mannitol medium on 12-hr viability of cells. Nalm-6 human lymphoid cells were shocked at 2.5 kV/cm (nine pulses of 10 μsec), followed by incubation for the times indicated, at 0°C (△), 22°C (◇), and 37°C (○), plus incubation at 22°C in preshocked medium (nine pulses of 10 μsec at 14 kV/cm) (□). All cells were then transferred to growth medium at 37°C for 12 hr, when residual viability was determined.

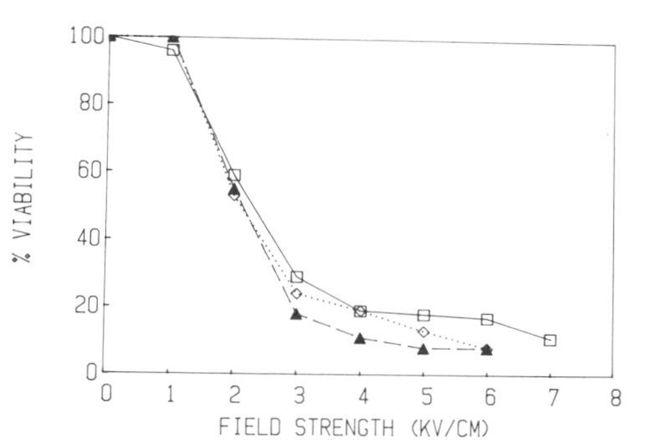

Figure 8. Effect of shocking temperature on 12-hr viability of shocked cells. HMy2 human lymphoblastoid cells were shocked (nine pulses of 10 μsec) at the field strengths indicated and transferred immediately to growth medium from low-conductivity mannitol medium at ice temperature (□), 22°C (▲), and 37°C (◇).

chamber for extended periods of time by varying the pulse interval from 1 to 80 sec, there was a noticeable decline in postshock viability. This was probably due to the gradual warming of the pipetting chamber toward ambient temperature, as a similar effect was obtained when a rapid pulse train (1 sec interval) was followed by an equivalent time during which the shocked sample was retained in the pipetting electrode chamber (data not shown). The shocks themselves should not warm the chamber significantly since joule heating effects are minimal in low-conductivity medium, such as that used here (conductivity = 0.027 S/m at 20°C).

4. DISCUSSION

The differences in electrosensitivity which we have documented in this study can be broadly summarized by stating that lymphoid cells were the most sensitive followed by epithelial and fibroblastic cells while endothelial cells were resistant to DC field intensities much higher than those required to permeabilize the cells (Table 2, Fig. 1).

Resistance or sensitivity to the long-term effects of shock is not determined by differences in permeabilization as such, which occurs when electrocompression of the membrane causes breakdown and transient pore formation (Zimmermann, 1982). This is apparent when electropermeabilization is visualized by flow cytometry (see Fig. 2). The patterns of permeant versus impermeant dye uptake obtained with flow cytometry are of considerable help in establishing the appropriate range of conditions for individual cell types. On their own, however, they do not establish the precise optimum conditions for, say, electrotransfection, or electrofusion, as these will be determined by a combination of effective permeabilization (which can be measured by flow cytometry) and long-term postshock viability. As we have seen, this varies from cell type to cell type even when they exhibit similar patterns of propidium iodide/fluorescein diacetate uptake.

Electrosensitivity is not directly determined by specific conditions *in vitro,* as is evident from experiments with cells adapted to different regimes. There appears, furthermore, to be no direct correlation between responsiveness to electroshocks and conventional criteria with respect to ease or difficulty of *in vitro* maintenance, or to the method used to harvest the cells. The shocking protocol itself must, however, be rigidly controlled to prevent extrinsic factors from biasing the results. The deleterious effect of postshock incubation in low-conductivity mannitol-based medium (Fig. 7) is a case in point.

As transfectant yields are actually increased at room temperature compared with ice (O'Hare *et al.*, 1987), the protocol described here in which cells are immediately transferred to warm growth medium must be explicitly followed if these results are to be reproduced. No advantage accrues from extending interpulse intervals, either with respect to transfectant yield (O'Hare *et al.*, 1987), or long-term viability (see above). Protocols for electrofusion which involve exposing the cells after shocking to a gradually reducing AC field using similar nonphysiological media (Zimmermann *et al.*, 1985) must therefore involve some, perhaps unnecessary, mortality which may reduce the absolute yield of hybrids. The use of an alternative shocking medium with variations in ionic composition and osmolarity might mitigate the deleterious response in terms of postshock viability, and thus improve results. However, we (O'Hare *et al.*, 1987) have not obtained any increase in transfectant yields by using hypotonic medium for shocking. Most workers have used saline-based media for electroporation due to limitations imposed by their apparatus (see O'Hare *et al.*, 1987, for references), although mixtures of intermediate composition have been used by some (Stopper *et al.*, 1985). Addition of Ficoll has been shown to improve viability under electrofusion conditions (Herzog *et al.*, 1986).

The differences noted in Table 2 could not be accounted for by any known variable in the procedure other than cell type itself. Laplace's equation (see Zimmermann, 1982) predicts that the voltage at the surface of a dielectric sphere between two electrodes is directly proportional to the radius of the sphere as well as the cosine of the angle between the field direction and any given membrane site. Thus:

$$V_m = V_{app} K \cos(\theta) r$$

where V_m is the voltage at the membrane, V_{app} is the applied field, and r is the cell radius; constant K is 1.5 for spherical cells.

Assuming that the threshold voltage for the breakdown of the plasma membrane is similar for different types of cell, as seems to be the case for quite disparate membranes (Sale and Hamilton, 1968; Smith and Cleary, 1983) and as our results show (Fig. 4), the corollary is that larger cells should require a smaller voltage for permeabilization. The flow cytometric data showed that all cells studied had a detectable threshold for permeabilization at an applied field strength of close to 1.5 kV/cm. Thus, the survival of resistant cells above this voltage was not simply due to differences in thresholds. Furthermore, when the theoretical effects of cell size on transmembrane voltage were taken into account by means of an electrosensitivity index (Table 2), the differences between individual cell types were enhanced rather than reduced. In fact, some of the smallest cells were the most sensitive (e.g., Nalm-6).

We therefore considered the possibility that heterogeneity in cell size might be more important than modal diameter in determining cell sensitivity. As expected, the more heterogeneous populations gave flatter viability curves (Fig. 1) but the observed variations in electrosensitivity were still too large to be explained by this effect.

In summary, therefore, it seems that the difference in electrosensitivity cannot be accounted for simply by differences in cell diameter or size distribution, or by differences in culture procedures or shocking techniques. They seem to depend on two attributes:

1. The ability of the cell to reseal the plasma membrane immediately after shocking. Flow cytometry showed that this was comparatively inefficient in lymphoid cells, compared with other types.
2. The ability of the resealed cell to recover from the imposed trauma. It is this which gives the endothelial cells their greater resistance, despite the ease with which they can be permeabilized.

Several alternatives present themselves as possible causes for the differences in these attributes. The membrane constituents of lymphoid, epithelial, fibroblastic, and endothelial cells clearly differ, at least insofar as their protein composition is concerned, and differences probably also exist in the membrane lipids. The effect of temperature on rates of resealing of synthetic lipid bilayers has been well documented (Benz and Zimmermann, 1981), with evidence for two exponential processes controlling resealing above 25°C. Although the effect of lipid composition on resealing was not investigated as only a limited range of synthetic mixtures survived electroshock, another study (Chernomordik *et al.*, 1983) has shown differences in conductance in synthetic membranes of differing composition. However, the lack of marked differences in viability curves obtained at widely differing temperatures above and below phase transition temperatures in the present study (Fig. 8) argues against major effects of differences in membrane lipids in determining long-term postshock viability. Nevertheless, other differences in the membranes or the constitution of the cortical cytoplasm could affect the efficiency of resealing after shocks and thus materially influence the resultant vitality. Furthermore, the cell membrane seldom presents a uniform surface. Many cells are covered with microvilli, lamellipodia, and other protrusions that may alter their true surface area and geometry and thus their response to electroshocks and ability to reseal the membrane.

The short-term survival of electroshocked cells might be determined at the level of the plasma membrane but in the longer term it is more likely to be governed by changes in the cytoplasm. Theory indicates that the threshold voltage of, for example, a mitochondrion or lysosome should be very large compared with the cell membrane itself, and thus these organelles should not be directly affected, due to their size. However, vital soluble constituents may leak out of the cells and it is probably the ability of the cell to replace these which ultimately determines the survival of a resealed cell.

Whether effects on the nucleus itself are involved remains speculative. The present results relate only to immediate postshock viability rather than clonogenicity or proliferative potential as the latter parameters differ widely between different cell types and are thus difficult to compare directly. Nevertheless, observations on these shocked cultures indicate that if they survive the immediate postshock phase (< 24hr), their long-term proliferative potential is not perceptibly impaired. Electroshocks do not, therefore, appear to result in long-term effects on replicatory potential, provided that cells survive the immediate postshock phase.

Whatever the precise cause or causes of the intrinsic differences that we have observed in this study, our results demonstrate that the effect of electroshocks on the viability of individual cell types cannot be predicted on the basis of presently known parameters and needs to be monitored empirically. We have previously shown that flow cytometry can be used to follow electropermeabilization of cells and the resealing of their membranes (O'Hare *et al.*, 1987). When attempting to optimize electrotransfection or electrofusion of mammalian cells, a combination of flow cytometry and the measurement of cell viability should be used. An arbitrary selection of electrical parameters may not suffice, particularly when a large number of independent variables are involved, and the cells themselves are particularly electrosensitive, as for example with transfection (Dembic *et al.*, 1987) or electrofusion (O'Hare and Edwards, 1986) of lymphocytes.

5. SUMMARY

The electrosensitivity of a variety of different types of mammalian cells has been examined using a dedicated electroporator which delivers a square wave pulse of predefined duration and height. Significant differences were found which could not be accounted for by either cell size, heterogeneity, thresholds for electropermeabilization, or culture conditions. Lymphoid cells were the most sensitive, followed by epithelial cells and fibroblasts, with a strain of endothelial cells the most resistant.

ACKNOWLEDGMENTS. This work was supported by grants to the Institute of Cancer Research (Royal Cancer Hospital) from the Cancer Research Campaign and the Medical Research Council. We are grateful to the workers indicated for the cell lines used in this study (Table 1) and to Mrs. S. Davies for assistance with tissue culture.

REFERENCES

Benz, R., and Zimmermann, U., 1981, The resealing process of lipid bilayers after reversible electrical breakdown, *Biochim. Biophys. Acta* **640:**169.

Borman, L. S., Swartzendruber, D. C., and Littlefield, L. G., 1972, Establishment of two parental cell lines and three clonal strains from rat colonic carcinoma, *Cancer Res.* **42:**5074.

Chernomordik, L. V., Sukharev, S. I., Abidor, J. G., and Chizmadzhev, Y. A., 1983, Breakdown of lipid bilayer membranes in an electric field, *Biochim. Biophys. Acta* **736:**203.

Chu, G., Hayakawa, H., and Berg, P., 1987, Electroporation for the efficient transfection of mammalian cells with DNA, *Nucleic Acids Res.* **15:**1311.

Danielson, K. G., Oborn, C. J., Durban, E. M., Butel, J. S., and Medina, D., 1984, Epithelial mouse mammary cell line exhibiting normal morphogenesis *in vivo* and functional differentiation *in vitro, Proc. Natl. Acad. Sci. USA* **81:**3756.

DeLarco, J. E., and Todaro, G. J., 1978, Epithelioid and fibroblastic rat kidney cell clones: Epidermal growth factor (EGF) receptors and the effects of mouse sarcoma virus transformation, *J. Cell Physiol.* **94:**335.

Dembic, Z., Haas, W., Zamoyska, R., Parnes, J., Steinmetz, M., and von Boehmer, H., 1987, Transfection of CD8 gene enhances T-cell recognition, *Nature* **326:**510.

Dunnington, D. J., Hughes, C. M., Monaghan, P., and Rudland, P. S., 1983, Phenotypic instability of rat mammary tumor epithelial cells, *J. Natl. Cancer Inst.* **71:**1227.

Edwards, P. A. W., Smith, C. M., Neville, A. M., and O'Hare, M. J., 1982, A human–human hybridoma system based on a fast-growing mutant of the ARH77 plasma cell leukaemia-derived line, *Eur. J. Immunol.* **12:**641.

Engel, L. W., Young, N. A., Tralka, T. S., Lippman, M. E., O'Brien, S. J., and Joyce, M. J., 1978, Establishment and characterization of three new continuous cell lines derived from human breast carcinomas, *Cancer Res.* **38:**3352.

Galfré, G., and Milstein, C., 1981, Preparation of monoclonal antibodies, strategies and procedures, *Methods Enzymol.* **173:**3–45.

Glassy, M. C., Handley, H. H., Hagiwara, H., and Royston, I., 1983, UC729-6, a human lymphoblastoid B-cell line useful for generating antibody-secreting human–human hybridomas, *Proc. Natl. Acad. Sci. USA* **80:** 6327.

Gospodarowicz, D., Mescher, A. L., and Birdwell, C. R., 1977, Stimulation of corneal endothelial cell proliferation *in vitro* by fibroblast and epidermal growth factors, *Exp. Eye Res.* **25:**75.

Hama-Inaba, H., Shiomi, T., Sato, K., Ito, A., and Kasai, M., 1986, Electric pulse-mediated gene transfer in mammalian cells grown in suspension culture, *Cell Struct. Funct.* **11:**191.

Hayflick, L., 1965, The limited *in vitro* lifetime of human diploid cell strains, *Exp. Cel! Res.* **37:**614.

Herzog, R., Müller-Wekensiek, A., and Voelter, W., 1986, Usefulness of Ficoll in electric field-mediated cell fusion, *Life Sci.* **39:**2279.

Hurwitz, R., Hozier, J., Le Bien, T., Minowada, J., Gajl-Peczalska, J., Kubonishi, I., and Kersey, J., 1979, Characterization of a leukemia cell line of the pre-B phenotype, *Int. J. Cancer* **23:**174.

Neumann, E., Schaefer-Ridder, M., Wang, Y., and Hofschneider, P. H., 1982, Gene transfer into mouse lyoma cells by electroporation in high electric fields, *EMBO J.* **7:**841.

O'Hare, M. J., and Edwards, P. A. W., 1986, Human monoclonal antibodies and the LICR-LON-HMy2 system, in: *Human Hybridomas: Diagnostic and Therapeutic Application* (A. Strelkauskas, ed.), Dekker, New York, pp. 44–63.

O'Hare, M. J., Asche, W., Williams, L., and Ormerod, M. G., 1987, Optimization of transfection of breast epithelial cells by electroporation, *DNA* (in press).

Sale, H., and Hamilton, W. A., 1968, Effects of high electric fields on micro-organisms. III. Lysis of erythrocytes and protoplasts, *Biochim. Biophys. Acta* **163:**37.

Smith, G. H., and Cleary, S. F., 1983, Effects of pulsed electric fields on mouse spleen lymphocytes *in vitro, Biochim. Biophys. Acta* **763:**328.

Stopper, B., Zimmermann, U., and Wecker, E., 1985, High yields of DNA-transfer into mouse L-cells by electropermeabilisation, *Z. Naturforsch.* **40:**929.

Taylor-Papadimitriou, J., Purkis, E., Lane, B. F., McKay, L. A., and Chang, S. E., 1982, Effects of SV40 transformation on the cytoskeleton and behaviour of human keratinocytes, *Cell Differ.* **11:**169.

Todaro, G. J., and Green, H., 1963, Quantitative studies of the growth of mouse embryo cells in culture and their development into established cell lines, *J. Cell Biol.* **17:**299.

Zimmermann, U., 1982, Electric field-mediated fusion and related electrical phenomena, *Biochim. Biophys. Acta* **694:**227.

Zimmermann, U., Vienken, J., Halfmann, J., and Emeis, C. C., 1985, Electrofusion: A novel hybridization technique, *Adv. Biotechnol. Processes* **4:**79.

Chapter 21

Molecular Genetic Applications of Electroporation

Huntington Potter

1. INTRODUCTION

Both academic and industrial applications of molecular biology depend on being able to efficiently express cloned genes in various eukaryotic cells. The traditional method of gene transfer, by uptake of calcium phosphate/DNA coprecipitates (Graham and Van der Eb, 1973), works well with fibroblasts but has proved difficult to apply to other differentiated mammalian cell types such as lymphocytes or neuronal cells, and is completely unsuited to plant cells or parasites. The solution to this problem has been provided by a completely new approach—electroporation. In studying the effect of high-voltage electric discharges on biological membranes, it was discovered that such shocks could induce cells to fuse via their plasma membranes, apparently by creating holes or pores in the cell membrane (Zimmermann *et al.*, 1976; Senda *et al.*, 1979; Scheurich *et al.*, 1980; Neumann *et al.*, 1980; for review, see Zimmermann and Vienken, 1982). Neumann and his colleagues (Neumann *et al.*, 1982; Wong and Neumann, 1982) then found that mouse fibroblasts (L cells) take up and express exogenous DNA when subjected to electric shock. However, because L cells are easily made to take up DNA by traditional methods, it was not clear that the procedure could be applied to any other type of cell. We extended and modified electroporation (Potter *et al.*, 1984) to allow the introduction of exogenous DNA into a broad spectrum of cell types, including neuronal cells, endocrine cells, primary animal cells, hepatoma cells, hematopoietic stem cells, and plant protoplasts (Fromm *et al.*, 1985; Igarashi *et al.*, 1986; Potter and Montminy, 1986; Montminy *et al.*, 1986; Ou-Lee *et al.*, 1986; Sureau *et al.*, 1986; Toneguzzo *et al.*, 1986; Potter, 1987). Electroporation yields a high frequency of permanent transfectants, has a high efficiency of transient gene expression, and is substantially easier to carry out than alternative techniques. Thus, for various reasons electroporation is becoming increasingly popular. Indeed, in the last few years, electroporation has moved out of a few developmental laboratories to become the method of choice for gene transfer in many situations (Fig. 1).

In essence, electroporation makes use of the fact that the cell membrane acts as an electrical capacitor which is unable (except through ion channels) to pass current. Subjecting membranes to a high-voltage electric field results in their temporary breakdown and the formation of pores that are

HUNTINGTON POTTER • Department of Neurobiology, Harvard Medical School, Boston, Massachusetts 02115.

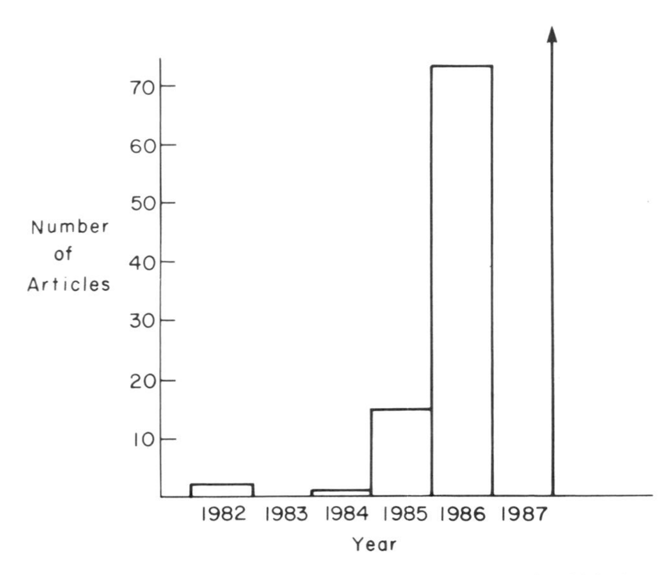

Figure 1. An indication of the recent increase in the number of publications in which electroporation has been used to carry out gene transfer.

large enough to allow macromolecules (as well as smaller molecules such as ATP) to enter or leave the cell. The reclosing of the membrane is a natural decay process which is delayed at 0°C. The reader is referred to other chapters in this book which consider the biophysics behind the mechanism of pore formation. It is the physical, rather than biochemical nature of electroporation that probably accounts for its wide applicability.

Comparison with Calcium Phosphate Transfection

The main difference between electroporation and, for instance, calcium phosphate coprecipitation procedures for transfecting eukaryotic cells is the state of the integrated DNA after selection in appropriate antibiotic media. In the case of calcium phosphate, the amount of DNA taken up and integrated into the genome of each transfected cell is in the range of 3 million base pairs. As a result, the transfected DNA often integrates as large tandem arrays containing many copies of the transfected DNA. The advantage of this procedure is evident primarily when one wishes to transfect whole genomic DNA into recipient cells and select for some phenotypic change such as malignant transformation. Here, a large amount of DNA integrated per recipient cell is essential. In contrast, electroporation can be adjusted to yield 1 to about 20 copies of an inserted gene (Potter *et al.*, 1984; Toneguzzo *et al.*, 1986, 1988) or a large amount of genomic DNA (Jastreboff *et al.*, 1987). For gene expression studies, low copy number is advantageous, since it allows one to know the particular copy responsible for the gene expression being observed. Even when few copies of an electroporated marker gene are introduced, cotransfection of a nonselectable gene of interest is very efficient (Toneguzzo *et al.*, 1988).

2. EXPERIMENTAL METHODS

2.1. Transfection by Electroporation

DNA may be transfected into various cells by applying a high-voltage pulse to a suspension of the cells and DNA as follows. Actively growing cells at about 10^6/ml medium are centrifuged,

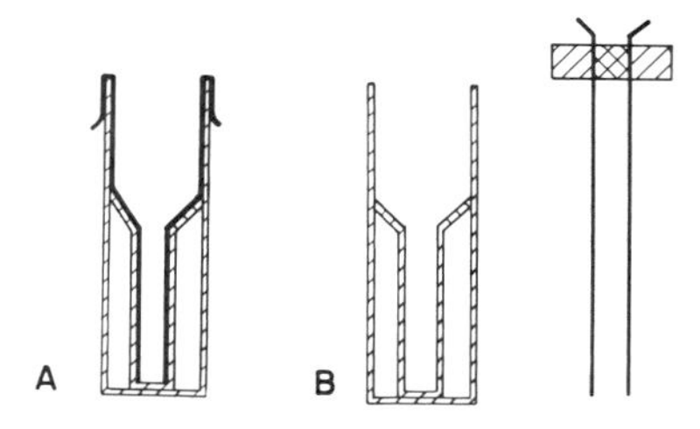

Figure 2. Two designs for electroporation chambers for delivering a high-voltage pulse to a suspension of cells and DNA. Each consists of a flat-sided, open-topped, plastic cuvette (from Sarstedt, Princeton, N.J.) having two metal (aluminum, stainless steel, or platinum) electrodes placed against the opposite walls (0.5 cm apart). A high-voltage pulse is applied to the electrodes and thus to the cell–DNA suspension.

suspended in ice-cold phosphate-buffered saline (without added $MgCl_2$ or $CaCl_2$), and centrifuged again. The pellet is resuspended in cold phosphate-buffered saline at a concentration of $1–2 \times 10^7$ cells/ml. Plasmid vector DNA (usually linearized by restriction enzyme digestion; see below) is added to the cell suspension to 20 μg/ml (although concentrations as low as 1 μg/ml are effective). The DNA and cells are allowed to sit for approximately 5 min at 0°C in one of the chambers shown in Fig. 2 and then an electric pulse is delivered.

The electric pulse may be applied to the electrodes of the electroporation chamber (Fig. 2) by any number of commercial or homemade apparatus. We have used primarily the Biorad Gene Pulser. This system stores charge in capacitors of various sizes which are then discharged into the cell–DNA suspension. Both voltage and capacitance (and hence decay half-time of the voltage pulse) can be varied to accommodate different cell types. After electroporation, the cells and DNA are allowed to sit for approximately 10 min at 0°C before being added to 10 ml of growth medium. Cells are grown for about 48 hr before transformants are selected in medium supplemented with the appropriate drug to which the plasmid confers resistance. Selection medium contains either G418 at 800 μg/ml (from GIBCO; true G418 concentration, ~370 μg/ml) or xanthine at 250 μg/ml, hypoxanthine at 15 μg/ml, mycophenolic acid at 1 μg/ml (from Lilly).

2.2. Critical Parameters

There are two critical parameters that are relevant to successful electroporation: the maximum voltage of the shock and the duration of the current pulse. Some power supplies have been designed to completely control these two parameters, but they are quite expensive and there is no evidence that they are preferable to the simple capacitor power supplies. The length of time of the discharge is determined by the capacitance and the resistance of the circuit. We have found with the Biorad Gene Pulser system that most mammalian cells are effectively transfected at a high voltage of 1.5 kV at a capacitance of 25 μF. If the cells are particularly sensitive, 2 kV or less at 3 μF should be tried. In general, the larger the diameter of the cell, the lower the voltage and capacitance that they will withstand before suffering unacceptable levels of cell death. However, each cell type can vary substantially in its sensitivity to the shock. In general, we aim for between 40 and 80% viability.

A large contributing factor to cell death appears to be the pH change that occurs due to electrolysis in the region of the electrodes. This problem can be alleviated by replacing some of the ionic strength of the phosphate-buffered saline with extra buffer, for instance, Hepes, or Tris at pH 7.5.

Recently, it has been discovered (Chu *et al.*, 1987) that lower voltage and higher capacitance—resulting in longer pulse duration—can be equally effective as high-voltage shocks. Under these conditions, 20°C was reported to result in better transfection than 0°C (see below).

The efficiency of transfection by electroporation is also dependent on cell type. For cells such as fibroblasts, which are easily transfected by traditional procedures, electroporation gives a frequency of permanent transfectants of one per 10^3 to 10^4 live cells—approximately that obtainable by calcium phosphate coprecipitation. For cells refractory to traditional methods, electroporation gives a frequency of permanent transfectants of between 10^{-4} and 10^{-5} for most cell types. Occasionally a cell line will transfect poorly under our standard conditions (10^{-6}), but these have not been optimized, and even this frequency is easily sufficient to obtain significant numbers of clones. In general, cells that transfect efficiently for permanent transfectants also do so for transient gene expression. Increasing the number of cells and the amount of DNA used in the electroporation for studying transient gene expression can circumvent problems of low transfection efficiency and low promoter/enhancer efficiency.

3. RESULTS

3.1. Gene Transfer by Electroporation

The results of transfecting DNA into several cell types under different conditions are shown in Table 1. Table 1 also indicates several important aspects of the transfection procedure. (1) Linearized plasmid DNA was at least 20 times more effective in transfecting cells than was circular DNA (lines 1 and 2). This result probably reflects the increased ease with which linear DNA is integrated into the cellular genome. (2) Carrying out the electroporation at 0°C was 6- to 16-fold more effective than at 20°C (compare lines 3 and 4). This result may be partly due to the slower closing of the membrane pores at 0°C (but see also Chu *et al.*, 1987). (3) Treatment of the cells with a sublethal concentration of colcemide for 18 hr prior to electroporation increased transfection 3- to

Table 1. Transfection Frequency by Electroporation[a]

Cells	Power supply settings (kV/μF)	Temperature (°C)	Transfectants/ 10^5 live cells
	Linear versus supercoiled DNA		
1. 1881 (ln)	1.2/25	20	1.5
2. 1881 (sc)	1.2/25	20	0.03
	Temperature and voltage effects		
3. M12	1.2/25	20	2.5
4. M12	1.2/25	0	15.0
5. M12/col	1.2/25	20	30.0
	Comparison of cell lines and species		
	Mouse		
6. L	1.5/25	0	15.0
7. 1881	1.5/25	0	3.0
8. M12	1.5/25	0	25.0
9. CTLL	1.5/25	0	20.0
	Human		
10. LAZ509	1.5/25	0	6.0
11. Ly65	1.5/25	0	1.0
12. BL18	1.5/25	0	10.0

[a]The efficiency of gene transfer by electroporation varies with different conditions and recipient cells. Mouse L cells are fibroblasts. M12 is a spontaneous mouse B-cell lymphoma, 1881 is an Abelson virus-transformed mouse pre-B-cell lymphoma, CTLL is a mouse cytotoxic T-lymphocyte line, LAZ509 is an Epstein–Barr virus-transformed human lymphoblastoid line, and Ly65 and BL18 are human Burkitt lymphoma cell lines (see Potter *et al.* 1984, for details). The vector used in these transfections was pSV7-neogpt. Line 5: cells were treated with colcemide (col) at 0.1 μg/ml for 24 hr prior to transfection.

10-fold (compare lines 3 and 5). This experiment was based on the assumption that, in cells temporarily halted in metaphase, the nuclear membrane would be either absent or more permeable and thus would pose less of a barrier to exogenous DNA. Table 1 also shows the results of transfecting several mouse and human cell lines under the same (optional) conditions.

3.2. The Immunoglobulin κ Gene Is Expressed in a Tissue-Specific Manner after Electroporation

A model system for demonstrating the use of electroporation for studying gene expression is provided by the human κ immunoglobulin gene (Potter *et al.*, 1984). Specifically, we wished to address the question of how lymphocytes of the B lineage are uniquely able to express Ig genes. 1881 cells show μ heavy chain production but no κ protein or mRNA. The two selectable plasmid vectors shown in Fig. 3A were therefore constructed, in which either the complete, or a specially deleted, human Ig κ gene was linked to the Ecogpt selectable marker gene. Following electroporation of the complete gene into mouse pre-B cells and mouse L cells, RNA from clones of permanently transfected cells was analyzed for κ-specific sequences by blot hybridization (Fig. 3B,D). The complete κ gene is seen to be transcribed in B cells to yield an mRNA of the same size as authentic human κ messenger from BL31 Burkitt lymphoma cells (Fig. 3C, right-hand lane). On the other hand, when the same normal human κ gene was introduced into mouse L cells, there was very little κ-specific transcription of any kind (Fig. 3D) and the transcripts that were produced were about 300 bases longer than the authentic human κ message. For comparison, the right-hand lane of Fig. 3E was loaded with 15 μg of RNA from one of the 1881 clones shown in Fig. 3B that received the normal human κ gene.

In sum, although the Ecogpt gene was transcribed in both L cells and pre-B cells (not shown), Ig gene transcription occurred only in the lymphoid cells. This indicates that the cell-type specificity of Ig gene expression is solely dependent on the DNA sequence in the area of the gene and not, for instance, on its chromosome location. Furthermore, it is evident that, although the 1881 pre-B cells have not yet undergone a DNA recombination event leading to a completed light chain, they are far enough along the path of B-cell development to be able to transcribe an exogenously added κ gene.

3.3. A Region of the J–C Intron of the κ Ig Gene Is Necessary for Effective Transcription

Because the human κ gene was actively transcribed in mouse B cells, but not mouse L cells, it seemed reasonable to ask which region of the DNA was responsible for this tissue-specific transcription. A likely candidate for such a sequence was a region of DNA in the J–C intron that serves no protein coding function, yet is highly conserved between mouse and human and thus might play a regulatory role in both species (Hieter *et al.*, 1980). The second construct in Fig. 3A contains an altered human κ gene from which this putative regulatory region was deleted. This vector was then introduced by electroporation into mouse 1881 pre-B cells and L cells as before.

The results of subjecting the RNA from transfected clones of mouse pre-B cells and fibroblasts to hybridization with the human κ constant region probe are shown in Fig. 3B–D. There was a marked decrease in the level of transcription of κ Ig mRNA in the 1881 clones that received the deleted gene (Fig. 3C) compared with those that received the normal gene (Fig. 3B). No further reduction in the low level of aberrant κ transcripts in L cells resulted from deleting the transcriptional enhancer sequence (compare Fig. 3E and 3D).

To quantitate the level of κ-specific mRNA in individual clones of mouse pre-B cells transfected with the κ Ig gene, with or without its enhancer, the autoradiograms of Northern blots were analyzed by densitometry. The amount of κ mRNA relative to the number of integrated κ genes (measured by Southern blot hybridization of *Hind*III-restricted DNA) was calculated and is displayed in Fig. 4. Ten individual 1881 clones that received the normal human κ gene are compared with 11 clones that received the "enhancer"-deleted human κ gene. On average, κ mRNA expression in pre-B cells is depressed by a factor of approximately 15 when the κ enhancer is deleted. In sum, the region of DNA deleted from the human κ intron must contain DNA necessary for efficient transcription in B lymphocytes—in fact, an enhancer sequence.

3.4. Transient Expression of Electroporated Genes

In addition to generating permanently transfected cell lines carrying genetically engineered genes of interest, electroporation can also be used to introduce DNA into cells for transient expression assays. In this application, the bacterial gene coding for chloramphenicol acetyltransferase (CAT) is linked to eukaryotic transcription promoter and enhancer sequences. Gene expression is assayed by measuring CAT activity by virtue of its ability to acetylate [^{14}C]chloramphenicol and change its mobility during thin-layer chromatography (Gorman *et. al.*, 1982). Figure 5 shows the application of this assay to studying the strength of different promoters in a human lymphoma cell line (M12) (H. Potter and E. Sinn, unpublished). The most active promoter-enhancer is seen to be that derived from the SV40 virus (PSV2CAT). The c-myc promoter alone is seen to be fairly weak (c-myc CAT), but is much enhanced after its translocation to the immunoglobulin locus in the BL22 Burkitt lymphoma (BL22-myc CAT). Evidently, a sequence in the immunoglobulin gene has boosted the strength of the normal promoter of the c-myc oncogene in the lymphoma cells, and may thus account for their malignant transformation.

4. APPLICATION OF ELECTROPORATION TO PROBLEMS OF GENE THERAPY

One long-term goal of many of the studies being carried out in modern molecular genetics is to be able to apply our newfound knowledge to the treatment of human disease. As more and more genes become identified and ultimately cloned that, when mutant, result in a disease state, it is possible that, in some instances, gene replacement therapy of some kind might be useful. However, several criteria must be realized before such a goal can be practically achieved. First, the gene must be available which, when introduced into a relatively few number of cells of the individual, can

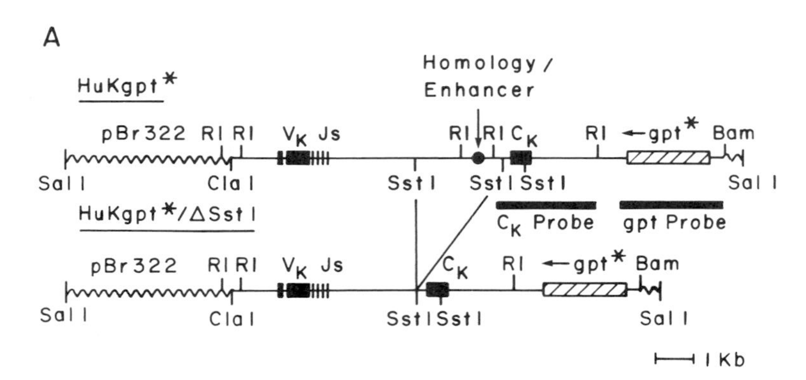

Figure 3. (A) Construction of the normal and deleted human Ig κ gene vectors. The complete human κ gene (top) contains a transcriptional enhancer sequence located on a 2-kb *Sst* fragment that was deleted to yield the plasmid shown on the bottom line. These plasmids were linearized with *Sal*I before transfection by electroporation. (B–E) Northern blot analysis of RNA from mouse L cells and pre-B cells transfected with the normal and deleted κ genes. Panels D and E were overexposed by a factor of 5–10 to reveal the low level of aberrant transcripts present in L cells transfected with the normal and deleted κ genes.

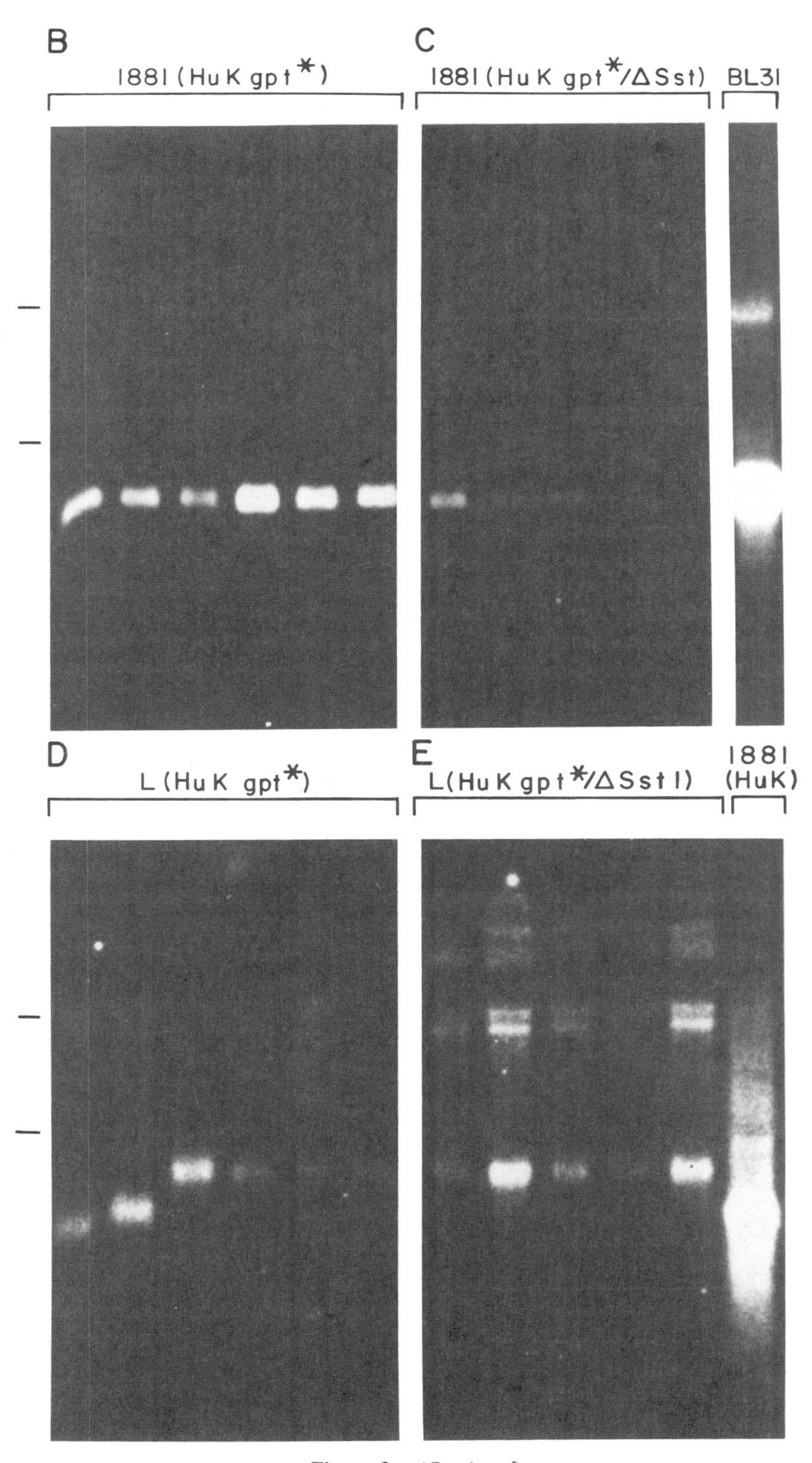

Figure 3. *(Continued)*

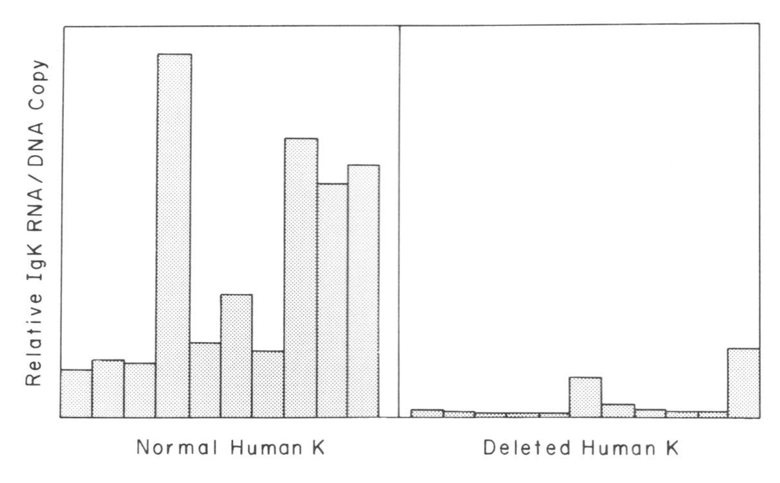

Figure 4. Relative amounts (normalized for gene dosage) of κ mRNA present in mouse 1881 pre-B cells transfected by electroporation with either the normal or deleted human Ig κ gene.

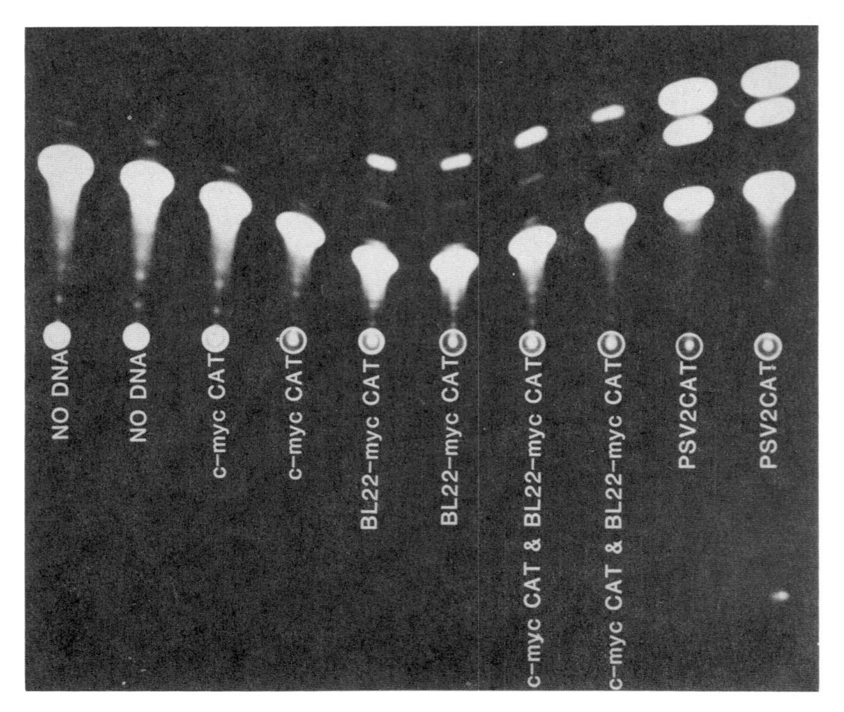

Figure 5. Expression of bacterial chloramphenicol acetyltransferase after transfection of human lymphoid cells by electroporation. Different transcriptional control elements have been linked to the CAT gene coding region to assay their relative strength. PSV2 CAT contains the SV40 promoter and enhancer; c-myc CAT contains the promoter from the human c-myc oncogene; BL22-myc CAT contains the c-myc promoter linked to an immunoglobulin gene by virtue of the chromosome translocation present in the BL22 Burkitt lymphoma. M12 is a cell line derived from a human lymphoma.

result in improved health. Second, there must be a way to introduce the cloned gene into appropriate cells without damaging them and then reintroduce the cells into the individual such that they can make use of their newly engineered genotype.

The first problem is rapidly being solved and already several genes are now candidates for such an application. The remaining problem is therefore to introduce the genes into the appropriate cells and reintroduce those cells into the organism without adverse effects. At this point there are essentially two ways in which foreign genes can be introduced into mammalian cells of a wide variety of tissue types. The first is electroporation, described here, and the second is the use of retroviral vectors which have been designed by Mulligan and his colleagues to infect any human cell (Williams *et al.*, 1984).

Although for many purposes in scientific research both electroporation and retroviral vectors are adequate for gene transfer, with certain advantages accruing to one or the other in different situations, gene therapy in humans poses special problems. First, the transfection technique should be efficient, so that most of the target cells receive and express the gene of interest. Both techniques are capable of delivering genes with high efficiency, approaching 100% for the retroviral vectors and potentially as much as 20 or 30% for electroporation (Sugden *et al.*, 1985). Furthermore, both methods can be used to transfect bone marrow cells which may then be reintroduced into a recipient animal (Williams *et al.*, 1984; Toneguzzo and Keating, 1986). The recent demonstration of transfection of human hematopoietic stem cells by electroporation (Tonneguzo and Keating, 1986, 1987) is particularly encouraging. However, the mechanisms by which these two techniques work strongly favor electroporation as a long-term solution to gene therapy. When a retrovirus enters a target cell, it integrates essentially randomly in the genome and thus has potential for introducing mutational damage by the mere fact of its insertion. In addition, the promoter generally carried on both long terminal repeats of an integrating retrovirus can result in the expression of nearby genes which, if they are oncogenes, can, at low frequency, lead to the malignant transformation of the target cell. Finally, because the retrovirus still retains the capability of excising from the genome and reintegrating or potentially even reinfecting other cells, provided it is supplied *in trans* with the appropriate proteins (as might occur if the cells are infected by a second retrovirus of a viable type), there is no guarantee that the retrovirus-transfected cells represent a safe and stable means of introducing an engineered gene into a living organism. In contrast, these drawbacks are lessened by using electroporation for gene transfer. First, there are no long terminal repeats associated with the introduced genes. Second, there is good evidence that it should be possible to direct electroporated genes to recombine with their homologous host gene (Smithies *et al.*, 1985; Thomas and Capecchi, 1987). The resultant cell actually acquires the wild-type normal version of its mutant gene at the exact chromosome location that it normally should have, thus reducing the potential mutagenetic effects of random insertion.

5. CONCLUSIONS

Electroporation is a simple, highly effective means of introducing cloned genes into a wide variety of cell types. It affords substantial benefits over alternative procedures, being easier to use, and is often more efficient and applicable to more different kinds of cells. As the parameters of electroporation become optimized, efficiencies of DNA entry into the recipient cell may approach 100%. Coupled with developing means for directing homologous recombination to replace a defective gene on the chromosome with its transfected wild-type counterpart, electroporation may play a key role in future gene replacement therapy for certain human diseases.

ACKNOWLEDGMENTS. I am grateful for the assistance of Stefan Cooke in the preparation of the manuscript and to Francis Toneguzzo for unpublished data. The work in our laboratory has been supported by NIH Grants AI21848 and GM35967.

REFERENCES

Chu, G., Hayakawa, H., and Berg, P., 1987, Electroporation for the efficient transfection of mammalian cells with DNA, *Nucleic Acids Res.* **15:**1311–1326.

Fromm, M., Taylor, L. P., and Walbot, V., 1985, Expression of genes transferred into monocot and dicot plant cells by electroporation, *Proc. Natl. Acad. Sci. USA* **82:**5824–5828.

Gorman, C. M., Moffat, L. F., and Howard, B. H., 1982, Recombinant genomes which express chloramphenicol acetyl transferase in mammalian cells, *Mol. Cell. Biol.* **2:**1044–1051.

Graham, F.L., and Van der Eb, A., 1973, A new technique for the assay of infectivity of human adenovirus 5 DNA, *Virology* **52:**456–467.

Hieter, P. A., Max, E. E., Seidman, J. G., Maizel, J. V., Jr., and Leder, P., 1980, Cloned human and mouse kappa immunoglobulin constant and J region genes conserve homology in functional segments, *Cell* **22:** 197–207.

Igarashi, T., Okazaki, T., Potter, H., Gaz, R., and Kronenberg, H. M., 1986, Cell-specific expression of the human parathyroid hormone gene in rat pituitary cells, *Mol. Cell. Biol.* **6:**1830–1833.

Jastreboff, M. M., Ito, E., Bertino, J.R., and Narayanan, R., 1987, Use of electroporation for high-molecular-weight DNA-mediated gene transfer, *Exp. Cell Res.* **171:**513–517.

Montminy, M. R., Sevarino, K. A., Wagner, J. A., Mandel, G., and Goodman, H., 1986, Identification of a cyclic-AMP responsive element within the rat somatostatin gene, *Proc. Natl. Acad. Sci. USA* **83:**6682–6686.

Neumann, E., Gerisch, G., and Opatz, K., 1980, Cell fusion induced by high electric impulses applied to Dictyostelium, *Naturwissenschaften* **67:**414–415.

Neumann, E., Schaefer-Ridder, M., Wang, Y., and Hofschneider, P. H., 1982, Gene transfer into mouse lyoma cells by electroporation in high electric fields, *EMBO J.* **1:**841–845.

Ou-Lee, T. M., Turgeon, R., and Wu, R., 1986, Expression of a foreign gene linked to either a plant virus or a Drosophila promoter, after electroporation of protoplasts of rice, wheat, and sorghum, *Proc. Natl. Acad. Sci. USA* **83:**6815–6819.

Potter, H., 1987, Electroporation: A general method of gene transfer in: *Proceedings of the Ninth Annual Conference of the IEEE Engineering in Medicine and Biology Society*, pp. 705–707.

Potter, H., and Montminy, M., 1986, Introduction of cloned genes into PC 12 pheochromocytoma cells by electroporation, in: *Discussions in Neurosciences,* Volume III, No. I (A. Bignami, L. Bolis, and D. C. Gadjusek, eds.), FESN, Geneva, pp. 138–143.

Potter, H., Weir, L., and Leder, P., 1984, Enhancer-dependent expression of human κ immunoglobulin genes introduced into mouse pre-B lymphocytes by electroporation, *Proc. Natl. Acad. Sci. USA* **81:**7161–7165.

Scheurich, P., Zimmerman, U., Mischel, M., and Lamprecht, I., 1980, Membrane fusion and deformation of red blood cells by electric fields, *Z. Naturforsch.* **35c:**1081–1085.

Senda, M., Takeda, J., Abe, S., and Nakamura, T., 1979, Cell fusion by electrical stimulation, *Plant Cell Physiol.* **20:**144–1443.

Smithies, O., Gregg, R. G., Boggs, S. S., Koralewski, M. A., and Kucherlapati, R. S., 1985, Insertion of DNA sequences into the human chromosome β-globin locus by homologous recombination, *Nature* **317:** 230–234.

Sugden, B., Marsh, K., and Yates, J., 1985, A vector that replicates as a plasmid and can be efficiently selected in B-lymphoblasts transformed by Epstein–Barr virus, *Mol. Cell Biol.* **5:**410–413.

Sureau, C., Romet-Lemonne, J.-L., Mullins, J. I., and Essex, M., 1986, Production of hepatitis-B virus by a differentiated human hepatoma cell line after transfection with a cloned circular HBV DNA, *Cell* **47:**37–47.

Thomas, K. R., and Capecchi, M. R., 1987, Site-directed mutagenesis by gene targeting in mouse embryo-derived stem cells, *Cell* **51:**503–512.

Toneguzzo, F., and Keating, A., 1986, Stable expression of selectable genes introduced into human hematopoietic stem cells by electric field-mediated DNA transfer, *Proc. Natl. Acad. Sci. USA* **83:**3496–3499.

Toneguzzo, F., and Keating, A., 1987, Mechanism of transfer and integration of genes introduced into hematopoietic cells by electroporation, in: *Proceedings of the Ninth Annual Conference of the IEEE Engineering in Medicine and Biology Society,* pp. 715–716.

Toneguzzo, F., Hayday, A. C., and Keating, A., 1986, Electric field mediated DNA transfer: Transient and stable gene expression in human and mouse lymphoid cells, *Mol. Cell. Biol.* **6:**703–706.

Toneguzzo, F., Keating, A., Lilly, S., and McDonald, K., 1988, Electric field-mediated gene transfer: Characterization of DNA transfer and patterns of integration in lymphoid cells, *Nucleic Acids Res.* **16:**5515–5532.

Williams, D. A., Lemischka, I. R., Nathan, D. G., and Mulligan, R. C., 1984, Introduction of new genetic material into pluripotent haematopoietic stem cells of the mouse, *Nature* **310**:476–480.
Wong, T. K., and Neumann, E., 1982, Electric field mediated gene transfer, *Biochem. Biophys. Res. Commun.* **107**:584–587.
Zimmermann, U., and Vienken, J., 1982, Electric field-induced cell-to-cell fusion, *J. Membr. Biol.* **67**:165–182.
Zimmermann, U., Riemann, F., and Pilwat, G., 1976, Enzyme loading of electrically homogeneous human red blood cell ghosts prepared by dielectric breakdown, *Biochim. Biophys. Acta* **436**:460–474.

Chapter 22

Plant Gene Transfer Using Electrofusion and Electroporation

James A. Saunders, Benjamin F. Matthews, and Paul D. Miller

1. INTRODUCTION

In recent years, the rapidly developing fields of electrofusion of somatic cells and electroporation of selected cell types have become an extremely important tool in the production and modification of hybrid somatic gene pools. In plants, these procedures have thus far been restricted to protoplasts in which the thick secondary cell wall has been stripped off usually through enzymatic digestion of the cellulosic and pectic components of the cell wall. Protoplasts with a diameter of 35–55 μm can be predictably isolated from normal mesophyll leaf tissue or from cell suspension cultures. These cells are ideally suited for the simple, rapid, and consistent procedures that have been developed for both electrofusion and electroporation (Bates *et al.*, 1987). These procedures are very effective for cells of this size and have made electromanipulation of cell membranes the method of choice for gene transfer due to the simplicity, convenience, and ease of operation.

1.1. Rationale for Electrofusion Techniques

Until the 1980s, protoplast fusion was achieved strictly by using chemical fusogens or mechanical stresses such as centrifugal force. The most widely used chemical fusogen, polyethylene glycol (PEG), is often used in combination with alkaline pH and high concentrations of calcium to achieve a 2–5% yield of fusion products. However, this system has several drawbacks including toxic effects on the cells, low fusion rates, and lack of specificity. As an alternative technique, electrofusion of plant protoplasts offers a nontoxic procedure with relatively high fusion rates. In our laboratory, fusion yields as high as 40–50% of the tobacco protoplasts have been obtained, although typically 30% represents a working average (Fig. 1). In most cases, the viability of these cells is maintained at greater than 85% throughout the fusion events (Fig. 2).

James A. Saunders • Germplasm Quality and Enhancement Laboratory, USDA, ARS, Beltsville, Maryland 20705. Benjamin F. Matthews • Plant Molecular Genetics Laboratory, USDA, ARS, Beltsville, Maryland 20705. Paul D. Miller • Vegetable Laboratory, USDA, ARS, Beltsville, Maryland 20705. *Present address:* DNA Plant Technology Corporation, Cinnaminson, New Jersey 08077.

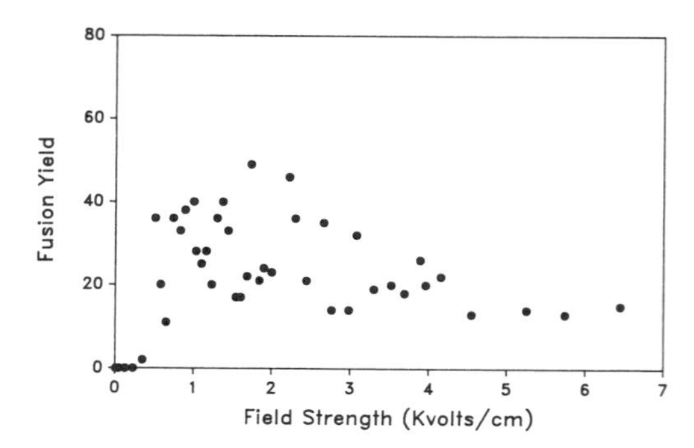

Figure 1. The effects of field strength on the yield of fusion products of two or more protoplasts isolated from mesophyll leaf tissue of *Nicotiana rustica*. A single square wave DC pulse of 40 μsec duration was applied and the number of fused cells was determined optically 20 min after the fusion pulse.

1.2. Rationale for Electroporation Techniques

The transfer of small numbers of well-characterized genes into a recipient plant variety offers the possibility of specific and controlled genetic engineering. It eliminates the transfer of unwanted, unnecessary, or deleterious genes into the recipient variety. Such extraneous genes complicate varietal development by producing adverse side effects. These genes must be deleted through conventional breeding efforts before the variety is commercially feasible. Thus, techniques such as protoplast fusion, which combines two genomes doubling the chromosome number, may have disadvantages over electroporation techniques for commercial applications.

Single gene transfer can be accomplished by several methods. Modified pathogens, such as *Agrobacterium tumefaciens*, have been used quite successfully in introducing foreign genes into plants (Fraley *et al.*, 1983; Barton *et al.*, 1983; Horsch *et al.*, 1984; Broglie *et al.*, 1984). However, the gene to be transferred must be cloned into the equivalent of T-DNA of *A. tumefaciens* before gene transfer can take place. Also, the host range of *A. tumefaciens* is limited and the response of different crop species to *A. tumefaciens* varies. Thus, it is much more difficult to transform some plant crops than it is others. For example, many important crops including members of the Gramineae are particularly recalcitrant to infection and gene transfer through *A. tumefaciens*.

Another method of gene transfer is microinjection. This method has been successful (Crossway *et al.*, 1986); however, it is difficult, tedious, and is not conducive to treating large numbers of plant cells or protoplasts. Microinjection also demands a high level of skill in micromanipulation and expensive manipulation equipment.

In contrast to these methods, electroporation can be used for the transfer of foreign genes into protoplasts made from monocots or dicots. Thus, the range of crop plants which can be used for gene transfer is not dependent on pathogen–host recognition, as in the case of *A. tumefaciens*. Furthermore, the genes to be transferred do not have to be recloned into T-DNA, bypassing an elaborate and time-consuming step. Electroporation can transfer genes efficiently into a large

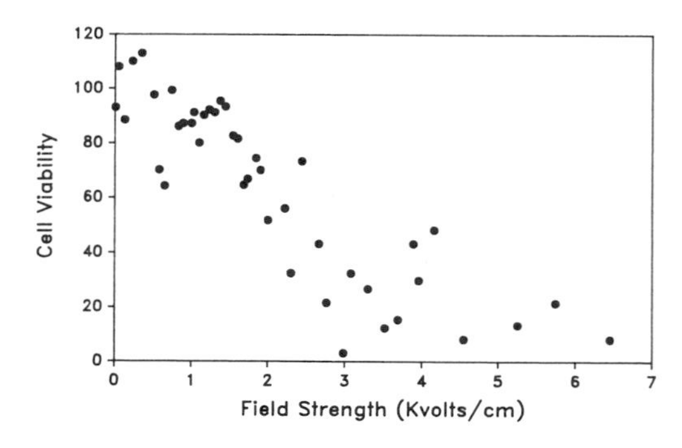

Figure 2. The effects of field strength on the viability of protoplasts isolated from mesophyll leaf tissue of *Nicotiana rustica*. Cell viability was determined by staining the cells with fluorescein diacetate 20 min after a 40 μsec DC pulse was applied.

population of plant protoplasts, and is not as tedious and time-consuming as is microinjection. Thus, large numbers of plant protoplasts can be transformed at one time with confidence that at least some of those protoplasts treated will be transformants.

Electroporation is a simple procedure. It does not require a high level of skill nor does it require equipment as expensive as an electrofusion apparatus. Moreover, equipment is readily available from commercial sources, or can be made in the laboratory (Mischke *et al.*, 1986). The application of electroporation is currently limited to crop plants which can be regenerated from protoplasts. However, it may be possible to circumvent this limitation by using intact cells or other plant material. This could increase the applicability of electroporation to include virtually any plant crop species.

2. OPTIMIZATION OF ELECTROFUSION

2.1. Alignment of Cells

The electrofusion of cells or protoplasts is a two-step procedure. It involves not only the perturbation of the cellular membrane to induce cytoplasmic fusion, but also manipulation of the cells to bring them into close physical contact to allow fusion to proceed. Protoplasts isolated from mesophyll leaf tissue of *Nicotiana tabacum* can be brought together through dielectrophoresis using an alternating sine wave current of 100 volts/cm. This process, which induces a polarity in individual cells, creates areas of high field strength between localized regions of a fusion chamber. The polarized cells are attracted to these areas of high field strength. This ultimately results in the migration of the cells toward each other and facilitates cell fusion. The procedure is rapid and has a negligible effect on the viability of the protoplasts exposed to frequencies above 10 kHz and below 6 MHz. In tobacco, the optimal frequency for dielectrophoresis was determined to be a sine wave of 1 MHz, although AC sine waves of other frequencies were effective in the movement of the isolated protoplasts (Saunders *et al.*, 1986).

2.2. Fusion of Cells

The fusion of cells using electrical currents typically occurs following the AC cell alignment phase of the process. However, Sowers and Kapoor (1987) have shown that red blood cell ghosts remain in a fusogenic condition for several minutes following a rapid DC pulse and that cells coming into contact with each other during this fusogenic period will coalesce.

The diameter of the cells to be fused appears to be an important consideration for fusion. An effective fusion field strength for large-diameter cells, such as those from plant protoplasts, is typically much lower than an effective field strength for a cell with a smaller diameter, such as a bacterial or animal cell. Typically, a 40-μsec square wave pulse of 1–3 kvolts/cm electrode gap is sufficient to fuse plant protoplasts (Fig. 1). In our laboratory we have shown that protoplast fusion can occur using both square wave pulses as well as exponentially decaying capacitor discharge waves. One of the most important differences between the wave forms appears to be the effect of the pulse on the viability of the protoplasts being fused. When a capacitor discharge pulse generator with an exponentially decaying wave form is used, it is often necessary to apply a pulse large enough to kill more than 50% of the cells in the fusion chamber before effective fusion yields are obtained. Using a square wave pulse generator with a 40-μsec pulse of 1–1.5 kvolts/cm, we can typically fuse 30% of the cells while maintaining cell viability at 85% of the untreated control protoplasts (Fig. 2). In addition, capacitor discharge waves can change their rate of discharge and thus the shape of the wave as the capacitor ages, which can affect both the yield of fusion products obtained from the treatment as well as the viability of the treated cells. These differences notwithstanding, both the square wave pulse generators as well as the capacitor discharge generators are effective in membrane fusion and membrane electroporation.

Excessively high field strengths can have undesirable effects both on the viability of the cells being treated as well as on the yield of fused cells produced by the electrofusion process (Figs. 1 and 2). It is often difficult and time-consuming to empirically determine the optimum field strength necessary for fusion by evaluating the fusion yield and the viability of each cell population which is being treated. Accurate information on effective fusion pulses can be obtained very rapidly by monitoring the change in conductivity which occurs in the fusion media induced by the DC pulse. During electroporation or electrofusion, the pores induced in the cell membrane not only allow for the uptake of foreign materials into the cell, but also allow cellular components to leak into the fusion media. The latter increases the conductivity of the media. This change in conductivity is only apparent if two conditions are met prior to the application of the fusion pulse. First, the fusion media must have a low conductivity relative to the ionic concentration of the protoplasts. Second, the volume of the protoplasts must be sufficiently large compared to the volume of the fusion chamber to influence the conductivity of the media by cellular leakage. A dramatic rise in conductivity of the fusion media indicates the effectiveness of the electropulse treatment. This can be determined by initially starting with low field strengths and gradually increasing the fusion pulse while monitoring the conductivity both before and after a pulse. This procedure, if repeated on freshly isolated protoplasts, can be utilized to rapidly determine effective field strengths for electrofusion of cells from a variety of sources.

3. SOMATIC HYBRID SELECTION AND IDENTIFICATION

3.1. Selection Strategies

Protoplast fusion results in the production of a mixed population of protoplasts consisting of somatic hybrids (heterokaryonic fusions), parental protoplasts, and some self-fusions (homokaryonic fusions). From this population the investigator tries to efficiently identify those cells which are somatic hybrids and to eliminate all of the others. Numerous selection and identification schemes have been devised to accomplish this. Some schemes take advantage of differential growth rates of the somatic hybrid versus the parental lines on culture medium. In these schemes, protoplasts from each parental line grow poorly or not at all on the protoplast culture medium, while the somatic hybrids grow well. For example, Power *et al.*, (1977) were able to select for interspecific somatic hybrids by developing a culture medium which allowed only the hybrid to grow but not the parental protoplasts. Using a *Nicotiana sylvestris* cell line resistant to streptomycin, Medgyesy *et al.*, (1980) obtained intraspecific and interspecific hybrids.

Other somatic hybrid selection schemes include the use of mutant cell lines resistant to growth inhibitory compounds, such as amino acid analogues (Horn *et al.*, 1983; Harms *et al.*, 1982). For example, two cell lines are selected for resistance to growth inhibition by amino acid analogues. When protoplasts from these cell lines are fused, the resultant somatic hybrid possesses dual resistance and can grow in the presence of both analogues, while the parental lines cannot. Harms *et al.*, (1981) were able to select dually resistant carrot somatic hybrids by this technique. One parental line of *Daucus carota* was preselected for resistance to the lysine analogue S(2-amino)-ethyl cysteine (AEC), while the other parental line was resistant to the tryptophan analogue, 5-methyltryptophan (5MT). The resultant somatic hybrids exhibited dual resistance which contained predominantly tetraploid chromosome numbers.

Complementation is another method for selecting somatic hybrids. One parental cell line is an albino resistant to growth inhibition by a specific compound while the other parent is sensitive to growth inhibition by the compound but has a normally green complement. The somatic hybrids can be identified by their growth in the presence of the inhibitory compound and their regeneration into green shoots. For example, protoplasts from a *Nicotiana tabacum* cell line deficient in nitrate reductase activity were fused with protoplasts from a cytoplasmic albino cell line (Glimelius and Bonnett, 1981). Plants representing somatic hybrids were obtained which were green and possessed active nitrate reductase. Complementation has also been used for the identification of somatic hybrids of *Nicotiana* using nitrate reductase-deficient cell lines without pigment selection (Marton *et al.*, 1982). Nine nitrate reductase-deficient cell lines were tested for complementation after hybridization. Allelic deficiencies would not complement, whereas apoenzyme and cofactor deficiencies complemented. Thus, somatic hybrids contained restored nitrate reductase activity. Similar corrections in nitrate-deficient cell lines of *Nicotiana* have been made by fusion of these lines with irradiated protoplasts of *Datura innoxia* (Gupta *et al.*, 1982). Dudits *et al.* (1977) conducted a pigment complementation experiment by fusing albino *D. carota* protoplasts with *D. capillifolius* protoplasts. The green regenerating shoots were examined for *D. carota* morphological features to identify somatic hybrids. These are only a few of the many selection schemes which have been developed over the past several years and which have been subjects of reviews (Evans, 1983; Saunders and Bates, 1987).

3.2. Cell Separation

Visual selection is another method of obtaining somatic hybrids. The most common method of visual selection is through the fusion of green mesophyll protoplasts with protoplasts from cell cultures and visually identifying hybrids using a microscope (Flick *et al.*, 1985). With care, the somatic hybrids are mechanically isolated and then cultured. Variations on this method would include the fusion of protoplasts from other colored plant organs such as the petals. There are several advantages to selecting hybrids visually. It is an efficient method, whereby the vast majority of those selected are indeed hybrids. Furthermore, the time-consuming task of selecting and characterizing mutations usable for somatic hybridization is eliminated. However, mechanical isolation of somatic hybrids can be quite tedious. To circumvent this problem, automated sorting systems are being developed which utilize a combination of fluorescence labeling of protoplasts, autofluorescence of chlorophyll, and laser technology. Protoplasts can be labeled using fluorescence dyes such as FITC (green) or RITC (red) and then fused (Galbraith and Mauch, 1980). The heterokaryons can be separated from parental protoplasts using a cell cytometer–cell sorter. *Nicotiana* protoplasts can be routinely sorted at a rate of 300 protoplasts/sec. If, after fusion, the heterokaryons represent between 0.1 and 1.0% of the total protoplast population, then approximately three heterokaryons could be identified and selected per second (Afonso *et al.*, 1985). The sorting procedure provides a population of protoplasts which can be from 40 to 90% heterokaryons. Although the possibilities for this method are intriguing, unfortunately the cost of the cell sorting apparatus is currently prohibitively expensive for many laboratories.

In an effort to bypass the use of biochemically selectable mutant cell lines, our laboratory attempted to separate mixtures of protoplasts from both *Nicotiana* and *Daucus* on the basis of inherent differences of surface charges on the cell membrane. In this procedure, the cells were subjected to a constant electrophoretic field of low voltage. The velocity of migration of the cells toward the opposite charged pole is a function of the surface charge on the cell membrane and is termed the electrophoretic mobility value of a population of cells. Our data indicate that the inherent differences in electrophoretic mobility values within both the *Nicotiana* and the *Daucus* genera are sufficient to separate mixed populations of parental cells by this technique. Further, cells which are the result of electrofusions between parental cell lines with differing electrophoretic mobility values, have characteristic intermediate mobility values. These intermediate populations of fused cells appear sufficiently resolved from the parental cell lines to allow the recovery and subsequent culture of specifically selected populations of fused cells. From our data, it appears that these techniques for the electrofusion and subsequent separation of the fused cells are applicable to a wide variety of plant sources.

3.3. Isoenzyme Patterns

Besides the conventional screens for selecting heterokaryons, several methods are used for confirming that the plants regenerated from the isolated protoplast are a result of somatic hybridization. Often karyotypes of the parental lines and the putative somatic hybrid are compared to determine if all chromosomes of both parents are present. Another method commonly used to confirm the isolation of a heterokaryon is evaluation of the expression of distinguishing enzyme isoforms of the somatic hybrid and its parents using electrophoretic gels. Peroxidase (Afonso *et al.*, 1985; Medgyesy *et al.*, 1980), alcohol dehydrogenase (Maliga *et al.*, 1977), malate dehydrogenase (Afonso *et al.*, 1985), esterase (Sidorov *et al.*, 1981; Medgyesy *et al.*, 1980), xanthine dehydrogenase (Gupta *et al.*, 1982), homoserine dehydrogenase (Matthews and Widholm, 1985), glucose-6-phosphate dehydrogenase (Dudits *et al.*, 1977; Maliga *et al.*, 1977), aspartate aminotransferase (Wetter and Kao, 1980), and RuBPCase (Chen *et al.*, 1977; Iwai *et al.*, 1980; Akada and Hirai, 1983) are some of the many enzymes which have been used to differentiate between somatic hybrids and parental lines. This method may have its drawbacks at times. It is possible that several enzymes screened will not distinguish between parent lines, or that one parental form of an enzyme is not strongly expressed in the hybrid. On the other hand, new isoenzymic forms may be present in the somatic hybrid which were not present in either parent (Matthews and Widholm, 1985). These new combinations may be interesting not only for their ability to distinguish somatic hybrids, but also because they may possess new and unique biochemical and catalytic properties.

3.4. Molecular Biology Techniques for Somatic Hybrid Identification

Nuclear, chloroplast, and mitochondrial DNA restriction patterns have recently been used to examine organelle genetics and heredity and to confirm somatic hybridization. Very few investigators have published nuclear DNA analyses of parental and somatic hybrid lines. Saul and Potrykus (1984) used repetitive DNA sequences to identify somatic hybrids of *Hyoscyamus muticus* and *Nicotiana tabacum*. This method has not been used often and relies on the use of non-cross-hybridizing DNA repetitive sequences from each species. Such probes become more difficult to find as the similarity of the parents increases.

Although chloroplast DNA restriction fragment patterns have been used to identify somatic hybrids, chloroplasts tend to sort during cell division. The analysis of chloroplast DNA may reveal the presence of chloroplast DNA from one parent and may not indicate that somatic hybridization has actually occurred. However, some plants obtained through somatic hybridization have contained mixed populations of chloroplasts from both parents (Fluhr *et al.*, 1983). Interspecific chloroplast recombination has also been reported (Medgyesy *et al.*, 1985) to occur in a somatic hybrid formed between *N. tabacum* and *N. plumbaginifolia*. In this case the parental cell lines differed in three different cytoplasmic genetic markers. The chloroplast DNA of the somatic hybrid contained at least six recombination sites. However, most analyses of chloroplast DNA from somatic hybrids have been much less successful in demonstrating somatic hybridization.

In contrast, mitochondrial DNA restriction fragment patterns have been used frequently to demonstrate somatic hybridization (Belliard *et al.*, 1979; Nagy *et al.* 1981; Boeshore *et al.*, 1983; Matthews and Widholm, 1985). Not only are portions of the mitochondrial genomes of both parents present in the somatic hybrid, but recombination often occurs, which generates novel DNA fragments not present in either parental line. Thus, identifying somatic hybrids is much easier using mitochondrial DNA as a genetic marker.

4. OPTIMIZATION OF ELECTROPORATION

Electroporation is the application of one or more short electrical pulses to a membrane system such that holes or pores form in the outermost membrane. Under the proper conditions, these pores are transient, and may exist for more than 10 min before resealing (Okada *et al.*, 1986a; Serpersu *et al.*, 1985). The size of the pore diameters can exceed 30 nm (Okada *et al.*, 1986a), and their duration is controllable. The physical parameters necessary for this control are dependent on the membrane composition and its immediate environment. When the membrane pores are open, molecules in the suspension media may enter or leave the cell. In the past few years, electroporation of DNA or RNA into cells has become a very popular method for direct gene transfer and transfection of cells. The high efficiency of this relatively easy technique was first demonstrated by Neumann *et al.*(1982) and Potter *et al.* (1984) with animal cells. Fromm *et al.* (1985) showed that the technique could be used to transfer genes into plant protoplasts. Since these early reports, electroporation has been used to study the effects and efficiencies of cell transformation (Shillito *et al.*, 1985; Schocher *et al.*, 1986), the expression of foreign genes in cells (Pietrzak *et al.*, 1986; Miller and Sinden, 1986; Reiss *et al.*, 1986), and the regulation of such expression (Ou-Lee *et al.*, 1986). Electroporation applied to whole plant cells (i.e., through the intact cell wall) has been reported and referred to as electroinjection (Morikawa *et al.*, 1986). Obviously, the transformation of whole cells or tissues will circumvent the problems associated with plant regeneration from protoplasts. This application is therefore an extremely important endeavor and promises to attract much attention.

In any given cell, the extent of poration is determined by the intensity and duration of the electrical impulses and the ionic concentration and constitution of the electroporation buffer. The electrical parameters which interact and generally define an electroporation system are the field strength of the pulse, the duration of the pulse, and the resistance of the fusion media. Hence, variation in results using the same cell variety is still common even when all electroporation parameters are held constant. Because there is such a wide variety of cell responses, there is a wide range of experimental treatments in the literature (Table 1).

Though existence of the pores is transient, several minutes may be required for their closing. We try to minimize the manipulation of the protoplasts for 10 min immediately following electroporation, not only to allow the nucleic acids time to move across the plasmalemma, but also to maintain viability of the protoplasts. The rate of pore closure may also be slowed by decreased temperature (Potter *et al.*, 1984; Kinosita and Tsong, 1977).

The majority of electroporation studies in plants have employed selectable genes conferring antibiotic (chloramphenicol or neomycin) resistance upon the host cells. The monitoring of chloramphenicol acetyltransferase (CAT) activity provides a relatively quick and easy verification of the uptake and expression of DNA by protoplasts. We electroporated the CAT gene into six varieties of potato protoplasts including: Russett Burbank, Kennebec, Atlantic, USW 2225 (a dihaploid *Solanum tuberosum*), *S. chocoense*, and a fusion hybrid (X_4) between *tuberosum* and *chocoense*. All six varieties have shown levels of transient CAT activity in a range easily assayed, from 0.07 to 1.0 unit of CAT activity per 10^5 protoplasts. We determined that replacement of the NaCl by LiCl in the electroporation buffer made a significant difference in the amount of DNA transferred, hence assayable CAT activity (Fig. 3). When 135 mM Na^+ was replaced by Li^+, there was a 4- to 70-fold increase in CAT activity. The increase in activity appears to be roughly linear with increasing Li^+ concentration. To help elucidate the effect of Li^+, we compared the effect of Li^+ with that of Na^+ on the activity of purified CAT and no difference was found. Ito *et al.* (1984) found efficient transformation of yeast cells possible when the latter were incubated in the presence of the monovalent alkali cations Na^+, K^+, Li^+, Cs^+, and Rb^+. They suggested a chemiosmotic-type mechanism of transport across the plasma membrane. If the process of poration occurs at a predisposed site in the membrane, e.g., a H^+ channel, then perhaps Li^+ increases this predisposition to channel opening. This might be accomplished by increasing membrane fluidity or by enhancing membrane electrical potentials making them more susceptible to an applied electrical pulse. Alternatively, Li^+ might enhance electroporation by maintaining pore stability.

Electroporation has also proven to be a very efficient means of inducing uptake of viruses (Okada *et al.*, 1986a; Nishiguchi *et al.*, 1986) and viroids (Miller and Sinden, 1986) into protoplasts and even into whole (intact) plant cells (Morikawa *et al.*, 1986). Our study screened for the presence of the virus or viroid by nucleic acid hybridization (Fig. 4). The number of protoplasts infected

Table 1. Partial List of Electroporation Parameters[a]

Tissue source	Fusion media salts	Field strength (kV/cm)	Selection marker	Efficiency	Reference
Carrot, tobacco, maize	150 mM NaCl 5 mM KCl	0.50–0.875	CAT	TA	Fromm *et al.* (1985)
Maize	140 mM NaCl 4 mM $CaCl_2$	0.50	NPT	0.8%	Fromm *et al.* (1986)
Potato, carrot, cabbage	135 mM LiCl 5 mM $CaCl_2$ 2.7 mM KCl	0.30–0.40	CAT PSTV-cDNA	TA > 23%	Miller and Sinden (1986)
Tobacco cells	1.3 mM $CaCl_2$ 0.7 mM $ZnSo_4$ 0.5 mM $MgSO_4$	0.65	TMV-RNA	50%	Morikawa *et al.* (1986)
Tobacco	0	10	TMV-RNA CMV-RNA	75%	Nishiguchi *et al.* (1986)
Tobacco, vinca	70 mM KCl 5 mM $CaCl_2$	0.75	TMV-RNA CMV-RNA	40–60% 80%	Okada *et al.* (1986a)
Tobacco	70 mM KCl	0.75	CAT NPT	TA $5/10^3$	Okada *et al.* (1986b)
Rice, wheat, sorghum	150 mM NaCl	10.0	CAT	TA	Ou-Lee *et al.* (1986)
Tobacco	6 mM $MgCl_2$	1.25–1.5	NPT, CAT	> 0	Pietrzak *et al.* (1986)
Tobacco	0	2.0	NPT	0.02%	Riggs and Bates (1986)
Tobacco	6 mM $MgCl_2$	1.25–1.5	APH	0.03%	Schocher *et al.* (1986)
Tobacco	6 mM $MgCl_2$	1.25–1.5	NPT	2%	Shillito *et al.* (1985)
HeLa cells, fibroblasts, kidney cells	137 mM NaCl, 5 mM KCl	0.53	CAT, GPT, NPT	1%	Chu *et al.* (1987)
Lymphoid	150 mM NaCl	7–9	Ig κ	0.005%	Falkner *et al.* (1984)
Mouse lyoma	150 mM NaCl	8	TK	0.01%	Neumann *et al.* (1982)

[a]Abbreviations used: APH, aminoglycoside phosphotransferase; CAT, chloramphenicol acetyltransferase; CMV, cucumber mosaic virus; GPT, xanthine-guanine phosphotransferase; Ig κ, immunoglobulin kappa gene; NPT, neomycin phosphotransferase; PSTV, potato spindle tuber viroid; TA, transient expression assay; TK, thymidine kinase; TMV, tobacco mosaic virus.

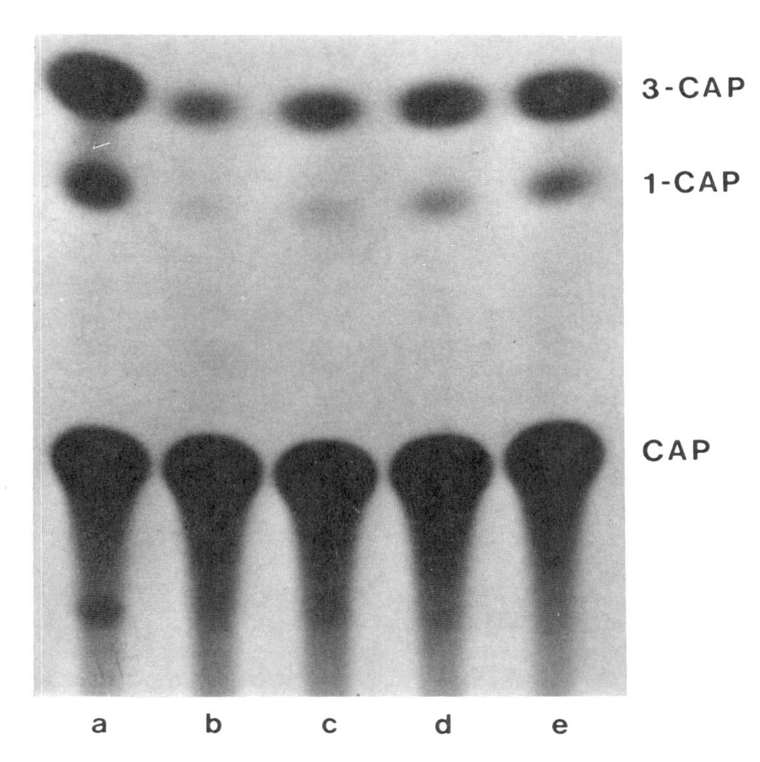

Figure 3. Li^+ effect on protoplast electroporation. Potato protoplasts were electroporated at 375 V/cm in electroporation buffer [10 mM Hepes, pH 7.2, 5 mM $CaCl_2$, 2.7 mM KCl, 0.2 M mannitol, 135 mM salt (NaCl, LiCl, or a combination of both)] containing 40 μg of pUC8CaMVCATN. CAT assays were performed on 10^5 protoplasts in each of lanes b–e. Lanes: a, 0.1 unit of purified chloramphenicol acetyltransferase (CAT) (Sigma Chemical Co.); b, 135 mM NaCl; c, 108 mM NaCl + 27 mM LiCl; d, 27 mM NaCl + 108 mM LiCl; e, 135 mM LiCl. CAP, chloramphenicol; 1-CAP, 1-acetylchloramphenicol; 3-CAP, 3-acetylchloramphenicol.

ranged from 23 to 80%. These values indicate the minimum number of cells which were successfully infused by at least one copy of the foreign plasmid or virus. Whether these pathogens are eventually used as gene-shuttle vectors or subcellular "vaccines," in the mutant or antisense form, their current use proves that electroporation is an extremely efficient delivery system and therefore a technology of great promise.

5. SUMMARY

Electrofusion of plant protoplasts is a promising technique for combining the genetic content of more than one cell. Electrofusion is rapid, reproducible, and gives predicted results. It is possible to fuse any two or more membrane-bound cells or vesicles of a similar diameter with this procedure. Its present limitations involve the genetic instability of hybrid cell lines with multiple ploidy levels. Electroporation represents a modification of the techniques used in electrofusion and can be used to transfer discrete genes or gene packages efficiently. These techniques represent powerful tools for the cellular and molecular biologists in gene transfer technologies.

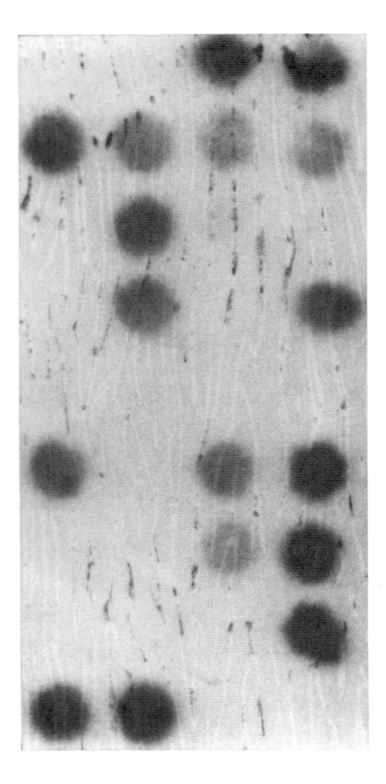

Figure 4. Dot-blot of nucleic acid extracted from calli derived from potato protoplasts electroporated with PSTV-cDNA. Seventeen of thirty-six calli are positive for the presence of the viroid. No positives were obtained from calli derived from protoplasts treated in an identical manner without the electrical pulse.

REFERENCES

Afonso, C. L., Harkins, K. R., Thomas-Compton, M. A., Krejci, A. E., and Galbraith, D. W., 1985, Selection of somatic hybrid plants in *Nicotiana* through fluorescence-activated sorting of protoplasts, *Bio/Technology* **3:**811–816.

Akada, S., and Hirai, A., 1983, Studies on the mode of separation of chloroplast genomes in parasexual hybrid calli. II. Heterogeneous distribution of two kinds of chloroplast genomes in hybrid callus, *Plant Sci. Lett.* **32:**95–100.

Barton, K. A., Binns, A. N., Matzke, A. J. M., and Chilton, M. D., 1983, Regeneration of intact tobacco plants containing full length copies of genetically engineered T-DNA, and transmission of T-DNA to R1 progeny, *Cell* **32:**1033–1043.

Bates, G. W., Saunders, J. A., and Sowers, A. E., 1987, Electrofusion principles and applications, in: *Cell Fusion* (A. E. Sowers, ed.), Plenum Press, New York, pp. 367–395.

Belliard, G., Vedel, F., and Pelletier, G., 1979, Mitochondrial recombination in cytoplasmic hybrids of *Nicotiana tabacum* by protoplast fusion, *Nature* **281:**401–403.

Boeshore, M. L., Lifshitz, I., Hanson, M. R., and Izhar, S., 1983, Novel composition of mitochondrial genomes in *Petunia* somatic hybrids derived from cytoplasmic male sterile and fertile plants, *Mol. Gen. Genet.* **190:**459–467.

Broglie, R., Coruzzi, G., Fraley, R. T., Rogers, S. G., Horsch, R. B., Niedermeyer, J. G., Fink, C. L., Flick, J. S., and Chua, N. H. 1984, Light-regulated expression of a pea ribulose-1, 5-bisphosphate carboxylase small subunit gene in transformed plant cells, *Science* **224:**838–843.

Chen, K., Wildman, S. G., and Smith, H. H. 1977, Chloroplast DNA distribution in parasexual hybrids as shown by polypeptide composition of fraction I protein, *Proc. Natl. Acad. Sci. USA* **74:**5109–5112.

Chu, G., Hayakawa, H., and Berg, P., 1987, Electroporation for the efficient transfection of mammalian cells with DNA, *Nucleic Acids Res.* **15:**1311–1326.

Crossway, A., Oakes, J., Irvine, J., Ward, B., Knauf, V., and Schumaker, C., 1986, Integration of foreign DNA following microinjection of tobacco mesophyll protoplasts, *Mol. Gen. Genet.* **202:**179–185.

Dudits, D., Hadlaczky, G., Levi, E., Fejer, O., Haydu, Z., and Lazar, G., 1977, Somatic hybridization of *Daucus carota* and *D. capillifolius* by protoplast fusion, *Theor. Appl. Genet.* **51:**127–132.

Evans, D. A., 1983, Protoplast fusion, in: *Handbook of Plant Cell Culture,* Volume 1 (D. A. Evans, W. R. Sharp, P. V. Ammirato, and Y. Yamada, eds.), Macmillan Co., New York, pp. 291–321.

Falkner, F. G., Neumann, E., and Zachau, H. G., 1984, Tissue specificity of the initiation of immunoglobulin κ gene transcription, *Z. Physiol. Chem.* **365:**1331–1343.

Flick, C. E., Kut, S. A., Bravo, J. E., Gleba, Y. Y., and Evans, D. A., 1985, Segregation of organelle traits following protoplast fusion in *Nicotiana, Bio/Technology* **3:**555–560.

Fluhr, R., Aviv, D., Edelman, M., and Galun, E., 1983, Hybrids containing mixed and sorted-out chloroplasts following interspecific somatic fusions in *Nicotiana, Theor. Appl. Genet.* **65:**289–294.

Fraley, R. T., Rogers, S. G., Horsch, R. B., Sanders, P. R., Flick, J. S., Adams, S. P., Bittner, M. L., Brand, L. A., Fink, C. L., Fry, J. S., Galluppi, G. R., Goldberg, S. B., Hoffmann, N. L., and Woo, S. C., 1983, Expression of bacterial genes in plant cells, *Proc. Natl. Acad. Sci. USA* **80:**4803–4807.

Fromm, M., Taylor, L. P., and Walbot, V., 1985, Expression of genes transferred into monocot and dicot plant cells by electroporation, *Proc. Natl. Acad. Sci. USA* **82:**5824–5828.

Fromm, M. E., Taylor, L. P., and Walbot, V., 1986, Stable transformation of maize after gene transfer by electroporation, *Nature* **319:**791–793.

Galbraith, D. W., and Mauch, T. J., 1980, Identification of fusion of plant protoplasts. II. Conditions for the reproducible fluorescence labelling of protoplasts derived from mesophyll tissue, *Z. Pflanzenphysiol.* **98:** 129–140.

Glimelius, K., and Bonnett, H. T., 1981, Somatic hybridization in *Nicotiana:* Restoration of photoautotrophy to an albino mutant with defective plastids, *Planta* **153:**497–503.

Gupta, P. P., Gupta, M., and Schieder, O., 1982, Correction of nitrate reductase defect in auxotrophic plant cells through protoplast-mediated intergeneric gene transfers, *Mol. Gen. Genet.* **188:**378–383.

Harms, C. T., Potrykus, I., and Widholm, J. M., 1981, Complementation and dominant expression of amino acid analogue resistance markers in somatic hybrid clones from *Daucus carota* after protoplast fusion, *Z. Pflanzenphysiol.* **101:**377–390.

Harms, C. T., Oertli, J. J., and Widholm, J. M., 1982, Characterization of amino acid analogue resistant somatic hybrid cell lines of *Daucus carota* L., *Z. Pflanzenphysiol.* **106:**239–249.

Horn, M. E., Kameya, T., Brotherton, J. E., and Widholm, J. M., 1983, The use of amino acid analog resistance and plant regeneration ability to select somatic hybrids between *Nicotiana tabacum* and *N. glutinosa, Mol. Gen. Genet.* **192:**235–240.

Horsch, R. B., Fraley, R. T., Rogers, S. G., Sanders, P. R., Lloyd, A., and Hoffmann, N., 1984, Inheritance of functional foreign genes in plants, *Science* **223:**496–498.

Ito, H., Murata, K., and Kimura, A., 1984, Transformation of intact yeast cells treated with alkali cations or thiol compounds, *Agric. Biol. Chem.* **48:**341–347.

Iwai, S., Nagao, T., Nakata, K., Kawashima, N., and Matsuyama, S., 1980, Expression of nuclear and chloroplastic genes coding for fraction-1 protein in somatic hybrids of *Nicotiana tabacum + rustica, Planta* **147:**414–417.

Kinosita, K., and Tsong, T. Y., 1977, Formation and resealing of pores of controlled sizes in human erythrocyte membrane, *Nature* **268:**438–441.

Maliga, P., Lazar, G., Joo, F., Nagy, A. H., and Menczel, L., 1977, Restoration of morphogenic potential in *Nicotiana* by somatic hybridization, *Mol. Gen. Genet.* **157:**291–296.

Marton, L., Sidorov, B., Baisini, G., and Maliga, P., 1982, Complementation in somatic hybrids indicates four types of nitrate reductase deficient lines in *Nicotiana plumbaginifolia, Mol. Gen. Genet.* **187:**1–3.

Matthews, B. F., and Widholm, J. M., 1985, Organelle DNA composition and isoenzyme expression in an interspecific somatic hybrid of *Daucus, Mol. Gen. Genet.* **198:**371–376.

Medgyesy, P., Menczel, L., and Maliga, P., 1980, The use of cytoplasmic streptomycin resistance: Chloroplast transfer from *Nicotiana tabacum* into *Nicotiana sylvestris,* and isolation of their somatic hybrids, *Mol. Gen. Genet.* **179:**693–698.

Medgyesy, P., Fejes, E., and Maliga, P., 1985, Interspecific chloroplast recombination in a *Nicotiana* somatic hybrid, *Proc. Natl. Acad. Sci. USA* **82:**6960–6964.

Miller, P. D., and Sinden, S. L., 1986, Effects of protoplast source and buffer on electroporation efficiency, *Mid-Atlantic Plant Molecular Biology Society, Proceedings,* p. 12.

Mischke, S., Saunders, J. A., and Owens, L., 1986, A versatile low-cost apparatus for cell electrofusion and other electrophysiological treatments, *J. Biochem. Biophys. Methods* **13:**65–75.

Morikawa, H., Iida, A., Matsuri, C., Ikegami, M., and Yamada, Y., 1986, Gene transfer into intact plant cells by electroinjection through cell walls and membranes, *Gene* **41:**121–124.

Nagy, F., Torok, I., and Maliga, P., 1981, Extensive rearrangements in the mitochondrial DNA in somatic hybrids of *Nicotiana tabacum* and *Nicotiana knightiana, Mol. Gen. Genet.* **183:**437–439.

Neumann, E., Schaefer-Ridder, M., Wang, Y., and Hofschneider, P. H., 1982, Gene transfer into mouse lyoma cells by electroporation in high electric fields, *EMBO J.* **1:**841–845.

Nishiguchi, M., Langridge, W. H. R., Szalay, A. A., and Zaitlin, M., 1986, Electroporation-mediated infection of tobacco leaf protoplasts with tobacco mosaic virus RNA and cucumber mosaic virus RNA, *Plant Cell Rep.* **5:**57–60.

Okada, K., Nagata, T., and Takebe, I., 1986a, Introduction of functional RNA into plant protoplasts by electroporation, *Plant Cell Physiol.* **27:**619–626.

Okada, K., Takebe, I., and Nagata, T., 1986b, Expression and integration of genes introduced into highly synchronized plant protoplasts, *Mol. Gen. Genet.* **205:**398–403.

Ou-Lee, T. M., Turgeon, R., and Wu, R., 1986, Expression of a foreign gene linked to either a plant-virus or a *Drosophila* promoter, after electroporation of protoplasts of rice, wheat, and sorghum, *Proc. Natl. Acad. Sci. USA* **83:**6815–6819.

Pietrzak, M., Shillito, R. D., Hohn, T., and Potrykus, I., 1986, Expression in plants of two bacterial resistance genes after protoplast transformation with a new plant expression vector, *Nucleic Acids Res.* **14:**5857–5868.

Potter, H., Weir, L., and Leder, P., 1984, Enhancer-dependent expression of human κ immunoglobulin genes introduced into mouse pre-B lymphocytes by electroporation, *Proc. Natl. Acad. Sci. USA* **81:**7161–7165.

Power, J. B., Berry, S. F., Frearson, E. M., and Cocking, E. C., 1977, Selection procedures for the production of inter-species somatic hybrids of *Petunia hybrids* and *Petunia parodii.* I. Nutrient media and drug sensitivity complementation selection, *Plant Sci. Lett.* **10:**1–6.

Reiss, M., Jarestreboff, M. M., Bertino, J. R., and Narayanan, R., 1986, DNA-mediated gene transfer into epidermal cells using electroporation, *Biochem. Biophys. Res. Commun.* **137:**244–249.

Riggs, C. D., and Bates, G. W., 1986, Stable transformation of tobacco by electroporation: Evidence for plasmid concentration, *Proc. Natl. Acad. Sci. USA* **83:**5602–5606.

Saul, M. W., and Potrykus, I., 1984, Species-specific repetitive DNA used to identify interspecific somatic hybrids, *Plant Cell Rep.* **3:**65–67.

Saunders, J. A., and Bates, G. W., 1987, Chemically induced fusion of plant protoplasts, in: *Cell Fusion* (A. E. Sowers, ed.), Plenum Press, New York, pp. 497–520.

Saunders, J. A., Roskos, L. E., Mischke, S., Aly, M. A. H., and Owens, L. D., 1986, Behavior and viability of tobacco protoplasts in response to electrofusion parameters, *Plant Physiol.* **80:**117–121.

Schocher, R. J., Shillito, R. D., Saul, M. W., Paszkowski, J., and Potrykus, I., 1986, Co-transformation of unlinked foreign genes into plants by direct gene transfer, *Bio/Technology* **4:**1093–1096.

Serpersu, E. H., Kinosita, K., and Tsong, T. Y., 1985. Reversible and irreversible modification of erythrocyte membrane permeability by electric field, *Biochim. Biophys. Acta* **812:**779–785.

Shillito, R. D., Saul, M. W., Paszkowski, J., Muller, M., and Potrykus, I., 1985, High efficiency direct gene transfer to plants, *Bio/Technology* **3:**1099–1103.

Sidorov, V. A., Menczel, L., Nagy, F., and Maliga, P., 1981, Chloroplast transfer in *Nicotiana* based on metabolic complementation between irradiated and iodoacetate treated protoplasts, *Planta* **152:**341–345.

Sowers, A. E., and Kapoor, V., 1987, The electrofusion mechanism in erythrocyte ghosts, in: *Cell Fusion* (A. E. Sowers, ed.), Plenum Press, New York, pp. 397–408.

Wetter, L. R., and Kao, K. N., 1980, Chromosome and isoenzyme studies on cells derived from protoplast fusion of *Nicotiana glauca* with *Glycine max–Nicotiana glauca* cell hybrids, *Theor. Appl. Genet.* **57:**273–276.

Chapter 23

Electric Field-Induced Fusion and Cell Reconstitution with Preselected Single Protoplasts and Subprotoplasts of Higher Plants

Hans-Ulrich Koop and Germán Spangenberg

1. INTRODUCTION

Since in higher plants—at least in a number of species—it is possible to regenerate the whole organism from single somatic cells, here cell fusion techniques can be regarded not only as a means of analysis at the cellular level, but also as a means of creating new organisms. Protoplast fusion for the production of somatic hybrids (for reviews see: Evans, 1983; Evans *et al.*, 1983; Gleba and Sytnik, 1984; Harms, 1985) or for the transfer of cytoplasmic traits (Galun and Aviv, 1983) is generally performed in a statistical manner, i.e., fusion is induced in a mixed population of the respective fusion partners. This approach allows only limited, if any, control of the type and numbers of cells participating in a particular fusion event and therefore requires identification and selection of putative fusion products either shortly after the fusion procedure (Kao, 1977) or during later stages of culture (Melchers and Labib, 1974). The availability of selectable markers may be a limiting factor for the applicability of cell fusion techniques in genotypes qualifying for use in breeding programs. Moreover, in the case of subprotoplast fusion for transfer of cytoplasmic traits, i.e., in the case of cell reconstitution experiments, unavoidable contamination of subprotoplast preparations with different subprotoplasts or intact protoplasts (Wallin *et al.*, 1978, 1979; Maliga *et al.*, 1982) may considerably reduce the value of this technique by obscuring the identity and composition of particular fusion products, cell lines, or plants regenerated thereof.

In order to overcome the problems encountered in fusion schemes using populations of cells or cellular fragments, we have designed experimental systems that allow for controlled fusion of single

HANS-ULRICH KOOP • Botanical Institute, University of Munich, D-8000 Munich 19, Federal Republic of Germany. GERMÁN SPANGENBERG • Max Planck Institute for Cell Biology, D-6802, Ladenburg, Federal Republic of Germany. In memory of Prof. Dr. H. G. Schweiger, the late director of the Max Planck Institute for Cell Biology, Ladenburg near Heidelberg, who not only helped to get the work described started, but also made a number of valuable suggestions and substantial contributions while the experiments were in progress.

selected protoplasts or subprotoplasts, respectively, and for the subsequent culture and regeneration of single fusion products.

2. TOWARD SOMATIC HYBRIDIZATION VIA FUSION OF SINGLE PROTOPLASTS

2.1. Selecting Single Cells

Various devices, largely differing with respect to complexity of construction, costs, reliability of operation, and skill needed for their successful use, have been developed for manual isolation of single cells from cell suspensions (Kao, 1977; Koop *et al.*, 1983a,b). The hydraulic system we designed (Fig. 1) uses modern technology in order to facilitate handling of single selected cells in high numbers within a reasonable time (Koop and Schweiger, 1985a; Spangenberg, 1986; Schweiger *et al.*, 1987). The degree of automatization is high enough to avoid the necessity of extreme manual skill without losing the capability of isolating defined single cells under microscopical control using high-resolution optics. In addition, it allows not only selection of cells but also precise control of the amount of liquid which is transferred along with a cell, a feature which is highly important for culturing individual cells (see Section 2.2).

Briefly, the instrumental components combined with an inverted microscope (Zeiss IM 35) are a capillary pipette which is connected to a dispenser/diluter–pump (Hamilton Microlab M), modified to produce volume changes in the nanoliter range, and a video recording system. The pump is operated by a step-motor, which is computer-controlled. Three more step-motors serve for vertical positioning of the capillary pipette and for adjusting the position of the microscope stage. Positioning is alternatively achieved in a preprogrammed fashion by the computer or manually by the use of a joystick. When operated under computer control, the pump and the positioning systems for capillary and stage are activated by the use of two footswitches. A cell is selected by adjusting it in the selection chamber (Fig. 1) underneath the opening of the capillary pipette, taking it up into the capillary by suction along with a few nanoliters of liquid in which the cells are suspended. The capillary is then raised out of the selection chamber, the stage moved to adjust the desired location of the selected cell underneath the capillary opening, the capillary lowered, and the pump activated to release the cell along with a few nanoliters of medium. Up to 200 cells can thus be selected and transferred in 1 hr (Koop and Schweiger, 1985a; Spangenberg, 1986; Schweiger *et al.*, 1987).

2.2. Culturing Single Cells

One of the most critical requirements for the culture of cells is to maintain an appropriate cell density, i.e., the cell number to volume of culture medium ratio may not exceed or fall below a certain value. In the case of higher plant cells, the required density is in the range of 10^4 to 10^5 cells per ml, which computes to 10 to 100 nl of culture medium per single cell. This was first achieved in the culture of *Datura innoxia* suspension cells and *Nicotiana tabacum* mesophyll protoplasts (Koop *et al.*, 1983a,b).

Later, following the same principle but using an improved microculture system, we were able to regenerate whole plants from up to 60% of individually selected and cultured mesophyll protoplasts of *N. tabacum* (Koop and Schweiger, 1985a). The microculture system consists of an array of microdroplets of culture medium on a microscope coverslip. Each microdroplet is protected from evaporation by a separate layer of mineral oil (1 μl). Oil layers are kept separated by impregnating the areas between the oil droplets with silane, as depicted in Fig. 2. Handling of the microculture system is simple; it combines excellent optical conditions for high-resolution light microscopy with good accessibility to the microdroplets and the selected cells. Withdrawals from or additions to the culture medium are easily performed, and harvesting of microcolonies grown from single cells is readily achieved. The microculture system has been successfully used for suspension cells, the regeneration of plants from protoplasts of *N. tabacum* (Koop and Schweiger, 1985a,b), *Brassica*

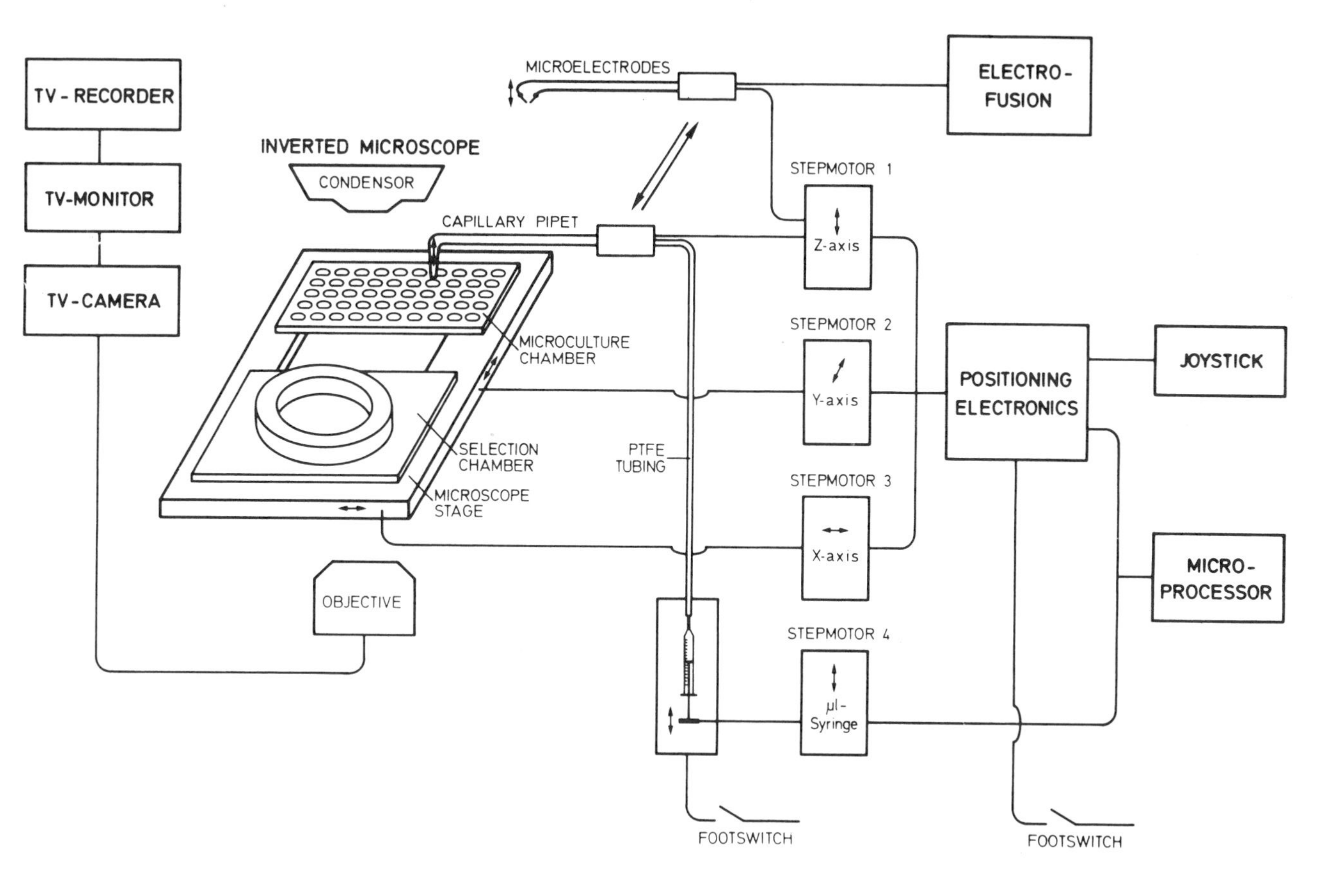

Figure 1. Experimental setup for individually selecting, culturing, and electrofusing single protoplasts.

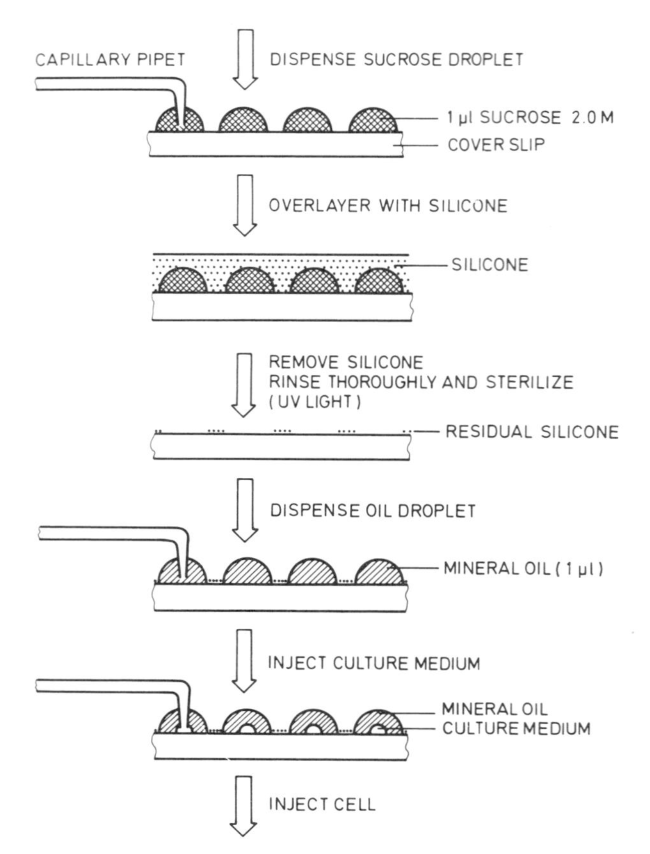

Figure 2. Preparing microculture chambers for individual culture of single cells. From Koop and Schweiger (1985a), with permission.

napus (Spangenberg, 1986; Schweiger *et al.*, 1987), and *Funaria hygrometrica* (Mejía *et al.*, 1988), the analysis of interactions between different types of cells (Spangenberg *et al.*, 1985), cell reconstitution experiments (Spangenberg *et al.*, 1986a,b), the culture of microinjected protoplasts of *B. napus* (Spangenberg, 1986; Spangenberg *et al.*, 1986b; Schweiger *et al.*, 1987) and *N. tabacum*, and more recently for cloning of hybridomas (O. Walla, personal communication).

2.3. Fusing Single Protoplasts

A system allowing the controlled fusion of defined single cells requires monitoring of the single fusion event, a condition which is best met by electric field-induced fusion (Zimmerman, 1982). Indeed, one of the earliest reports on electrically induced fusion (Senda *et al.*, 1979) describes controlled fusion of single protoplasts. The method used did not allow the fusion products to be separated from nonfused protoplasts since the experiments were performed in a suspension. Also, it was very cumbersome, requiring the use of micromanipulators. The system we designed uses the same positioning device as for single cell selection with the only difference being that microelectrodes are substituted for the capillary pipette (Figs. 1 and 3). The electrodes consist of two rectangular loops of platinum wire (diameter 50 µm, 10 mm long, 0.3 mm wide) and leave a gap of 200 to 300 µm (Koop and Schweiger, 1985b). Fusion is performed with pairs of protoplasts preselected into separate drops of mannitol (500 nl, covered with mineral oil) (Figs. 3 and 4). After

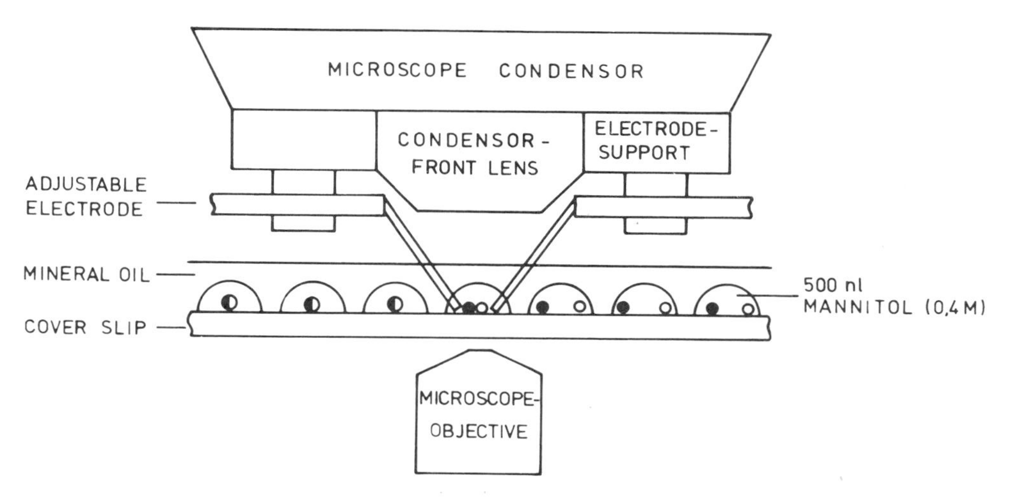

Figure 3. Arrangement of microelectrodes and microdroplets containing selected pairs of protoplasts for electrofusion of defined pairs of cells. From Koop and Schweiger (1985b), with permission.

selecting pairs of protoplasts, the capillary pipette is replaced by the microelectrodes, which are then sequentially lowered into the droplets. Dielectrophoretic alignment of the two cells, induced at 1 MHz and 60 to 80 V/cm, is complete after a few seconds and a single or a series of a few DC pulses (50 μsec, 0.9 to 1.8 kV/cm) is applied to induce fusion. The alternating field is turned off once fusion has started, the electrodes are removed from the mannitol droplet, lowered into the next one, and so forth (Figs. 3 and 4). About 50 preselected pairs of protoplasts can be fused and transferred into microculture in 1 hr. Efficiency of fusion varies with the types of protoplasts to be fused and was found to be 45 to 60% in the case of *B. napus* (Spangenberg *et al.*, 1986a,b; compare Section 3.1), 50 to 60% in the moss *Physcomitrella patens* (Hansen *et al.*, unpublished), and up to 100% in *N. tabacum* (Koop and Schweiger, 1985b).

2.4. Regenerating Plants from Single Fusion Products

One of the most significant advantages of electric field-induced fusion over chemically induced fusion is supposed to be that it does not require the use of potentially toxic agents. We have investigated whether conditions applied during electric field-induced fusion reduce the viability and regenerative capacity of the treated cells. As can be seen from the comparison with nonfused controls (Table 1), there is a slight but nonsignificant reduction of viability and division frequency in fused samples (Koop and Schweiger, 1985b). No difference has been found in the morphogenetic capacity of microcalli derived through individual culture of fusion products and nonfused controls (Table 2).

No quantitative data on the frequencies of nuclear fusion and, in the case of suitable deficient lines (Melchers and Labib, 1974; Schieder, 1977; Glimelius *et al.*, 1978; Kohn *et al.*, 1985), of complementation are available from experiments on controlled fusion of single protoplasts. In our experiments, we found chromosome doubling in 60% of the plants regenerated from fusion products of two defined protoplasts of *N. tabacum* (Table 2). Chromosome counting in plants regenerated from control unfused protoplasts and microcultured in the same way, revealed the normal chromosome set ($2n = 48$). Thus, nuclear fusion rather than ploidy level alternations induced during the *in vitro* culture can accomplish the chromosome doubling observed in the plants regenerated from microcultured fusion products.

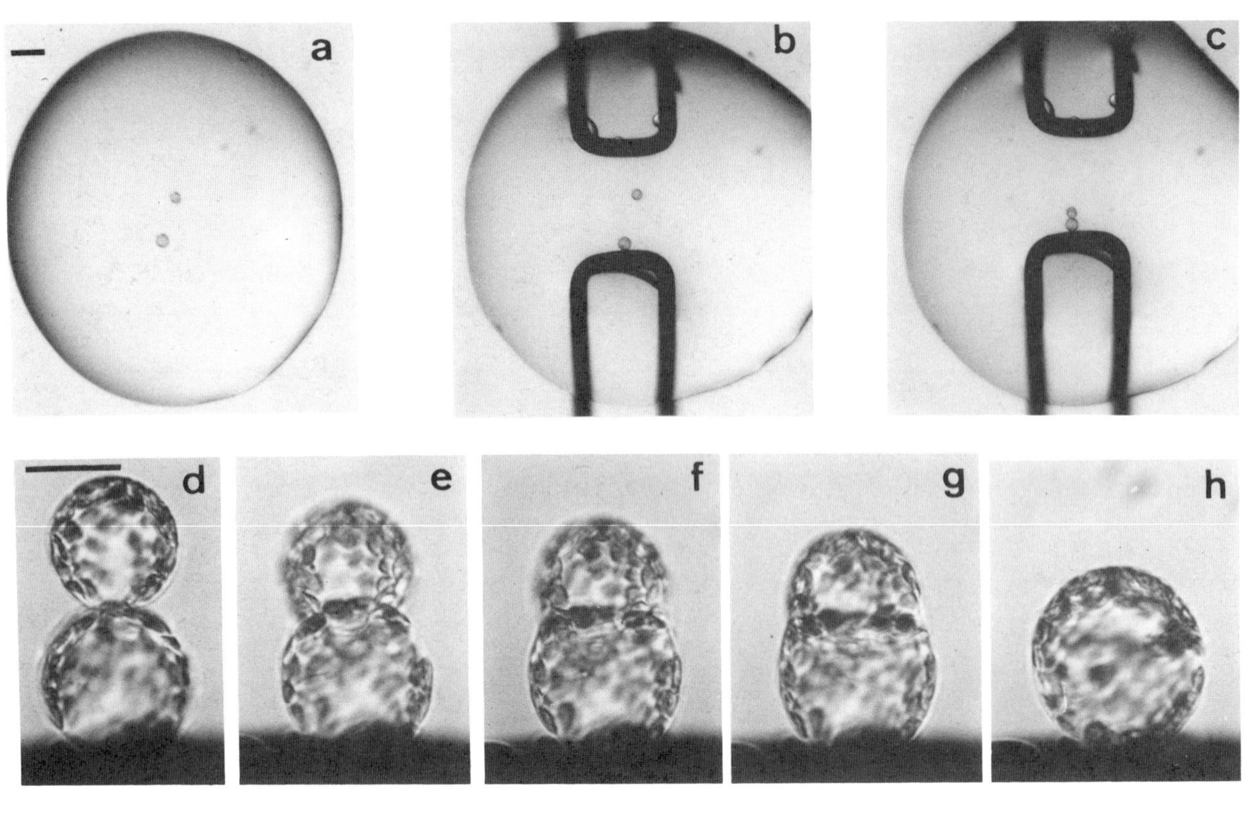

Figure 4. Dielectrophoretic alignment and fusion of two selected single protoplasts. (a) Two protoplasts in a microdroplet of 0.4 M mannitol, (b) attachment of one of the protoplasts to one of the electrodes, and (c) alignment of the protoplasts in a "pearl chain." Time interval between a and c is less than 1 min. Bar = 100 μm. (d–h) The protoplasts shown in a–c before (d) and 10 (e), 80 (f), 110 (g), and 180 sec (h) after application of the DC pulse. Bar = 25 μm. From Koop *et al.* (1983b), with permission.

Table 1. Survival and Multiple Divisions after 8 Days of Individual Culture of Nonfused Controls and Fusion Products from Single Selected Pair of Protoplasts of *Brassica napus*, *Nicotiana tabacum*, and *Physcomitrella patens*

	Brassica napus	Fusion	*Nicotiana tabacum*	Fusion	*Physcomitrella patens*	Fusion
	Control	Products	Control	Products	Control	Products
Survival (%)	50	58	86	77	52	37
Multiple divisions (%)	46	37	56	47	52	37

3. CELL RECONSTITUTION BY FUSION OF SINGLE SUBPROTOPLASTS

The possibility of mixing cytoplasmic genophores and thus achieving a more general system for biparental inheritance in higher plants is one of the main new advantages of protoplast fusion over sexual reproduction (Gleba and Sytnik, 1984). In this respect, the transfer of alien cytoplasm by protoplast fusion represents an important means of increasing the genetic diversity of the extranuclear genomes of plants (Pelletier *et al.*, 1985). In particular, plant cell reconstitution using subprotoplasts—cytoplasts and karyoplasts—as one or both of the parents in the one-to-one electrofusion process (see Section 2.3) allows *a priori* a definition of the subcellar compartments that will participate in the fusion event. This *cell engineering* represents a unique feature of the one-to-one electrofusion described in this chapter and is not achievable by mass fusion performed at the populational level even if combined with a microisolation of the fusion products *a posteriori*. The main reason is that all of the techniques available for subprotoplast isolation (Archer *et al.*, 1982; Lörz *et al.*, 1981; Bradley, 1978; Vatsya and Bhaskaran, 1981; Wallin *et al.*, 1978; Lesney *et al.*, 1986) yield preparations of subprotoplasts contaminated with protoplasts or other subprotoplasts. In addition, protoplasts contaminating cytoplast preparations are known to be more stable or efficient in the fusion process. For instance, while dealing with organelle transfer by performing mass fusions of protoplast–cytoplast populations, experimental evidence for the preferential participation of the protoplasts contaminating the cytoplast preparation in the fusion events, has been presented by Maliga *et al.* (1982). This would probably mean that most of the results on organelle transfer by protoplast–cytoplast mass fusions so far reported could be interpreted as normal protoplast–protoplast fusion followed by nuclear segregation rather than as true protoplast–cytoplast *ab initio* fusions.

Only the individual production, culture, and analysis of fusion products originating from a cytoplast and a protoplast, a karyoplast and a protoplast, or a karyoplast and a cytoplast (by using, for example, the one-to-one electrofusion approach herein presented) allow a thorough study of the transmission genetics underlying protoplast–subprotoplast fusion products.

Table 2. Morphogenesis in Calli and Chromosome Doubling in Plants Regenerated through Individual Culture of Fusion Products from Single Selected Pairs of Protoplasts of *Nicotiana tabacum*

Number of	Experiment 1	Experiment 2	Control
Calli transferred to morphogenesis medium	14	20	35
Calli regenerating shoots	14	19	28
Calli regenerating plants	14	18	28
Plants showing doubled chromosome numbers	8	14	—

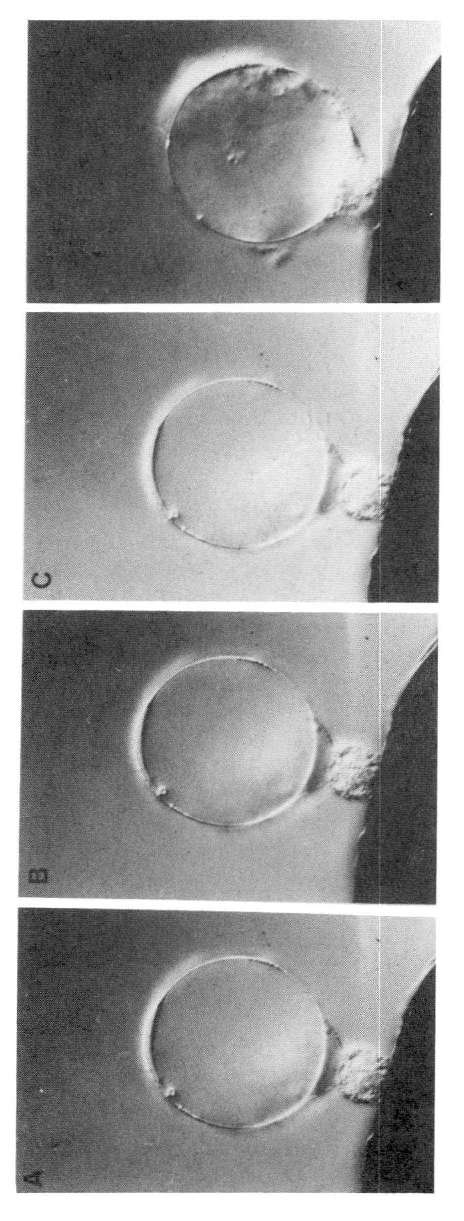

Figure 5. Cell reconstitution mediated by electrofusion of a karyoplast and a cytoplast. (From Spangenberg and Schweiger, 1986, with permission.)

3.1. Preparing and Fusing Cytoplasts and Karyoplasts

Different protocols for the isolation of enucleated subprotoplasts (cytoplasts) and subprotoplasts containing mainly the cell nucleus and very little cytoplasm (karyoplasts) have been reported (Bradley, 1978; Wallin *et al.*, 1978; Lörz *et al.*, 1981; Archer *et al.*, 1982; Vatsya and Bhaskaran, 1981; Lesney *et al.*, 1986). Protoplast fragmentation by plasmolysis of elongated cells, centrifugation of plant protoplasts causing the extrusion of the nucleus, collapse of the tonoplast in highly vacuolated protoplasts, or budding of subprotoplasts due to the increase in volume of protoplasts are different mechanisms for the production of higher plant subprotoplasts (Bradley, 1983).

For our purposes, the subprotoplast isolation by enucleation of a cytochalasin B- and DMSO-treated population of protoplasts ultracentrifuged through discontinuous density gradients (Wallin *et al.*, 1978) proved to be useful. The cytoplasts and karyoplasts could be easily recovered from the distinct bands obtained and after washing, could be individually placed into electrofusion microchambers for the one-to-one fusion.

Cell reconstitutions by fusing defined pairs of karyoplasts and cytoplasts of *B. napus* could be routinely achieved in 30% of the cases (Fig. 5). Similar efficiencies were obtained while performing

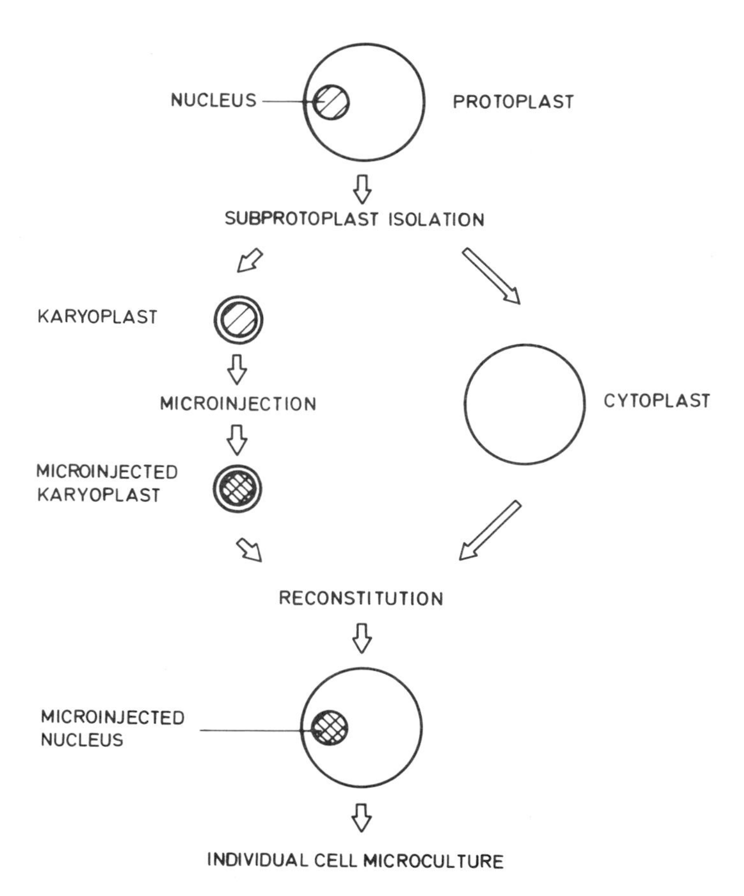

Figure 6. Electrofusion-mediated cell reconstitution from a cytoplast and a microinjected karyoplast. (From Spangenberg, 1986.)

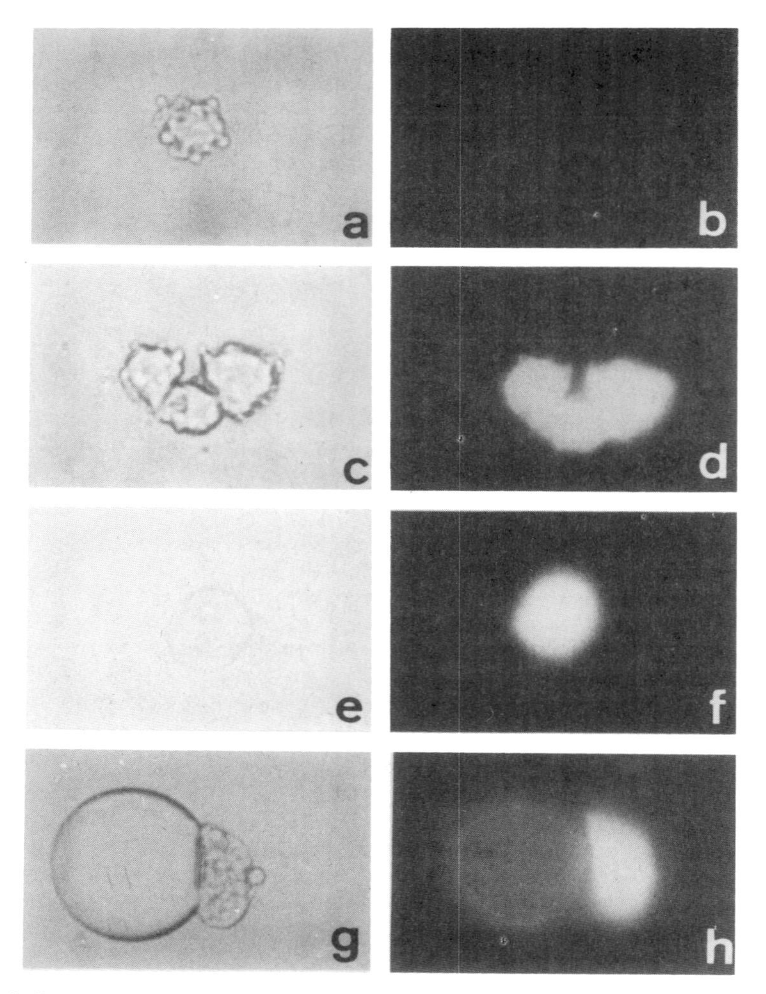

Figure 7. Indirect immunoflourescence analysis of protoplasts. Protoplasts derived either from one microinjected protoplast (c, d) or karyoplast (e, f) or from the fusion product of a microinjected karyoplast with a cytoplast (g, h) were tested. The cells were microinjected either with pLGV23neo (c, d) or pSV2neo (e, h) or with pU12 (a, b) for control. The left side (a, c, e, g) represents brightfield images, the right side (b, d, f, h) the corresponding fluorescence images. (From Spangenberg *et al.*, 1986b, with permission.)

electrofusion of a protoplast and a subprotoplast to achieve nuclear transplantation (protoplast–karyoplast fusion) or organelle transfer (protoplast–cytoplast fusion) (Spangenberg and Schweiger, 1986).

3.2. Reconstituting Cells from Cytoplasts and Microinjected Karyoplasts

The analysis of gene expression in different cellular compartments and the interaction between them represents an interesting field of research in plant cell biology. An elegant approach aimed at getting a better understanding of the interrelationships between nuclear and cytoplasmic compartments would be to manipulate one of them and analyze gene expression in the other.

The combination of some micromanipulation techniques, such as intranuclear microinjection of protoplasts and karyoplasts, one-to-one electrofusion-mediated cell reconstitution, and microculture of manipulated cells, allows this direct genetic modification of a cellular compartment under controlled conditions and therefore an unequivocal interpretation of the results obtained on gene expression.

Conditions for controlled subprotoplast isolation of *B. napus,* microinjection of karyoplasts, electrofusion-mediated cell reconstitution starting from a microinjected karyoplast and a cytoplast, as well as expression of the microinjected foreign DNA were established (Fig. 6). More than 20% of the selected pairs of cytoplasts and microinjected karyoplasts were successfully electrofused and microcultured. Between 70 and 90% of the individually selected karyoplasts and protoplasts survived the microinjection and transfer to microculture (Spangenberg *et al.*, 1986b).

After microinjection of two different clones containing the neomycin phosphotransferase II gene (pSV2neo and pLGV23neo) into either protoplasts, karyoplasts, or reconstructed cells with microinjected karyoplasts, the expression of the NPT II gene could be assessed by indirect immunofluorescence (Fig. 7) in 60 to 70% of the cases (Spangenberg *et al.*, 1986b). The integration of the full-length gene into high-molecular-weight plant genomic DNA could be shown by Southern blot analysis.

3.3. Culturing Reconstituted Cells

After electrofusion of selected pairs of a subprotoplast and a protoplast or a karyoplast and a cytoplast, 10 to 20% of the fusion products obtained formed microcalli. By transferring them through intermediate microculture steps, further growth allowed the establishment of some hybrid cell clones. No significant difference in behavior in individual culture between reconstituted cells and unfused control protoplasts of *B. napus* was detected (Spangenberg and Schweiger, 1986; Spangenberg, 1986).

REFERENCES

Archer, E. K., Landgren, C. R., and Bonnett, H. T., 1982, Cytoplast formation and enrichment from mesophyll tissues of *Nicotiana* spp., *Plant Sci. Lett.* **25:**175–185.

Bradley, P. M., 1978, Production of enucleated plant protoplasts of *Allium cepa, Plant Sci. Lett.* **13:**287–290.

Bradley, P. M., 1983, The production of higher plant subprotoplasts, *Plant Mol. Biol. Rep.* **1**(3)**:**117–123.

Evans, D. A., 1983, Protoplast fusion, in: *Handbook of Plant Cell Culture,* Volume 1 (D. A. Evans, W. R. Sharp, P. V. Ammirato, and Y. Yamada, eds.), Macmillan Co., New York, pp. 291–321.

Evans, D. A., Bravo, J. E., and Gleba, Y. Y., 1983, Somatic hybridization: Fusion methods, recovery of hybrids, and genetic analysis, *Int. Rev. Cytol. Suppl.* **16:**143–159.

Galun, E., and Aviv, D., 1983, Cytoplasmic hybridization: Genetic and breeding applications, in: *Handbook of Plant Cell Culture,* Volume 1 (D. A. Evans, W. R. Sharp, P. V. Ammirato, and Y. Yamada, eds.), Macmillan Co., New York, pp. 385–392.

Gleba, Y. Y., and Sytnik, K. M., 1984, *Protoplast Fusion: Genetic Engineering of Higher Plants,* Springer-Verlag, Berlin.

Glimelius, K., Eriksson, T., Grafe, R., and Müller, A. J., 1978, Somatic hybridization of nitrate reductase-deficient mutants of *Nicotiana tabacum* by protoplast fusion, *Physiol. Plant.* **44:**273–277.

Harms, C. T., 1985, Hybridization by somatic cell fusion, in: *Plant Protoplasts* (L. C. Fowke, and F. Constabel, eds.), CRC Press, Boca Raton, Fla.

Kao, K. N., 1977, Chromosomal behaviour in somatic hybrids of soybean—*Nicotiana glauca, Mol. Gen. Genet.* **150:**225–230.

Kohn, H., Schieder, R., and Schieder, O., 1985, Somatic hybrids in tobacco mediated by electrofusion, *Plant Sci.* **38:**121–128.

Koop, H. U., and Schweiger, H. G., 1985a, Regeneration of plants from individually cultivated protoplasts using an improved microculture system, *J. Plant Physiol.* **121:**245–257.

Koop, H. U., and Schweiger, H. G., 1985b, Regeneration of plants after electrofusion of selected pairs of protoplasts, *Eur. J. Cell Biol.* **39:**46–49.

Koop, H. U., Weber, G., and Schweiger, H. G., 1983a, Individual culture of selected single cells and protoplasts of higher plants in microdroplets of defined media, *Z. Pflanzenphysiol.* **112:**21–34.

Koop, H. U., Dirk, J., Wolff, D., and Schweiger, H. G., 1983b, Somatic hybridization of two selected single cells, *Cell Biol. Int. Rep.* **7:**1123–1128.

Lesney, M. S., Callow, P. W., and Sink, K. C., 1986, A technique for bulk production of cytoplasts and miniprotoplasts from suspension culture-derived protoplasts, *Plant Cell Rep.* **5:**115–118.

Lörz, H., Paszkowşki, J., Dierks-Ventling, C., and Potrykus, I., 1981, Isolation and characterization of cytoplasts and miniprotoplasts derived from protoplasts of cultured cells, *Physiol. Plant.* **53:**385–391.

Maliga, P., Lörz, H., Lázár, G., and Nagy, F., 1982, Cytoplast–protoplast fusion for interspecific chloroplast transfer in *Nicotiana, Mol. Gen. Genet.* **185:**211–215.

Mejía, A., Spangenberg, G., Koop, H. U., and Bopp, M., 1988, Microculture and electrofusion of defined protoplasts of the moss *Funaria hygrometrica, Botanica Acta* **101:**166–173.

Melchers, G., and Labib, G., 1974, Somatic hybridization of plants by fusion of protoplasts, I. Selection of light resistant hybrids of "haploid" light sensitive varieties of tobacco, *Mol. Gen. Genet.* **135:**227–294.

Pelletier, G., Vedel, F., and Belliard, G., 1985, Cybrids in genetics and breeding, *Hereditas Suppl.* **3:**49–56.

Schieder, O., 1977, Hybridization experiments with protoplasts from chlorophyll-deficient mutants of some solanaceous species, *Planta* **137:**253–257.

Schweiger, H. G., Dirk, J., Koop, H. U., Kranz, E., Neuhaus, G., Spangenberg, G., and Wolff, D., 1987, Individual selection, culture and manipulation of higher plant cells, *Theor. Appl. Genet.* **73:**769–783.

Senda, M., Takeda, J., Abe, S., and Nakamura, T., 1979, Induction of cell fusion of plant protoplast by electrical stimulation, *Plant Cell Physiol.* **20:**1441–1443.

Spangenberg, G., 1986, Manipulation individueller Zellen der Nutzpflanze *Brassica napus L.* mit Hilfe von Elektrofusion, Zellrekonstruktion und Mikroinjektion, PhD thesis, University of Heidelberg.

Spangenberg, G., and Schweiger, H. G., 1986, Controlled electrofusion of different types of protoplasts and subprotoplasts including cell reconstitution in *Brassica napus L., Eur. J. Cell Biol.* **41:**51–56.

Spangenberg, G., Koop, H. U., and Schweiger, H. G., 1985, Different types of protoplasts from *Brassica napus L.:* Analysis of conditioning effects at the single-cell level, *Eur. J. Cell Biol.* **39:**41–45.

Spangenberg, G., Koop, H. U., Lichter, R., and Schweiger, H. G., 1986a, Microculture of single protoplasts of *Brassica napus, Physiol. Plant.* **66:**1–8.

Spangenberg, G., Neuhaus, G., and Schweiger, H. G., 1986b, Expression of foreign genes in a higher plant cell after electrofusion-mediated cell reconstitution of a microinjected karyoplast and a cytoplast, *Eur. J. Cell Biol.* **42:**236–238.

Vatsya, B., and Bhaskaran, S., 1981, Production of sub-protoplasts in *Brassica oleracea var. capitate*—A function of osmolarity of the media, *Plant Sci. Lett.* **23:**277–282.

Wallin, A., Glimelius, A., and Eriksson, T., 1978, Enucleation of plant protoplasts by cytochalasin B, *Z. Pflanzenphysiol.* **87:**333–340.

Wallin, A., Glimelius, A., and Eriksson, T., 1979, Formation of hybrid cells by transfer of nuclei via fusion of miniprotoplasts from cell lines of nitrate reductase deficient tobacco, *Z. Pflanzenphysiol.* **91:**89–94.

Zimmermann, U., 1982, Electric field-mediated fusion and related electrical phenomena, *Biochim. Biophys. Acta* **694:**227–277.

Chapter 24

Critical Evaluation of Electromediated Gene Transfer and Transient Expression in Plant Cells

M. Pröls, J. Schell, and H.-H. Steinbiß

1. INTRODUCTION

Applied genetics can benefit from direct gene transfer in two important research areas: (1) transformation of protoplasts followed by regeneration procedures resulting in fertile stably transformed plants, and (2) gene transfer to plant protoplasts leading to a transient expression of the marker gene(s). In both cases, very simple plasmids have been used to develop improved techniques. In our laboratory, we preferred a construction made by Jeff Velten which was designated pCAP 212 by Velten and Schell (1985). This vector carries a dual plant promoter fragment from the Ti plasmid of *Agrobacterium tumefaciens*. It is fused, in both orientations: once to the neomycin phosphotransferase (NPT II) gene of the transposon Tn*5*, and secondly to the chloramphenicol acetyltransferase (CAT) gene. Expression of NPT II and CAT may be readily assayed in plant cells and tissue. Therefore, this vector is very useful for optimizing transformation procedures as well as for measuring transient gene expression. Additionally, we made all our experiments with protoplasts derived from a carrot suspension culture (exception: Fig. 6). The culture conditions (Hain *et al.*, 1985) as well as the parameters of the enzyme assay were kept constant. Only in this way did a comparison of different techniques seem possible.

2. PEG-MEDIATED GENE TRANSFER

PEG is the active substance in most of the chemical methods for gene uptake into plant protoplasts. Such a treatment is harmful to living cells and results in a high number of damaged cells and a reduced regeneration capacity of the protoplasts. Kao and Saleem (1986) improved the viability of PEG–high pH–Ca^{2+}-treated mesophyll protoplasts. However, protoplast formation itself influences cell metabolism by inducing stress proteins or ethylene biosynthesis (Fleck *et al.*,

M. Pröls, J. Schell, and H.-H. Steinbiß • Max Planck Institute for Plant Breeding, D-5000 Köln 30, Federal Republic of Germany.

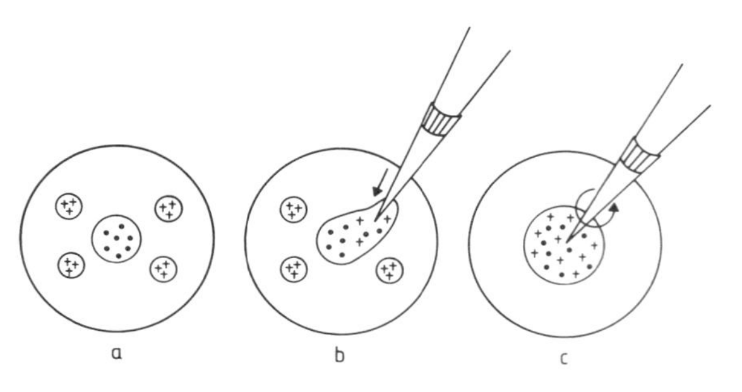

Figure 1. Chemical method for gene uptake. The protoplast fusion method of Hein *et al.* (1983) was adapted for DNA uptake. Protoplasts of *Daucus carota* were isolated from a suspension culture according to Hain *et al.* (1985). About 10^6 protoplasts were washed in MS medium (Murashige and Skoog, 1962) supplemented with 0.4 M mannitol and resuspended in a final volume of 180 μl of the same medium. Ten to thirty micrograms of plasmid DNA (pCAP 212; Velten and Schell, 1985) dissolved in 20 μl of sterile water were added. (a) This mixture was dropped in the center of a petri dish surrounded by four drops of PEG solution [50 μl, 25% polyethyleneglycol 1500, 100 mM $Ca(NO_3)_2$, 0.45 M mannitol, pH9]. (b) The protoplasts are now blended with the PEG droplets successively, and (c) the resulting protoplast–DNA–PEG solution is mixed carefully. After 20 min, 5 ml of wash solution [0.275 M $Ca(NO_3)_2$, 0.45 M mannitol, pH 6] is added dropwise. After 10 min, the protoplasts are transferred into centrifuge tubes, pelleted, and resuspended in culture medium (Hain *et al.*, 1985). Two days later, regenerating cells are harvested, and a CAT assay is carried out according to Gorman *et al.* (1982). Briefly, a protocol of a transformation experiment made in Cologne by U. Wirtz (personal communication): about 10^6 tobacco protoplasts; 60–80 μg recombinant DNA; from 250,000 microcolonies entering selection, about 200–400 colonies are resistant to kanamycin; efficiency: 0.08–0.16%.

1982; Cress, 1982; Hartung *et al.*, 1983; Mussel *et al.*, 1986). Additionally, heat-shock pretreatments (Potrykus *et al.*, 1985b,c) result in a drastic alteration of the protein synthesis pattern and appearance of a new set of proteins termed heat-shock proteins (e.g., Kimpel and Key, 1985; Sinibaldi and Turpen, 1985; Belanger *et al.*, 1986; Vierling *et al.*, 1986; Neumann *et al.*, 1987). These phenomena must be taken into account. Thus, interpretation of transient gene expression, as well as stable integration of foreign genes in a plant genome, becomes more and more complicated. Taken together, protoplast formation and PEG-mediated gene transfer are not harmless to plant cells. To date, there have been no reports that PEG is unsuitable for DNA-uptake experiments. In mammalian cells there are lines in which PEG or other chemicals failed to transfer DNA into recipient cells. In such cases, an urgent need for methods without restriction to cell types is obvious (Chu *et al.*, 1987) and "electroporation" might be the method of choice.

PEG-mediated gene transfer can be applied to essentially any plant species, in contrast to *Agrobacterium*-mediated gene transfer with its host range limitation (De Cleene, 1985). The oldest and most commonly used method is the so-called "Krens method" (Krens *et al.* 1982, 1985; Paszkowski *et al.*, 1984; Lörz *et al.*, 1985; Potrykus *et al.*, 1985a; Uchimiya *et al.*, 1986a,b). We preferred a procedure described by Hein *et al.* (1983). A typical protocol is given in Fig. 1.

Gene transfer experiments lead to a transient expression of genes shortly after their introduction. The majority of the DNA is maintained as extrachromosomal copies within the plant cells (Werr and Lörz, 1986). The first visible CAT activity appears in carrot protoplasts between 7 and 12 hr after the plasmid DNA is transferred. We found the maximum CAT expression at day 3. The signal disappeared completely after 2 weeks (Fig. 2). The sensitivity of the CAT assay can be determined by mixing untreated protoplasts with DNA-treated ones. The expression signal of 10^4 protoplasts cannot be detected in a background of 2×10^6 untreated ones. Therefore, initiation of the transient CAT expression and the end of expression cannot be determined precisely due to the limited sensitivity of the CAT assay.

Using the transient CAT-expression signal, it is possible to estimate the specific influence of different technical parameters. PEG alone or in combination with high pH is not sufficient, but a mixture of PEG and Ca^{2+} gives rise to a strong expression signal. A significant difference between linear and circular plasmid DNA was not detected.

Addition of PEG to the droplet of protoplasts leads to tight agglutination. Globular products are formed consisting of two or more protoplasts.

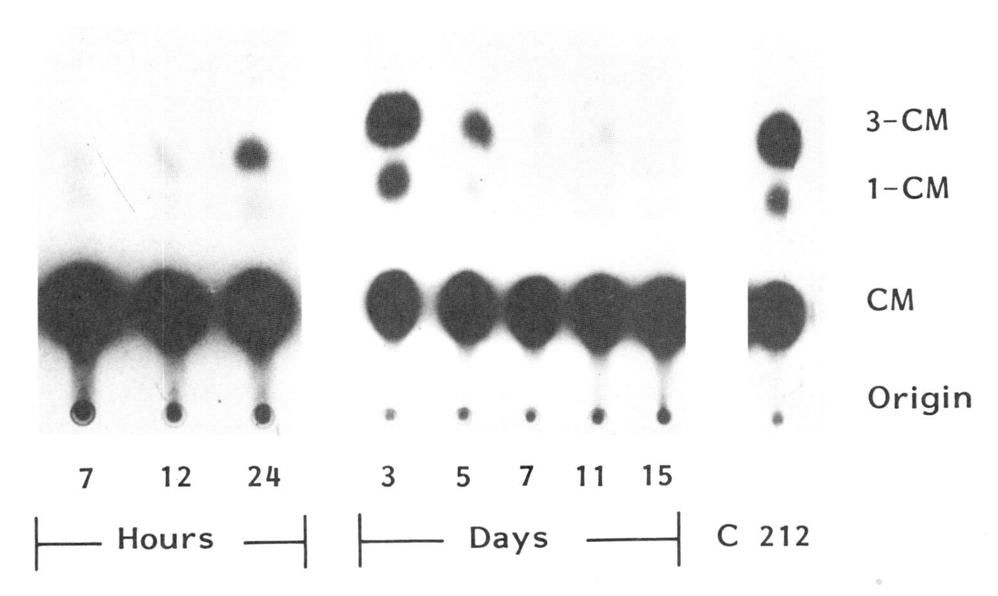

Figure 2. Transient expression of the chloramphenicol acetyltransferase (CAT) gene in cultured carrot cells. Protoplasts of a carrot suspension culture were isolated and cultured according to Hain *et al.* (1985). A recombinant plasmid in which the bacterial CAT gene is under the control of a dual-promoter DNA fragment, originally isolated from the T-DNA of *Agrobacterium tumefaciens* (pCAP 212; Velten and Schell, 1985), was introduced into the protoplasts using the PEG-mediated gene transfer system, described in Fig. 1. At different time intervals, cells were harvested and mechanically broken. Lysates were clarified by centrifugation and the supernatant was assayed for CAT activity according to the procedure of Gorman *et al.* (1982). The first detectable signal with the CAT assay was observed 7 hr after transfection. The CAT activity increased up to day 3 and declined afterwards. The CAT signal was detectable for at least 2 weeks. The plasmid construction was also transferred to tobacco protoplasts and a stably transformed callus has been obtained (C212) and used as a positive control. Abbreviations: CM, chloramphenicol; 1-CM, 1-acetate chloramphenicol; 3-CM, 3-acetate chloramphenicol.

The plasmalemma surface preferentially exhibits negative charges. Positively charged cations like Ca^{2+} and Mg^{2+} are assumed to link PEG and DNA molecules with protoplasts (Kao and Michayluk, 1974). Thus, the integrity of the plasmalemma and electrostatic forces seem to be necessary for PEG-induced agglutination (Constabel and Kao, 1974) and binding of DNA to the protoplast's surface. The final uptake mechanism is not yet clear and several theories have been advanced to explain the function of PEG (e.g., Krähling *et al.*, 1978; Boss and Mott, 1980). After adding the washing solution, the aggregates disappear and some fusion products are visible. Wong and Capecchi (1985) have shown that DNA inside the cytoplasm is 1000 times less efficient for getting stably transformed mouse cells than DNA microinjected directly into the nucleus. Once inside the nucleus, integration of foreign DNA into the host genome is independent of the cell cycle. The nuclear membrane acts as a natural barrier, but we found that transient gene expression can appear in a nondividing cell system (*Valerianella locusta;* Steinbiß, 1978) indicating that PEG can somehow facilitate the entry of DNA into the nucleus.

The driving force to develop electrical systems as well as other methods of direct gene transfer is the lack of effective gene vectors for cereals like wheat, rice, maize, and barley. It has been shown by Graves and Goldman (1986) that the gene vector *Agrobacterium tumefaciens* is able to interact with *Zea mays*. These findings have to be confirmed and investigated in more detail.

Additionally, other simple gene transfer methods, like injection of recombinant DNA into

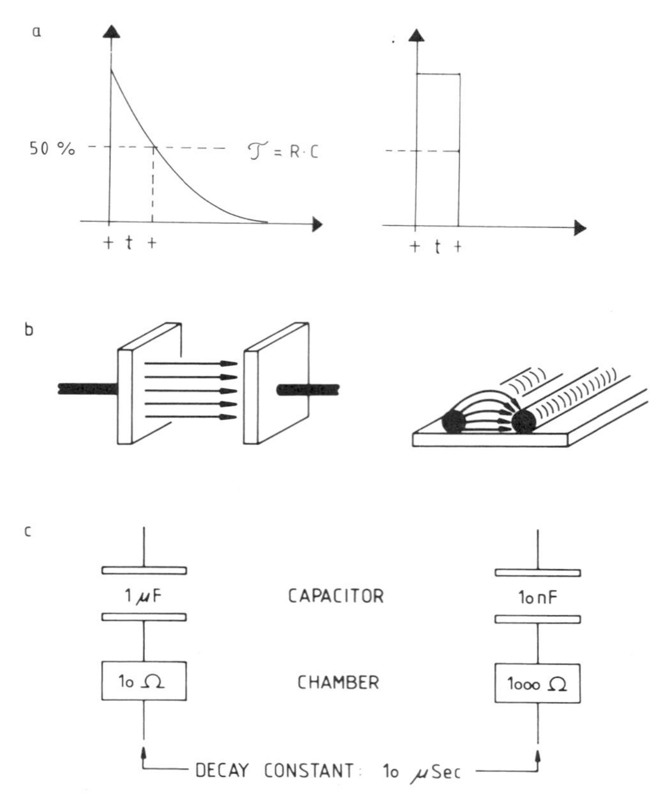

Figure 3. Divergences in "electroporation" setups. (a) In the case of a capacitor discharge, the voltage decays exponentially. The field-decay time constant is $\tau = R\ C$, where C is the capacitance of the discharge capacitor and R is the resistance of the discharge unit. Some commercial voltage generators provide a rectangular pulse. (b) The electric pulses can be delivered to a chamber containing two plane-parallel platinum or stainless steel electrodes providing a homogeneous electric field. The "GCA/Precision electroporation vessel" (Langridge *et al.*, 1985) contains two platinum wires which give rise to a slightly inhomogeneous field. (c) The resistance of the sample solution inside the pulse chamber determines the total circuit resistance. We preferred a salt-rich pulse medium (about 10 ohms) similar to that of Fromm *et al.* (1985), whereas Shillito *et al.* (1985) adjust the resistance to a value of about 1000 ohms. The decay constant of the pulses is on the order of 10 μsec, but the energy delivered to the chambers is different in both cases.

young floral tillers of rye (De la Pena *et al.*, 1987), might be of more practical importance for improving crops than a direct gene transfer system if regeneration of cereal protoplasts remain unsuccessful.

It is very difficult to compare experimental results from different publications because of their experimental diversity. Figure 3 summarizes some obvious differences. Additionally, most investigators prefer a group-specific plasmid construction. Moreover, the vitality of the used plant system shows large differences. Two extreme examples: The tobacco cell line BY2 of Okada *et al.* (1986) needs about 12 hr to undergo the first cell division after protoplast formation and the mesophyll protoplasts sometimes require up to 1 week to finish their first mitosis.

A very simple but important divergence is the scheme to estimate the transformation efficiency. There is no international agreement, but the best compromise seems to be a calculation on the basis of growing colonies which could be recovered without selection. Thus, a compensation of divergences regarding the vitality of selected cell lines might be possible.

3. ELECTROPULSES IN THE MICROSECOND RANGE

The first setup to achieve electrical breakdown of animal cell membranes was published by Neumann and Rosenheck (1972). In 1978, it was demonstrated that the electroshock technique could also be applied to plant protoplasts (Steinbiß, 1978). A typical homemade setup is given in Fig. 4. Several commercial apparatus are available today. The setup has to be adapted to the protoplasts of the selected cell line. This can be done properly with the vital dye trypan blue (Fig. 5).

By changing field strength, number, and length of the pulses, we found the following compromise for carrot protoplasts: four successive electrical pulses, a field strength of 1400 volts/cm, a pulse length of 12 μsec, and a 5 sec delay between pulses. Thus, about 45% of treated protoplasts are permeabilized to the vital stain trypan blue. They can reestablish the semipermeable properties of their plasmalemma within 40 min by means of excluding this normally nonpermeable pigment; 30% have lost this capability completely. Such an electrical stress has a strong negative influence over the regeneration ability of the protoplasts (Fig. 6). Therefore, we combined the parameters which resulted in a half-maximal uptake of low-molecular-weight substances like [^{14}C]mannitol (Fig. 6), in the highest percentage of reversibly permeabilized protoplasts and in a division ratio not beyond 50% of untreated controls.

Several experiments have been performed under optimal conditions and we never succeeded in any measurable transient CAT-gene expression. In comparison, with a Ca^{2+}–PEG-mediated gene transfer, we got a strong expression signal. We cannot conclude from these negative results that no plasmids were transfected into the protoplasts, but it may be that the copy number was not sufficient to produce a detectable CAT signal. Langridge *et al.* (1985) obtained transformed carrot somatic embryos using a commercial transfection setup (BCA, electroporation vessel Z1200, and a rectangular pulse). This result documents that short electric pulses are suitable for transfection of functional DNA molecules as well as viral RNA (e.g., Nishiguchi *et al.*, 1986; Riggs and Bates, 1986; Hibi *et al.*, 1986) into plant protoplasts.

Since we did not obtain any transient expression signal of the CAT gene by applying short electric pulses alone, the carrot protoplasts were additionally subjected to pretreatments such as heat shock and/or PEG as suggested by Shillito *et al.* (1985). Electric pulses in combination with PEG resulted in strong transient CAT expression. However, more detailed studies revealed that it was the PEG pretreatment alone which was the crucial step to obtain measurable expression signals.

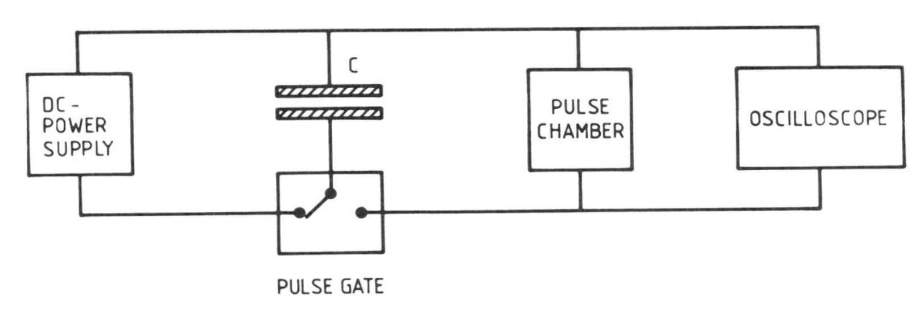

Figure 4. Schematic representation of the electrotransfection setup. A high-voltage power supply provides the electrical energy, which at first is stored in a high-voltage capacitor (C). The capacitor is discharged via a pulse gate and thereby the voltage is applied to the flat platinum electrodes of the pulse chamber. The distance between the electrodes is 0.5 cm. The voltage across the conducting protoplast suspension between the electrodes decays exponentially. The field-decay time constant is $\tau = R\,C$, where C is the capacitance of the discharge capacitor and R is the resistance of the discharge unit. Time courses of the electric field can be controlled by an oscilloscope.

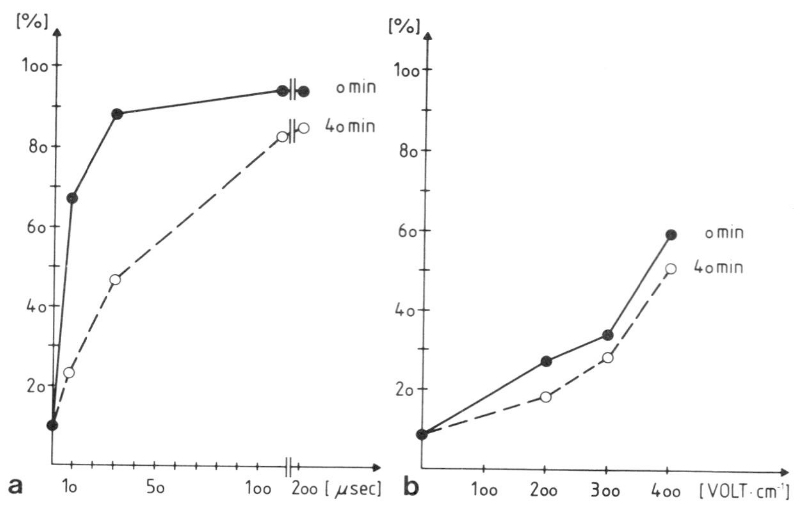

Figure 5. Viability of "electroporated" carrot protoplasts. (a) Short-time pulses. Protoplasts of *Daucus carota* were exposed to four exponentially decaying electric impulses with an initial intensity of 1400 V/cm. One part of the protoplasts was exposed immediately to the nonpermeating pigment trypan blue (final concentration 0.05%). The other part was allowed to recover for 40 min before it was exposed to the dye in the same fashion. Two hours later, the blue-stained protoplasts could be counted easily with a microscope. The percentage of stained cells is plotted as a function of increasing pulse length. About 90% of untreated protoplasts retain their semipermeable properties and therefore exclude trypan blue, whereas electrically treated cells become pigmented as the dye penetrates. The dashed lines represent the percentage of "dead" protoplasts. Consequently, the difference of the curves represents the vitality of the "electroporated" protoplasts. (b) Long-time pulses. The percentage of stained cells is plotted as a function of the electric field strength. The pulse length was 20 msec. In comparison to short electric pulses, the percentage of protoplasts which could restore their semipermeable properties is much lower. On the other hand, a very large percentage (about 40%) of protoplasts retain an intact plasmalemma and cannot be "electroporated" at up to 400 V/cm.

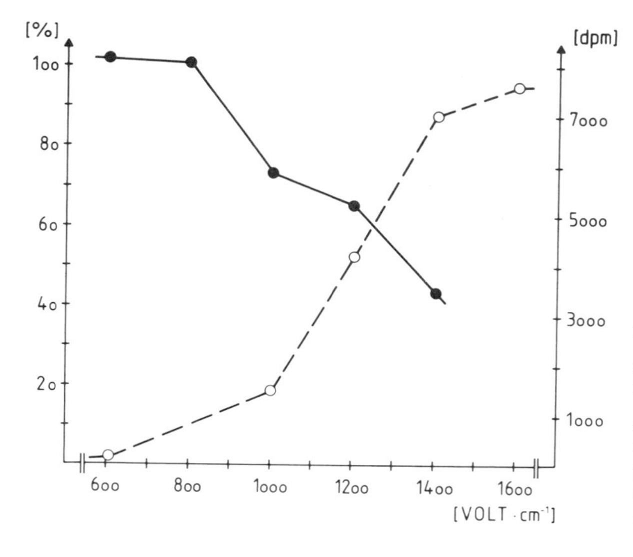

Figure 6. (●) Tobacco protoplasts were exposed to a single electric pulse. The field-decay time constant was 36 μsec. After 1 week, the number of cells that finished their first mitosis were scored and served as control (100%). The reduction of the mitosis rate in comparison to the control is plotted as a function of the external field strength. (○) Protoplasts of *Valerianella locusta* were treated with a single electric pulse. The pulse length was 26 μsec. Uptake of [^{14}C]mannitol is plotted as a function of the external field strength. Modified after Steinbiß (1978).

Taken together, PEG-mediated gene transfer gave rise to a strong transient gene expression signal in carrot protoplasts whereas electropulses in the microsecond range failed completely. With respect to efficiencies of transformation experiments, the combination of heat shock, Mg^{2+}–PEG, and electrotransfection (Shillito *et al.*, 1985) was as efficient as PEG-mediated gene transfer (Meyer *et al.*, 1985; Uchimiya *et al.*, 1986b).

Electrotransfection bears two significant advantages. In contrast to chemical techniques, it can be applied to complete cells or tissues. As far as we are aware, the culture of cereal and some important grain legume protoplasts beyond a callus has been unsuccessful. Why not use meristems or other pieces of tissues which guarantee plant regeneration? The main problem seems to be the transport of nucleic acids across the cell wall of recipient cells. The limiting diameter of pores, through which molecules can pass freely, is about 50 Å (Carpita *et al.*, 1979). This is large enough for dyes and sugar molecules, but not for proteins or nucleic acids except where there are specific secretion channels. We added the vital stain trypan blue (0.05%) to the carrot suspension culture. A low percentage of dead cells and calli were stained rapidly. An electrical treatment appropriate for carrot protoplasts yielded a high percentage of blue calli. Surprisingly, all cells of the stained calli were blue, not simply the outer ones.

Additionally, we have to consider the unique possibility that electric fields (e.g., alternating electric fields) might disrupt the integrity of coat membranes of specific cell organelles selectively and thus mediate gene transfer to mitochondria or plastids.

4. ELECTROPULSES IN THE MILLISECOND RANGE

Transient expression of the CAT gene in carrot protoplasts was achieved with a 20-msec pulse (Fromm *et al.*, 1985). Any pretreatment with PEG destroyed the majority of the cells. The CAT-expression signal varied highly from experiment to experiment. We obtained also stably transformed calli; on one occasion, 16 resistant calli were obtained from 10^4 calli entering the kanamycin selection procedure (Fromm *et al.*, 1985, 1986).

Just as did Chu *et al.* (1987), we observed pulse-induced gas formation in the chamber. Furthermore, the viability of the treated cells is strongly decreased in comparison to that of protoplasts exposed to electric pulses in the microsecond range (Fig. 5). Not more than 10% of the protoplasts excluded trypan blue again after 40 min of recovery. The results obtained with short pulses (1400 V, 16 μsec) and long pulses (400 V, 20 msec) suggest that two different transfection processes may occur.

REFERENCES

Belanger, F. C., Brodl, M. R., and Ho, T. D., 1986, Heat shock causes destabilization of specific mRNAs and destruction of endoplasmic reticulum in barley aleurone cells, *Proc. Natl. Acad. Sci. USA* **83:**1354–1358.

Boss, W. F., and Mott, R. L., 1980, Effects of divalent cations and PEG on the membrane fluidity of protoplasts, *Plant Physiol.* **66:**835–837.

Carpita, N., Sabularse, D., Montezinos, D., and Delmer, D. P., 1979, Determination of the pore size of cell walls of living plant cells, *Science* **205:**1144–1147.

Chu, G., Hayakawa, H., and Berg, P., 1987, Electroporation for the efficient transfection of mammalian cells with DNA, *Nucleic Acids Res.* **15:**1311–1326.

Constabel, F., and Kao, K. N., 1974, Agglutination and fusion of plant protoplasts by polyethylene glycol, *Can. J. Bot.* **52:**1603–1606.

Cress, D. E., 1982, Osmotic stress inhibits thymidine incorporation into soybean protoplast DNA, *Plant Cell Rep.* **1:**186–188.

De Cleene, M., 1985, The susceptibility of monocotyledons to *Agrobacterium tumefaciens*, *Phytopathol. Z.* **113:**81–89.

De la Pena, A., Lörz, H., and Schell, J., 1987, Transgenic rye plants obtained by injecting DNA into young floral tillers, *Nature* **325**:274–276.

Fleck, J., Durr, A., Fritsch, C., Vernet, T., and Hirth, L., 1982, Osmotic-shock "stress-proteins" in protoplasts of *Nicotiana sylvestris, Plant Sci. Lett.* **26**:159–165.

Fromm, M. E., Taylor, L. P., and Walbot, V., 1985, Expression of genes transferred into monocot and dicot plant cells by electroporation, *Proc. Natl. Acad. Sci. USA* **82**:5824–5828.

Fromm, M. E., Taylor, L. P., and Walbot, V., 1986, Stable transformation of maize after electroporation, *Nature* **319**:791–793.

Gorman, C. M., Moffat, L. F., and Howard, B. H., 1982, Recombinant genomes which express chloramphenicol-acetyltransferase in mammalian cells, *Mol. Cell. Biol.* **2**:1044–1051.

Graves, A. C. F., and Goldman, S. L., 1986, The transformation of *Zea mays* seedlings with *Agrobacterium tumefaciens, Plant Mol. Biol.* **7**:43–50.

Hain, R., Stabel, P., Czernilofsky, A. P., Steinbiß, H. H., Herrera-Estrella, L., and Schell, J., 1985, Uptake, integration, expression and genetic transmission of a selectable chimaeric gene by plant protoplasts, *Mol. Gen. Genet.* **199**:161–168.

Hartung, W., Kaiser, W. M., and Burschka, C., 1983, Release of abscisic acid from leaf strips under osmotic stress, *Z. Pflanzenphysiol.* **112**:131–138.

Hein, T., Prezewozny, T., and Schieder, O., 1983, Culture and selection of somatic hybrids using an auxotrophic cell line, *Theor. Appl. Genet.* **64**:119–122.

Hibi, T., Kano, H., Sugiura, M., Kazami, T., and Kimura, S., 1986, High efficiency electro-transfection of tobacco mesophyll protoplasts with tobacco mosaic virus RNA, *J. Gen. Virol.* **67**:2037–2042.

Kao, K. N., and Michayluk, M. R., 1974, A method for high frequency inter-generic plant protoplast fusion, *Planta* **115**:355–367.

Kao, K., and Saleem, M., 1986, Improved fusion of mesophyll and cotyledon protoplasts with PEG and high pH–Ca solution, *J. Plant Physiol.* **122**:217–225.

Kimpel, J.A., and Key, J. L., 1985, Heat shock in plants, *Trends Biochem. Sci.* **10**:353–357.

Krähling, H., Schinkewitz, U., Barker, A., and Hülser, D. F., 1978, Electron microscopial and electrophysiological investigations on PEG induced cell fusions, *Eur. J. Cell Biol.* **17**:51–61.

Krens, F. A., Molendijk, L., Wullems, G. J., and Schilperoort, R. A., 1982, In vitro transformation of plant protoplasts with Ti-plasmid DNA, *Nature* **296**:72–74.

Krens, F. A., Mans, R. M. W., van Blogteren, T. M. B., Hoge, J. H. C., Wullems, G. J., and Schilperoort, R. A., 1985, Structure and expression of DNA transferred to tobacco via transformation of protoplasts with Ti-plasmid DNA: Co-transfer of T-DNA and non T-DNA sequences, *Plant Mol. Biol.* **5**:223–234.

Langridge, W. H. R., Li, B. J., and Szalay, A. A., 1985, Electric field mediated stable transformation of carrot protoplasts with naked DNA, *Plant Cell Rep.* **4**:355–359.

Lörz, H., Baker, B., and Schell, J., 1985, Gene transfer to cereal cells mediated by protoplast transformation, *Mol. Gen. Genet.* **199**:178–182.

Meyer, P., Walgenbach, E., Bussmann, K., Hombrecher, G., and Saedler, H., 1985, Synchronized tobacco protoplasts are efficiently transformed by DNA, *Mol. Gen. Genet.* **201**:513–518.

Murashige, T., and Skoog, F., 1962, A revised medium for rapid growth and bioassays with tobacco tissue cultures, *Physiol. Plant.* **15**:473–494.

Mussel, H., Earle, E., Campbell, L., and Batts, L. A., 1986, Ethylene synthesis during protoplast formation from leaves of *Avena sativa, Plant Sci. Lett.* **47**:207–214.

Neumann, D., zur Nieden, U., Manteuffel, R., Walter, G., Scharf, K.-D., and Nover, L., 1987, Intracellular localization of heat-shock proteins in tomato cell cultures, *Eur. J. Cell Biol.* **43**:71–81.

Neumann, E., and Rosenheck, K., 1972, Permeability changes induced by electric impulses in vesicular membranes, *J. Membr. Biol.* **10**:279–290.

Nishiguchi, M., Langridge, W. H. R., Szalay, A. A., and Zaitling, M., 1986, Electroporation-mediated infection of tobacco leaf protoplasts with tobacco mosaic virus RNA and cucumber mosaic virus RNA, *Plant Cell Rep.* **5**:57–60.

Okada, K., Takebe, I., and Nagata, T., 1986, Expression and integration of genes introduced into highly synchronized plant protoplasts, *Mol.Gen. Genet.* **205**:398–403.

Paszkowski, J., Shillito, R. D., Saul, M., Mandak, V., Hohn, T., Hohn, B., and Potrykus, I., 1984, Direct gene transfer to plants, *EMBO J.* **3**:2717–2722.

Potrykus, I., Paszkowski, J., Saul, M. W., Petruska, J., and Shillito, R. D., 1985a, Molecular and general genetics of a hybrid foreign gene introduced into tobacco by direct gene transfer, *Mol. Gen. Genet.* **199**: 169–177.

Potrykus, I., Saul, M. W., Petruska, J., Paszkowski, J., and Shillito, R., 1985b, Direct gene transfer to cells of a graminaceous monocot, *Mol. Gen. Genet.* **199:**183–188.

Potrykus I., Shillito, R. D., Saul, M. W., and Paszkowski, J., 1985c, Direct gene transfer state of the art and future potential, *Plant Mol. Biol. Rep.* **3:**117–128.

Riggs, C. D., and Bates, G. W., 1986, Stable transformation of tobacco by electroporation: Evidence for plasmid concatenation, *Proc. Natl. Acad. Sci. USA* **83:**5602–5606.

Shillito, R. D., Saul, M. W., Paszkowski, J., Müller, M., and Potrykus, I., 1985, High efficiency direct gene transfer to plants, *Biotechnology* **3:**1099–1103.

Sinibaldi, R. M., and Turpen, T., 1985, A heat shock protein is encoded within mitochondria of higher plants, *J. Biol. Chem.* **260:**15382–15385.

Steinbiß, H.-H., 1978, Dielektrischer Durchbruch des Plasmalemmas von *Valerianella locusta* Protoplasten, *Z. Pflanzenphysiol.* **88:**95–102.

Uchimiya, H., Hirochika, H., Hashimoto, H., Hara, A., Masuda, T., Kasumimoto, T., Harada, H., Ikeda, J.-E., and Yoshioka, M., 1986a, Co-expression and inheritance of foreign genes in transformants obtained by direct DNA transformation of tobacco protoplasts, *Mol. Gen. Genet.* **205:**1–8.

Uchimiya, H., Fushimi, T., Hashimoto, H., Harada, H., Syono, K., and Sugawara, Y., 1986b, Expression of a foreign gene in callus derived from DNA-treated protoplasts, *Mol. Gen. Genet.* **204:**204–207.

Velten, J., and Schell, J., 1985, Selection-expression plasmid vectors for use in genetic transformation of higher plants, *Nucleic Acids Res.* **13:**6981–6998.

Vierling, E., Mishkind, M. L., Schmidt, G. W., and Key, J. L., 1986, Specific heat-shock proteins are transported into chloroplasts, *Proc. Natl. Acad. Sci. USA* **83:**361–365.

Werr, W., and Lörz, H., 1986, Transient gene expression in a Gramineae cell line, *Mol. Gen. Genet.* **202:**471–475.

Wong, E. A., and Capecchi, M. R., 1985, Effect of cell cycle position on transformation by microinjection, *Somatic Cell Mol. Genet.* **11:**43–51.

Chapter 25

Transformation Studies in Maize and Other Cereals

Barbara Junker and Horst Lörz

1. INTRODUCTION

Electroporation and electrofusion have proven to be powerful tools for gene transfer studies in model plants (Potrykus *et al.*, 1985). In order to apply these techniques to cereals, which are by far the most important crops for human nutrition, knowledge of tissue culture is needed. This is especially true for regeneration from protoplasts to plants.

Using maize (*Zea mays* L.) as an example, the present level of cell and tissue culture will be discussed. The situation in maize is representative of most of the major cereals. We will discuss the techniques for transformation at the level of protoplasts, including electroporation and electrofusion. We include alternative strategies for the transformation of cereals.

2. CELL AND TISSUE CULTURE OF MAIZE

Maize tissue culture began about four decades ago with the observation of callus formation of plated endosperm (La Rue, 1949). Reports on callus induction on explants such as stem segments (Mascarenhas *et al.*, 1965), immature inflorescences (Linsmeier-Bedner and Bedner, 1972), and mature embryos (Green *et al.*, 1974) followed. However, no plants could be regenerated from these cultures. The breakthrough with plant regeneration was reached in 1975 (Green and Phillips, 1975), when immature embryos were used as explants for callus formation. At the same time, suspension cultures could also be established from callus originated leaf base explants (Sheridan, 1975). Plant regeneration from these suspensions was not possible, but protoplasts could be isolated from such suspensions and cultured successfully up to the callus stage (Choury and Zuwarski, 1981).

Although considerable knowledge has accumulated on tissue culture of maize and other Gramineae, the basic situation remains unchanged. Callus cultures originating from multicellular explants exhibit plant regeneration while protoplasts from suspensions develop only to the callus stage. Immature embryos, excised 10 to 14 days after pollination, have proven to be the most suitable

Barbara Junker and Horst Lörz • Max Planck Institut für Züchtungsforschung, D-5000 Köln 30, Federal Republic of Germany.

source of explant for the induction of totipotent maize *in vitro* cultures. Plants could also be regenerated from other explants such as mesocotyl (Harms *et al.*, 1976), young leaves (Chang, 1983; Santos *et al.*, 1984), and mature embryos (Vasil *et al.*, 1983). Haploid maize plants can be obtained by culturing ovaries (Ao *et al.*, 1982) and anthers (Brettell *et al.*, 1981; Genovesi and Collins, 1982). The regenerated plants do not always phenotypically and genotypically resemble the donor plants. The phenomenon of "somaclonal variation" (Larkin and Scowcroft, 1981) is especially significant in maize (Göbel *et al.*, 1986) and is probably due to tissue culture stress (Lörz and Brown, 1986).

Different types of callus can develop from explants mostly depending on the genotype of the donor plant and the medium used (Fig. 1). This was originally described by Green *et al.* (1983) for immature embryo-derived cultures of the line A 188:

- Type I: Compact, yellow-white, nodular, slow-growing callus, from which plants can develop by organogenesis (Springer *et al.*, 1979) or somatic embryogenesis (Lu *et al.*, 1983). Long-term suspension cultures could not be established from this type of callus.
- Type II: Friable, white-yellow callus, which is relatively fast-growing and is covered with somatic embryos. This type of callus can give rise to plants by embryogenesis (Green, 1982, 1983) and can be used to establish suspensions (Green *et al.*, 1983).
- Type III: Translucent, gray-yellow, granular callus, from which plants cannot be regenerated. However, the material is very well suited for the establishment of cell suspension cultures. All protoplasts giving rise to callus described thus far, were isolated from this type of cell (Potrykus *et al.*, 1979; Polikarpochkina *et al.*, 1979; Choury and Zuwarski, 1981) (Fig. 2).

Induction of type I callus and plant regeneration thereof are efficient and reproducible for a large number of genotypes (Duncan *et al.*, 1985), including commercially important lines (Fahey *et al.*, 1986). By genetic analysis, it could be shown that dominant, nucleus-coded genes are of importance for the efficiency of callus initiation and plant regeneration (Hodges *et al.*, 1986). There are only a few genotypes known which can be used to establish type III callus and suspensions (Potrykus *et al.*, 1979; Polikarpochkina *et al.*, 1979; Choury and Zuwarski, 1981). There are still difficulties in obtaining callus and especially totipotent suspensions of type II cultures. By optimizing media (Armstrong and Green, 1985) and crosses of type I and type III genotypes, progress was reported for callus induction (Lupotto, 1986) and the initiation of embryogenic suspensions (Kamo and Hodges, 1986). Protoplasts isolated from this specific type of suspension should be a promising way to bridge the gap which still exists between maize protoplasts and plants derived thereof. Another approach is the isolation and culture of protoplasts from type I and type II callus. Thus far, only divisions up to a microcallus stage have been obtained (Imbrie-Mulligan and Hodges, 1986).

In maize, as in most plant species, protoplasts can be isolated directly from different plant tissues, i.e., leaf, stem, roots, or microspores (Ozias-Akins and Lörz, 1984). These protoplasts can be kept in culture for some time, but reproducible, sustained divisions have not been reported. It has been concluded that protoplasts from cells which cannot divide in plants anymore have lost the capacity for division and totipotency *in vivo* as well (Lörz *et al.*, 1987). In most other cereals, the situation of tissue culture is comparable to the one reported in maize (for reviews see Bright and Jones, 1985). Only in rice has regeneration of plants from protoplasts been achieved using embryogenic suspensions of Japonica-type rice for the isolation of protoplasts (Abdullah *et al.*, 1986; Fujimura *et al.*, 1986; Yamada *et al.*, 1986).

3. TRANSFORMATION OF PROTOPLASTS

Protoplasts isolated from nonembryogenic suspensions of maize have been used for gene transfer experiments. After gene transfer to protoplasts by electroporation, integration and ex-

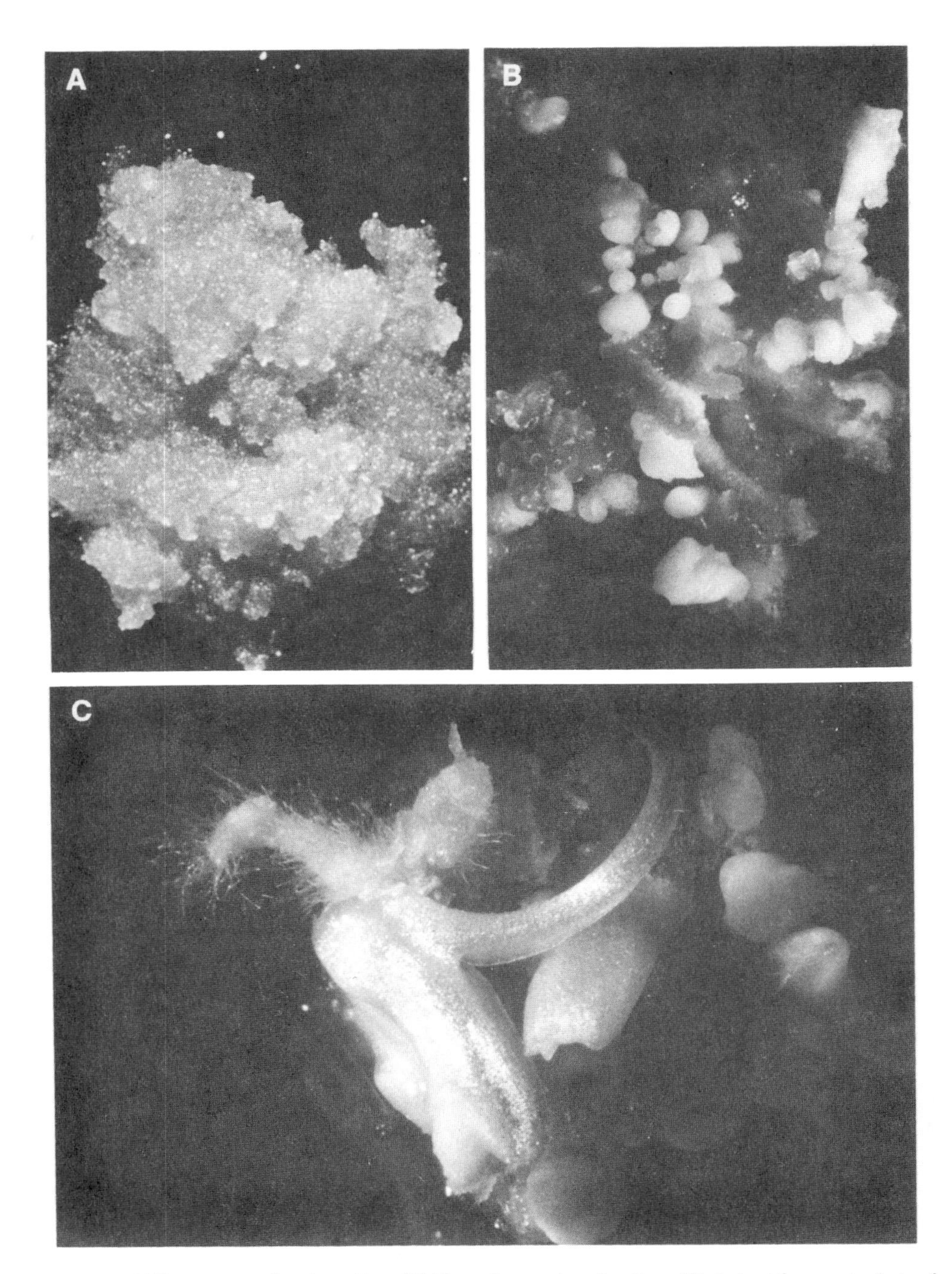

Figure 1. Different types of maize callus. (A) Nonembryogenic callus (type III) derived from protoplasts of the line Black Mexican Sweet. (B) Embryogenic callus (type II) which originated from excised immature embryos and has been cultured for 4 weeks. (C) Embryogenic callus (type I) which originated from immature embryos and has already been transferred to regeneration medium. One of the embryos is germinating.

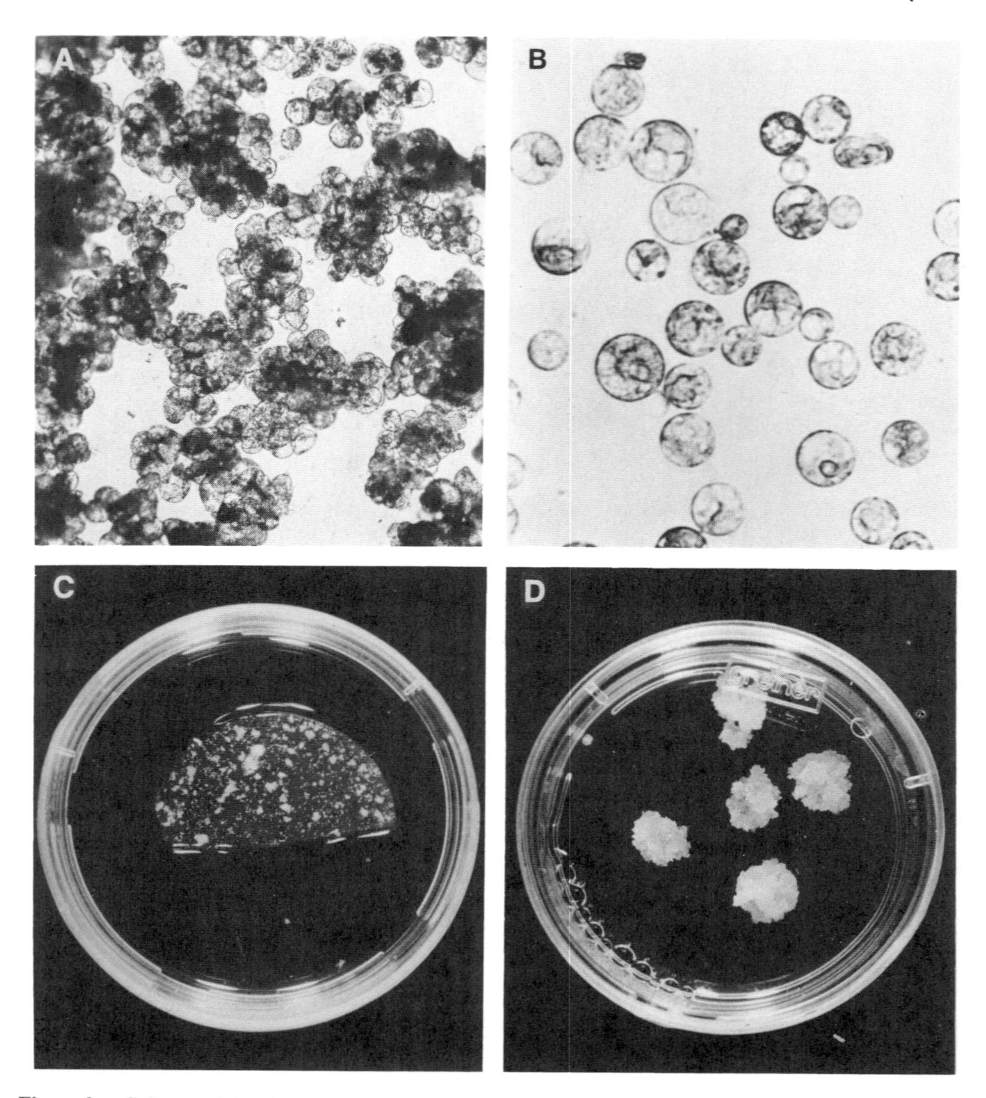

Figure 2. Culture and development of suspension-derived protoplasts of maize. (A) Cell suspension of maize, line Black Mexican Sweet consisting of small cell clusters comprising about 5–30 cells. (B) Freshly isolated protoplasts from a suspension as shown in A. (C) Development of microcalli from protoplasts plated in agarose after 3 weeks of culture. (D) Protoplast-derived calli of maize 7 weeks after culture initiation.

pression of a plasmid-coded selectable marker gene could be shown in the protoplast-derived calli (Fromm *et al.*, 1986). Comparable results were obtained with suspension-derived protoplasts of other cereals, e.g., *Triticum monococcum* L. (Lörz *et al.*, 1985) and rice (Uchimiya *et al.*, 1986). In both cases, DNA uptake was achieved by PEG treatment.

Independent of integration, the uptake and expression of chimeric genes can be monitored by assaying for the enzymatic activity of the foreign gene product a few days after transformation. This phenomenon of "transient gene expression" has been described for maize suspension protoplasts after electroporation (Fromm *et al.*, 1985) as well as after PEG treatment (Junker *et al.*, 1986). In addition to "dividing protoplasts," "nondividing protoplasts" isolated from leaves and roots of

Table 1. Stable Genetic Transformation of Cereal Cells Originating from Protoplasts

Species	Procedure	Reference
Panicum maximum	Electroporation	Ozias-Akins *et al.* (1987)
Oryza sativa	PEG	Uchimiya *et al.* (1986)
Triticum monococcum	PEG	Lörz *et al.* (1985)
	Electroporation	Ozias-Akins *et al.* (1987)
Zea mays	Electroporation	Fromm *et al.* (1986)

maize plants could express foreign genes after PEG-mediated DNA uptake (Junker *et al.*, 1987). The results published to date for transient gene expression and stable genetic transformation of cereal protoplasts are listed in Tables 1 and 2.

In somatic cell genetics, DNA-mediated transformation is the most direct way to transfer a defined gene into plant cells. Genetic modification of plant cells can also be achieved by transfer of isolated cell organelles or by somatic hybridization and cybridization resulting in newly combined organellar and nuclear genomes (Lörz, 1985).

Cereal protoplasts have been used for fusion experiments. Both chemical (PEG) and electrical fusion methods have been applied. Oat (*Avena sativa* L.) leaf protoplasts have been fused in the electric field with protoplasts of *Kalanchoe daigremontiana* (Vienken *et al.*, 1981). PEG-induced fusion was reported for maize and barley mesophyll protoplasts (Brar *et al.*, 1980). A few cell divisions have occurred after such fusion experiments, but no regeneration has been achieved. Mostly cytogenetic and biochemical analyses were used to confirm the hybrid character of these hybrid cell lines.

A comparison of efficiencies of chemical versus electrical transformation and fusion methods is difficult at present, since technical equipment, gene constructions, and especially the protoplast cultures differ in the various studies. However, all these experiments show clearly that protoplasts of cereals are as suitable for transformation and fusion experiments as the ones of model plants. The lack of tissue culture knowledge, especially how to regenerate plants from protoplasts, remains the limiting step in the application of these techniques to crop plants.

Table 2. Transient Expression Studies with Cereal Protoplasts

Species	Source of protoplasts	Procedure	Reference
Hordeum vulgare	Suspension	PEG	Junker *et al.* (1987)
Panicum maximum	Suspension	Electroporation	Ozias-Akins *et al.* (1987)
Pennisetum purpureum	Suspension	Electroporation	Ozias-Akins *et al.* (1987)
Oryza sativa	Leaves	Electroporation	Ou-Lee *et al.* (1986)
	Leaves	PEG	Junker *et al.* (1987)
	Suspension	PEG	Junker *et al.* (1987)
Secale cereale	Leaves	PEG	Junker *et al.* (1987)
Sorghum bicolor	Callus	Electroporation	Ou-Lee *et al.* (1986)
Triticum monococcum	Suspension	PEG	Werr and Lörz (1986)
	Suspension	Electroporation	Ou-Lee *et al.* (1986)
Zea mays	Suspension	Electroporation	Fromm *et al.* (1985)
	Suspension	PEG	Junker *et al.* (1986)
	Leaves	PEG	Junker *et al.* (1987)
	Roots	PEG	Junker *et al.* (1987)

Figure 3. Scheme of natural life cycle of maize and different steps of *in vitro* culture. (1) Mature seed; (2) germinating seedling (2–4 days); (3) vegetative growth; (4) early stages of development of inflorescences; (5) flowering plant; (6) embryogenic callus derived from excised immature embryos; (7) *in vitro* regenerated plantlet; (8) embryogenic or nonembryogenic suspension; (9) protoplasts isolated from suspension cultures; (10) protoplast-derived nonembryogenic callus; (11) protoplasts isolated directly from the plant. Transformation studies in maize and other cereals have been performed using *in vitro* cultured cells as well as intact plants: (2) treatment of young seedlings with *Agrobacterium tumefaciens;* (4) injection of plasmid DNA into young inflorescences; (5) pollen as vector for DNA-mediated transformation; (9) electroporation or PEG-mediated direct gene transfer for stable transformation; (9 + 11) electroporation or PEG-mediated direct gene transfer for transient expression studies of introduced genes.

4. ALTERNATIVE TRANSFORMATION METHODS

Methods for gene transfer, independent of *in vitro* culture, have been studied most intensively in cereals. Strategies investigated include pollen transformation, micro- and macroinjection, and gene transfer with *Agrobacterium tumefaciens* as a possible vector system (see Fig. 3).

Maize pollen has been treated with isolated DNA of a ''donor'' maize plant and used for pollination of the ''recipient'' plant. In the progeny, phenotypic modifications correlating to genes of the donor strain were found (Ohta, 1986). The phenotypes analyzed include colorless aleurone, shrunken and waxy endosperm. Molecular evidence for gene transfer has not been obtained. It may be that DNA treatment of pollen is mutagenic to the gametophytic cells, giving rise to phenotypic and genetic changes (Lörz *et al.,* 1987; Negrutiu *et al.,* 1986).

Transformation of numerous dicotyledonous plants and a few monocotyledonous species is performed efficiently and routinely using the natural vector system of the Ti plasmid from *A. tumefaciens*. The natural host range of *A. tumefaciens* and *A. rhizogenes* does not include the Gramineae. Therefore, cereals were considered not to be amenable to *Agrobacterium*-mediated gene transfer (De Cleene, 1984). However, very recent and preliminary results indicate the possible use of *A. tumefaciens* or *A. rhizogenes* as natural vectors for the transformation of cereals. After infection of young maize seedlings with *A. tumefaciens*, specific compounds such as opines normally formed in plant cells after infection with *Agrobacterium*, were also found in developing plants (Graves and Goldman, 1986). Thus far, no molecular evidence for the integration of T-DNA into the maize genome was obtained. Opines could be detected in nontransformed tissue of several plant species, including maize (Christou *et al.*, 1986). We have observed, after infection of maize seedlings with *A. tumefaciens* strains carrying genetically modified T-DNA, the expression of neomycin phosphotransferase in the tissue of treated plants and no expression in nontreated controls. *Agrobacterium*-mediated delivery of infectious maize streak virus into maize plants (Grimsley *et al.*, 1987) can also be taken as further evidence that *Agrobacterium* might be developed in the near future as a natural and suitable vector system for cereals.

Another approach which does not involve any tissue culture techniques was taken by injection of DNA into young floral tillers of rye (*Secale cereale* L) plants (De la Pena *et al.*, 1987). Three transformed rye plants, subsequently yielding about 6000 seeds, were obtained when plasmid DNA carrying a gene coding for kanamycin resistance was injected about 14 days prior to the first meiotic metaphase into about 350 floral tillers. Experiments are in progress to determine whether this rather simple transformation procedure can also be applied to other graminaceous species.

5. CONCLUSION

Considerable progress has been made in the last few years with *in vitro* culture and plant regeneration of maize and other cereals. Plant regeneration from *in vitro* cultures originating from multicellular explants has been developed into an efficient and routine procedure. Progress in the culture and regeneration of protoplasts isolated from maize, wheat, and other cereals is still limited and plant regeneration from protoplasts has been achieved thus far only for rice.

Stable genetic transformation of cultured cells of maize, wheat, and rice has been obtained by transforming protoplasts with plasmid DNA containing a selectable marker gene. DNA uptake was stimulated by chemical treatment with PEG or electroporation. Protoplasts isolated from numerous graminaceous species and from different tissues have proven to be a useful tool to study chimeric gene constructions and the function of promoters in transient gene expression. The present difficulties of regenerating plants from protoplasts stimulated the development of other transformation techniques applicable to cereals. These approaches include the transfer of genes via microinjection into embryogenic cells, via macroinjection into developing plants, or via the natural vector system of *Agrobacterium*. Electroporation and electrofusion are of special interest in the context of genetic manipulation of protoplasts. Basic findings made with model plant species will contribute to the progress with crop plants such as cereals.

NOTED ADDED IN PROOF

Since the completion of this chapter, plant regeneration from protoplasts and direct DNA-mediated transformation of maize protoplasts have been reported from several laboratories (Prioli and Sondahl, 1988; Rhodes *et al.*, 1988a,b; Shillito *et al.*, 1988).

REFERENCES

Abdullah, R., Cocking, E. C., and Thompson, J. A., 1986, Efficient plant regeneration from rice protoplasts through somatic embryogenesis, *Biotechnology* **4:**1087–1090.

Ao, G. M., Zhao, S. X., and Li, G.H., 1982, In vitro induction of haploid plantlets from unpollinated ovaries of corn (*Zea mays* L.), *Acta Genet. Sin.* **9:**281–283.

Armstrong, C. L., and Green, C. E., 1985, Establishment and maintenance of friable, embryogenic maize callus and the involvement of L-proline, *Planta* **164:**207–214.

Brar, D. S., Rambold, S., Constable, F., and Gamborg, O. L., 1980, Isolation, fusion and culture of sorghum and corn protoplasts, *Z. Pflanzenphysiol.* **96:**269–275.
Brettell, R. I. S., Thomas, E., and Wernicke, W., 1981, Production of haploid maize plants by anther culture, *Maydica* **26:**101–111.
Bright, S. W. J., and Jones, M. G. K. (eds.), 1985, *Cereal Tissue and Cell Culture,* Junk, The Hague.
Chang, Y. F., 1983, Plant regeneration in vitro from leaf tissues derived from cultured immature embryos of *Zea mays* L., *Plant Cell Rep.* **2:**183–185.
Choury, P. S., and Zuwarski, D. B., 1981, Callus formation from protoplasts of a maize culture, *Theor. Appl. Genet.* **59:**341–344.
Christou, P. S., Platt, S. G., and Ackermann, M. C., 1986, Opine synthesis in wild-type plant tissue, *Plant Physiol.* **82:**218–221.
De Cleene, M., 1984, The susceptibility of monocotyledonous plants to *Agrobacterium tumefaciens, Phytopathol. Z.* **113:**81–89.
De la Pena, A., Lörz, H., and Schell, J., 1987, Transgenic rye plants obtained by injecting DNA into young floral tillers, *Nature* **325:**274–276.
Duncan, D. R., Williams, M. E., Zehr, B. E., and Widholm, J. M. 1985, The production of callus capable of plant regeneration from immature embryos of numerous *Zea mays* genotypes, *Planta* **165:**322–332.
Fahey, J. W., Reed, J. N., Readdy, T. L., and Pace, G. M. 1986, Somatic embryogenesis from three commercially important inbreds of *Zea mays, Plant Cell Rep.* **5:**35–38.
Fromm, M., Taylor, L. P., and Walbot, V., 1985, Expression of genes transferred into monocot and dicot plant cells by electroporation, *Proc. Natl. Acad. Sci. USA* **82:**5824–5828.
Fromm, M. E., Taylor, L. P., and Walbot, V., 1986, Stable transformation of maize after gene transfer by electroporation, *Nature* **319:**791–793.
Fujimura, T., Sakurai, M., Negishi, T., and Hirose, A., 1986, Regeneration of rice plants from protoplasts, *Plant Tissue Culture Lett.* **2:**74–75.
Genovesi, A. D., and Collins, G. B., 1982, In vitro production of haploid plants of corn via anther culture, *Crop Sci.* **22:**1137–1144.
Göbel, E., Brown, P. T. H., and Lörz, H., 1986, In vitro culture of *Zea mays* and analysis of regenerated plants, in: *Nuclear Techniques and In Vitro Culture for Plant Improvement,* International Atomic Energy Agency, Vienna, pp. 21–27.
Graves, A. C. F., and Goldman, S. L., 1986, The transformation of *Zea mays* seedlings with *Agrobacterium tumefaciens, Plant Mol. Biol.* **7:**43–50.
Green, C. E., 1982, Somatic embryogenesis and plant regeneration from the friable callus of *Zea mays,* in: *Plant Tissue Culture* (A. Fujiwara, ed.), Maruzen, Tokyo, pp. 107–108.
Green, C. E., 1983, New developments in plant tissue culture and plant regeneration, in: *Basic Biology of New Developments in Biotechnology* (A. Holländer, A. J. Larkin, and P. Rogers, eds.), Plenum Press, New York, pp. 195–205.
Green, C. E., and Phillips, R. L., 1975, Plant regeneration from tissue cultures of maize, *Crop. Sci.* **15:**417–421.
Green, C. E., Phillips, R. L., and Kleese, R. A., 1974, Tissue cultures of maize (*Zea mays* L.): Initiation, maintenance and organic growth factors, *Crop Sci.* **14:**54–58.
Green, C. E., Armstrong, C. L., and Anderson, P. C., 1983, Somatic cell genetic systems in corn, in: *Molecular Genetics of Plants and Animals* (K. Downey, R. W. Vollmy, A. Fazelahmad, and J. Schmitz, eds.), Academic Press, New York, pp. 147–159.
Grimsley, N., Hohn, T., Davis, J. W., and Hohn, B., 1987, Agrobacterium-mediated delivery of infectious maize streak virus into maize plants, *Nature* **325:**177–179.
Harms, C. T., Lörz, H., and Potrykus, I., 1976, Regeneration of plantlets from callus cultures of *Zea mays* L., *Z. Pflanzenzücht.* **77:**347–351.
Hodges, T. K., Kamo, K. K., Imbrie, C. W., and Becwar, M. R. 1986, Genotype specificity of somatic embryogenesis and regeneration in maize, *Biotechnology* **4:**219–223.
Imbrie-Milligan, C. W., and Hodges, T. K., 1986, Microcallus formation from maize protoplasts prepared from embryogenic callus, *Planta* **168:**395–401.
Junker, B., Baker, B., and Lörz, H., 1986, Culture and transformation studies with maize protoplasts, in: *Genetic Manipulation in Plant Breeding* (W. Horn, C. J. Jensen, W. Odenbach, and O. Schieder, eds.), de Gruyter, Berlin, pp. 827–830.
Junker, B., Zimny, J., Lührs, R., and Lörz, H., 1987, Transient expression of chimaeric genes in dividing and non-dividing cereal protoplasts after PEG-induced DNA uptake, *Plant Cell Rep.* **6:**329–332.

Kamo, K. K., and Hodges, T. K., 1986, Establishment and characterization of long-term embryogenic maize callus and cell suspension cultures, *Plant Sci.* **45:**111–117.

Larkin, P. J., and Scowcroft, W. R. 1981, Somaclonal variation—A novel source of variability from cell cultures for plant improvement, *Theor. Appl. Genet.* **60:**197–214.

La Rue, C. D., 1949, Cultures of the endosperm of maize, *Am. J. Bot.* **36:**798.

Linsmeier-Bedner, E. M., and Bedner, T. W., 1972, Light and hormonal control of root formation in *Zea mays* callus cultures, *Dev. Growth Differ.* **14:**165–174.

Lörz, H., 1985, Isolated cell organelles and subprotoplasts: Their roles in somatic cell genetics, in: *Plant Genetic Engineering* (J. H. Dodds, ed.), Cambridge University Press, London, pp. 25–29.

Lörz, H., and Brown, P. T. H., 1986, Variability in tissue culture derived plants—Possible origins, drawbacks and advantages, in: *Genetic Manipulation in Plant Breeding* (W. Horn, C. J. Jensen, W. Odenbach, and Schieder, eds.), de Gruyter, Berlin, pp. 513–534.

Lörz, H., Baker, B., and Schell, J., 1985, Gene transfer to cereal cells mediated by protoplasts transformation, *Mol. Gen. Genet.* **199:**178–182.

Lörz, H., Junker, B., Schell, J., and De la Pena, A., 1987, Gene transfer in cereals, in: *Plant Tissue and Cell Culture* (D. Somers, ed.), Liss, New York pp. 303–316.

Lu, C., Vasil, V., and Vasil, I. K. 1983, Improved efficiency of somatic embryogenesis and plant regeneration in tissue cultures of maize (*Zea mays* L.), *Theor. Appl. Genet.* **66:**285–289.

Lupotto, E., 1986, In vitro culture of isolated somatic embryos of maize (*Zea mays* L.), *Maydica* **21:**193–201.

Mascarenhas, A. F., Sayagover, B. M., and Jagannathan, V., 1965, Studies on the growth of callus cultures of *Zea mays* L., in: *Tissue Culture* (C. V. Ramakrishnan, ed.), Junk, The Hague, pp. 283–291.

Negrutiu, I., Heberle-Bors, E., and Potrykus, I., 1986, Attempts to transform for kanamycin-resistance in mature pollen of tobacco, in: *Biotechnology and Ecology of Pollen* (J. L. Mulcahy, G. Bergamini-Mulcahy, and E. Ottariano, eds.), Springer-Verlag, Berlin, pp. 65–76.

Ohta, Y., 1986, High-efficiency genetic transformation of maize by a mixture of pollen and exogenous DNA, *Proc. Natl. Acad. Sci. USA* **83:**715–719.

Ou-Lee, T. M., Turgeon, R., and Wu, R., 1986, Expression of a foreign gene linked to either a plant-virus or a *Drosophila* promoter, after electroporation of protoplasts of rice, wheat and sorghum, *Proc. Natl. Acad. Sci. USA* **83:**6815–6819.

Ozias-Akins, P., and Lörz, H., 1984, Progress and limitations in the culture of cereal protoplasts, *Trends Biotechnol.* **2:**119–123.

Ozias-Akins, P., Hauptmann, R., Vasil, V., Tabaeizadeh, Z. Horsch, R., Rogers, S., Fraley, R., and Vasil, J., 1987, Stable transformation of *Panicum maximum* and *Triticum monococcum:* More efficient selectable markers for the *Gramineae, J. Cell. Biochem. Suppl.* **11B:**28.

Polikarpochkina, R. T., Gamburg, K. Z., and Khavkin, E. E., 1979, Cell-suspension culture of maize (*Zea mays* L.), *Z. Pflanzenphysiol.* **95:**57–67.

Potrykus, I., Harms, C. T., and Lörz, H., 1979, Callus formation from cell culture protoplasts of corn (*Zea mays* L.), *Theor. Appl. Genet.* **54:**209–214.

Potrykus, I., Saul, M. W., Petruska, I., Paszkowski, L., and Shillito, R. D., 1985, Direct gene transfer to cells of a graminaceous monocot, *Mol. Gen. Genet.* **199:**183–188.

Prioli, L. M., and Sondahl, M. R., 1988, Plant regeneration from protoplasts of a tropical maize (*Zea mays,* L.) inbred line, Abstracts of the International EUCARPIA Congress on Genetic Manipulation in Plant Breeding, Elsinore, 1988 (C. J. Jensen, ed.), EUCARPIA, Roskidde, p. 94.

Rhodes, C. A., Lowe, K. I. S., and Ruby, K. L., 1988a, Plant regeneration from protoplasts isolated from embryogenic maize cell cultures, *Biotechnology* **6:**56–60.

Rhodes, C. A., Pierce, D. A., Mettler, I. J., Mascarenhas, D., and Detmar, J. J., 1988b, Genetically transformed maize plants from protoplasts, *Science* **240:**204–207.

Santos, M. A., Torne, J. M., and Blanco, J. L., 1984, Methods of obtaining maize totipotent tissue, *Plant Sci. Lett.* **33:**309–315.

Sheridan, W. F., 1975, Growth of corn cells in culture, *J. Cell Biol.* **67:**396a.

Shillito, R. D., Carswell, G. K., Johnson, C. M., and Harms, C. T., 1988, Plant regeneration from protoplasts of a tropical maize (*Zea mays* L.) inbred line, Abstracts of the International EUCARPIA Congress on Genetic Manipulation in Plant Breeding, Elsinore, 1988 (C. J. Jensen, ed.), EUCARPIA, Roskidde, p. 72.

Springer, W. D., Green, C. E., and Kohn, K. A., 1979, A histological examination of tissue culture initiation from immature embryos of maize, *Protoplasma* **101:**269–281.

Uchimiya, H., Fushimi, T., Hashimoto, H., Harada, H., Syono, K., and Sugawara, Y., 1986, Expression of a foreign gene in callus derived from DNA-treated protoplasts of rice (*Oryza sativa* L.), *Mol. Gen. Genet.* **204:**204–207.

Vasil, V., Lu, C. Y., and Vasil, I. K., 1983, Proliferation and plant regeneration from the nodal region of *Zea mays* L. (maize, *Graminaceae*) embryos, *Am. J. Bot.* **70:**951–954.

Vienken, J., Ganser, R., Hampp, R., and Zimmermann, U., 1981, Electric field-induced fusion of isolated vacuoles and protoplasts of different developmental and metabolic provenance, *Physiol. Plant.* **53:**64–70.

Werr, W., and Lörz, H., 1986, Transient gene expression in a *Gramineae* cell line, *Mol. Gen. Genet.* **202:**368–373.

Yamada, Y., Yang, Z. Q., and Tang, D. T., 1986, Plant regeneration from protoplast derived callus of rice (*Oryza sativa*), *Plant Cell Rep.* **5:**85–88.

Part V

Methods and Equipment

Chapter 26

Cells in Electric Fields

Physical and Practical Electronic Aspects of Electro Cell Fusion and Electroporation

Gunter A. Hofmann

1. INTRODUCTION

Pulses of an electric field can induce cell fusion (electrofusion) or produce micropores in the cell membrane allowing materials in the media to enter (electroporation). These two techniques have wide application in science, medicine, and biotechnology (Zimmermann, 1982, 1986; Zimmermann and Pilwat, 1982; Hofmann and Evans, 1986). This chapter discusses the physical aspects and the design of instrumentation for the two techniques. It includes the electrical requirements for electrofusion and electroporation, the requirements of the electric generator, and the design of chambers for large-scale electrofusion and electroporation, hereafter referred to as electromanipulation of cells. Not all biological aspects such as characteristics of various cell lines (which are the subject of electromanipulation), genetic selection systems, optimal conditions for fusion and electroporation have been well characterized, but additional information will be found elsewhere in this volume.

An electro cell manipulation system can be divided into the following three main subsystems: (1) the electronic generator or power supply, which provides the electric signals, (2) the chamber which contains the cells and in which the electric signals from the generator create electric fields, and (3) the biological subsystem encompassing the cells and membranes and purposes of the electromanipulation. The interaction of the three subsystems can lead to two basic goals:

- Electroporation—which allows materials to be introduced into the cytoplasm following the dielectric breakdown of the cell membrane (alternatively, the contents can be induced to leak out).
- Fusion—if cells or membranes are first aligned into close contact by dielectrophoresis, then the electrical signal which induces electroporation may also induce fusion.

Gunter A. Hofmann • BTX-Biotechnologies and Experimental Research Inc., San Diego, California 92109.

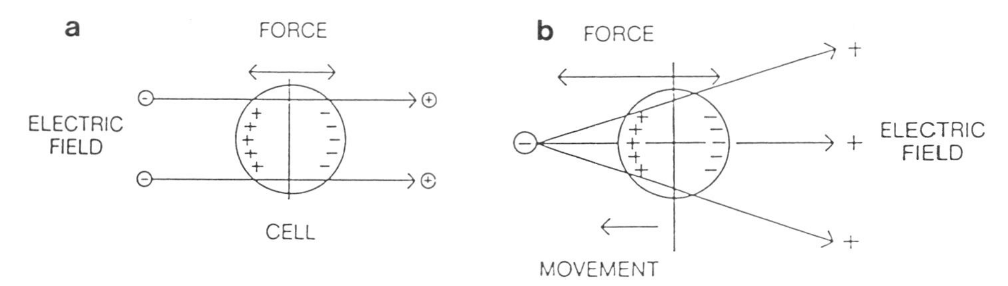

Figure 1. Charge separation and force on a cell in a homogeneous (a) and an inhomogeneous electric field (b).

2. DIELECTROPHORESIS

When cells are exposed to alternating electric fields, they experience a separation of electrical charges, forming an oscillating dipole. If the electric field is homogeneous (Fig. 1a) and the cell density low ($< 10^7/cm^3$) the cell will remain in place. If the cell density is high (10^7–$10^8/cm^3$), dipole interactions between adjacent cells occur, leading to mutual attraction. In the case of an inhomogeneous electric field, the force of the field on the side of the cell closer to the field concentration point (higher field strength) is larger than the force on the opposite side. As a result, the cell moves toward the field concentration area (Fig. 1b). The force on the cell is independent of the polarity of the electric field and always points toward the field concentration area. Therefore, an alternating (AC) electric field can be used to induce cell movement. The waveshape of the alternating field does not appear to be critical to the cell movement. We observed dielectrophoretic movement of cells with sinusoidal, square, and sawtooth-like AC waveforms. As cells approach the area of field concentration, they are attracted to each other and form pearl chains of two or more cells (alignment).

The following two derived relationships govern the movement and alignment of cells in AC electric fields (Pohl, 1978).

The force experienced by a cell is given as:

$$F = r^3 - E\ grad\ E \cdot 2\pi\epsilon_0 \frac{\epsilon - \epsilon_0}{\epsilon + 2\epsilon_0}$$

where F is the force on the cell in newtons, r is the radius of the cell in meters, ϵ_0 is the dielectric constant of the liquid in farads per meter, ϵ is the dielectric constant of the cell in farads per meter, E is the electric field strength in volts per meter, and grad E is the measure for inhomogeneity of the electric field in volts per square meter. As a general rule, ϵ decreases with increasing frequency; if $\epsilon > \epsilon_0$, then cells are drawn to higher fields (positive dielectrophoresis); if $\epsilon < \epsilon_0$, then cells are drawn to lower fields (negative dielectrophoresis); if $\epsilon = \epsilon_0$, there is no force on the cells (this is typical for dead cells, where the contents leaked out. Larger cells experience a stronger force (line up faster), and smaller cells need a higher field to be collected. The lower cell radius limit for collection is 0.3 μm.

The relative length L of a pearl chain is derived as (Pohl, 1978):

$$L \sim \sqrt{t}\, V\, c\, r^4$$

where L is the pearl chain length in relative units (micrometers or number of cells), t is the collection (alignment) time, V is the voltage, c is the concentration, and r is the radius of the cells.

If the desire is to create dimers (pairs), the following guidelines should be observed: short

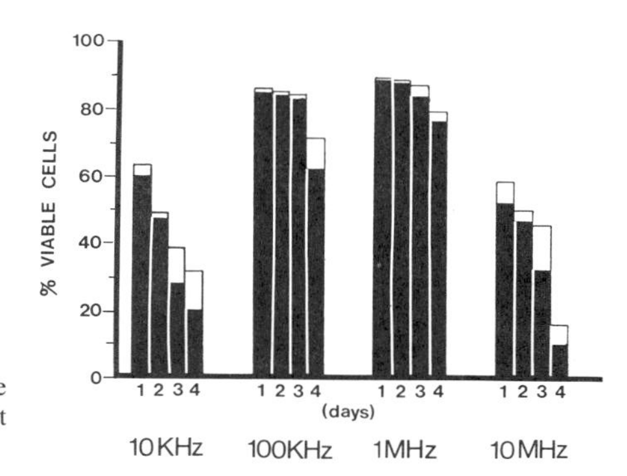

Figure 2. Effect of the alignment AC field on the viability of UC 729-6 (white bar) and PBL (black bar) at different frequencies.

alignment time, low alignment voltage, low cell concentration, moderate divergence of field. Small cells require longer time, higher voltage, higher concentration.

The viability of cells generally appears to be highest if the frequency is between 100 kHz and a few megahertz, as shown for the UC 729-6 cell line in Fig. 2.

3. DIELECTRIC BREAKDOWN—FUSION/ELECTROPORATION

Cells suspended in a liquid can be modeled as a structure consisting of a nonconducting membrane with aqueous solutions on both sides. Exposure to an electric field leads to charge separation in the membrane similar to the charge separation in the dielectric layers of an electrical capacitor. This results in a transmembrane potential difference. Opposite electrical charges on the membrane attract each other, exerting pressure on the membrane, which can induce membrane thinning. Zimmermann (1982) proposed that at a critical potential difference (field strength), localized breakdown of the membrane occurs and pores are formed, allowing a flow of medium or cytoplasm. Based on freeze-fracture electron microscopy, it appears that a large number of adjacent pores might lead to macroscopic cracks in the membrane which allow permeation of large molecules (Stenger and Hui, 1986). However, more work needs to be done before a model evolves which is sufficiently supported by experimental evidence. Removal of the field presumably leads to resealing of the pores provided that the field strength and pulse width are not excessive. If two cells touch each other during the process of pore formation, adjacent pores may form channels which permit communication between cytoplasmic compartments. This process can thus lead to the formation of a new spherical hybrid cell. Excessive field strength and/or duration of the pulses can lead to irreversible changes in the cell by damaging intracellular proteins and DNA.

Fusion yield depends on the following qualitative guidelines (Zimmermann, 1982): Smaller cells require higher fields; excessively high fields or long pulse duration may result in irreversible mechanical breakdown of the membrane; higher temperatures require lower fields; precompression of the cell membrane by pressure gradients (turgor pressure or hydrostatic pressure) leads to a decrease of the required breakdown field; rounding-off time (formation of a complete new spherical cell) may take up to 60 min and is a function of temperature; optimum time for rounding off after fusion is 3 to 5 min; rounding off is aided by adding small amounts of $CaCl_2$ or $MgCl_2$ after fusion; fusion efficiency is size dependent; pretreatment of larger cells with pronase stabilizes against high fields; for hybridoma production, it is recommended that the ratio of lymphocytes to myeloma cells in solution be 3–4 to 1; reapplying AC fields for a few minutes after the fusion pulse may aid the formation of spherical cells.

If the intensity of the electric field is excessive, the cell may lyse due to failure of membrane pores to reseal or osmotic swelling which leads to excess tension in the plasma membrane. Efficient introduction of DNA into cells by means of electroporation was found to occur at a field strength

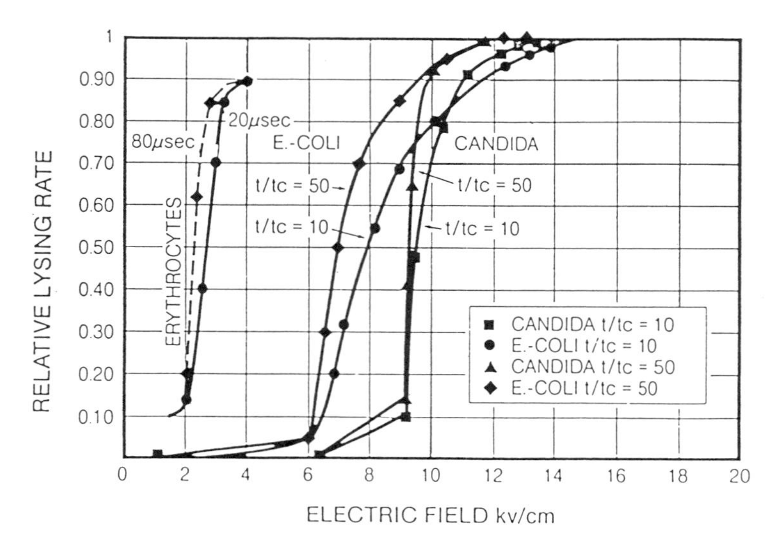

Figure 3. Lysing of cells in an electric field.

where substantial cell lysis takes place (Neumann *et al.*, 1982). The onset of lysing is quite sudden above a critical field strength. An empirical model for the lysis-induced killing of cells that shows some important relationships between experimental parameters has been developed (Hulsheger *et al.*, 1981). The relative rate of lysing can be expressed by the empirical relationship:

$$L = 1-(t/t_c)\exp[-(E-E_c)/K]$$

where L is the relative rate of lysing, t is the treatment time (this is the product of pulse width times the number of pulses, in microseconds), t_c is the threshold value for the treatment time in microseconds, E is the applied electric field strength in kilovolts per centimeter, E_c is the threshold value for the field strength in kilovolts per centimeter, and K is a model constant. t_c, E_c, and K are species-dependent constants. Typical values for bacteria are: t_c = 35 μsec, E_c = 6kV/cm, K = 6.3 cm/kV.

Figure 3 shows a typical lysing curve as a function of applied field strength. The onset of lysing is quite steep and if the field strength is selected below or above the optimum value, no lysing or complete lysing is observed, respectively. A 5 to 80% lysing rate appears to give good electroporation yields. The field strength for high-yield electroporation is critical if the chamber provides a uniform field. If the chamber provides a highly nonuniform field, some part of the chamber volume may be exposed to the optimum field strength even if the applied voltage is not optimal.

Pretreatment of cells changes the field strength at which lysing occurs (Table 1). Figure 4

Table 1. Pretreatment of Cells Changes the Critical Field Strength at Which 50% Lysing Occurs[a]

Cell line	Pretreatment	Field strength, kV/cm (pulse length 30 μsec)
UC 729-6	Trypsin	0.61
	Pepsin	1.27
	Papain	1.58
PBL	None	0.95

[a]Data from Glassy (personal communication).

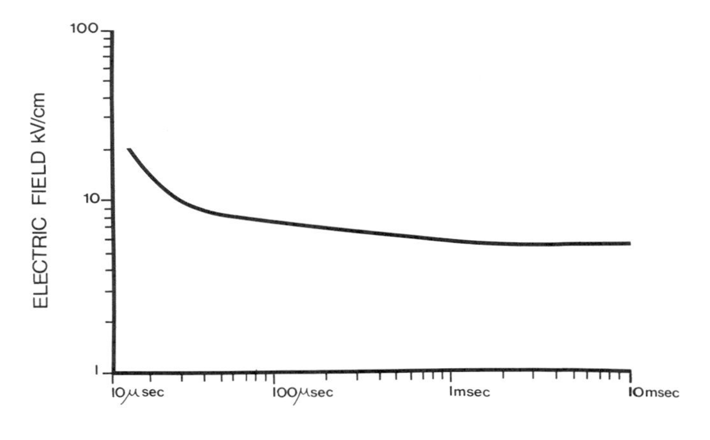

Figure 4. Fifty percent lysing curve of 16 hr *E. coli* as a function of field strength and pulse length. Below a critical field strength, no lysis is observed even with very long pulses.

shows the trade-off between field strength and pulse width to obtain a specific percentage (50%) of lysing, calculated from Eq. (3).

4. THE ELECTRIC GENERATOR

4.1. Frequency

The electric generator provides generally two electric signals: (1) continuous alternating high-frequency electric fields (AC) to align cells into pearl chains, and (2) short, unipolar pulses (often called DC pulses, signals which show only one polarity) of high amplitude. Figure 5 shows a typical sequence of electric signals for a fusion experiment. For electroporation, only the short, unipolar high electric field pulses are required. The parameters which describe the alternating fields are

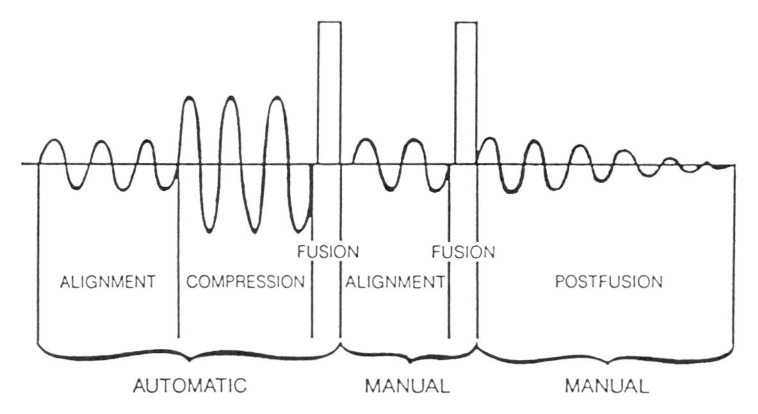

Figure 5. Typical electrical waveforms of a fusion sequence. First applied is the AC alignment field, typically for many seconds. Optional is a short (1 sec) increase in the alignment amplitude (compression) to increase cell proximity. The brief (microseconds) high-amplitude fusion pulse follows the AC signals within milliseconds. It is desirable to be able to set the sequence in an automatic mode.

frequency, amplitude, and waveshape. Cells can be influenced by an AC field from the kilohertz to the megahertz frequency range. The viability of the cells after exposure to the AC field must remain high for successful cell manipulation. A frequency around 1 MHz appears to work well for a wide variety of cells (Fig. 2).

Cells were found to rotate at certain resonance frequencies (Holzapfel *et al.*, 1982; Zimmermann *et al.*, 1981). This is an undesirable effect for fusion because it interferes with the alignment process. This effect can be minimized by selecting an AC frequency different from the resonant frequency. If the generator has only a fixed frequency, this solution is not available. In this case it is advantageous to use a nonsinusoidal waveshape. Sharp resonance of rotation is more probable with sinusoidal waveshapes than with other waveshapes exhibiting a high content of harmonics. A basically symmetrical square wave of 1 MHz gives good alignment without measurably changing the viability as compared to a sinusoidal waveshape.

The alignment effect is even stronger with a square wave of the same amplitude as a sine wave, because the cells are exposed to the maximum amplitude for a longer period of time. The AC voltage, however, still needs to be symmetrical. A unipolar square wave was found to lead to strong turbulence in the chamber (Glassy and Hofmann, personal communication).

4.2. Power Requirements

Some definitions of typical electrical parameters encountered in the electrofusion/electroporation literature appear to be appropriate. One can choose two different ways to look at an electrical system consisting of a generator and a chamber filled with a cell solution: a macroscopic or a microscopic view.

Macroscopic view: The generator provides a potential difference, V, between the electrodes of a chamber, which is measured in volts. The chamber exhibits a certain electrical resistance, R, which is measured in ohms. The current, I, flowing through the chamber is measured in amperes and can be calculated as $I = V/R$, which is the macroscopic Ohm's law. The electrical power, P, into the chamber is $P = IV$ and is measured in watts. If the power is delivered to the chamber for t seconds, then the energy, Q, delivered into the chamber is $Q = IVt$ and is measured in joules. The cells in the solution do not experience any of these macroscopic parameters directly. If we want to understand what is happening to the cells, we need to take a microscopic view.

Microscopic view: The macroscopic potential difference between the electrodes of the chamber creates a local electric field in the solution which is measured in volts per centimeter. (For a simple parallel plate geometry, it can be calculated by dividing the voltage by the electrode gap.) The resistive properties of the solution can be described by the specific resistivity, r (ohm-cm), which is the inverse of the specific conductivity (mho/cm). Note that the specific resistivity is a medium property and does not depend on the geometry of the chamber. The local electric field, E, generates a local current density, i (A/cm^2), which is determined by the microscopic Ohm's law: $i = E/r$. Now all electrical parameters are defined as local properties to which the cells are subjected.

We can now calculate the local power density and total power requirements for the AC and DC signal from the required field strength E (V/cm) in the chamber, the medium specific resistivity r (ohm-cm), and the chamber volume W(cm^3):

$$\text{Power density (in watts/cm}^3\text{): } p = E^2/r$$

$$\text{Total power (in watts): } P = E^2 \cdot W/r$$

which is equal to VI from the generator. A typical range for the required power is shown in Table 2. For larger volumes the power requirements are quite substantial, especially for the high electric field pulse required for fusion or poration. It becomes a nontrivial engineering task to economically provide high-frequency AC power and DC pulses.

The DC pulse needs to be of much higher amplitude than the AC signal because the necessary

Table 2. Power Requirements for AC and DC Signals in Typical Small and Large Chambers for a High- and a Low-Resistivity Medium

Chamber type and volume	R (ohms)	Field (V/cm)	Specific resistivity (ohm-cm)	Power (W)
Microslide, 20 μl	850	200	10,000[a]	0.08 (AC)
	8.5	200	100[b]	8.0 (AC)
	850	8000	10,000	3.2 (DC)
	8.5	8000	100	320 (DC)
Production, 50 ml	11	200	10,000	200 (AC)
	0.1	200	100	20,000 (AC)
	11	8000	10,000	8,000 (DC)
	0.1	8000	100	800,000 (DC)

[a]0.3 M mannitol.
[b]Electrolytes such as PBS.

field strength for electroporation is typically between 0.2 kV/cm for plant protoplasts and 10 kV/cm for bacteria. The momentary power during the pulse can be as high as 800,000 watts or more. However, the average power can be much lower (80 watts) because the electroporation pulse (typically 100 μsec in length) may be applied only once per second. In practice a storage device for electrical energy (a capacitor) is charged slowly with low power and then discharged very rapidly, delivering the required high-power pulse.

4.3. Circuit Configurations

The circuit configurations which are generally being used to generate pulses of high power can be grouped into two different types: (1) circuits which require ON–OFF switches (closing and opening a circuit when current is still flowing), resulting in a partial discharge of the energy storage device (mostly a capacitor), and (2) circuits which require only ON switches (i.e., discharging the energy storage device completely: both the current and the voltage go to zero). This distinction is important because it is rather difficult to switch off large currents against the high voltages which might still remain in the capacitor and this difficulty expresses itself in higher costs.

An example of a circuit which requires ON–OFF switches is the partial discharge of a capacitor shown in Fig. 6. The resulting waveshape resembles a square wave. Typical switching elements are vacuum tubes, high-power transistors, or force-commutated (injection of countercurrent to create an artificial zero current) thyristors, also called silicon-controlled rectifiers.

The following circuits require only ON switches: the complete discharge of a capacitor, resulting in an exponentially decaying waveshape (Fig. 7) with a time constant (the time of the maximum voltage to drop to $1/e = 1/2.7$) equal to the product of RC where C is the size of the storage capacitors and R is the sum of all resistances in the circuit (mainly internal generator resistance plus chamber resistance). Another circuit consists of the discharge of a pulse-forming network (an array of capacitors and inductors, or a long coaxial cable, Fig. 8), resulting in a more or less square pulse with some ripple. Useful switches for these circuits are triggered vacuum gaps, triggered spark gaps, thyratrons, thyristors, or even mechanical switches (relays) for some special applications. Fruengel (1965) will be helpful for further study.

The AC circuit needs to be electrically separated from the pulse circuit so there is no uncontrolled addition of the amplitudes nor any destructive interaction between the circuit components. A convenient method is the use of a fast, mechanical, high-voltage relay. This allows switch-over times between typically 30 msec for high-power versions and as short as 1 msec for lower-power requirements. After being aligned and in contact, it appears that cells can tolerate such a pause before the DC pulse is applied. If shorter pauses are desired, electronic switching needs to be

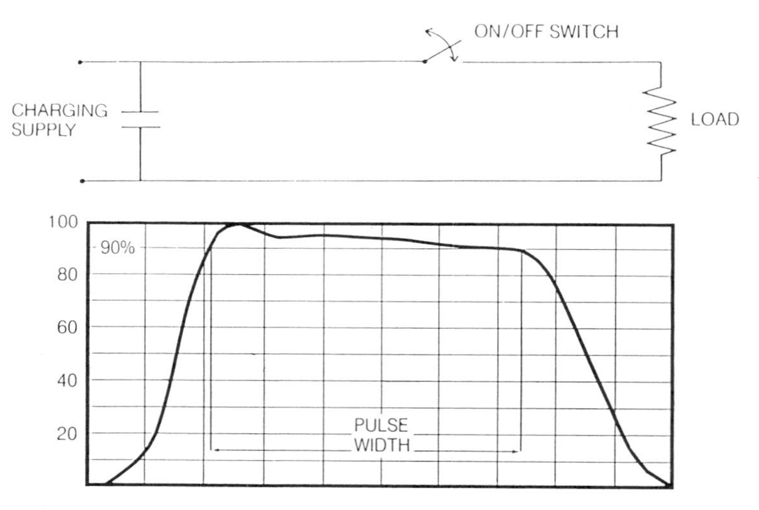

Figure 6. Waveform of a partial capacitor discharge. A convenient way of defining width is the time between the 90% points of maximum amplitude.

employed, but this is not an inexpensive task at high power levels. AC can also be reapplied after the fusion pulse because the membrane stays in a fusogenic state for quite a long time, on the order of many seconds at lower temperatures (Sowers, 1985, 1986).

4.4. Load Matching

The chambers, in which the cell manipulation takes place, are filled with a liquid medium in which the cells are suspended. This medium can exhibit a wide range of specific ohmic resistivities

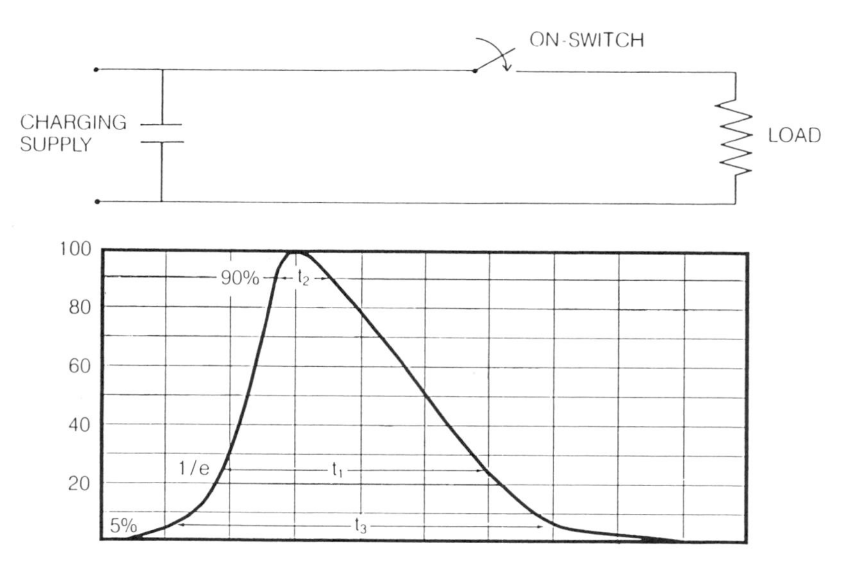

Figure 7. Waveform of a complete capacitor discharge with various definitions of pulse lengths. Most convenient is the $1/e$ time constant which is the product of the storage capacity C and the circuit resistance R (sum of generator and chamber resistance).

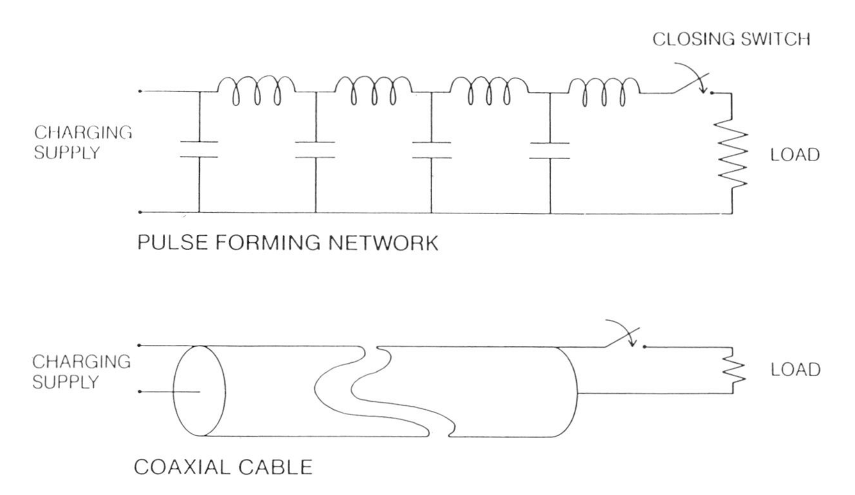

Figure 8. Generation of square pulses with a pulse-forming network (top) or a coaxial cable (bottom). An approximate square pulse will only result if the impedance of the generator (the charged-up pulse-forming network or the coaxial cable) is the same as the impedance of the load (the chamber). In practice, this is unachievable most of the time because of the widely varying medium conductivity. In order to get pulses of microsecond length, the coaxial cable needs to be thousands of feet long.

(ohm-cm), depending on the type of medium and type and amount of additives. Values can range from 10,000 ohm-cm for 0.3 M mannitol to 100 ohm-cm for phosphate-buffered saline solution (PBS) which results in a widely varying chamber resistance (ohm). In either case, medium-filled chambers exhibit a resistive behavior up to frequencies of about 10 MHz. At higher frequencies, capacitive effects start to play a role. Here, we will be concerned with frequencies below 10 MHz and therefore will treat the medium-filled chambers as ohmic resistors. It is important to understand how power is transferred from a generator to a resistive load. If the "match" between generator and load is poor, then the actual voltage on the chamber is substantially lower than what is normally assumed to be the generator output voltage.

Every generator has a certain internal resistance, R_i (measured in ohms), which is defined as the open circuit voltage V_O (no load connected) divided by the short circuit current (the maximum current the generator is capable of delivering). The voltage V on the load is expressed as:

$$V = V_O \cdot R_L/(R_L + R_i)$$

Figure 9 shows the relative voltage on the load for different ratios between load and generator resistance. If the resistance of the chamber, R_L, is much larger than the internal resistance, R_i, of the power supply, then the full open circuit voltage, V_O, appears across the load. This situation is typical for small-volume chambers such as microslides. If both resistances are equal, then half of the open circuit power supply voltage appears across the load. If the load resistance is much smaller than the internal resistance, then only a fraction of the open circuit power supply voltage appears across the load:

$$V = V_O \cdot R_L/R_i$$

This situation represents a poor match between power supply and chamber and is often encountered when working with a highly conductive medium such as PBS in larger volumes. The following is a good example for the importance of understanding the interaction between generator and chamber. Many early experiments in electroporation utilized continuous DC electrophoresis power supplies. Such supplies can be used as a pulse source because most have internal capacitors of capacitance C

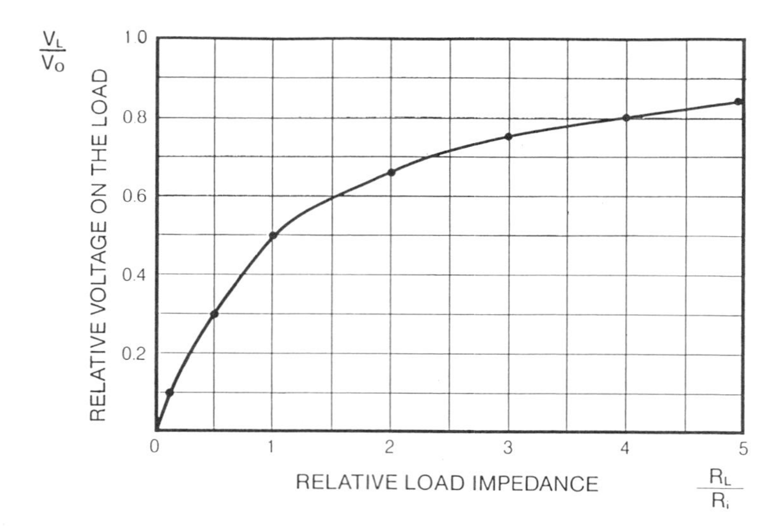

Figure 9. Variation of the load voltage (V_L) with the ratio between loan (chamber) resistance (R_L) and the internal generator resistance (R_i). V_O is the open circuit (no load) voltage of the generator. This is the maximum generator output voltage.

across the output in order to generate a smooth DC output voltage. If a closing switch is put in series with the output, these capacitors will discharge and provide a pulse depending on R_i, R_L, and C. However, this is a very marginal setup as a pulse generator because the power supply most likely exhibits a high internal impedance. One specific power supply had an internal resistance of about 230 ohms. As a result, the voltage across production-size chambers of low resistance was much lower than the voltage set on the power supply dial. Figure 10 shows the actual voltage on a load for a full power supply setting of 2000 V. If the chamber used with the power supply is a cuvette with a 4-mm gap and the cuvette is filled with PBS or a similar medium, then a pulse of only about 90 V

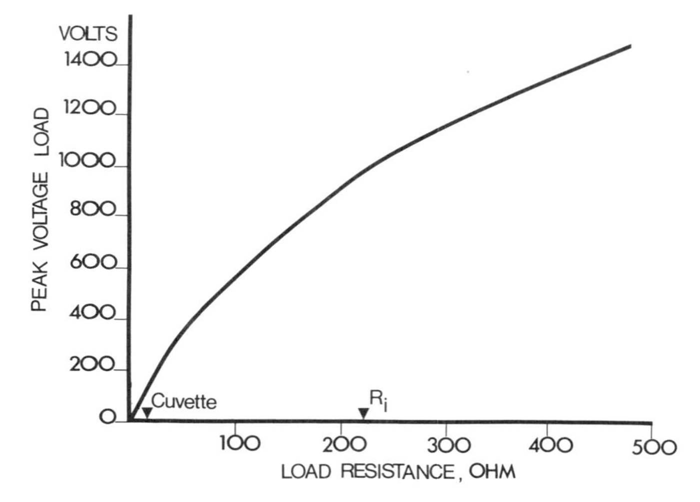

Figure 10. Output voltage of a typical electrophoresis power supply for different load resistances. If the power supply is used with a cuvette filled with PBS, then only a very small fraction of the open circuit voltage appears across the cuvette. This is a result of the high internal impedance (230 ohms) of the power supply.

will appear across the electrodes. The resulting field strength is only about 0.24 kV/cm. The pulse decay time constant is about 21 msec. If the cuvette is filled with mannitol, 1700 V will be across the cuvette and the field strength is 4.26 kV/cm. The pulse decay time constant is about 60 msec. In either case there is a follow-on DC output after the pulse of typically one-third the pulse amplitude. This will lead to a substantial, uncontrolled heating of the medium (25°C, if left on for 1 sec). Potter *et al.* (1984) have reported success with such a power supply. However, it is not designed specifically for electroporation purposes, and does not work well with PBS or other high-conductivity media. The values for field strength cited in the literature are by more than one order of magnitude too high, making it very difficult to exactly reproduce experiments with other systems. The actual field strength in the chamber is dependent on the medium conductivity and reproducible results are difficult to achieve. There is no choice of pulse length unless some easily controlled variable is changed. In practice, this is not available.

4.5. Practical Design and Safety Considerations

The output voltages of low-internal-resistance electro cell manipulation systems are of dangerous amplitude. Improper handling and carelessness with high-voltage systems can lead to serious injuries or even fatalities. We caution biological researchers attempting to assemble their own power supply. Besides the electrical danger, it will be difficult to duplicate exact conditions as cited in reliable literature unless the biological researcher is prepared to establish a quite sophisticated and expensive electronics laboratory with oscilloscope, wide-band high-voltage probes, and current transformers or shunts. Systems for small chamber volumes such as microslides (20 μl typically) can be assembled from high-impedance (50 ohm), low-voltage commercial pulse and AC signal generators. For most biologists without training in physics and electronics, it is impractical to custom design and build the necessary equipment for large-volume applications.

5. CHAMBERS

5.1. General Requirements

Chambers for fusion and transfection need to create the required field strength and geometry, contain the desired volume, and be easily sterilizable, filled, emptied, and cleaned.

The electric field between electrodes is caused by a potential difference (voltage) and is expressed in volts or kilovolts per centimeter. If the field is homogeneous, as it is between infinite parallel plane electrodes, the field strength can be calculated by dividing the voltage by the electrode gap (cm). The electric field between wire electrodes is higher at the electrode surface than farther away. This increase in field strength over a homogeneous field is called the field enhancement factor and is a function of the specific electrode geometry. It can have values of 5 or more, if the electrodes have sharp edges (Alston, 1968).

Dielectrophoresis (alignment, pearl-chaining) is a consequence of cells being exposed to an inhomogeneous field. The intensity of this effect depends on the electric field strength and gradient and not directly on the voltage between the electrodes. A certain electric field can be achieved either by a high voltage across a large gap or a lower voltage across a smaller gap. For safety reasons, it is preferable to operate with lower voltages and small gaps.

The ohmic resistance of a chamber is a function of the geometry and the specific resistivity, r (ohm-cm), of the solution in the chamber. For the simple case of a parallel plate electrode chamber with a gap between the electrodes of d centimeters and a cross section (electrode area) of q square centimeters, the resistance is:

$$R = r \cdot d/q \quad \text{(ohms)}$$

The resistance can also be expressed as:

$$R = r \cdot d^2/W$$

where W is the chamber volume in cubic centimeters. The specific resistivity of chamber solutions ranges from 10,000 ohm-cm (0.3 M mannitol) to 100 ohm-cm (salt solutions such as PBS), gaps typically vary from 0.05 cm to 0.4 cm, and the volume from 10 μl to more than 50 ml. The chamber resistance varies accordingly over a very wide range typically from 0.1 ohm or less to 1000 ohms or more. This wide range poses a considerable design challenge for power supplies which are generally efficient only over a narrow range of load resistance (see Section 4.4).

The resistance of the load as it appears to the generator, can be changed by inserting a transformer between the generator and the load. The apparent resistance of the load changes with the square of the turn ratio of the transformer secondary to primary windings. In this way a better match between generator and load can be achieved. However, care needs to be taken in the transformer design so that the high-frequency alignment signal can pass through.

The unipolar (the voltage does not change polarity) fusion/electroporation pulse presents another transformer design requirement, which can be characterized by the product of peak pulse voltage times the pulse duration. If the actual pulse exceeds the design value, the magnetic material of the transformer saturates before the full pulse has passed and cuts the pulse off. This transformer behavior, on the other hand, can be used to advantage to intentionally shorten long exponential pulses from a complete capacitor discharge (Fig. 7). It makes it look like a short square pulse by using a pulse transformer with a volt-second design value substantially below the exponential pulse volt-second value. The magnetic material of the transformer acts in this case like a magnetic off switch.

5.2. Uniform Fields

Chambers with uniform electric fields subject all cells to the same electric field. This is advantageous for electroporation where no pearl chaining is required. High transformation yields are possible provided that the field strength is optimal. The optimum value is not always known, and transformation yields can drop substantially above or below the optimum field strength (Fig. 11).

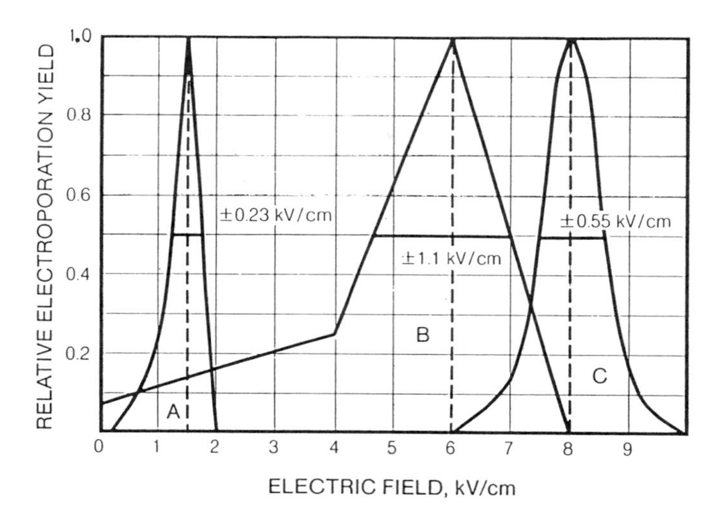

Figure 11. Relative electroporation yield as a function of applied electric field. (A) *Nicotiana tabacum* (Shillito *et al.*, 1985). (B) TK mammalian cells (Zerbib *et al.*, 1985). (C) Mouse L cells (Neumann *et al.*, 1982).

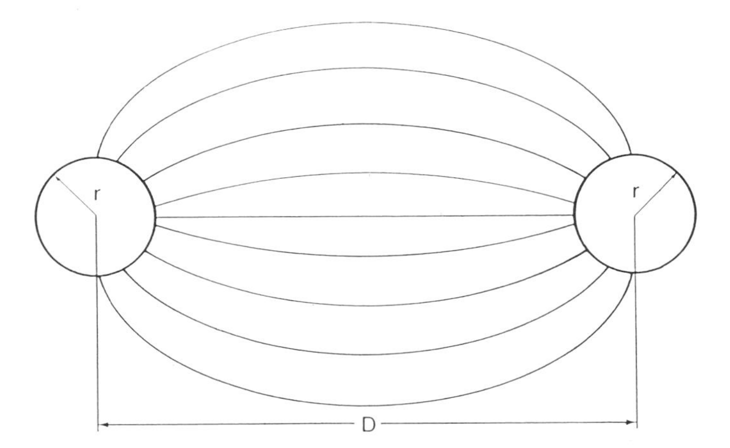

Figure 12. Field between parallel wires, a typical configuration for microslide chambers.

Practical realizations of uniform field chambers are chambers with flat parallel electrodes or to some degree coaxial chambers with a diameter many times larger than the electrode gap (see Fig. 13B). Alignment and fusion are still possible in uniform AC fields provided the cell density is sufficiently high for mutual dielectrophoresis (attraction) to take place. Without an AC field, fusion can be achieved using the avidin/biotin method (Lo *et al.*, 1984), by using high cell density and allowing the cells to settle into a multilayer on the chamber bottom, or by using PEG or other chemicals for cell agglutination.

5.3. Chambers with Nonuniform Fields

Chambers with nonuniform fields allow alignment at low cell densities with an AC field, and reduce the need to have exactly the right field strength in order to get high transfection yields. Generally, any electrode configuration besides an infinite parallel plate configuration provides a nonuniform field. It is desirable to use configurations which are practical to manufacture and have well-defined field shapes.

For small volumes (10 to 100 μl), two simple well-proven geometries are available:

1. Microslides with thin metal sheets separated by a small gap. The field inhomogeneity stems from the effect of adjacent electrode edges and depends on the fine structure of the edges which cannot be easily controlled.
2. A better geometry for small volumes is one which utilizes parallel wires (Fig. 12). The maximum electric field is well defined for this case: $E = V/2r \ln(D/r)$ if $D >> 2r$, where D is the distance between wire centers and r is the wire diameter.

For medium-size volumes of a few milliliters, coaxial configurations can be easily manufactured and give well-defined field configurations. They can be built with high or low divergence or field inhomogeneity (Fig. 13). For a highly divergent field the center electrode can be a wire or thin tubing; for a low-divergence field both electrodes can be tubes with a diameter much larger than the gap between them. The field strength between coaxial electrodes is: $E = V/r \ln(R_2/R_1)$. The chamber volume varies linearly with the chamber length.

Large volumes of 10 ml and more with inhomogeneous fields require a different design approach in order to provide multiple field enhancement points throughout the volume. One chamber design makes use of the field enhancement properties of layers of wire mesh, kept apart by insulating spacers. This "sandwich" construction method can be extrapolated to very large vol-

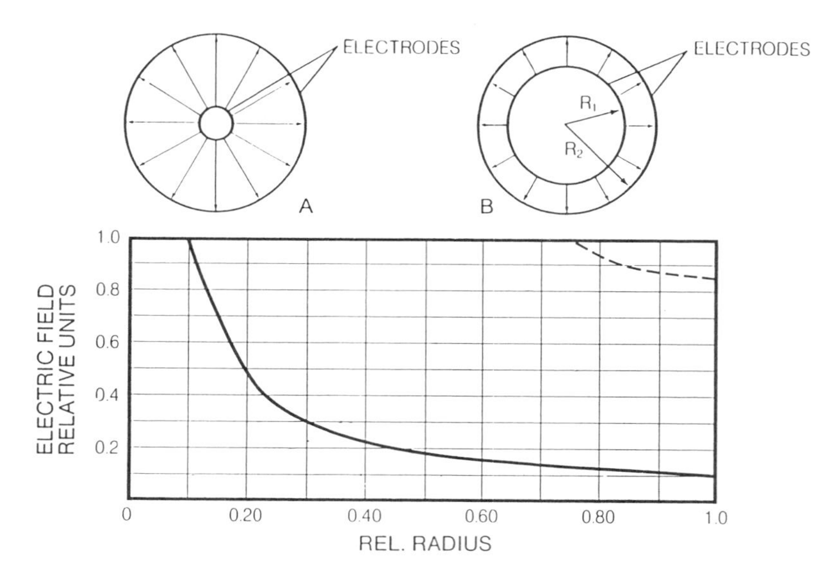

Figure 13. Cross section of coaxial chambers with high-divergence (A) and low-divergence (B) field.

umes. Figure 14 shows the sandwich structure of a chamber with a maximum volume of 75 ml which can be autoclaved and used as a flowthrough chamber. The power supply has to be matched to a chamber of such a large size so that the desired fields for cell manipulation can be achieved.

5.4. Chamber Materials

Preferred isolating materials with good mechanical, electrical, and thermal (for possible autoclaving) properties are polysulfone and nylon. The electrodes need not be made out of noble

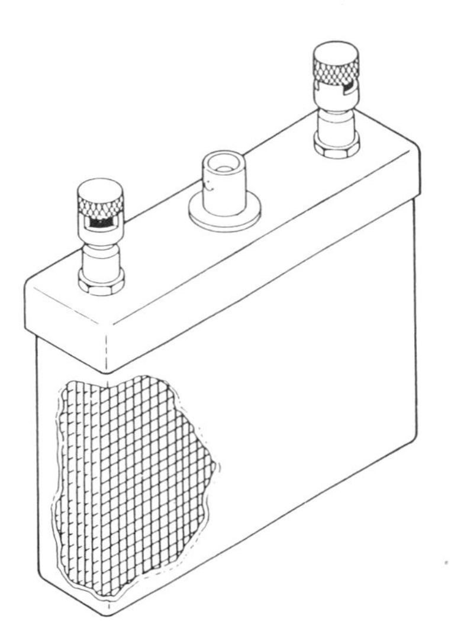

Figure 14. Large-volume (75 ml) chamber with inhomogenous fields. The electrodes consist of alternating layers of wire mesh held apart by insulating spacers. This design lends itself to extrapolation to very large volumes.

materials such as gold or platinum. We found no detrimental effect on cell viability when stainless steel 316 or 304 was used. Aluminum foil is not recommended because the oxide layer which forms rapidly on an aluminum surface in air is an electrical insulator. Applying a voltage between oxidized aluminum electrodes in a cell suspension liquid might lead to local breakdown of this oxide layer in the form of microscopic arcs which pit the electrode surfaces and eject electrode material into the liquid. This effect is known as the "Malter effect" (Malter, 1936a,b) and can be looked upon as "electroporating the electrodes." Presently available on the market are disposable (sterile) cuvette chambers with aluminum foil electrodes and reusable noble metal or stainless steel electrodes using low-cost disposable plastic cuvettes as the container. Sterilization of the electrodes with alcohol and/or brief flaming was found to be sufficient to prevent cross-contamination (Evans, personal communication).

5.5. Chamber Medium

The medium, in which the cells are suspended, is subjected to the electric field generated by the voltage across the chamber electrodes. The medium will carry an electric current according to its specific resistivity, following Ohm's law. For frequencies of up to several megahertz, the chamber/medium system behaves like a purely ohmic resistor; for higher frequencies, capacitive effects begin to play a role and voltage and current are out of phase with each other. Because the optimum frequencies for alignment with high cell viability in electro cell fusion were found to be around 1 MHz, we can treat the chamber/medium system as a purely ohmic resistance. The specific resistivity of the medium is determined by the concentration of charge carriers (ions). Sugar (e.g., mannitol or sucrose) solutions have a low charge carrier concentration and exhibit a high specific resistivity of typically 10,000 ohm-cm. They are also relatively compatible with cells at least for a limited amount of time if the sugar solution is isosmotic or hyperosmotic with respect to the cytosol of the cell. They are, therefore, a preferred medium for electrofusion and electroporation in cases where heating effects need to be minimized. Solutions with high ionic concentration such as PBS exhibit a low specific resistivity of typically 60 to 150 ohm-cm. For most electroporation experiments, PBS appears to be the preferred medium; however, the electrical power which is required to generate an electric field is about 100 times higher in PBS than in a typical sugar solution. An electrolyte such as PBS exhibits nonlinear electrical characteristics. Its specific electrical resistivity decreases with temperature and with increasing applied field. Fields of up to 18 kV/cm can be generated in an electrolyte (Bishop and Edmonds, 1985). Local inhomogeneities such as air or gas bubbles in a liquid cause a local field enhancement which can lead to arcing, where the current flow restricts to a thin filament, and a possible catastrophic failure or destruction of the chamber.

5.6. Chamber Medium Temperature Rise

The temperature rise of the medium as a consequence of the electric field application needs to be considered as one important system parameter. The temperature influences the critical field strength for membrane poration: it is lower at higher temperatures (Zimmermann, 1982). It should be pointed out that a heat shock might be beneficial to the electroporation process as was demonstrated with a separately applied heat shock (Shillito *et al.*, 1985). We can speculate that a yield improvement might occur if the heat shock is applied directly through the electric field pulse. For this reason and to calculate excessive temperature rise of the medium, we derive general formulas for the temperature rise. Earlier we derived the local power input into the medium as $p = Ei$ or $p = E^2/r$ (where p is in watts/cm^3, E is in V/cm, i is in A/cm^2, and r is in ohm-cm). Integrating the power over time gives the energy e deposited in a volume element (energy density). For a square wave this means multiplying the power by the pulse length t: $e = pt$ (watt-sec/cm^3 or joule/cm^3). This energy is being transformed into thermal energy leading to a rise in temperature T (K). For a square pulse with pulse length t (sec), $T = E^2\, t/4.2\, r$, r being the specific resistivity (ohm-cm). The

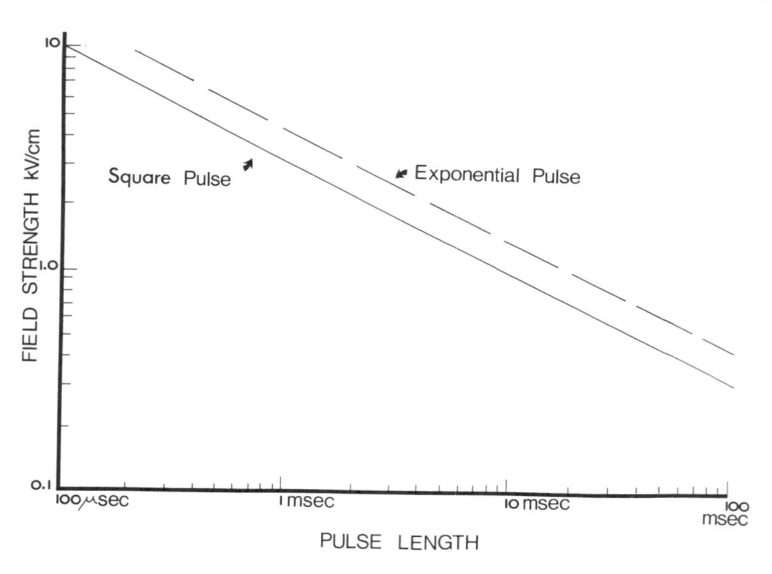

Figure 15. 25 K temperature rise in PBS as a function of field strength and pulse length for square and exponential pulses. For a given field strength the graph shows the pulse length which will result in a 25 K temperature rise of the medium; for a given pulse length, it allows determination of the field strength resulting in such a temperature rise. Pulse length for the exponential pulse is defined as the $1/e$ time constant.

factor 4.2 expresses the fact that 4.2 joules = 1 cal. We assumed that the density (1 g/cm^3) and the specific heat (1 cal/g per K = 4.2 J/g per K) of the medium are close to those of water.

A similar calculation can be performed for exponentially decaying pulses by integrating the momentary power input over the full duration of the pulse: $T = E^2\, t/8.4\, r$, where $t = RC$, the $1/e$ decay time constant.

Figure 15 shows a graph of a 25 K temperature rise of the medium as a function of field strength and pulse length for square and exponentially decaying pulses.

6. SYSTEM MONITORING

6.1. Electro Cell Fusion

The most convenient approach is the direct observation of the alignment, fusion, and rounding-off process under a microscope in a microslide chamber of typically 10–20 μl volume. Once good electrical parameters are established by direct observation, similar conditions in larger or different chambers can be achieved by adjusting the generator output voltages so that identical electric fields are realized (see Section 6.3). Optimum yield of viable cell hybrids can be achieved by optimizing the fusion pulse amplitude (Fig. 16).

6.2. Electroporation

The electroporation process cannot be observed under a microscope. However, good electroporation yields are usually achieved with electric field/pulse length conditions which result in lysis of some of the cells (5–80%). This can be utilized to establish good electrical operating parameters by constructing a lysis/viability curve: percentage of cell survival versus field strength. Lysis/viability can be observed either visually, by trypan blue exclusion, methylene blue, or plating and observation of growth. Scale up to different or larger chambers can be done by maintaining the electric field amplitude and pulse length.

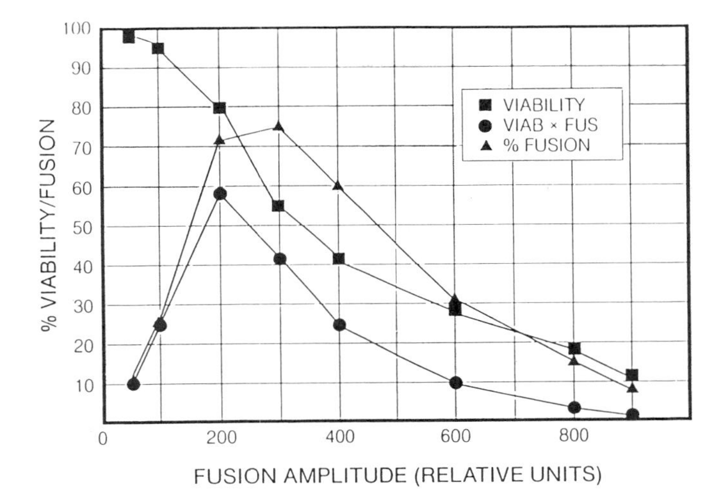

Figure 16. Viability and fusion yield of tobacco protoplasts as a function of applied electric field (Saunders, personal communication).

6.3. Determining the Actual Electrical Parameters in a Chamber

6.3.1. Calculations

The following steps will determine the chamber electrical field strength:

1. Measure the specific resistivity of the medium with a laboratory conductivity meter.
2. Calculate chamber resistance from known chamber geometry and specific resistivity.
3. Calculate the actual voltage on the chamber from the generator output voltage setting, internal resistance of the generator, and chamber resistance (see Section 4.4).
4. Calculate the electric field in the chamber from the actual voltage on the chamber and the geometry.

This process is quite tedious and it still does not yield the shape of the pulse.

6.3.2. Oscilloscope

An electronic oscilloscope and a high-voltage probe can be used to measure the actual voltage on the chamber, eliminating steps 1 to 3 above. The field can then be calculated from the voltage and the chamber geometry.

The pulse shape can be observed with the oscilloscope if it has a wide enough bandwidth. Such oscilloscopes are expensive and require operational knowledge. A commercial instrument is on the market which combines the functions of a digital storage scope with a computer to measure and calculate peak voltage and chamber field and resistance of the chamber as well as the medium specific resistivity.

7. RESULTS AND DISCUSSION

Electro cell manipulation is rapidly becoming an accepted technique in genetic engineering. The literature is growing steadily with hundreds of papers describing experiments in electrofusion or

Table 3. Comparison of the Techniques of Electromanipulation of Cells

Parameter	Exponentially decaying pulse	Square pulse
Heating of medium	More severe. Beneficial heat shock possible	Minimized
Electric field amplitude	Varies widely during pulse decay	Well defined
Available pulse length	Long	Short
Generator impedance	Lower	Higher
Acquisition costs	Lower	Higher
Efficacy	Can be high	Can be high
Required electric field strength	Lower	Higher
Size of stable pores	Smaller	Larger
Number of stable pores	Larger	Smaller
Possible size of transfectant	Smaller	Larger

electroporation. In some instances the values given for the electric field strength might not represent the actual field strength in the chamber because no electrical measurements were performed. However, it is important to realize that many different cell lines have already been fused or transfected with good yields. Most fusion experiments have been performed in nonionic media. The AC alignment signal, which is applied for a relatively long time (seconds), would lead to substantial heating otherwise. The unipolar fusion pulse is in most cases of short duration (microseconds) and high amplitude (many kV/cm). The electroporation literature falls into two distinctive categories: high fields (many kV/cm)–short pulses (microseconds, often square waveshape) and low fields (few hundred V/cm)–long pulses (milliseconds, exponentially decaying waveshape). Good transfection yields have been reported with both techniques. A systematic side-by-side comparison of the two regimes would be desirable. Some general trade-offs are suggested in Table 3. Biological methods and results are the subject of other chapters in this volume.

ACKNOWLEDGMENTS. I thank Eve Morris for her continuous encouragement and support, and John Fewster for valuable suggestions.

REFERENCES

Alston, L. L., 1968, *High Voltage Technology,* Oxford University Press, London.

Bishop, A. E., and Edmonds, G. D., 1985, Electrolytic resistors in plasma physics research, *J. Nucl. Energy Part C* **7**:423–426.

Fruengel, F., 1965, *High Speed Pulse Technology,* Volume 1, Academic Press, New York.

Hofmann, G. A., and Evans, G. A., 1986, Electronic genetics—Physical and biological aspects of cellular electromanipulation, *IEEE Eng. Med. Biol.* **5**:6–25.

Holzapfel, C., Vienken, J., and Zimmermann, U., 1982, Rotation of cells in an alternating electric field: Theory and experimental proof, *J. Membr. Biol.* **67**:13–26.

Hulsheger, H., Niemann, E. G., and Potel, J., 1981, Killing bacteria with electric pulses of high field strength, *Radiat. Environ. Biophys.* **20**:53–65.

Lo, M. M. S., Tsong, T. Y., Conrad, M. K., Strittmatter, S. M., Hester, L. D., and Snyder, S. H., 1984, Monoclonal antibody production by receptor-mediated electrically induced cell fusion, *Nature* **310**:792–794.

Malter, L., 1936a, Anomalous secondary electron emission: A new phenomenon, *Phys. Rev.* **49**:478.

Malter, L., 1936b, Thin film field emission, *Phys. Rev.* **50**:48–58.

Neumann, E., Schaefer-Ridder, M., Wang, Y., and Hofschneider, P. H. 1982, Gene transfer into mouse lyoma cells by electroporation in high electric fields, *EMBO J.* **1**:841–845.

Pohl, H. A., 1978, *Dielectrophoresis,* Cambridge University Press, London.

Potter, H., Weir, L., and Leder, P., 1984, Enhancer-dependent expression of human kappa immunoglobulin gene introduced into mouse pre-B lymphocytes by electroporation, *Proc. Natl. Acad. Sci. USA* **81:**7161–7165.

Shillito, R. D., Saul, M. W., Paszkowski, J., Muller, M., and Potrykus, I., 1985, High efficiency direct gene transfer to plants, *Biotechnology* **3:**1099–1103.

Sowers, A. E., 1985, Movement of a fluorescent lipid label from a labeled erythrocyte membrane to an unlabeled erythrocyte membrane following electric field-induced fusion, *Biophys. J.* **47:**519–525.

Sowers, A. E., 1986, A long-lived fusogenic state is induced in erythrocyte ghosts by electric pulses, *J. Cell Biol.* **103:**1358–1362.

Stenger, D. A., and Hui, S. W., 1986, Kinetics of ultrastructural changes during electrically induced fusion of human erythrocytes, *J. Membr. Biol.* **93:**43–53.

Zerbib, D., Amalric, F., and Teissie, J., 1985, Electric field mediated transformation: Isolation and characterization of a TK^+ subclone, *Biochem. Biophys. Res. Commun.* **129:**611–618.

Zimmermann, U., 1982, Electric field-mediated fusion and related electric phenomena, *Biochim. Biophys. Acta* **694:**227–277.

Zimmermann, U., 1986, Electrical breakdown, electropermeabilization and electrofusion, *Rev. Physiol. Biochem. Pharmacol.* **105:**176–256.

Zimmermann, U., and Pilwat, G., 1982, Electric field-mediated cell fusion, *J. Biol. Phys.* **10:**43–50.

Zimmermann, U., Vienken, J., and Pilwat, G., 1981, Rotation of cells in an alternating electric field: The occurrence of a resonance frequency, *Z. Naturforsch.* **36:**173–177.

Chapter 27

External Electric Field-Induced Transmembrane Potentials in Biological Systems

Features, Effects, and Optical Monitoring

Daniel L. Farkas

1. INTRODUCTION

1.1. Electrical Phenomena in Biological Systems

In living systems, a high level of organizational heterogeneity forms the structural basis for a wide variety of functions (energy transduction, ion and metabolite transport, excitability, development) and their regulation. At a fundamental level, practically all events and interactions involved are ultimately electrical in nature, due to the ubiquity of charges and the suitability for processing of electrical signals, as well as to evolutionary reasons. Most phenomena of interest involve biological membranes, shown to be highly specialized molecular assemblies and not mere proteolipid barriers between the inside and outside media. As a consequence, the electrical properties of membranes have been extensively investigated.

It has been found that the membrane electrical capacitance changes very little both throughout the wide biological spectrum and between various functional states of the cells (Cole, 1968). On the other hand, the membrane electrical conductivity depends on environmental parameters and is sensitive to (and indeed involved in) functional events. Ion-specific channels have a key role in these changes, and their study has received a new impetus in recent years by the advent of the patch-clamp technique. Another electrical characteristic of interest is the potential difference between the inner and outer domains that most cells (and subcellular organelles) maintain, at the expense of metabolic energy. The relevance of this membrane potential to cellular events has long been recognized, but the quantitative investigation of its extent, spatial distribution, and time course has been hindered by the relative scarcity of experimental approaches available, as well as the intrinsic limitations of the methods employed. Thus, the use of microelectrodes has severe physical constraints, while chemical methods are notoriously slow.

Daniel L. Farkas • Department of Physiology, University of Connecticut Health Center, Farmington, Connecticut 06032.

1.2. Optical Probes of Membrane Potential

The use of optical probes as indicators of membrane potential (Cohen and Salzberg, 1978; Loew, 1982) relies on the methodological advantages of these techniques (noninvasiveness, sensitivity, fast response) and has already yielded an impressive array of information during the past few years (DeWeer and Salzberg, 1986), with most of the studies concentrating on excitable cells. Some of the shortcomings of optical probes (such as pharmacological effects, photodynamic damage, bleaching, nonspecific responses, and low signal-to-noise) can be alleviated by proper choice of dye, recording equipment, and experimental conditions. Other problems, such as interpretational difficulties from measurements on cell suspensions, can be tackled by combining the dye technology with either quantitative light microscopy or miniaturized detectors of subcellular size. What remains to be contended with is the disconcerting dissimilarity of the quantitative aspects of dye response in different systems of interest. Calibration of probe response in various settings depends on the availability of a fast, reliable, and nondestructive method for inducing membrane potentials. Externally applied electric fields constitute one of the best answers to this need.

1.3. Aim and Scope of This Review

It would be unrealistic to attempt a complete review of membrane electrical phenomena and their optical monitoring. Of the various possible sources for membrane potentials–electrogenesis, diffusion, nonuniform charge distribution, and electrical polarization–only the last will be discussed due to its origin in externally applied electric fields and its direct relevance to electroporation and electrofusion. A succinct theory of external field-induced transmembrane potential differences and enumeration of known effects is followed by an analysis of three applications, drawing understandably on our own experience and dealing with an intrinsic probe in photosynthetic vesicles, an extrinsic probe in a model system, and the same probe in cell membranes, respectively. The additional bias in choosing these examples comes from the following considerations: A mechanistic explanation for electric field effects on membranes seems thoroughly needed. Quantitative kinetic information is of great relevance in such cases, and we believe that the systems described do afford the necessary sensitivity and time response.

2. EXTERNAL ELECTRIC FIELD-INDUCED TRANSMEMBRANE POTENTIAL DIFFERENCES IN VESICLES

2.1. Field Enhancement in the Membrane—Qualitative Description

Externally applied electric fields of a few hundred volts per centimeter are capable of inducing significant effects in cell suspensions, unlike in chemical electric field-jump experiments. The reason is a remarkable local enhancement of the applied fields at the membrane level. A qualitative explanation follows (see Fig. 1): If an electrolyte of conductivity λ_o is placed between two electrodes to which an electric field of, e.g., $E = 1000$ V/cm is applied, the potential drop within the solution, on a distance of 2 μm, is 200 mV. If a dielectric sphere (Fig. 1, left) of radius $R = 1$ μm and conductivity $\lambda_m << \lambda_o$ is placed in the solution, the corresponding potential drop on the same 2 μm distance is, by the laws of electrostatics, $\Delta\psi = 3/2\ (2RE) = 300$ mV. If, finally, a thin dielectric spherical shell of radius $R = 1$ μm and thickness $d << R$ is placed in the field, the corresponding potential difference is again $\Delta\psi = 300$ mV (as $\lambda_m << \lambda_o$), with the whole drop occurring within the nonconductive shell (Fig. 1, right) in two equal steps of 150 mV; clearly, by this mechanism the R/d ratio is now scaling the field E_m within the membrane. Since for cells R/d is typically on the order of 10^3, the fields resulting within the membrane are more than three orders of magnitude higher than the applied field intensity. Therefore, the main prerequisite for the local field enhancement is the existence of a topologically closed, low-conductivity thin boundary for the cell—its membrane. These considerations can be translated in more quantitative terms.

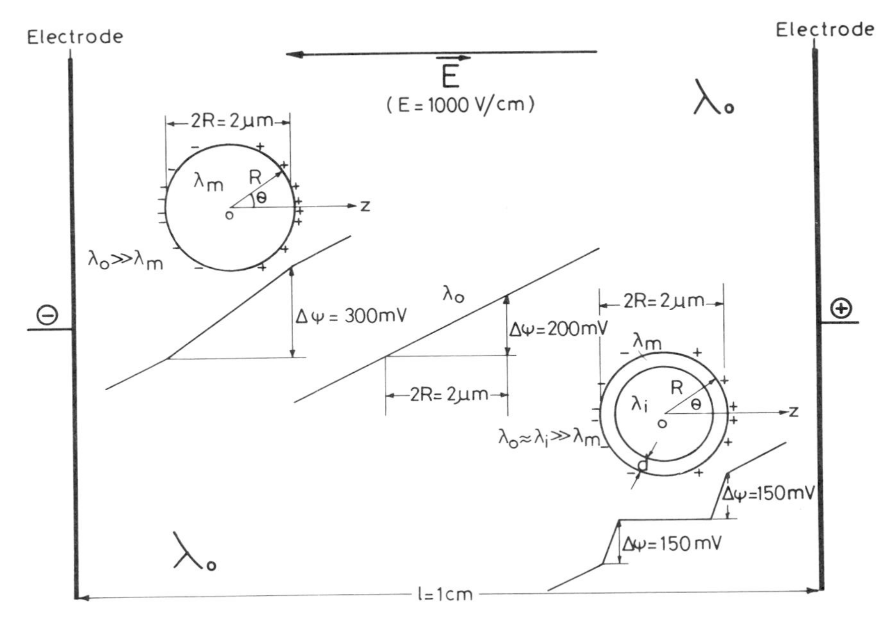

Figure 1. A qualitative model for electric field enhancement in a spherical membrane. Adapted from Witt (1979). Symbols and further details are given in the text.

2.2. Spatial and Temporal Variation of Field-Induced Membrane Potentials—Theory

The procedure for calculating the electrical potential difference across a spherical dielectric shell of thickness d and radius R induced by an externally applied field E was given by Maxwell (1873), and can be outlined as follows: The electrostatic potential U obeys the Laplace equation

$$\Delta U = 0 \tag{1}$$

With the appropriate choice of a spherical coordinate system, the solution can be written as a sum of Legendre polynomials for the three regions of the membrane (m), inner medium (i), and outer medium (o). Solving Eq. (1) requires the boundary conditions for the potential at the origin, infinity, and the interfaces separating the three media (i, m, and o), coupled with the conditions equating the rate of surface charge density change with the radial currents at the interfaces–expressed with the aid of the specific ion conductivities λ_j (J = i, m, o) in the three regions—for the time-dependent solution.

For most cases of interest, $R >> d$ and $\lambda_{\mathrm{i}} = \lambda_{\mathrm{o}} >> \lambda_{\mathrm{m}}$. With these assumptions, the electric field ($E_{\mathrm{m}} = \Delta\psi/d$) induced within the membrane is given by

$$E_{\mathrm{m}} = \frac{3}{2}\ \frac{R}{d} E \cos \theta \left[1-e^{-\frac{2\lambda_0 t}{3RC_{\mathrm{m}}}}\right] \tag{2}$$

where C_{m} is the membrane electrical capacitance and θ is the azimuthal angle between the applied field direction and the normal to the membrane at the locus of interest. Equation (2) underscores several important points:

a. The membrane field is proportional to the applied field E and enhanced by a factor of $3R/2d$, which can be 10^3 or more for cells.

b. The field has a strong angular dependence, being maximal (and opposite) at the "poles" ($\theta = 0°$, $180°$) facing the electrodes and zero at the "equator" ($\theta = 90°$).
c. Since the membrane capacitance C_m does not change much, the kinetics of the field increase within the membrane mainly depend on the medium conductivity λ_o.

For the more general conditions (Farkas *et al.*, 1984a,b; Ehrenberg *et al.*, 1987) in which the membrane conductivity λ_m is not negligible (e.g., in the presence of functional, induced, or extraneously added ion channels), as well as $\lambda_i \neq \lambda_o$ and d not negligible compared to R, the expression for E_m becomes

$$E_m = \frac{9\lambda_0\lambda_i[1-e^{-\frac{t}{RC_m}\left(\frac{2\lambda_0\lambda_i}{2\lambda_0+\lambda_i}+\frac{R}{d}\lambda_m\right)}]\,E\cos\theta}{(2\lambda_0+\lambda_m)(2\lambda_m+\lambda_i)-2(\lambda_0\lambda_m)(\lambda_i-\lambda_m)\left(\frac{R-d}{R}\right)^3} \quad (3)$$

In the more frequently encountered case of $\lambda_0 = \lambda_i$ and $R >> d$, this reduces to

$$E_m = \frac{3/2\,R/d\,E\cos\theta\,[1-e^{-\frac{2\lambda_0 t}{3RC_m}\left(1+\frac{3}{2}\frac{R}{d}\frac{\lambda_m}{\lambda_0}\right)}]}{1+3/2\,R/d\,\lambda_m/\lambda_0} \quad (4)$$

Equation (4) shows, in addition to points (a)–(c) above, that:

d. The membrane conductivity λ_m also controls field amplitudes and kinetics. An increase in λ_m is expected to simultaneously decrease (by the same factor) the intensity and rise-time of the induced field.
e. A large R/d ratio in an experimental system would accentuate the influence of λ_m, making it attractive for the study of membrane conductivity effects.

Clearly, the spherical model is the simplest and most widely used one. But, since cell shapes vary considerably, the results obtained do not always apply. To our knowledge, the only other geometry for which the electrostatic problem has been solved analytically is the ellipsoidal one (Bernhardt and Pauly, 1973). The solution is similar to Eq. (2) above, lacking somewhat the predictive value of, e.g., Eq. (4).

Moreover, the existence of any significant surface potentials should influence the values and time course of membrane fields obtained. Surprisingly, this aspect has only very recently been considered quantitatively (Gross, 1988).

2.3. The Electric Field as a Stimulus—Features and Merits

As a perturbation method in general, and as a source of membrane potentials in particular, externally applied electric fields have a number of methodological merits:

a. Intensity range: it is extremely wide (mV/cm to thousands of V/cm), with excellent reproducibility.
b. Kinetic profile: the duration and time course are precisely variable, within a wide continuum. Onset and offset times can be shorter than 1 μs, orders of magnitude lower than for other stimuli, allowing detection of fast responses.

c. Nonspecificity, transience: at the end of the perturbation, the sample is essentially the same as before, except for field-induced effects, if any.
d. Repetitivity: in view of (b) and (c) above, a sample can undergo multiple electrical stimulation, with field pulse duration, spacing, and number being variable. This is particularly important in cases where a poor signal-to-noise ratio warrants averaging, or where an effector pulse is to be followed by a probing one.
e. Directionality: the applied field has a fixed orientation with respect to a laboratory system of coordinates, resulting in axially rather than radially symmetric membrane potentials of known topology.
f. Sense (polarity): both positive and negative field pulses are obtainable, being useful for symmetry reasons, special applications, and artifact minimization.

Most of these features are of obvious importance in any systematic study of induced membrane potentials and related phenomena. On the other hand, since energetically applied fields constitute a strong perturbation, a number of possible unwanted effects can be associated with them; the most notable is joule heating, with its own potential consequences, both general (shift in chemical equilibria, shock waves) and specific to cell suspensions (temperature gradients across membranes, thermo-osmosis). However, conditions are usually carefully chosen to minimize artifacts, and only "pure" field effects will be discussed here.

2.4. Effects of Externally Applied Electric Fields

External electric fields have been applied to study the electrical parameters of cells in suspension ever since the beginning of the century. As a rule, early investigations used AC fields of variable frequency and were aimed at extracting rather general information, such as the average resistance and capacitance (Cole, 1968).

In recent years, external electric fields were used in a variety of new ways to obtain effects or information of direct biological relevance. One classification—based on field intensities—includes:

a. Low fields, resulting in potentials of a few millivolts across a cell membrane, usually having long application times with the field mainly acting upon regulatory mechanisms within the cell. The most convincing results show clear effects on growth and development, while other aspects require assumptions of strongly nonlinear effects and amplification mechanisms.
b. Intermediate to high fields, with induced potentials of tens to hundreds of millivolts, which, in turn, elicited a number of responses: (1) activation of membrane-bound ATPases, resulting in stimulation of transport (Tsong, 1983), or net ATP synthesis (Witt, 1979; Vinkler *et al.*, 1982, 1983; Kagawa and Hamamoto, 1986); (2) absorption (DeGrooth *et al.*, 1980) and emission (Ellenson and Sauer, 1976; Farkas, 1983) changes (see Section 3); (3) action potentials in excitable cells (Chan and Nicholson, 1986); and (4) electrophoresis of particles within the membrane (Poo, 1981). In these studies, the field constitutes usually the driving force and/or trigger for the events monitored.
c. Very high fields (corresponding to 0.5 V or more membrane potential), leading to electrical breakdown that can be either irreversible or reversible, the latter opening the exciting new field of electroporation and electrofusion to which this book is dedicated.

A number of interesting effects, not included in the classification above but discussed in other chapters, can be induced by homogeneous fields (electrophoresis, orientational movement), nonhomogeneous fields (dielectrophoresis; Pohl, 1978), or special field geometries and sequences (particle rotation; Schwan, this volume). Still other effects (e.g., acceleration of bone healing) are clinical in nature, poorly understood, and will not be discussed here.

3. ELECTROPHOTOLUMINESCENCE

3.1. The Photosynthetic Pigments as Intrinsic Optical Probes

The membranes of thylakoids (in green plants) and chromatophores (in photosynthetic bacteria) constitute the main sites for harnessing the sunlight's energy. This is achieved through a finely tuned sequence of energy-transducing events, starting with an ultrafast, light-induced charge separation followed by charge delocalization and secondary electron and ion transport. The main protagonists are the photosynthetic pigments (chlorophylls, carotenoids), undergoing significant function-related optical changes (in absorption, fluorescence and luminescence) and thus capable of yielding valuable information on photosynthetic functionality. Since most of the events involve charge movement, these "reporter groups" can also be regarded as fast, voltage-sensitive, intrinsic optical probes for membrane electrical phenomena, with a number of advantages (fixed membrane concentration and localization, direct functional involvement) over extrinsic ones.

The linear electrochromic absorbance changes of carotenoids (Witt, 1979) have been linked by kinetic, spectroscopic, and electrical evidence to photosynthetic membrane potentials, but their quantitative interpretation and calibration has been difficult due to the existence of complicating factors (Kakitani *et al.*, 1982). Nevertheless, the subnanosecond time response uniquely qualifies the technique for monitoring the fast initial events in photosynthesis; conversely, very few other investigations have used the carotenoid shift for electrical measurements, except the elegant work of Junge and co-workers dealing with ionophore action in the membrane (Schmid and Junge, 1975) and proton flow through the F_0F_1 ATPase of chloroplasts (Lill *et al.*, 1987). Quadratic carotenoid electrochromism has been demonstrated by two groups (DeGrooth *et al.*, 1980; Schlodder and Witt, 1980) upon application of external electric fields to chloroplast suspensions. The extent and kinetics of the induced changes were in accordance with Eq. (2) above, while their spectral dependence was as predicted by classical electrochromic theory. The main experimental problem consisted in the extensive averaging necessitated by low signal-to-noise ratios.

3.2. The Phenomenon of Electrophotoluminescence

In photosynthetic systems, the light-induced charge separation is an extremely efficient process, as the recombination is prevented by kinetic, energetic, and steric considerations. When the low-probability recombination does occur, one of the possible outcomes is the regeneration of the singlet excited state of chlorophyll, followed by its radiative decay, giving rise to delayed luminescence (DL). This emission is weak, multiphasic, and can be significantly modified by a change in the membrane electrochemical potential because the reactants are membrane-embedded and the rate of recombination contains a Boltzmann factor that includes the potential difference in the exponent.

A particularly efficient method of transiently stimulating DL is the application of an external electric field pulse to a suspension of preilluminated chloroplasts. The resulting enhanced emission is termed *electrophotoluminescence* (EPL) (Ellenson and Sauer, 1976) and was first described by Arnold and Azzi (1971). Its salient features and range of parameters are illustrated in Fig. 2; the most striking is its high intensity (EPL/DL $\sim 10^3$) and related excellent signal-to-noise ratio. Another interesting aspect of mechanistic relevance is the so-called frequency doubling (not shown): an applied AC field elicits EPL that has emission peaks for both positive and negative half-waves of the applied field.

The emission intensity and kinetics depend on photosynthetic, medium, and membrane electrical parameters, and can be used to prove the latter if the former two are controlled. Complicating factors, such as the existence of two kinetic emission components (R and S in Fig. 2), have recently been clarified by linking them to the two photosystems and topological considerations (Symons *et al.*, 1985, 1987).

Most EPL experiments (summarized below), as well as the carotenoid electrochromism studies

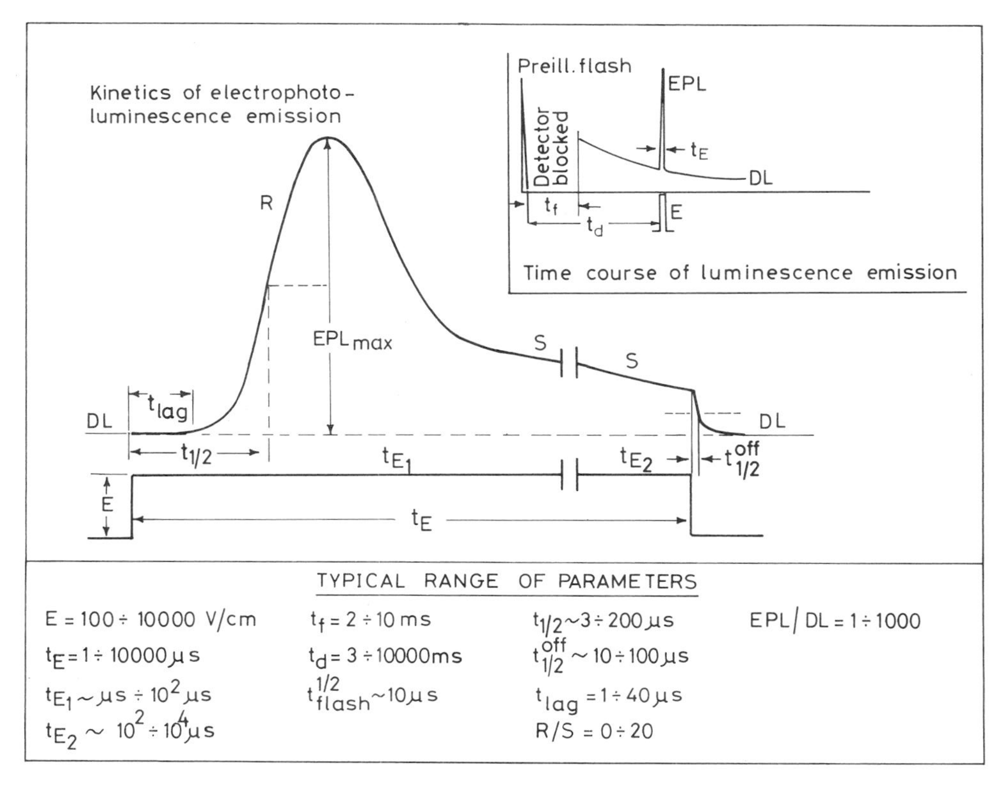

Figure 2. Schematized kinetics and typical range of parameters for electrophotoluminescence (EPL) emission. Insert: time course of delayed luminescence (DL). The applied electric field (E) has onset/offset times of less than 1 μsec. R and S denote two kinetic components of EPL, not discussed in the text (see Symons *et al.*, 1985, 1987). The dashed line indicates the level of nonstimulated luminescence (DL).

mentioned above, have been carried out on chloroplasts suspended hypotonically and yielding single-membrane-bounded, quasi-spherical, fully functional, cell-size entities called photosynthetic blebs. Indeed, from the mechanism of field enhancement described in Section 2.1, one can see the importance of (1) a large R/d ratio; (2) the outermost membrane being the pigment-containing one (unlike in, e.g., intact chloroplasts, in which there is no EPL), since the potential drop only occurs within it; and (3) spherical shape, for easy modeling. All these constraints are met by swelling of chloroplasts in low-salt media, which is also beneficial for minimizing joule heating during the field pulses and other artifacts.

The mechanism of EPL production for a bleb (see Fig. 3) is the following (Ellenson and Sauer, 1976; Farkas, 1983). The light-induced charge separation (thin arrows in Fig. 3) occurs from primary donors (some of which are marked D) to primary acceptors (marked A). Several lines of structural and functional evidence support the location of the donors toward the inside and of the acceptors toward the outside of the membranes; thus, the charge separation has a marked vectorial transmembranal character, and helps preserve some of the impinging photon energy in the form of a radially symmetric membrane potential. Upon application of an electric field E, the local field E_m induced within the membrane (thick arrows in Fig. 3) is given by

$$E_m = KE \cos \theta \tag{5}$$

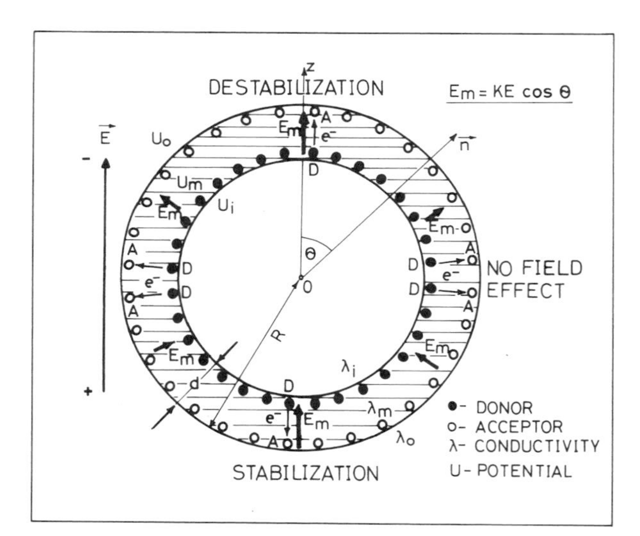

Figure 3. Model for electrophotoluminescence production in a bleb. The external electric field (E) induces in the membrane (hatched region) a field E_m, the direction, sense, and approximate size of which is shown by thick arrows. Increased emission (EPL) results, by a mechanism described in the text. Some of the notations are described in the insert and are used in Eqs. (1) and (4) of Section 2. Reproduced from *Biophysical Journal* (1984) **45:**363–373, with permission.

with K an enhancement factor that can be expressed from Eqs. (2) to (4). This simple form is used in order to stress the main point of mechanistic relevance: due to the cos θ dependence, E_m will be parallel or antiparallel to the light-induced charge separation in the two hemispheres, resulting in the latter's destabilization or stabilization, respectively (Fig. 3). In the hemisphere facing the negative electrode, the destabilization induces enhanced charge recombination; for high enough fields, the rates can be increased by orders of magnitude. Since in the opposite hemisphere charge stabilization can only lead to a minimum of no emission, the situation is not symmetrical and the net result is an intensely stimulated luminescence (EPL). The frequency doubling of AC field-stimulated EPL is thus easy to understand, since each hemisphere will contribute to emission when acted upon by a field of appropriate polarity.

The theoretical expectation for the applied field intensity dependence of EPL is that the EPL/DL ratio increases as sinh AE, with A a constant (Ellenson and Sauer, 1976; Farkas *et al.*, 1981a,b). This is borne out only within a limited range of field values, but throughout the investigated range, the peak EPL is either a linear or a supralinear function of the applied field intensity E (Farkas, 1983).

The same mechanism applies for EPL production in nonspherical vesicles. However, an analytical expression for EPL versus E is not possible in this case. A variety of other findings substantiate the model sketched here. The resulting usefulness of EPL as a probe of membrane electrodynamics is based on:

1. Its exquisite membrane field (E_m) sensitivity
2. Its excellent signal-to-noise ratio
3. Its very fast response to changes in E_m
4. The unique topology and symmetry involved in its generation

These points will be illustrated in the applications discussed below.

A typical experimental setup and various cell designs used are shown in Figs. 4 and 5, respectively.

3.3. Membrane Capacitance and Electrical Charging

Due to its origin in the recombination of charges trapped at different energetic levels, the emission of EPL appears to be a threshold phenomenon: below a certain applied field value E^* (and

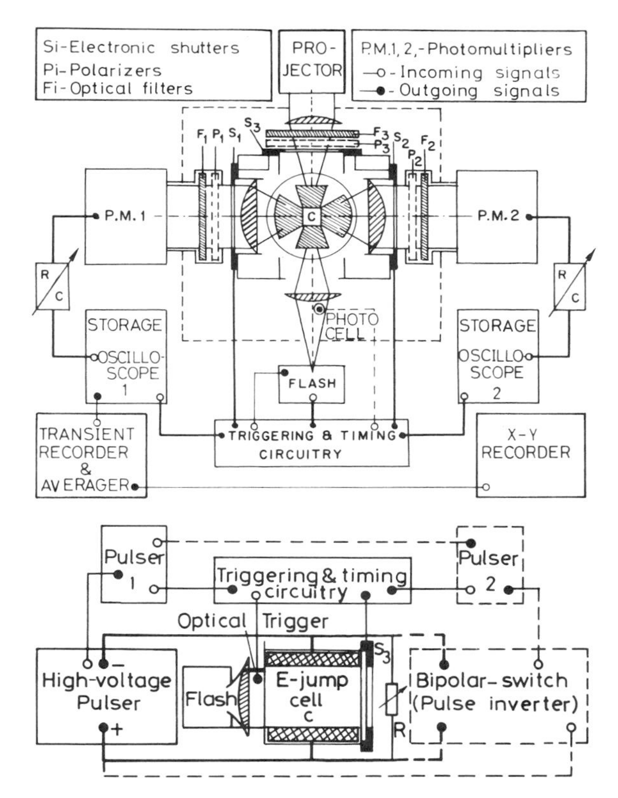

Figure 4. Experimental setup for electrophotoluminescence production and detection. The lower part of the figure represents the electrical design for producing the applied field with the appropriate timing and kinetics in the E-jump cell C. The upper part represents the optical and monitoring portion of the instrumentation, and the external electric field is applied to the suspension in the same cell C, perpendicularly to the plane of the drawing (for more details, see Farkas, 1983, and Farkas *et al.*, 1984a). Reproduced from *Biophysical Journal* (1984) **45:**363–373, with permission.

a corresponding membrane field $E_m{}^*$), no emission can be observed. We analyzed (Farkas *et al.*, 1984a) the characteristic onset lag (t_{lag} in Fig. 2) of EPL as reflecting the process of charging the bleb membranes as capacitors to this threshold value.

From Eq. (2), one obtains, with this assumption,

$$c_m = \frac{2Z\lambda_o}{3R} \frac{t_{lag}}{\ln \dfrac{E}{E-E^*}} \tag{6}$$

where Z is a constant depending on the ratio between the inner and outer ionic conductivities λ_i and λ_o ($Z = 1$ for $\lambda_i = \lambda_o$) and the other symbols are as previously defined. Direct measurement of E^* and of t_{lag} versus the applied field intensity E, for different chosen medium conductivities λ_o (varied by adding KCl to the suspension), yielded a value of $C_m = 1.2 \pm 0.3$ μF/cm² for the membrane capacitance of a population of blebs size-selected ($R = 4 \pm 1$ μm) with a fluorescence-activated cell sorter. This expected result (see Cole, 1968) was confirmed using non-size-selected bleb suspensions and by AC field-induced EPL experiments (Farkas *et al.*, 1984a; Farkas, 1983). The measurements—made possible by the unusually fast rise time and variable profile of the stimulus (E)—do not depend significantly on (light-induced) precursor availability and can be extended to investigate the relationship between the field- and light-induced membrane potentials: In the dark time (t_d) following preillumination, the latter decays during tens to hundreds of milliseconds. The lag time appears shortened by the existence of such a potential, to the extent that t_{lag} versus t_d resembles the mirror image of the potential indicating carotenoid shift's time decay. The additivity thus suggested (since, at constant applied field, less time is needed to reach the threshold potential for emission) could make EPL a tool for measuring light-induced potentials (Farkas *et al.*, 1984a).

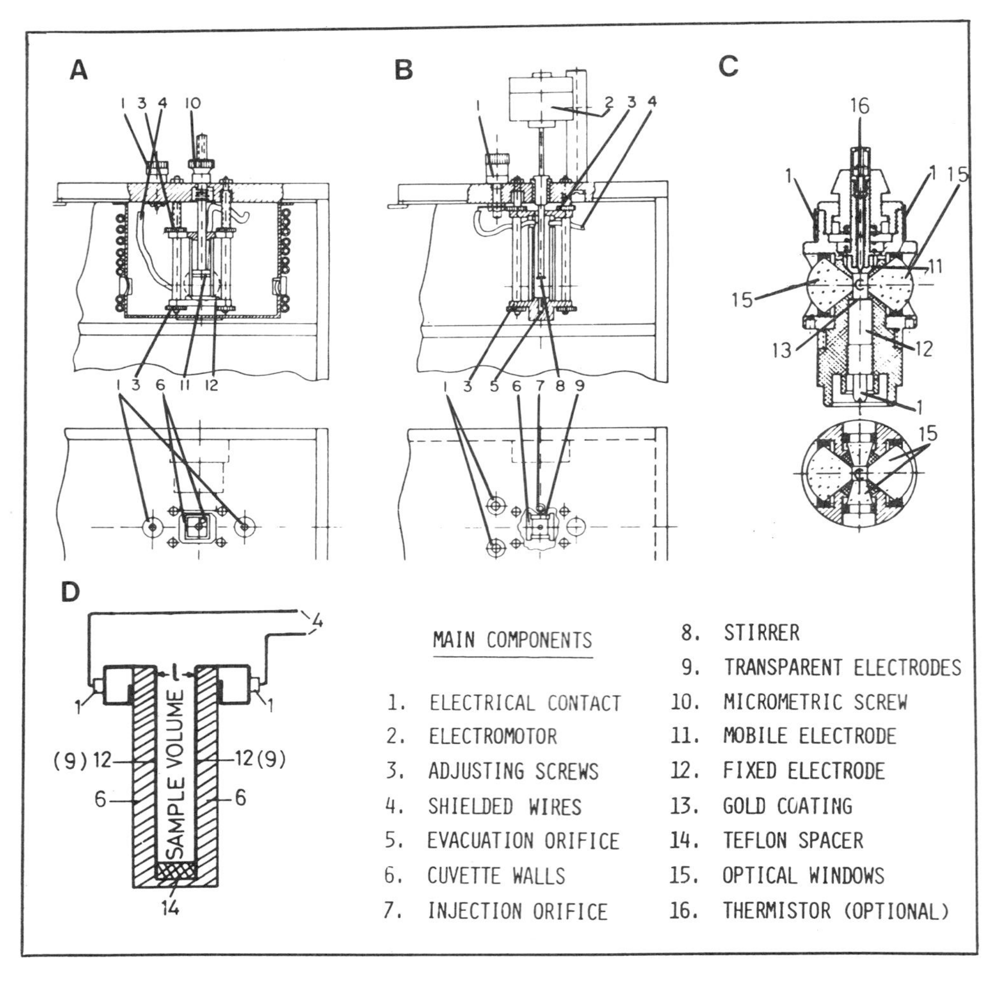

Figure 5. Cell designs for electrophotoluminescence experiments. The main components are identified in the insert. The variable interelectrode distance cell (A) was used to verify that electrode phenomena do not play a role; the transparent electrode cell (B) allowed EPL monitoring along the field direction (see Section 3.4); cell C facilitated application of high fields (up to kV/cm) and is the one appearing in Fig. 4, but all cells are interchangeable; the simplest design (D) is based on a disposable plastic cuvette, and is similar to most cells used in electroporation experiments described in the literature.

3.4. Electroselection

The local variation of the electric field E_m within the bleb membrane—which ultimately determines the intensity of EPL emission—is described by Eqs. (2) to (4). As a consequence, the relative contribution of different membrane regions to the overall emission is weighted by the cos θ dependence of E_m and potentiated by the supralinear dependence of EPL on E_m: In blebs, maximal emission from the polar regions ($\cos\theta = 1$), and no emission from regions with $\cos\theta \leqslant 0$ are expected. We called this topological selection by the field electroselection (Farkas *et al.*, 1980). If one assumes that the light-emitting molecules (specialized chlorophylls of the reaction center) have a preferential orientation in the membrane, electroselection should result in polarized emission. This

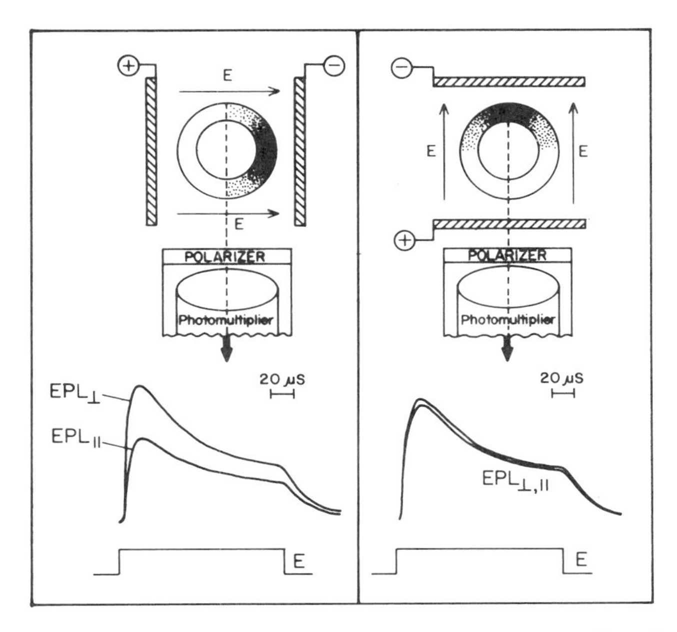

Figure 6. Polarization of electrophotoluminescence—role of detection geometry. The cell used (B in Fig. 5) had transparent electrodes. The detection geometries are sketched in the top part of the figure. The polarizers are rotated around the dotted axes, their axes being perpendicular or parallel to the applied field direction for detection of $EPL_{\perp}$ and $EPL_{\parallel}$, respectively.

prediction was borne out experimentally, and we were able to determine, by study of the polarization angles, that the emission transition moment of the primary donor involved in EPL emission lies in the plane of the membrane (Farkas *et al.*, 1980, 1981b). A typical set of measurements is shown in Fig. 6, which also underscores the importance of detection geometry in obtaining emission polarization. The conclusions regarding emitter orientation were strengthened by measurements of natural DL polarization in magnetic-field-oriented chloroplasts (Farkas *et al.*, 1981a).

The EPL polarization kinetics and dependence of experimental conditions were studied in detail, allowing the first clear distinction (Farkas, 1983) between the potential origins of the kinetic components R and S (see Fig. 2). More importantly, from the point of view of probing electrical events in the bleb membrane by EPL, one can conclude from electroselection that a change in emission polarization (not necessarily related to changes in emission intensity) is likely to be indicative of localized electrical changes. This point will be taken up in Section 3.6.3.

3.5. Ionophore Studies

3.5.1. General Considerations

For large enough R/d ratios, a significant increase in the membrane conductivity λ_m can make the second term in the denominator of Eq. (4) nonnegligible. Since the sensitivity of EPL to variation in E_m is likely to amplify any such effects, we attempted the use of emission changes to characterize induced variations of membrane conductivity in blebs (Farkas *et al.*, 1982, 1983).

The ionic conductivity of membranes can be selectively modified by a number of compounds, generically known as ionophores (the term is used in the widest sense, to include carriers and channels). Their chemical structure is usually known and their mechanism of action in artificial systems has been extensively investigated. In natural membranes, their behavior has mostly been assessed by functional effects, while kinetic and mechanistic details were usually extrapolated from model systems (McLaughlin and Eisenberg, 1975). Although acceptable in principle (and verified in practice), such an approach is likely to fail when it comes to details, due to inherent differences (of, e.g., complexity, protein content, thickness, fluidity) between biological and artificial membranes.

We first applied our approach to some of the best-characterized ionophores, representative of their particular classes: gramicidin (a channel), valinomycin (a rheogenic carrier), and monensin (a nonrheogenic carrier). In control experiments, their functional action on the thylakoid membrane was verified by their ability to inhibit photophosphorylation, produce diffusion-potential-stimulated DL, and accelerate the decay of the carotenoid electrochromic shift.

The typical EPL experiment consisted in measurement of intensity decrease and kinetic change in the presence of an added ionophore, keeping all other relevant features (applied field intensity, suspension medium) unchanged. All experiments had dark times exceeding 300 msec, to ensure the negligibility of light-induced membrane potentials.

The ionophore-induced changes in EPL were not only quite significant, but also differed markedly from one type of ionophore to another. Some of the findings were as follows.

3.5.2. Channel Ionophores

Gramicidin D induced a sizable emission decrease and a kinetic change in the emission pattern. In particular, the emission lag (see Section 3.3) was considerably lengthened, reflecting the longer time needed to reach a threshold voltage at increased λ_m. The effects depended on ionophore and monovalent cation concentrations. The selectivity of the channel has been tested by changing (at constant osmolarity and pH) the nature of the cation in the suspension medium. Based on the proportional diminishment of EPL, the intercationic selectivity sequence as Cs > Rb > K > Na >> Li, corresponding to the sequence obtained for artificial membranes (Eisenmann sequence I).

3.5.3. Carrier Ionophores

Valinomycin also diminished EPL, but to a lesser extent than gramicidin. The reason is likely to be a kinetic one, since turnover times for carriers are expected to be in the microsecond range; therefore, the mediated ion transfer cannot always compete efficiently with the buildup of external field-induced membrane potential. Moreover, for a short time following the fast (<0.2 μsec) onset of the external field *E*, the field in the membranes should be the same in the presence and absence of the carrier, for at least the time it takes for a transported ion to appear on the other side of the membrane. Thus, we propose, the time t_t after which the emission in the presence of the ionophore starts to diverge from the control should be an indicator of carrier turnover. For valinomycin, we found t_t = 1.3 μsec, the first such determination—to our knowledge—for a biological membrane; the value is lower than that (5–20 μsec) for artificial membranes (Lauger *et al.*, 1981), but not surprising in view of the high fluidity of the thylakoid membrane. The intercationic selectivity was Rb > Cs > K > Na >> Li, similar to Eisenmann sequence II and to results derived by Schmid and Junge (1975) by monitoring the carotenoid shift decay in chloroplasts. It is worth mentioning that the K/Na selectivity ratio was only about two orders of magnitude, as compared to about four in artificial membranes.

With the same approach, we could also detect nonspecific membrane damage induced by high (10^{-5} M) concentrations of valinomycin, as expected (D. Waltz, personal communication).

Monensin, which, as opposed to valinomycin, is a nonrheogenic carrier, did not have any influence on EPL, since it mediated no net charge movement. Interestingly, nigericin, another nonrheogenic carrier, behaved as monensin at low concentrations, but more like valinomycin above 1 μM. We conclude that it is a carrier of concentration-dependent rheogenicity.

3.5.4. Other Conductivity Modifiers

Besides other commonly used ionophores, we applied the method described above to a class of synthetic ionophores prepared by the group of A. Shanzer (Shanzer *et al.*, 1983). The high selectivity for the lithium ion, and the relative ionophoric efficiency within the series were found to be in good agreement with those obtained for liposomes and planar bilayer membranes (Farkas *et al.*, 1983).

A number of permeant ions (e.g., sodium tetraphenylboron) have also been tested and found to act similarly to slow-turnover carriers (Farkas, 1983).

3.5.5. Significance

EPL changes provide a method of investigating ionophore action in a biological membrane with the following advantages and ensuing conclusions: (1) microsecond time resolution; (2) a distinction between channels and carriers, and between rheogenic and nonrheogenic carriers; (3) an assessment of carrier turnover times; (4) a convenient study of intercationic selectivities and voltage dependence; and (5) existence of complementary functional checks for ionophore action within the same membrane.

3.6. Electrical Breakdown and Recovery

3.6.1. Rationale and Choice of Conditions

One of the reasons ionophore studies were discussed in some detail is the direct relevance of the approach to the study of electrical breakdown by EPL changes. The same formula [Eq. (4)] applies, with the change in membrane conductivity being caused by a reversible electrical breakdown of the bleb membrane induced by supercritical applied fields. Thus, the simplest experiment would consist in measuring EPL produced by a subthreshold (probing) field pulse preceded—or not, for comparison—by a breakdown-inducing pulse. However, pulse pairs of asymmetrical length and/or intensity are not easy to obtain for technical reasons. Moreover, a more elegant method is available, due to the special symmetry of the experimental system.

If, following a single preillumination, EPL emission is elicited by a succession of external field pulses, the peak EPL usually decreases and the kinetics change from pulse to pulse, depending on: (1) field intensity; (2) pulse duration; (3) pulse spacing; (4) pulse polarity; and (5) dark time between preillumination and field application. All these parameters can be controlled at will in our experiment, with remarkable precision. Factors (1) to (4) are related to precursor depletion and possibly to electrical breakdown, while (5) is due to the decay of EPL precursors in the dark. By a certain choice of experimental conditions, one can separate the electrical effects from the precursor-related ones: In a spherical bleb, as discussed in Section 3.2, a DC field pulse of a certain polarity induces EPL emission only from the hemisphere closer to the negative electrode. If a second pulse is applied to the suspension, without additional preillumination, it will induce emission from the same hemisphere (if it is of the same polarity as the first pulse) or from the opposite one (if the pulse sequence is bipolar). Thus, in the case of bipolar pulses, the electric field acts on the light-created precursors in the two hemispheres practically independently, the only condition being that the pulse spacing be short (typically milliseconds) compared to the bleb rotational diffusion time (which it is in the seconds range). Moreover, if the time elapsed from illumination (t_d) is also much longer than the spacing (t^*, see Fig. 7) between the pulses, the amount of precursors should be practically the same in the two hemispheres and identical amounts and kinetics of induced EPL are expected in the two pulses. Any decrease of EPL in the second pulse (as compared to the first) should be indicative of electrical "damage" induced by the first pulse. We assume this phenomenon to be electrical breakdown.

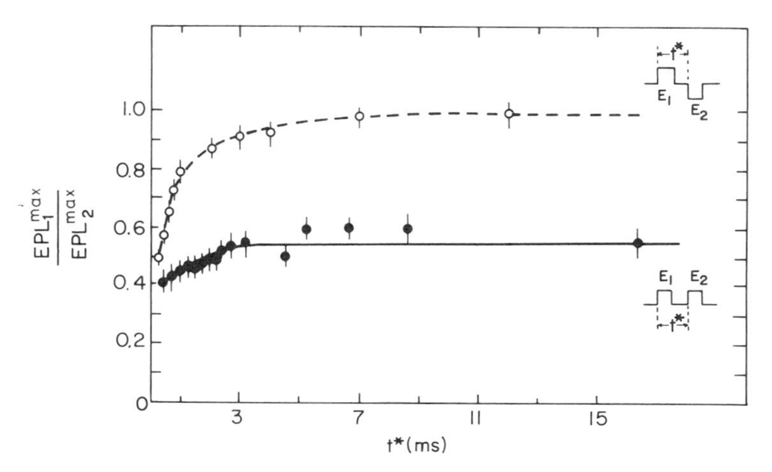

Figure 7. Electrical breakdown and recovery kinetics, monitored by electrophotoluminescence changes. The ratios of electrophotoluminescence emission peaks (EPL_1^{max}/EPL_2^{max}) obtained in a bipolar (○) or unipolar sequence (●) are plotted as a function of pulse spacing t^*. Conditions: E = 2500 V/cm; t_d = 300 msec; t_E = 300 μsec (see Fig. 2 for notations), corresponding to peak membrane potentials in excess of 1.2 V. Reproduced from *Biochimica et Biophysica Acta* (1984) **767**:507–514, with permission.

3.6.2. Breakdown and Recovery Kinetics (Farkas *et al.*, 1984b)

The results of a typical bipolar pulse experiment are shown in Fig. 7. The diminished EPL induced by the second pulse should reflect reversible breakdown, as full recovery of EPL_2 to the level of EPL_1 occurs upon increasing the pulse spacing t^* (subscripts indicate pulse order, and the superscript denoting emission peaks has been omitted). If, however, one further increases the pulse duration, especially over 500 μsec (see Fig. 8), full recovery is not obtained, indicative of irreversible breakdown of at least some of the larger blebs in the suspension. We confirmed this directly by

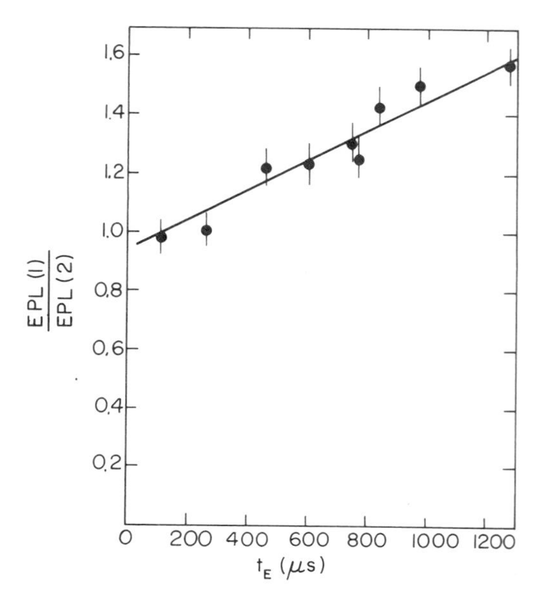

Figure 8. Applied pulse length dependence of electrical breakdown in blebs, measured by electrophotoluminescence changes. The experiment was similar to that illustrated in Fig. 7, with E = 2000 V/cm, t_d = 300 msec, and t^* = 6 msec. The increase of the peaks' ratio is mainly due to the appearance of irreversible breakdown, although a slight dependence of recovery times on pulse length (t_E) was also observed. Reproduced from *Biochimica et Biophysica Acta* (1984) **767**:507–514, with permission.

microscopy, revealing irreversible vesicle destruction upon pulsing by fields of the same intensity and duration.

The extent and type of damage also depended on local field intensity (the lowest potentials for breakdown being around 880 mV), either directly (applied field intensity) or indirectly (through bleb size distribution), as well as the structural/functional state of the membrane, since aging considerably reduced the breakdown threshold.

Longer-lasting, slow-rising AC fields induced breakdown which was more difficult to assess quantitatively, but the recovery (resealing) times were certainly longer (Farkas, 1983). An additional element, not stressed in our previous publications on the subject (Farkas *et al.*, 1984b; Korenstein *et al.*, 1984), is that the rise time of EPL induced by the second pulse—under breakdown conditions, with λ_m increased by the first pulse—is always shorter, in accordance with the provisions of Eq. (4).

3.6.3. Breakdown Localization by Electroselection

From the angular dependence of the transmembrane field, E_m, one concludes that breakdown should first occur at the blebs' poles, where the fields are maximal. Indeed (Farkas, 1983; Korenstein *et al.*, 1984) in a typical sequence the polarization of peak EPL drops from $q = (\mathrm{EPL}_{\perp}/\mathrm{EPL}_{\parallel}) = 1.43$ to approximately $q = 1.33$ for the second pulse, if the first is sufficiently intense to induce breakdown, but regains its initial value upon recovery at long enough pulse spacings.

3.6.4. Environmental Effects

The temperature of the suspension medium is expected to strongly influence breakdown voltages and recovery times. We investigated this problem and found that the breakdown recovery is not a monophasic process, but rather a multiexponential one. In heat-treated chloroplasts, one set of measurements (at $E = 2500$ V/cm, $t_E = 40$ μsec, and $t_d = 350$ msec) showed time constants of 95, 2, and 0.4 msec for 5°C and 8.5, 0.8, and < 0.2 (nonresolved) msec for 25°C (Farkas and Korenstein, unpublished). These results are somewhat preliminary, but illustrate the tremendous sensitivity and time resolution of the proposed approach for the study of membrane breakdown.

4. THE HEMISPHERICAL LIPID BILAYER AS A MODEL SYSTEM

The hemispherical lipid bilayer (HLB) was introduced as a model system for membrane studies 20 years ago (Pagano and Thompson, 1967). It is a millimeter-size, quasi-spherical, single bilayer that can be obtained from a variety of lipids, with or without proteins incorporated. It can be used either alone or in pairs (Brewer and Thomas, 1984) in contact, and the measurements can be electrical (Asami and Irimajiri, 1984) or optical (Loew and Simpson, 1981). Under conditions of voltage clamp, the HLB has been most successful in investigating voltage-sensitive dyes (Fluhler *et al.*, 1985).

Recently we used one of the better-characterized charge-shift styryl dyes, di-4-ANEPPS (1-(3-sulfonatopropyl)-4-[β] [2-(di-*n*-butylamino)-6-naphtyl-vinyl] pyridinium betaine), as an optical indicator to determine the extent and kinetics of the transmembrane potential rise induced in a hemispherical bilayer (Ehrenberg *et al.*, 1987). The experimental arrangement is shown in Fig. 9. We were able to demonstrate that the dye response—for both the relative transmittance change and fluorescence variation—is in accordance with Eq. (4). The response was linear with the membrane potential—produced by application of external electric field pulses rather than voltage clamp—while the rate of development of the potential was linearly related to the conductance of the bathing solution; the membrane capacitance to fit the data was $C_m = 0.71$ μF/cm^2. Testing the predictions of Eq. (4) further, we have examined how the kinetics and magnitude of the membrane potentials induced by the external field are modulated upon increasing λ_m by introduction of channels into the

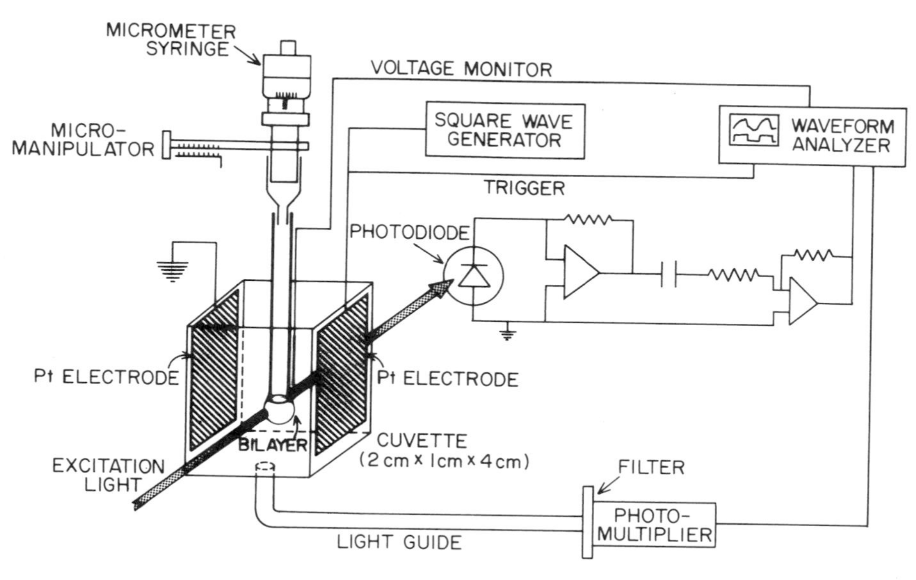

Figure 9. Experimental setup for monitoring the external electric field-induced membrane potential in a hemispherical lipid bilayer (HLB). Light from a monochromator is focused on one side ($\theta = \pm 90°$) of the bilayer, so that less than 10% of its area is illuminated. The transmittance changes are detected with a photodiode, and the AC-coupled signal sent to a waveform analyzer, together with the fluorescence signal excited by the same beam and detected by a photomultiplier with a blocking filter, via a light guide. Triggering of the waveform analyzer is by the square pulse generator that supplies the applied electric field to the membrane-containing solution. The bilayer radius can be determined either by moving it perpendicularly to the beam, from one side to the other and measuring $2R$ with the micromanipulator, by measuring the HLB volume with the micrometric syringe upon bilayer formation, or by observation through a specially calibrated stereomicroscope.

hemispherical bilayer. Addition of gramicidin, a channel-forming peptide, decreased both the time constant and the amplitude of the induced transmembrane potential equally (up to threefold; compare Figs. 10 and 11 for the change in charging time). This effect reached saturation for about 5 μM gramicidin (Fig. 12). The linear relationship of the kinetics with the bathing solution conductance was retained in the presence of the ionophore, but with a steeper slope, as predicted by Eq. (4). The observability of these effects is likely due to the unusually large R/d ($>10^5$) ratio for the hemispherical bilayer.

5. CELLS IN ELECTRIC FIELDS—MONITORING MEMBRANE POTENTIALS IN INDIVIDUAL CELLS BY A FAST OPTICAL PROBE

Interesting and informative as studies on model systems may be, the subject of real consequence in the area of applied electric field effects is their influence on cells in suspension, particularly since very recently reversible electrical breakdown—induced in the cell membranes by field pulses in excess of a certain critical intensity—has found a host of important applications, based on the exchange of materials between the inside and the outside of the vesicles through the field-produced pores (electroporation) and the fusion of two or more vesicles (similar or different, and in contact) upon their simultaneous breakdown (electrofusion). Some of these applications (see Zimmermann *et al.,* 1981; Neumann *et al.,* 1982; Tsong, 1983; and most chapters in this book) are: monoclonal antibody production by receptor-mediated electrofusion; production of red cell ghosts, giant erythrocytes, and giant culture cells; cell-membrane permeation by anticancer drugs; fusion of

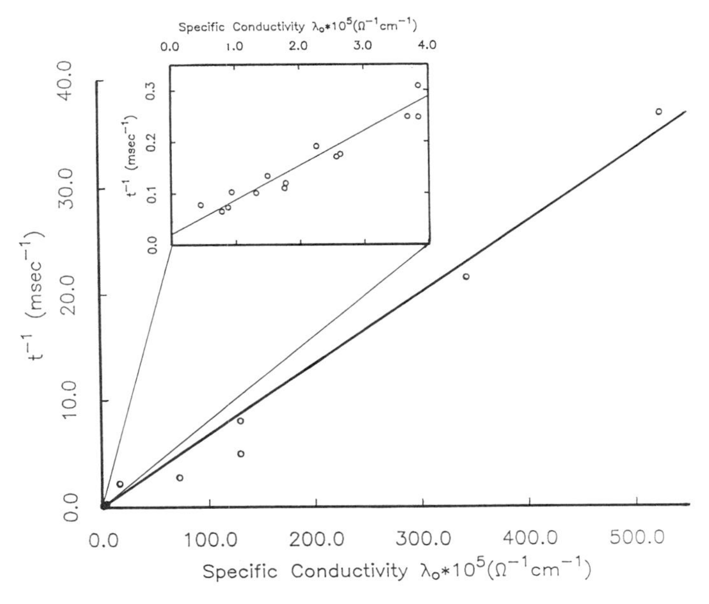

Figure 10. Dependence of bilayer charging rate constant on the electrical conductivity of the bathing solution. Results obtained on a number of bilayers of varying radii were normalized to $R = 1.5$ mm. Reproduced with permission from Z. K. Lojewska, D. L. Farkas, and L. M. Loew (unpublished results).

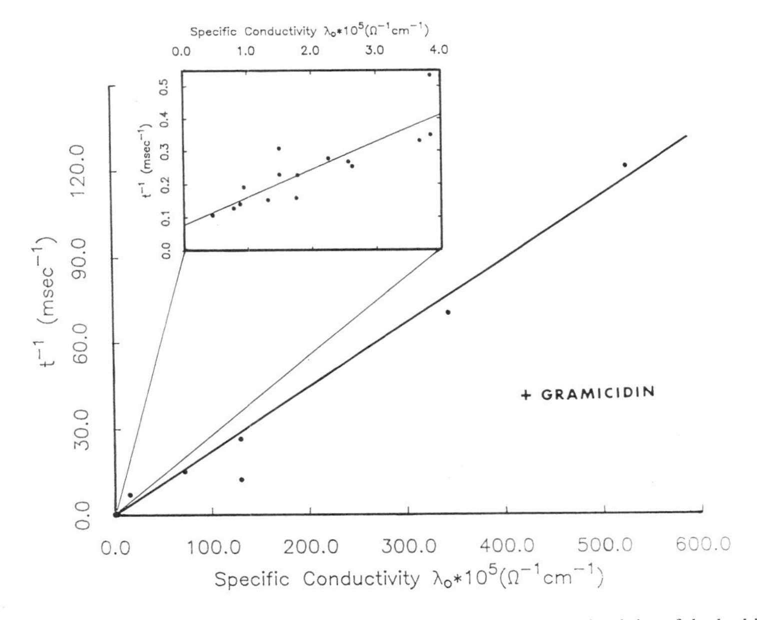

Figure 11. Dependence of bilayer charging rate constant on the electrical conductivity of the bathing solution, in presence of gramicidin. Conditions were the same as in Fig. 10, with 5 μM gramicidin added to the medium. Note the roughly threefold difference in the abscissa compared with that in the absence of ionophore (see Fig. 10). Reproduced with permission from Z. K. Lojewska, D. L. Farkas, and L. M. Loew (unpublished results).

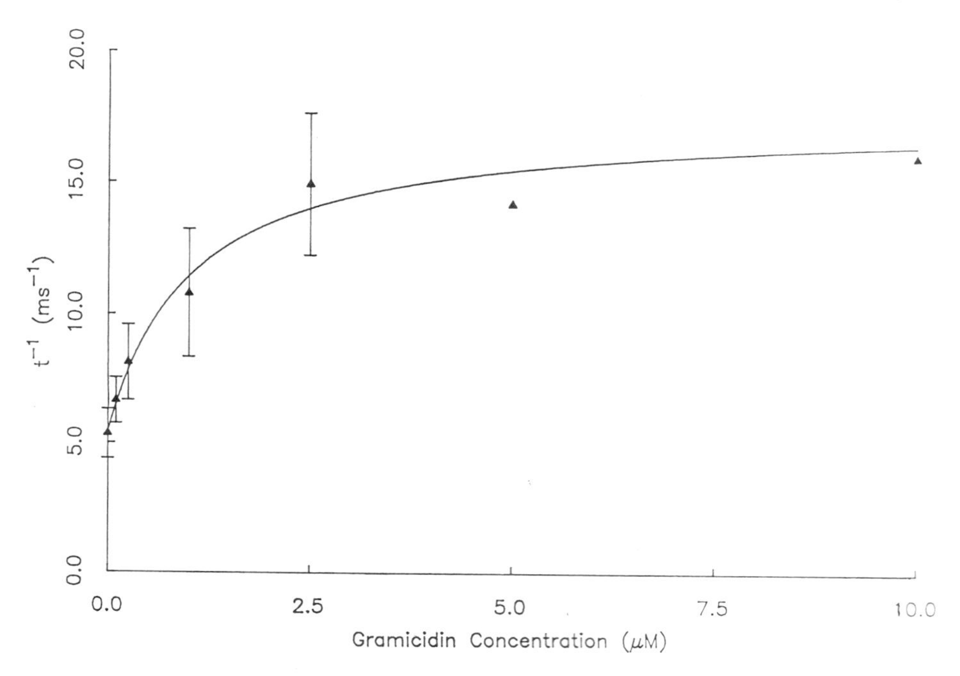

Figure 12. Rate constant for bilayer charging, as a function of added gramicidin concentration. As for Figs. 10 and 11, results were corrected for the variability in bilayer radius. Average values are shown, with their standard deviations (where applicable). Conditions: 5 mM KCl, 3.0 μM di-4-ANEPPS. Reproduced with permission from Z. K. Lojewska, D. L. Farkas, and L. M. Loew (unpublished results).

bacterial spheroplasts, yeast cells, mammalian cells, and plant protoplasts; direct gene transfer into mouse lyoma cells, monocot and dicot plant cells, including economically important cereals, leading to stable transformants; use of erythrocytes, lymphocytes, and liposomes as drug carrier and targeting systems. Thus, the horizons this technique opens in the biotechnological field and its scientific and practical importance could hardly be overstated. However, what should be stressed—besides the obvious potential—is that: (1) in only very few of these studies have the conditions (e.g., electric field parameters, medium composition, temperature) been considered and optimized carefully, and (2) if this was achieved, the respective assays (e.g., for optimal somatic hybrid production) were functional and thus both very time-consuming and reflecting a cumulative effect of a number of variables.

The fact that recent studies have shown (Tsong, 1983; Farkas *et al.*, 1984b; Sowers, 1985; Sowers and Lieber, 1986) that a simple model for electroporation, followed by electrofusion cannot explain certain phenomena (e.g., vastly different recovery times for BLMs versus biological membranes, long-lived fusogenic states, nonsymmetrical breakdown at the vesicle poles), underscores the need for a more systematic study of external field-induced breakdown phenomena.

Use of the fast, voltage-sensitive electrochromic probes described above (e.g. di-4-ANEPPS) in conjunction with computer-assisted quantitative fluorescence microscopy to monitor individual cells imparts to membrane potentials of induced measurements a spatial resolution limited only by the microscope optics and a temporal one limited only by the detection system bandwidth.

In recently reported work, Gross *et al.* (1986) were able to image with excellent resolution the spatial distribution of transmembrane potential changes induced by applied electric fields in nonexcitable cells using digital intensified video microscopy. The measurements confirmed the predictions of Eq. (4) for spherical cells and could be extended to ellipsoidal and aggregated cells. Results of our ongoing work, aimed at extending this approach, are illustrated in Fig. 13.

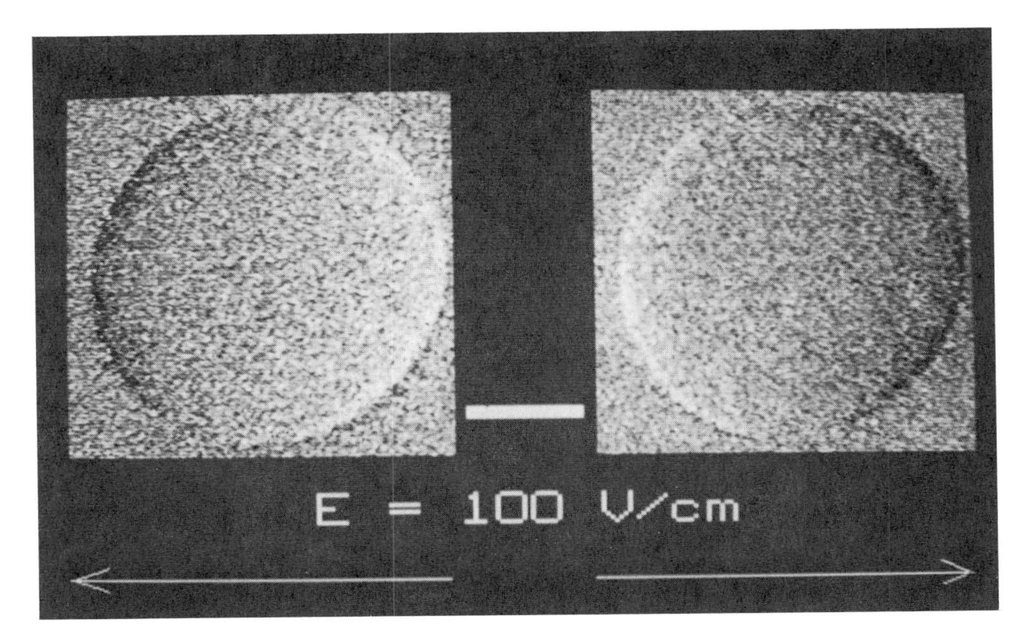

Figure 13. Effect of electric fields on single cells, quantitated by video-enhanced fluorescence microscopy. The details of the experimental arrangement will be described elsewhere (Farkas and Loew, in preparation). HeLa cells were grown on 22 × 30 mm coverslips, cold-incubated with 5 μM di-4-ANEPPS and 0.2% of the surfactant Pluronic 127 (which induces size increase, cell rounding and improved staining), and exposed to fields of E = 100 V/cm, for 0.5 seconds. The image results from subtracting the fluorescence (excited at 546 nm and detected above 590 nm) in the absence of the field from that obtained in the presence of the field, at 23°C. Analysis shows the predicted cosine dependence (see Eqs. 2–5) and about 9%/100 mV for the potential-indicating relative fluorescence change in the polar regions (where V = ± 225 mV), reversing sign with a change in the sense of the applied field (compare left vs. right in the figure). Upon further increase in the applied field intensities, the extent of relative fluorescence changes in the polar regions lags behind that of nearby areas and theoretical predictions, indicative of localized, symmetrical electric breakdown.

In order to improve our understanding of the sequence of events leading to and following electroporation and electrofusion with the ultimate aim of finding the conditions necessary to increase their efficiency and reproducibility, it would be important to use the technique described above under conditions leading to electrical breakdown of the cell membranes. While the efficiency of electroporation-mediated transfection can approach 1%, as recently reported (Chu *et al.*, 1987; Stopper *et al.*, 1987), and electrofusion can be enhanced by various treatments (Herzog *et al.*, 1986; Nea *et al.*, 1987; Ohno-Shosaku and Okada, 1984), optimizing the electrical part of the experiments is still incomplete, at best. For instance, (1) square-wave field pulses (rather than exponentially decaying ones) seem far superior, for various reasons, and yet their use is very scarce; (2) the persistence of membrane leakiness as a function of field energy and duration and medium parameters—a critical issue in all studies—is poorly understood; (3) the membrane field distribution during breakdown pulses and immediately following them is not known. Moreover, the electrical techniques for genetic manipulation seem superior to other techniques, not only in overall efficiency, but, according to recent reports (Bertling *et al.*, 1987; Boggs *et al.*, 1986), also in their direct consequences (e.g., single-site, single-copy insertion of DNA into the host genome).

Based on the reemergence of light microscopy as a main tool in cellular research due to the novel capabilities brought to it by computer-based digital data analysis and storage, we suggest the following (solely illustrative) list of experiments.

1. Application of superthreshold potentials to roughly spherical cells in a protocol similar to that in Gross *et al.* (1986) and Ehrenberg *et al.* (1987). The fluorescence change—measured at the pole, e.g.—should deviate from linearity with the applied potential, as already visible in some of the results in the papers above. The breakdown potential, its dependence on pulse duration and profile and the medium characteristics (e.g., temperature, divalent cation concentration) could thus be determined for a number of cells of interest.

2. The recovery of breakdown could be investigated by double pulse experiments, either similar to those with the blebs (Section 3.5) or with a lower intensity measuring pulse following a brief, high-intensity pulse. The temperature dependence of recovery seems the most important to address, but other factors (pH, cation nature and concentrations, osmolarity) could also play significant roles.

3. The location and spread (delocalization) of breakdown could be followed by monitoring—for the same applied field intensity—the potential-indicating fluorescence response at various points (corresponding to different θ angles) on the cells.

4. The size of electropores can be studied by loading the cells with fluorescent molecules of various sizes, and monitoring their field-induced efflux following breakdown. With the same technique (see Sowers and Lieber, 1986), one can address the question of whether electroporation is simultaneous at both poles of the cells. The dependence of any asymmetry in this case can be investigated as a function of a preexisting resting potential, which can be independently measured in the same setup by Nernstian dye distribution.

In order to compare natural membranes with pure lipid ones, essentially the same experiments (1–4) could be carried out on the HLB (see above).

5. By dielectrophoresis (Pohl, 1978), one can orient ellipsoidal cells with their axes at desired angles to the field, and measure the resulting potentials and breakdown-reflecting changes, upon field application.

6. Very recently, Bryant and Wolfe (1987) suggested that mechanical stresses induced in the membranes by the applied fields may (partly) be responsible for cell lysis previously attributed solely to irreversible breakdown. Field-induced cell shape changes that would precede such events and support this contention could easily be followed microscopically.

7. The electrofusion of a larger vesicle with a smaller one can be conveniently monitored in a video microscopy setup. Investigation should focus on the dependence of such events on the applied field intensity, the relative cell sizes, and the nature and pretreatment of their membranes.

8. In experiments of great relevance—in our view—for the use of breakdown phenomena in gene transfer (and thus genetic engineering), one could tag DNA molecules by specific fluorescent labels, and follow their ability to penetrate cells through electropores, under a variety of imposed conditions. Such an investigation would also explore the importance of the medium characteristics *following* breakdown for optimal penetration of the DNA molecules into the cells, since it could well be that one of the first (and fast) events of recovery is the partial closure of electropores to a size that precludes movement of relatively large molecules.

6. CONCLUSIONS AND OUTLOOK

A direct, universal, and versatile method for obtaining relevant information about bioelectrical properties is the application of external electric fields. In order to fully take advantage of the methodological merits of the technique, use of optical probes is recommended. The right choice of probe is crucial, and the potential value of the somewhat neglected intrinsic indicators has been stressed here. Moreover, the availability of very fast extrinsic probes–used within adequate model systems or monitored by microscopic imaging in cells—opens the way to the simultaneous assessment of the spatial distribution and time course of membrane electrical potential changes. The entire approach is easily extendable into the range of superthreshold applied fields that lead to electroporation and electrofusion. The topological and kinetic information thus obtained should contribute to expanding and optimizing existing techniques, as well as providing insights into underlying mechanisms. This should bring us closer to achieving the dual goal of having a simple, rapid, nonspecific, and versatile method for gaining access to the interior of cells, and improving our understanding of their membranes in the process, by use of molecular probes.

REFERENCES

Arnold, W., and Azzi, J., 1971, The mechanism of delayed light production by photosynthetic organisms and a new effect of electric fields on chloroplasts, *Photochem. Photobiol.* **14:**233–240.

Asami, K., and Irimajiri, A., 1984, Dielectric dispersion of a single spherical bilayer membrane in suspension, *Biochim. Biophys. Acta* **769:**370–376.

Bernhardt, J., and Pauly, H., 1973, On the generation of potential differences across the membranes of ellipsoidal cells in an alternating electrical field, *Biophysik* **10:**89–98.

Bertling, W., Hunger-Bertling, K., and Cline, M. J., 1987, Intranuclear uptake and persistence of biologically active DNA after electroporation of mammalian cells, *J. Biochem. Biophys. Methods* **14:**223–232.

Boggs, S. S., Gregg, R. G., Borenstein, N., and Smithies, O., 1986, Efficient transformation and frequent single-site, single-copy insertion of DNA can be obtained in mouse erythroleukemia cells transformed by electroporation, *Exp. Hematol.* **14:**988–994.

Brewer, G. J., and Thomas, P. D., 1984, Role of gangliosides in adhesion and conductance changes in large spherical model membranes, *Biochim. Biophys. Acta* **776:**279–287.

Bryant, G., and Wolfe, J., 1987, Electromechanical stresses produced in plasma membranes of suspended cells by applied electric fields, *J. Membrane Biol.* **96:**129–139.

Chan, C. Y., and Nicholson, C., 1986, Modulation by applied electric fields of Purkinje and stellate cell activity in the isolation turtle cerebellum, *J. Physiol. (London)* **371:**89–114.

Chu, G., Hayakawa, H., and Berg, P., 1987, Electroporation for the efficient transfection of mammalian cells with DNA, *Nucleic Acid Res.* **15:**1311–1326.

Cohen, L. B., and Salzberg, B. M., 1978, Optical measurement of membrane potential, *Rev. Physiol. Biochem. Pharmacol.* **83:**35–88.

Cole, K. S., 1968, *Membranes, Ions and Impulses,* University of California Press, Berkeley.

DeGrooth, B. G., Van Gorkom, H. J., and Meiburg, R. F., 1980, Electrochromic absorbance changes in spinach chloroplasts induced by an external electrical field, *Biochim. Biophys. Acta* **589:**299–314.

DeWeer, P., and Salzberg, B. M. (eds.), 1986, *Optical Methods in Cell Physiology,* Wiley–Interscience, New York.

Ehrenberg, B., Farkas, D. L., Fluhler, E. N., Lojewska, Z., and Loew, L. M., 1987, Membrane potential induced by external electric field pulses can be followed with a potentiometric dye, *Biophys. J.* **51:**833–837.

Ellenson, J. L., and Sauer, K., 1976, The electrophotoluminescence of chloroplasts, *Photochem. Photobiol.* **23:** 113–123.

Farkas, D. L., 1983, External electric field induced phenomena in the photosynthetic membrane: Electrophotoluminescence and phosphorylation, Ph.D. thesis, The Weizmann Institute of Science, Rehovot, Israel.

Farkas, D. L., Korenstein, R., and Malkin, S., 1980, Electroselection in the photosynthetic membrane: Polarized luminescence induced by an external electric field, *FEBS Lett.* **120:**236–242.

Farkas, D. L., Neeman, M., and Malkin, S., 1981a, Polarization of delayed luminescence in magneto-oriented chloroplasts, *FEBS Lett.* **134:**221–224.

Farkas, D. L., Korenstein, R., and Malkin, S., 1981b, External electric field induced delayed luminescence in the photosynthetic membrane, in: *Photosynthesis,* Volume I (G. Akoyunoglou, ed.), Balaban International Science Services, Philadelphia, pp. 627–636.

Farkas, D. L., Korenstein, R., and Malkin, S., 1982, Ionophore-mediated ion transfer in a biological membrane: A study by electrophotoluminescence, in: *Transport in Biomembranes: Model Systems and Reconstitution* (R. Antolini, A. Gliozzi, and A. Gorio, eds.), Raven Press, New York, pp. 215–226.

Farkas, D. L., Korenstein, R., and Malkin, S., 1983, Electrophotoluminescence in the photosynthetic membrane: A novel approach to the characterization of ionophore action in a biological system, in: *Physical Chemistry of Transmembrane Ion Motions* (G. Spach, ed.), Elsevier, Amsterdam, pp. 307–317.

Farkas, D. L., Korenstein, R., and Malkin, S., 1984a, Electrophotoluminescence and the properties of the photosynthetic membrane. I. Initial kinetics and the charging capacitance of the membrane, *Biophys. J.* **45:** 363–373.

Farkas, D. L., Malkin, S., and Korenstein, R., 1984b, Electrophotoluminescence and the electrical properties of the photosynthetic membrane. II. Electric field-induced electrical breakdown of the photosynthetic membrane and its recovery, *Biochim. Biophys. Acta* **767:**507–514.

Fluhler, E., Burnham, V. G., and Loew, L. M., 1985, Spectra, membrane binding and potentiometric responses of new charge shift probes, *Biochemistry* **24:**5749–5755.

Gross, D., 1988, Electromobile surface charge alters membrane potential changes induced by applied electric fields, *Biophys. J.* **54:**879–884.

Gross, D., Loew, L. M., and Webb, W. W., 1986, Optical imaging of cell membrane potential changes induced by applied electric fields, *Biophys. J.* **50:**339–348.

Herzog, R., Muller-Wellensiek, A., and Voelter, W., 1986, Usefulness of Ficoll in electric field mediated cell fusion, *Life Sci.* **39:**2279–2288.

Kagawa, Y., and Hamamoto, T., 1986, ATP formation in mitochondria, submitochondrial particles, and F_0F_1 liposomes driven by electric pulses, *Methods Enzymol.* **126:**640–643.

Kakitani, T., Honig, B., and Crofts, A. R., 1982, Theoretical studies of the electrochromic response of carotenoids in photosynthetic membranes, *Biophys. J.* **39:**57–63.

Korenstein, R., Farkas, D. L., and Malkin, S., 1984, Reversible electrical breakdown of swollen thylakoid membrane vesicles—A study of electrophotoluminescence, *Bioelectrochem. Bioenerg.* **13:**191–197.

Lauger, P., Benz, R., Stark, G., Bamberg, E., Jordan, P. C., Fahr, A., and Brock, W., 1981, Relaxation studies of ion transport systems in lipid bilayer membranes, *Q. Rev. Biophys.* **14:**513–598.

Lill, H., Althoff, G., and Junge, S., 1987, Analysis of ionic channels by a flash spectrophotometric technique applicable to thylakoid membranes: CF_0, the proton channel of the chloroplast ATP synthase, and, for comparison, gramicidin, *J. Membr. Biol.* **98:**69–78.

Loew, L. M., and Simpson, L. L., 1981, Charge-shift probes of membrane potential—A probable electrochromic mechanism for p-aminostyrylpyridinium probes on a hemispherical lipid bilayer, *Biophys. J.* **34:**353–365.

McLaughlin, S., and Eisenberg, M., 1975, Antibiotics and membrane biology, *Annu. Rev. Biophys. Bioeng.* **4:** 335–366.

Maxwell, J. C., 1873, *A Treatise on Electricity and Magnetism,* Volume I, Clarendon Press, Oxford, pp. 362–366.

Nea, L. J., Bates, G. W., and Gilmer, P. J., 1987, Facilitation of electrofusion of plant protoplasts by membrane-active agents, *Biochim. Biophys. Acta* **897:**293–301.

Neumann, E., Schaefer-Ridder, M., Wang, Y., and Hofschneider, P. H., 1982, Gene transfer into mouse lyoma cells by electroporation in high electric fields, *EMBO J.* **1:**841–845.

Ohno-Shosaku, T., and Okada, Y., 1984, Facilitation of electrofusion of mouse lymphoma cells by the proteolytic action of proteases, *Biochim. Biophys. Res. Commun.* **120:**138–143.

Pagano, R., and Thompson, T. E., 1967, Spherical lipid bilayer membranes, *Biochim. Biophys. Acta* **144:**666–669.

Pohl, H. A., 1978, *Dielectrophoresis,* Cambridge University Press, London.

Poo, M.-m., 1981, In situ electrophoresis of membrane components, *Annu. Rev. Biophys. Bioeng.* **10:**245–276.

Schlodder, E., and Witt, H. T., 1980, Electrochromic absorption changes of a chloroplast suspension induced by an external electric field, *FEBS Lett.* **112:**105–112.

Schmid, R., and Junge, W., 1975, Current–voltage studies on the thylakoid membranes in the presence of ionophores, *Biochim. Biophys. Acta* **394:**76–92.

Shanzer, A., Samuel, D., and Korenstein, R., 1983, Lipophilic lithium ion carriers, *J. Am. Chem. Soc.* **105:** 3815–3818.

Sowers, A. E., 1985, Movement of a fluorescent lipid label from a labeled erythrocyte membrane to an unlabeled erythrocyte membrane following electric field-induced fusion, *Biophys. J.* **47:**519–525.

Sowers, A. E., and Lieber, M. R., 1986, Electropore diameters, numbers, and locations in individual erythrocyte ghosts, *FEBS Lett.* **205:**179–184.

Stopper, H., Jones, H., and Zimmermann, U., 1987, Large scale transfection of mouse L-cells by electropermeabilization, *Biochim. Biophys. Acta* **900:**38–44.

Symons, M., Korenstein, R., and Malkin, S., 1985, External electric field effects on photosynthetic vesicles: The relationships of the rapid and slow phases of electrophotoluminescence in hypotonically swollen chloroplasts to PSI and PSII activity, *Biochim. Biophys. Acta* **806:**305–310.

Symons, M., Malkin, S., and Farkas, D. L., 1987, Photochemical properties of PSI studied by electrical field induced luminescence, *Biochim. Biophys. Acta* **894:**578–582.
Tsong, T. Y., 1983, Voltage modulation of membrane permeability and energy utilization in cells, *Biosci. Rep.* **3:**487–505.
Vinkler, C., Korenstein, R., and Farkas, D. L., 1982, External electric field driven ATP synthesis in chloroplast: A slow, ATP-synthase dependent reaction, *FEBS Lett.* **145:**235–240.
Vinkler, C., Korenstein, R., and Farkas, D. L., 1983, External electric field driven ADP phosphorylation (EFP) in thylakoid membranes, in: *Biological Structures and Coupled Flows* (A. Oplatka and M. Balaban, eds.), Academic Press, New York and Balaban International Science Services, Philadelphia, pp. 113–121.
Witt, H. T., 1979, Energy conversion in the functional membrane of photosynthesis—Analysis by light pulse and electric pulse methods: The central role of the electric field, *Biochim. Biophys. Acta* **505:**355–427.
Zimmermann, U., Scheurich, P., Pilwat, G., and Benz, R., 1981, Cells with manipulated functions: New perspectives for cell biology, medicine and technology, *Angew. Chem. Int. Ed. Engl.* **20:**325–344.

Index